S0-CBI-361

Methods in Enzymology

Volume 63
ENZYME KINETICS AND MECHANISM
Part A
Initial Rate and Inhibitor Methods

METHODS IN ENZYMOLOGY

EDITORS-IN-CHIEF

Sidney P. Colowick   Nathan O. Kaplan

Methods in Enzymology

Volume 63

Enzyme Kinetics and Mechanism

Part A
Initial Rate and Inhibitor Methods

EDITED BY

Daniel L. Purich
DEPARTMENT OF CHEMISTRY
UNIVERSITY OF CALIFORNIA
SANTA BARBARA, CALIFORNIA

ACADEMIC PRESS   New York   San Francisco   London   1979
A Subsidiary of Harcourt Brace Jovanovich, Publishers

COPYRIGHT © 1979, BY ACADEMIC PRESS, INC.
ALL RIGHTS RESERVED.
NO PART OF THIS PUBLICATION MAY BE REPRODUCED OR
TRANSMITTED IN ANY FORM OR BY ANY MEANS, ELECTRONIC
OR MECHANICAL, INCLUDING PHOTOCOPY, RECORDING, OR ANY
INFORMATION STORAGE AND RETRIEVAL SYSTEM, WITHOUT
PERMISSION IN WRITING FROM THE PUBLISHER.

ACADEMIC PRESS, INC.
111 Fifth Avenue, New York, New York 10003

United Kingdom Edition published by
ACADEMIC PRESS, INC. (LONDON) LTD.
24/28 Oval Road, London NW1 7DX

Library of Congress Cataloging in Publication Data

Main entry under title:

Enzyme kinetics and mechanism.

(Methods in enzymology ; v. 63, pt. A)
Includes bibliographical references and index.
1. Enzymes. 2. Chemical reaction, Rate of.
I. Purich, Daniel L. II. Series. [DNLM: 1. Enzymes--
Metabolism. 2. Kinetics. W1 ME9615K v. 63 pt. A /
QU135.3 E61]
QP601.M49 [QP601] 574.1'925'08s
ISBN 0–12–181963–9 vol. 63, pt. A [574.1'925] 79–18746

PRINTED IN THE UNITED STATES OF AMERICA

79 80 81 82   9 8 7 6 5 4 3 2 1

Table of Contents

CONTRIBUTORS TO VOLUME 63 vii

PREFACE ix

VOLUMES IN SERIES xi

Section I. Initial Rate Methods

1. Practical Considerations in the Design of Initial Velocity Enzyme Rate Assays — R. DONALD ALLISON AND DANIEL L. PURICH — 3

2. Techniques in Coupled Enzyme Assays — FREDERICK B. RUDOLPH, BENNETT W. BAUGHER, AND ROBERT S. BEISSNER — 22

3. Summary of Kinetic Reaction Mechanisms — HERBERT J. FROMM — 42

4. Derivation of Initial Velocity and Isotope Exchange Rate Equations — CHARLES Y. HUANG — 54

5. Computer-Assisted Derivation of Steady-State Rate Equations — HERBERT J. FROMM — 84

6. Statistical Analysis of Enzyme Kinetic Data — W. WALLACE CLELAND — 103

7. Plotting Methods for Analyzing Enzyme Rate Data — FREDERICK B. RUDOLPH AND HERBERT J. FROMM — 138

8. Kinetic Analysis of Progress Curves — BRUNO A. ORSI AND KEITH F. TIPTON — 159

9. Effects of pH on Enzymes — KEITH F. TIPTON AND HENRY B. F. DIXON — 183

10. Temperature Effects in Enzyme Kinetics — KEITH J. LAIDLER AND BRANKO F. PETERMAN — 234

11. Approaches to Kinetic Studies on Metal-Activated Enzymes — JOHN F. MORRISON — 257

12. Stability Constants for Biologically Important Metal–Ligand Complexes — WILLIAM J. O'SULLIVAN AND GEOFFREY W. SMITHERS — 294

13. Cryoenzymology: The Study of Enzyme Catalysis at Subzero Temperatures — ANTHONY L. FINK AND MICHAEL A. GEEVES — 336

14. Anomeric Specificity of Carbohydrate-Utilizing Enzymes — STEPHEN J. BENKOVIC — 370

Section II. Inhibitor and Substrate Effects

15. Reversible Enzyme Inhibition — JOHN A. TODHUNTER — 383

16. Product Inhibition and Abortive Complex Formation — FREDERICK B. RUDOLPH — 411

17. The Kinetics of Reversible Tight-Binding Inhibition — JEFFREY W. WILLIAMS AND JOHN F. MORRISON — 437

18. Use of Competitive Inhibitors to Study Substrate Binding Order — HERBERT J. FROMM — 467

19. Use of Alternative Substrates to Probe Multisubstrate Enzyme Mechanisms — CHARLES Y. HUANG — 486

20. Substrate Inhibition — W. WALLACE CLELAND — 500

AUTHOR INDEX 515

SUBJECT INDEX 527

Contributors to Volume 63

Article numbers are in parentheses following the names of contributors.
Affiliations listed are current.

R. DONALD ALLISON (1), *Department of Chemistry, University of California, Santa Barbara, California 93106*

BENNETT W. BAUGHER (2), *Department of Biochemistry, Rice University, Houston, Texas 77001*

ROBERT S. BEISSNER (2), *Department of Biochemistry, Rice University, Houston, Texas 77001*

STEPHEN J. BENKOVIC (14), *Department of Chemistry, Pennsylvania State University, University Park, Pennsylvania 16802*

W. WALLACE CLELAND (6, 20), *Department of Biochemistry, University of Wisconsin, Madison, Wisconsin 53706*

HENRY B. F. DIXON (9), *Department of Biochemistry, Cambridge University, Cambridge CB2 1QW, England*

ANTHONY L. FINK (13), *Division of Natural Sciences, University of California, Santa Cruz, California 95064*

HERBERT J. FROMM (3, 5, 7, 18), *Department of Biochemistry and Biophysics, Iowa State University, Ames, Iowa 50011*

MICHAEL A. GEEVES (13), *Division of Natural Sciences, University of California, Santa Cruz, California 95064*

CHARLES Y. HUANG (4, 19), *Laboratory of Biochemistry, National Institute of Heart, Lung, and Blood, National Institutes of Health, Bethesda, Maryland 20205*

KEITH J. LAIDLER (10), *Department of Chemistry, University of Ottawa, Ottawa, Ontario KIN 9B4, Canada*

JOHN F. MORRISON (11, 17), *Department of Biochemistry, John Curtin School of Medical Research, Australian National University, Canberra, ACT 2601, Australia*

BRUNO A. ORSI (8), *Department of Biochemistry, Trinity College, Dublin, Ireland*

WILLIAM J. O'SULLIVAN (12), *School of Biochemistry, University of New South Wales, Kensington, New South Wales, 2033, Australia*

BRANKO F. PETERMAN (10), *Department of Chemistry, University of Ottawa, Ottawa, Ontario KIN 9B4, Canada*

DANIEL L. PURICH (1), *Department of Chemistry, University of California, Santa Barbara, California 93106*

FREDERICK B. RUDOLPH (2, 7, 16), *Department of Biochemistry, Rice University, Houston, Texas 77001*

GEOFFREY W. SMITHERS (12), *School of Biochemistry, University of New South Wales, Kensington, New South Wales, 2033, Australia*

KEITH F. TIPTON (8, 9), *Department of Biochemistry, Trinity College, Dublin, Ireland*

JOHN A. TODHUNTER (15), *Department of Biology, The Catholic University of America, Washington, D. C. 20064*

JEFFREY W. WILLIAMS (17), *Department of Biochemistry, John Curtin School of Medical Research, Australian National University, Canberra, ACT 2601, Australia*

Preface

In the early years of chemistry the identification of new reactions preceded serious consideration of reaction kinetics, and it was not until Berthelot derived the bimolecular rate equation in 1861 that chemical kinetics offered any real value to the practicing chemist. Fortunately, biochemistry, which had its roots in the late nineteenth century, experienced the benefit of developments in kinetic theory. In fact, kinetic arguments have played a major role in defining the metabolic pathways, the mechanistic action of enzymes, and even the processing of genetic material. Nevertheless, it is amusing to witness the disdain of many investigators toward mechanistic conclusions drawn from kinetic data. After all, kinetic arguments are frequently tediously detailed with algebra and calculus, and so many refuse to believe that such abstract constructs truly apply to real systems. For those of us who derive much fascination, excitement, and satisfaction from the combination of chemical and kinetic probes of enzyme mechanism and regulation, the statement that "kinetics never proves anything" is especially amusing. When one views the definition of the word "proof" as an operation designed to test the validity of a fact or truth, the preceding statement serves only to demonstrate that we have failed to communicate the power and scope of kinetic arguments. The purpose of this volume is to initiate those who are interested in an advanced treatment of enzyme kinetic theory and practice. Indeed, this area of biochemistry is rich in information and experimental diversity, and it is the only means to examine the most fundamental characteristic of enzymes—catalytic rate enhancement.

Parts A (Volume 63) and B (Volume 64) are the first of a series of volumes to treat enzyme kinetics and mechanism, and the chapters presented have been written to provide practical as well as theoretical considerations. However, there has been no attempt on my part to impose a uniform format of symbols, rate constants, and notation. Certainly, uniformity may aid the novice, but I believe that it would also present a burden to those wishing to examine the literature. There, the diversity of notation is enormous, and with good reason, because the textural meaning of particular terms must be considered. In this respect, the practice of utilizing a variety of notations should encourage the student to develop some flexibility and thereby ease the entry into the chemical literature of enzyme dynamics and mechanism. Each of the contributors is an expert in the literature, and I have been especially pleased by the constant reference to key sources of experimental detail.

I wish to acknowledge with pleasure and gratitude the cooperation and ideas of these contributors, and I am indebted in particular to Professors Fromm and Cleland for many suggestions during the initial stages of developing the scope of this presentation. My students, certainly R. Donald Allison, also deserve much praise for surveying the literature and convincing me that a balanced view of the field may be presented in the confines of this series. The staff of Academic Press has also provided great encouragement and guidance, and to them I am deeply indebted. Finally, I wish to acknowledge the wisdom and friendship offered to me by Sidney Colowick and Nathan Kaplan.

DANIEL L. PURICH

METHODS IN ENZYMOLOGY

EDITED BY

Sidney P. Colowick and Nathan O. Kaplan
VANDERBILT UNIVERSITY
SCHOOL OF MEDICINE
NASHVILLE, TENNESSEE

DEPARTMENT OF CHEMISTRY
UNIVERSITY OF CALIFORNIA
AT SAN DIEGO
LA JOLLA, CALIFORNIA

  I. Preparation and Assay of Enzymes
 II. Preparation and Assay of Enzymes
III. Preparation and Assay of Substrates
 IV. Special Techniques for the Enzymologist
  V. Preparation and Assay of Enzymes
 VI. Preparation and Assay of Enzymes *(Continued)*
     Preparation and Assay of Substrates
     Special Techniques
VII. Cumulative Subject Index

METHODS IN ENZYMOLOGY

EDITORS-IN-CHIEF

Sidney P. Colowick   Nathan O. Kaplan

VOLUME VIII. Complex Carbohydrates
Edited by ELIZABETH F. NEUFELD AND VICTOR GINSBURG

VOLUME IX. Carbohydrate Metabolism
Edited by WILLIS A. WOOD

VOLUME X. Oxidation and Phosphorylation
Edited by RONALD W. ESTABROOK AND MAYNARD E. PULLMAN

VOLUME XI. Enzyme Structure
Edited by C. H. W. HIRS

VOLUME XII. Nucleic Acids (Parts A and B)
Edited by LAWRENCE GROSSMAN AND KIVIE MOLDAVE

VOLUME XIII. Citric Acid Cycle
Edited by J. M. LOWENSTEIN

VOLUME XIV. Lipids
Edited by J. M. LOWENSTEIN

VOLUME XV. Steroids and Terpenoids
Edited by RAYMOND B. CLAYTON

VOLUME XVI. Fast Reactions
Edited by KENNETH KUSTIN

VOLUME XVII. Metabolism of Amino Acids and Amines (Parts A and B)
Edited by HERBERT TABOR AND CELIA WHITE TABOR

VOLUME XVIII. Vitamins and Coenzymes (Parts A, B, and C)
Edited by DONALD B. MCCORMICK AND LEMUEL D. WRIGHT

VOLUME XIX. Proteolytic Enzymes
Edited by GERTRUDE E. PERLMANN AND LASZLO LORAND

VOLUME XX. Nucleic Acids and Protein Synthesis (Part C)
Edited by KIVIE MOLDAVE AND LAWRENCE GROSSMAN

VOLUME XXI. Nucleic Acids (Part D)
Edited by LAWRENCE GROSSMAN AND KIVIE MOLDAVE

VOLUME XXII. Enzyme Purification and Related Techniques
Edited by WILLIAM B. JAKOBY

VOLUME XXIII. Photosynthesis (Part A)
Edited by ANTHONY SAN PIETRO

VOLUME XXIV. Photosynthesis and Nitrogen Fixation (Part B)
Edited by ANTHONY SAN PIETRO

VOLUME XXV. Enzyme Structure (Part B)
Edited by C. H. W. HIRS AND SERGE N. TIMASHEFF

VOLUME XXVI. Enzyme Structure (Part C)
Edited by C. H. W. HIRS AND SERGE N. TIMASHEFF

VOLUME XXVII. Enzyme Structure (Part D)
Edited by C. H. W. HIRS AND SERGE N. TIMASHEFF

VOLUME XXVIII. Complex Carbohydrates (Part B)
Edited by VICTOR GINSBURG

VOLUME XXIX. Nucleic Acids and Protein Synthesis (Part E)
Edited by LAWRENCE GROSSMAN AND KIVIE MOLDAVE

VOLUME XXX. Nucleic Acids and Protein Synthesis (Part F)
Edited by KIVIE MOLDAVE AND LAWRENCE GROSSMAN

VOLUME XXXI. Biomembranes (Part A)
Edited by SIDNEY FLEISCHER AND LESTER PACKER

VOLUME XXXII. Biomembranes (Part B)
Edited by SIDNEY FLEISCHER AND LESTER PACKER

VOLUME XXXIII. Cumulative Subject Index Volumes I-XXX
Edited by MARTHA G. DENNIS AND EDWARD A. DENNIS

VOLUME XXXIV. Affinity Techniques (Enzyme Purification: Part B)
Edited by WILLIAM B. JAKOBY AND MEIR WILCHEK

VOLUME XXXV. Lipids (Part B)
Edited by JOHN M. LOWENSTEIN

VOLUME XXXVI. Hormone Action (Part A: Steroid Hormones)
Edited by BERT W. O'MALLEY AND JOEL G. HARDMAN

VOLUME XXXVII. Hormone Action (Part B: Peptide Hormones)
Edited by BERT W. O'MALLEY AND JOEL G. HARDMAN

VOLUME XXXVIII. Hormone Action (Part C: Cyclic Nucleotides)
Edited by JOEL G. HARDMAN AND BERT W. O'MALLEY

VOLUME XXXIX. Hormone Action (Part D: Isolated Cells, Tissues, and Organ Systems)
Edited by JOEL G. HARDMAN AND BERT W. O'MALLEY

VOLUME XL. Hormone Action (Part E: Nuclear Structure and Function)
Edited by BERT W. O'MALLEY AND JOEL G. HARDMAN

VOLUME XLI. Carbohydrate Metabolism (Part B)
Edited by W. A. WOOD

VOLUME XLII. Carbohydrate Metabolism (Part C)
Edited by W. A. WOOD

VOLUME XLIII. Antibiotics
Edited by JOHN H. HASH

VOLUME XLIV. Immobilized Enzymes
Edited by KLAUS MOSBACH

VOLUME XLV. Proteolytic Enzymes (Part B)
Edited by LASZLO LORAND

VOLUME XLVI. Affinity Labeling
Edited by WILLIAM B. JAKOBY AND MEIR WILCHEK

VOLUME XLVII. Enzyme Structure (Part E)
Edited by C. H. W. HIRS AND SERGE N. TIMASHEFF

VOLUME XLVIII. Enzyme Structure (Part F)
Edited by C. H. W. HIRS AND SERGE N. TIMASHEFF

VOLUME XLIX. Enzyme Structure (Part G)
Edited by C. H. W. HIRS AND SERGE N. TIMASHEFF

VOLUME L. Complex Carbohydrates (Part C)
Edited by VICTOR GINSBURG

VOLUME LI. Purine and Pyrimidine Nucleotide Metabolism
Edited by PATRICIA A. HOFFEE AND MARY ELLEN JONES

VOLUME LII. Biomembranes (Part C: Biological Oxidations)
Edited by SIDNEY FLEISCHER AND LESTER PACKER

VOLUME LIII. Biomembranes (Part D: Biological Oxidations)
Edited by SIDNEY FLEISCHER AND LESTER PACKER

VOLUME LIV. Biomembranes (Part E: Biological Oxidations)
Edited by SIDNEY FLEISCHER AND LESTER PACKER

VOLUME LV. Biomembranes (Part F: Bioenergetics)
Edited by SIDNEY FLEISCHER AND LESTER PACKER

VOLUME LVI. Biomembranes (Part G: Bioenergetics)
Edited by SIDNEY FLEISCHER AND LESTER PACKER

VOLUME LVII. Bioluminescence and Chemiluminescence
Edited by MARLENE A. DELUCA

VOLUME LVIII. Cell Culture
Edited by WILLIAM B. JAKOBY AND IRA H. PASTAN

VOLUME LIX. Nucleic Acids and Protein Synthesis (Part G)
Edited by KIVIE MOLDAVE AND LAWRENCE GROSSMAN

VOLUME LX. Nucleic Acids and Protein Synthesis (Part H)
Edited by KIVIE MOLDAVE AND LAWRENCE GROSSMAN

VOLUME 61. Enzyme Structure (Part H)
Edited by C. H. W. HIRS AND SERGE N. TIMASHEFF

VOLUME 62. Vitamins and Coenzymes (Part D)
Edited by DONALD B. MCCORMICK AND LEMUEL D. WRIGHT

VOLUME 63. Enzyme Kinetics and Mechanism (Part A: Initial Rate and Inhibitor Methods)
Edited by DANIEL L. PURICH

VOLUME 64. Enzyme Kinetics and Mechanism (Part B: Isotopic Probes and Complex Enzyme Systems) (in preparation)
Edited by DANIEL L. PURICH

VOLUME 65. Nucleic Acids (Part I) (in preparation)
Edited by LAWRENCE GROSSMAN AND KIVIE MOLDAVE

VOLUME 66. Vitamins and Coenzymes (Part E) (in preparation)
Edited by DONALD B. MCCORMICK AND LEMUEL D. WRIGHT

VOLUME 67. Vitamins and Coenzymes (Part F) (in preparation)
Edited by DONALD B. MCCORMICK AND LEMUEL D. WRIGHT

VOLUME 68. Recombinant DNA (in preparation)
Edited by RAY WU

VOLUME 69. Photosynthesis and Nitrogen Fixation (Part C) (in preparation)
Edited by ANTHONY SAN PIETRO

Methods in Enzymology

Volume 63
ENZYME KINETICS AND MECHANISM
Part A
Initial Rate and Inhibitor Methods

Section I
Initial Rate Methods

[1] Practical Considerations in the Design of Initial Velocity Enzyme Rate Assays

By R. Donald Allison and Daniel L. Purich

Developing a reliable initial velocity enzyme assay procedure is of prime importance for achieving a detailed and faithful analysis of any enzyme. This objective is quite different from the use of enzyme assays in enzyme purification or clinical chemistry, where the focus is on estimates of the enzyme content of various samples. In that case, one is particularly concerned with optimizing assay conditions by including substrates, cofactors, and activators at optimal (often saturating) levels and with minimizing interfering agents. Thus, the emphasis is on determining enzyme concentration in a routine, easy, and reproducible fashion. On the other hand, the enzyme kineticist must often work at subsaturating substrate and effector levels to evaluate the rate-saturation behavior. When two or more substrates are involved, the problem of obtaining initial velocity data becomes more considerable. This chapter treats of the practical aspects of initial rate enzyme assay.

General Experimental Design

The initial rate phase of an enzymic reaction typically persists for 10 sec to several hundred seconds. Thus, various methods including spectrophotometry, radioactive assay, and pH-stat procedures may be used along with manual mixing and manipulation of samples. Prior to addition of the enzyme (or one of the substrates) to initiate the reaction, the assay sample (usually in 0.05–3.0 ml volumes) is preincubated at the reaction temperature for several minutes to achieve thermal equilibration, and a small aliquot of enzyme is added to initiate the reaction. The increase in the product concentration or the drop in substrate concentration may then be measured. The basic goals are to initiate the reaction in a manner that leads to immediate attainment of the initial velocity phase and to obtain an accurate record of the reaction progress.

For most enzyme rate equations to apply to real systems, one must be certain that the conditions placed upon the mathematical derivation are satisfied in the experiment. Since rate equations become quite complex as product accrual becomes significant, the initial rate assumption is frequently taken to linearize the equations. Experimentally, one draws the tangent to the reaction progress curve as shown in Fig. 1. The best estimates of the slope of this line will be obtained from the most complete

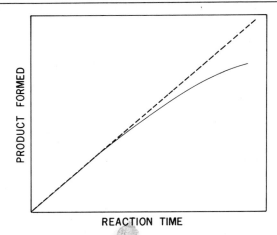

FIG. 1. Plot of product formation versus reaction time for an enzyme-catalyzed reaction. The solid line represents the reaction progress, and the dashed line is the tangent to the curve at low product formation. This tangent is the initial velocity, and it is expressed in units of molarity per minute.

record of the initial rate phase, and continuous assays are thus preferable to single-point assays (see below). The duration of the linear initial rate phase depends upon many factors, including the equilibrium constant, the fractional saturation of the enzyme with substrate(s) and product(s), the buffering capacity of the medium, and the concentration ratio of the least abundant substrate relative to the enzyme. Below a $[S]_{total}/[E]_{total}$ value of 100, the steady state may not persist for long, and nonlinear initial rates will frequently be observed. In some cases, the rate may appear linear, but virtual linearity should not be the only criterion used in establishing reaction conditions. With conditions such as a very favorable equilibrium constant, no product inhibition, and a high $[S]_{total}/[E]_{total}$ value, the initial reaction velocities can be maintained for a considerable period of time.

The Initial Rate Condition

As a general guideline, one assumes that the initial rate persists for a period of time during which the substrate(s) concentration is within 10% of the initial value. This is probably true only for reactions that are thermodynamically quite favorable, and even so it is best to choose an assay method that is safely within this range. Nonetheless, there is no *a priori* guarantee that product inhibition will not account for a significant error in the estimation of initial rates, and tests should be made for even the most favorable reactions (see below). Since the equilibrium constant for the particular reaction will presumably be known, one may estimate the extent of the reaction by the following simple expression:

$$K'_t = \frac{(P_0 + x)(Q_0 + x)(R_0 + x) \ldots}{(A_0 - x)(B_0 - x)(C_0 - x) \ldots} \tag{1}$$

where K'_t is the apparent product:substrate ratio (or "mass action ratio") at time t, x is the concentration change measured at the midpoint of the experimental assay (i.e., time t), and A_0, B_0, C_0, P_0, Q_0, and R_0 are the initial substrate and product concentrations. If it is found that K'_t is not much different from the apparent equilibrium constant (K') for the reaction, then one must reduce x by use of a more sensitive assay. Let us consider the case of yeast hexokinase, where the apparent equilibrium constant ($K' = 4.9 \times 10^3$ at pH 7.5) is quite favorable. Assuming we have both the glucose and ATP concentrations at 0.1 mM (i.e., near their Michaelis constants), then a 5% conversion would yield a K'_t of 2.8×10^{-3}, suggesting that the system is quite far from equilibrium. On the other hand, we may consider the acetate kinase reaction (written in the direction of acetyl phosphate formation where $K' = 3.3 \times 10^{-4}$ at pH 7.4). At an acetate concentration of 10 mM and an ATP level of 1 mM, a 5% conversion of substrate to products would yield a K'_t of 2.7×10^{-4}, not far away from the equilibrium value. In this case, it would be advisable to reduce the percentage of substrate conversion in the rate assay. An obvious extension of these comments is that the deviation from initial rates may be a greater problem in some product inhibition studies.

It is not certain that product accumulation during the "initial rate" will lead to insignificant error even when substrate conversion is quite low. Indeed, for some systems the inhibition constants for a particular product may be quite low. Consider the brain hexokinase reaction where $K_{\text{glucose-6-P}}$ is approximately 10^{-5} M[1] but the K_{glucose} is about 5-fold higher. Another case in point is the PRPP:ATP phosphoribosyltransferase from *Salmonella typhimurium;* here N^1-phosphoribosyl-ATP has a dissociation constant of 3.8×10^{-6} M, but the affinity for either substrate is considerably lower.[2] The reduced coenzyme in NAD^+-dependent dehydrogenases is frequently a potent inhibitor as well. One strategy around the problem of product accumulation is to remove the product by use of an auxiliary enzyme system (see this volume [2]). This can be especially useful when the auxiliary system also serves to regenerate one of the substrates. For example, the pyruvate kinase/lactate dehydrogenase coupled assay for kinases maintains the initial ATP concentration, and it also provides for a

---

[1] J. Ning, D. L. Purich, and H. J. Fromm, *J. Biol. Chem.* **244**, 3840 (1969).
[2] J. E. Kleeman and S. M. Parsons, *Arch. Biochem. Biophys.* **175**, 687 (1976).

convenient method of assay. Another example is provided by the use of adenylate kinase (AK) and AMP deaminase (AD) in the assay for nucleoside diphosphate kinase (NDPK):

$$\text{ATP} + \text{GDP} \xrightleftharpoons{\text{NDPK}} \text{ADP} + \text{GTP}$$
$$2\text{ADP} \xrightleftharpoons{\text{AK}} \text{ATP} + \text{AMP}$$
$$\text{AMP} \xrightleftharpoons{\text{AD}} \text{IMP} + \text{NH}_3 \qquad (2)$$

Here, ATP is partially regenerated by the auxiliary enzyme system.

Finally, one may be forced to restrict the assay period to prevent the occurrence of competitive reactions. This is particularly true with many transglycosylases, where a number of competing processes may proceed in parallel. In other cases, there are kinases that catalyze slow hydrolysis of ATP in the absence of a phosphoryl acceptor substrate. For this reason, it is advisable to rigorously establish the reaction stoichiometry by characterization and quantitation of product formation.

When the initial rate phase is particularly short, the observed velocity will fall off quickly. This can lead to a situation that yields false cooperativity in plots of velocity versus substrate concentration. The reason for this is that the duration of the initial rate period is short at low substrate concentration and longer at high levels of substrate. Thus, the velocities will be depressed at low concentrations and a sigmoidal curve will be observed. This problem becomes most serious with single-point determinations of velocity where a continuous record of the reaction progress is not available. For example, glutamine synthetase may be assayed by the γ-glutamyl transfer reaction.

$$\text{Glutamine} + \text{NH}_2\text{OH} \xrightleftharpoons[\text{Me}^{2+}]{\text{ADP, As}_\text{i}} \gamma\text{-glutamyl hydroxamate} + \text{NH}_3 \qquad (3)$$

The hydroxamate concentration may be estimated by a colorimetric assay with ferric chloride in strongly acidic solutions. The assay is fast and convenient, but, as shown in Fig. 2, care must be exercised to validate the linear formation of hydroxamate with time. In this particular case, the reaction was linear for a short time (10 min), and the plots are based upon values plotted at 10-min and 20-min incubations. A false cooperativity with respect to L-glutamine becomes pronounced and evident because the initial rate phase has disappeared.

Enzyme Purity and Stability

Ideally, one wants a pure, stable enzyme preparation. Although enzyme purity permits one to evaluate numerically various kinetic parameters that depend on a knowledge of the enzyme concentration, a more important aspect of purity deals with elimination of contaminating activities.

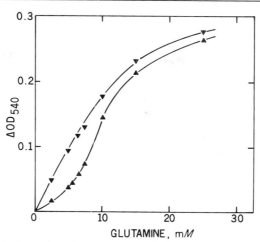

FIG. 2. Plot of the formation of γ-glutamyl hydroxamate (measured at 540 nm) versus the concentration of glutamine. ADP, hydroxylamine, arsenate, magnesium chloride, HEPES (pH 7.8), and KCl were present at 1.5 mM, 100 mM, 30 mM, 30 mM, 40 mM, and 100 mM, respectively (temperature 37°). Glutamine was varied from 2.5 to 25 mM. The final concentration of *Escherichia coli* glutamine synthetase (state of adenylylation, 1.5) was 4 μg/ml, and the velocity was measured by single-point assays at 10 (▼) and 20 min (▲), respectively. D. L. Purich, unpublished findings.

It is imperative to demonstrate that such minor activities do not unfavorably interfere with the particular assay under consideration. For multisubstrate reactions, the stability of each substrate, product, and effector may be examined separately in the absence and in the presence of the enzyme preparation. The enzyme's presence should not cause any change in the substrate, product, or effector concentrations in the absence of the complete reaction mixture. For example, one might consider the liver pyruvate kinase reaction, which is activated by fructose 1,6-bisphosphate. ADP should be unaffected by the enzyme in the absence of phosphoenol pyruvate (PEP), and this may be confirmed by chromatographic analysis. Likewise, the PEP should not be affected by the enzyme in the absence of ADP, and the fructose 1,6-bisphosphate should be stable with the enzyme in the absence or presence of ADP and PEP. In this case, the presence of contaminating adenylate kinase, enolase, fructose-1,6-bisphosphatase, or phosphofructokinase activity can easily be determined. Adenylate kinase is a frequent contaminant in nucleotide-dependent phosphotransferases; fortunately, in this case, one may purchase specific and potent inhibitors including P^1,P^4-di(adenosine-5')-tetraphosphate and P^1,P^5-di(adenosine-5')-pentaphosphate to block this activity.[3,4]

[3] D. L. Purich and H. J. Fromm, *Biochim. Biophys. Acta* **276**, 563 (1972).
[4] G. E. Lienhard and I. I. Secemski, *J. Biol. Chem.* **248**, 1121 (1973).

Sometimes it is impossible to eliminate a side reaction that arises from the intrinsic catalytic properties of a particular enzyme. For example, there are several transglycosylation reactions that are catalyzed by sucrose phosphorylase, and it was such evidence that provided insight into the possible role of a glucosyl-enzyme intermediate. Likewise, *E. coli* acetate kinase also catalyzes a purine nucleoside-5'-diphosphate kinase reaction in the absence of acetyl phosphate. In this case, it was possible to use inactivation studies to demonstrate parallel loss of both activities when the enzyme was first treated with acetyl phosphate and subsequently submitted to hydroxaminolysis.[5] Unfortunately, the loss of some enzymic activities by such treatment does not prove that a common active site was responsible for both activities.

Another check of enzyme purity may be the use of coupled dye assays with polyacrylamide gels. The presence of isozymes may be frequently uncovered by such procedures. This is a powerful probe that is recommended wherever practicable. One problem with using coupled assays to generate NADH or NADPH is that the auxiliary enzymes must diffuse into the gel. Newer methods, especially flat-bed isoelectric focusing, permit the use of starch, agarose, and very fine dextrans. Such support materials allow uniform entry of auxiliary enzymes into the matrix of the support.

Enzyme instability is also a problem often encountered in enzyme assays. One must search out conditions that minimize or eliminate time-dependent changes in the enzyme's catalytic power. Often the presence of stabilizing agents such as glycerol, certain salts, one substrate, or an effector may render the enzyme more stable. Indeed, instability is common with very pure enzyme preparations where adsorption to glass walls, the absence of protective effects by other proteins, or high dilution into the assay mixture may be encountered. With regulatory enzymes this may be a particular problem since protein conformational mobility may lead to conformational states that are kinetically controlled. Simple heat treatments during enzyme purification or cold exposure during storage may lock the enzyme into an inactive (or less active) conformation.

If inactivation of enzyme during the course of a series of assays is simply due to loss of activity, one may correct for this by establishing an activity decay curve. Here, a standard reference assay is intermittently used to measure the variation in enzyme activity during the experiment. But one must be certain that the inactivation is just the result of a decrease in the fraction of total active enzyme. If there is a change to a less active state, either the Michaelis constant or the maximum velocity will

---

[5] B. C. Webb, J. A. Todhunter, and D. L. Purich, *Arch. Biochem. Biophys.* **173**, 282 (1976).

change and might jeopardize the validity of the kinetic study. (This possibility is frequently ignored in many experiments.) A change in Michaelis constant with time can be detected by using two reference assays: one at subsaturating substrate levels and another at saturating levels.

Another useful approach is to examine the effect of preincubation of the enzyme with one of the substrates in a multisubstrate enzyme-catalyzed reaction. Substrate binding frequently acts to stabilize an enzyme to thermal denaturation and proteolysis. Likewise, some enzymes have remarkable conformational mobility, and reversibly inactive forms may occur in the absence of a stabilizing effect from a substrate. Incidentally, it might be added that there are situations where an inhibitor might actually "snap" an inactive enzyme into its active conformation. The *Salmonella typhimurium* PRPP:ATP phosphoribosyltransferase[6] provides one such an example. Histidine is a potent allosteric feedback inhibitor of this enzyme, and it was observed that prior incubation with the inhibitor can eliminate lags in the assay. In this case, the investigators made good advantage of the cooperativity of histidine binding by exposing the concentrated enzyme to 0.4 mM histidine and then diluting the enzyme by a factor of 315 to a final histidine level where its inhibitory effect on the initial rate was negligible. Under these conditions, the active enzyme form could be studied without a lag occurring. Likewise, the *Escherichia coli* acetate kinase enzyme undergoes reversible cold denaturation, and brief incubation with ATP, a product, can restore the activity.[5]

Some enzymes require reduction of critical thiol groups for activity, and the inclusion of 2-mercaptoethanol, dithiothreitol, or dithioerythritol is useful to restore activity. With *E. coli* coenzyme A-linked aldehyde dehydrogenase, it was observed that omission of 2-mercaptoethanol resulted in a lag in the activity in both directions. The lag was eliminated by prior incubation for 15 min with enzyme, 2-mercaptoethanol, NAD^+, and either CoA or acetaldehyde. In this case, the reaction was then initiated by addition of the substrate omitted in the preincubation. It was possible to demonstrate a requirement for the thiol in both directions of the reaction, and this eliminates a trivial explanation of the thiol effect in terms of reducing coenzyme A. Other cases of lags are described in Vol. 64 [8]. Generally dithiothreitol (DTT) and dithioerythritol (DTE) are preferred as reducing agents.[7] The intramolecular displacement of reduced enzyme is more facile than intermolecular reduction of mixed disulfides, and DTT and DTE may be used at lower concentrations.

---

[6] R. M. Bell, S. M. Parsons, S. A. Dubravac, A. G. Redfield, and D. E. Koshland, Jr., *J. Biol. Chem.* **249**, 4110 (1974).

[7] W. W. Cleland, *Biochemistry* **3**, 480 (1964).

Finally, the enzyme dilution should be minimized. The presence of 1–2 mg/ml of serum albumin or another protein may frequently afford the enzyme greater stability, and this approach has been gainfully exploited in many cases. It is advisable to test several different proteins if this method is used, especially in light of the tendency of serum albumin to bind fatty acids and other metabolites.

Substrate Purity

Many biochemical substances are fairly unstable, and impurities in each substrate must be considered as a possible source of experimental error. Although chromatographic and spectral analyses are among the best tools for establishing substrate purity, enzymic analysis is probably one of the most powerful tests of purity. Here, advantage is made of the stereochemical specificity of certain enzymes, but contaminating alternative substrates might give misleading results.

Since substrates and competitive inhibitors are generally sufficiently similar to bind to the same active site, it is not surprising that the similar physical properties of some substrates and inhibitors prevent facile and complete purification. With a competitive inhibitor present in a constant ratio to substrate, the observed kinetic parameters may be affected. This can be shown by rearranging the competitive inhibition expression [Eq. (4)] to account for this contamination.

$$\frac{1}{v} = \frac{1}{V_m} + \frac{K_m}{V_m} \frac{1}{[S]} \left(1 + \frac{[I]}{K_i}\right) \quad (4)$$

When an inhibitor is present in the substrate (i.e., $[I] = \alpha[S]$), we get

$$\frac{1}{v} = \frac{1}{V_m} + \frac{K_m \alpha}{V_m K_i} + \frac{K_m}{V_m} \frac{1}{[S]} \quad (5)$$

The form of this equation is indistinguishable from the Michaelis–Menten equation, and the double-reciprocal plots will yield the wrong estimates of the kinetic parameters. The occurrence of vanadate ions in some commercial yeast ATP preparations is a notable example of this situation. Another similar situation may occur when substrates are contaminated by alternative substrates. Depending on whether the alternative substrate has a different V_{max} or K_m, or both, a variety of nonlinear reactions may be observed. A corollary situation also occurs with inhibitors or effectors containing substrates, and incomplete inhibition may be observed in such cases. For example, the validity of product inhibition studies of nucleoside diphosphate kinase may be compromised if the nucleoside triphosphate contains significant levels of the corresponding nucleoside diphosphate. In this respect, the stability of substrates and effectors can be extremely important.

Sometimes, substrate and inhibitor instability is so serious that detectable decay may occur during the initial rate assay. This may be especially true for impure enzyme preparations containing contaminating enzyme activities. For multisubstrate cases, it is fairly easy to examine the stability of each substrate separately to check for this possibility. By using identical concentrations of two substrates containing different isotopic labels, one may verify the stoichiometry of bisubstrate reactions by the maintenance of the ratio of radioactivity. For example, [^{3}H]glutamate and [^{14}C]ATP may be used in the glutamine synthetase reaction to examine the presence of contaminating enzymes acting on either substrate. If identical concentration of ATP and glutamate are used, then the [^{3}H]:[^{14}C] ratio of the products should be identical to the same ratio of the substrates. While such tests appear a bit tedious, there is much merit in preventing the accumulation of false data. When effectors are added to an enzyme system, it may also prove to be advantageous to demonstrate their stability by reisolation or direct assay.

Range of Substrate Levels

For many one-substrate systems, choice of the substrate concentration range is not particularly a problem. In preliminary trials, one merely chooses the widest range about the Michaelis constant with due care to avoid substrate inhibition. A rough value of K_m may then be estimated, and the range can be refined. Since the greatest velocity change occurs in the region of the K_m, it is frequently satisfactory to vary the concentration from about 0.2 to 5.0 times this constant. This changes the fractional attainment of maximal velocity from 0.14 to 0.83, and reasonable estimates of K_m can be made. (Consult this volume [6] for statistical treatment of rate data.) Since reciprocal plots are commonly used to analyze the rate data, it is best to choose substrate concentrations that yield an equal spacing across the reciprocal plot. By initially making up the most concentrated sample, one may dilute with buffer to get dilutions of 1/1, 1/3, 1/5, 1/7, 1/9, 1/11, etc., and these will give values of 1, 3, 5, 7, 9, 11, etc. on the abscissa of a Lineweaver–Burk plot.

For one-substrate systems requiring a nonconsumed cofactor (e.g., monovalent or divalent metal ions, an essential activator) and for all multisubstrate cases, the choice of a suitable reactant concentration may be more tedious. Let us consider the case of a bisubstrate enzyme to illustrate the problem. Now, the velocity is a function of two components, and the relative contribution of each to velocity is determined by the value of the rate constants in the initial rate expression. If both substrates are below one-third to one-half of the corresponding substrate dissociation constants, the fractional attainment of V_m will be small, and velocity measurements may have considerable error. Above concentrations of substrates corresponding to five times their respective dissociation constants, the change in velocity will be relatively small. Some investigators work in a very narrow range, but then the chance for obscuring slope changes can be sizable. Fromm[8] has outlined a useful method for obtaining rate data using a five-by-five matrix of substrate concentrations. Three solutions (A, containing one substrate at the highest level to be employed; B, containing the second substrate also at the highest level; and C, containing buffer, nonvaried cofactors, and auxiliary enzymes) are prepared. Then, five dilutions of solution A (A/1, A/3, A/5, A/7, and A/9) and five dilutions of solution B (as solution A was prepared) are combined with solution C added to each. Thus, twenty-five velocity measurements are made in a single experimental trial, and one may make plots of v^{-1} vs $[A]^{-1}$ at five constant [B] levels and likewise v^{-1} vs $[B]^{-1}$ at five constant [A] levels.

For three-substrate enzymic systems the problems of data gathering become more cumbersome. To use the approach outlined above, 125 velocity measurements for each experiment would be required if A, B, and C were each varied at five concentrations. Even with a continuous-assay protocol, the amount of time needed to carry out a single experiment would be considerable. Thus, two basic procedures have been used in such investigations. In the Frieden method[9] one substrate is constant while the other two are treated like solutions A and B described in the previous paragraph. In this way the three-substrate system reduces to a pseudo-two-substrate mechanism, but care must be taken to keep the nonvaried substrate at a level above its respective Michaelis constant but nonsaturating. As noted by Fromm,[8] there is always the possibility that a high concentration of the nonvaried substrate can lead to apparent parallel-line data in double-reciprocal plots. The careful investigator will repeat experiments that yield parallel-line plots, but at a lower level of the nonvaried substrate. An example of the Frieden protocol is presented in Fig. 3 for the sheep brain glutamine synthetase system.[10] A second method was proposed and implemented by Fromm and co-workers.[8,11-13]

---

[8] H. J. Fromm, "Initial Rate Enzyme Kinetics." Springer-Verlag, Berlin and New York, 1975.
[9] C. Frieden, *J. Biol. Chem.* **234**, 2891 (1959).
[10] R. D. Allison, J. A. Todhunter, and D. L. Purich, *J. Biol. Chem.* **252**, 6046 (1977).
[11] H. J. Fromm, *Biochim. Biophys. Acta* **139**, 221 (1967).
[12] F. B. Rudolph, D. L. Purich, and H. J. Fromm, *J. Biol. Chem.* **243**, 5539 (1968).
[13] F. B. Rudolph and H. J. Fromm, *J. Biol. Chem.* **244**, 3832 (1969).

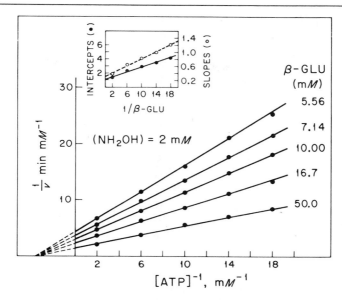

FIG. 3. Plot of the reciprocal of the initial reaction velocity (in units of min/mM) versus the reciprocal of ATP concentration at varying levels of β-glutamate (β-GLU) and a constant level of NH$_2$OH (2 mM). The amount of ovine brain glutamine synthetase used was 0.2 μg. Assays were performed at 37° in 50 mM HEPES (pH 7.2), 100 mM KCl, with free, uncomplexed Mg^{2+} held at 1 mM and a final reaction volume of 0.135 ml. *Inset:* A replot of the slopes and intercepts versus the reciprocal of β-glutamate concentration. From R. D. Allison, J. A. Todhunter, and D. L. Purich, *J. Biol. Chem.* **252**, 6046 (1977).

Here, one substrate is varied while the other two substrates are maintained constant in the general concentration range of their Michaelis constants. The experiment is then repeated; however, a different concentration of fixed substrates is chosen, care being exercised to maintain the ratio of fixed substrates constant in both experiments. This procedure is then repeated until all substrates are varied. It is noteworthy that all mechanisms involving quaternary complexes (i.e., EABC complexes) will give Lineweaver-Burk plots that intersect to the left of the v^{-1} axis. On the other hand, a Ping Pong mechanism will yield one or more sets of data that result in parallel-line plots. This procedure is illustrated in Fig. 4 by experiments of Rudolph *et al.*[12] on the *E. coli* CoA-linked aldehyde dehydrogenase reaction. The major limitation of the latter approach is that values for kinetic parameters are difficult to obtain from such plots.

Finally, a common assumption in many experiments is that the concentration of enzyme-bound substrate is negligible relative to the total substrate concentration, and this eliminates various quadratic terms from rate expressions. In most cases, this assumption will be valid, but it is

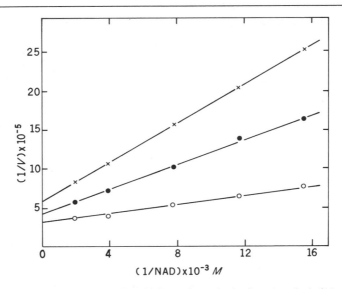

FIG. 4. Plot of the reciprocal of the initial reaction velocity (in units of min/M) versus the reciprocal of NAD$^+$ concentration (in units of M^{-1}). The respective concentrations of CoA and acetaldehyde were: ○, 9.75 μM and 5.50 mM; ●, 4.87 μM and 2.76 mM; ×, 2.44 μM and 1.38 mM. The NAD$^+$ concentration was varied from 5.2×10^{-4} M to 6.5×10^{-5} M. For other experimental conditions for the assay of the *Escherichia coli* coenzyme A-linked aldehyde dehydrogenase, see F. B. Rudolph, D. L. Purich, and H. J. Fromm, *J. Biol. Chem.* **243**, 5539 (1968).

advisable to determine that the total true substrate concentration exceeds the active enzyme concentration by 50- to 100-fold. The emphasis on a true "substrate" refers to the fact that many enzymes act on only one conformer of the substrate, which if in low concentration may violate the above assumption. For example, this may occur in cases where the open chain of a monosaccharide is the true substrate. With very low K_m substrates, some problems may also be encountered. Many enzymes acting on macromolecules (e.g., protein and polynucleotide kinases, various polymerases and depolymerases) may have exceptionally low Michaelis constants for the macromolecules, and their turnover numbers may not be large. In such cases, the substrate and enzyme concentrations required to yield an adequate rate determination may be in the same neighborhood, and the velocity dependence will be complex. Here, the initial rate assumption becomes impractical, and computer fitting may be required.

Analytical Methods

Every activity assay relies upon an analytical method to properly represent the progress of reaction. The possibility of side reactions during the analysis must be adequately controlled to maintain experimental validity. With enzymic systems, it is convenient to distinguish two different analytical methods: continuous and periodic sampling. The choice depends upon the availability of some reliable method to continuously follow the extent of reaction, and such methods generally rely upon some spectral change attending the reaction. Spectrophotometry, radioactivity methods, polarimetry, fluorimetry, polarography, and pH measurements represent common methods for analytically determining concentration changes. Only the first is discussed here since they are probably the most frequently applied methods. Periodic sampling is frequently utilized when the number of assays is quite large or when an adequate continuous assay is unavailable. Basically, one estimates product formation over a fixed period of reaction.

The spectrophotometric method thrives as a result of the wide variety of naturally occurring chromophores (including nucleotides, coenzymes, thiol esters, and hematoporphyrins) and many synthetic alternative substrates containing any of a variety of chromophoric groups. The most common assays involve NAD^+-linked dehydrogenases by taking advantage of the large molar extinction of NADH at 340 nm ($\epsilon_M = 6.22 \times 10^3\ M^{-1}\ cm^{-1}$). Their use as auxiliary enzyme assay systems is described in this volume [2].

By far the best spectrophotometers use a double monochromator system for "purifying" the light. This is especially important at high absorbances, where deviation from the Beer–Lambert relation may occur. It cannot be overemphasized that one must work under conditions where the absorbance change is a *linear* measure of the concentration change of the absorbing reaction component. In the presence of a significant stray light component leaking through the monochromator:

$$A = \log \left(\frac{I_0 + I_s}{I + I_s} \right) \neq \epsilon c l \tag{6}$$

where A, I_0, I, I_s, ϵ, c, and l are absorbance, incident light intensity, transmitted light intensity, stray light intensity, the molar absorptivity ($M^{-1}\ cm^{-1}$), molar concentration, and light path length (cm). At high sample absorbance, the value of I may approach I_s and the error will be considerable. Other sources of nonlinearity may result from micellar formation, stacking of aromatic chromophores, and sample turbidity. The first two may be relieved by addition of certain nonionic detergents, provided no unfavorable effects on the enzyme occur. The latter may be prevented by centrifugation or filtration. For highest sensitivity of measurement, the wavelength should be adjusted to yield the greatest difference between substrate and product absorbances. This may not necessarily be the so-called λ_{max} for the product, since the spectra of substrate and product may overlap. If the λ_{max} can be conveniently used, it is probably the region that best obeys Beer–Lambert's law. Of course, a record of the entire spectrum would be desirable in some special cases, but single-wavelength monitoring is probably the most common method. In some cases, changes at several wavelengths may be followed, and a simultaneous equation treatment of the rate data can be applied.

The most sensitive direct assay of product accumulation can be made with radionuclides. Assays can be made continuously or by periodic sampling, and the latter is most frequently employed. Product formation, rather than substrate depletion, provides the most sensitivity and accuracy. In periodic sampling protocols, it is absolutely necessary to verify that the reaction is thoroughly quenched at the time of sampling. Likewise, the stability of labeled substrate and product during the quenching of the reaction's progress must be demonstrated. These two problems are frequently the source of published errors in enzyme rate data. Periodic sampling requires one to stop reactions completely at desired times. Many agents (including ethanol, other organic solvents, mercurials, phenol, acids, bases, EDTA, EGTA) may be employed to quench the progress of enzyme-catalyzed reactions. In addition, heating can often effect complete denaturation of the enzyme. The problem that is frequently encountered is the failure to rapidly achieve complete quenching. This is especially true for low molecular weight enzymes which have very stable protein structures (e.g., ribonuclease or adenylate kinase). Nonetheless, one may generally rely on pH changes to stop many reactions, provided the conditions are suited to substrate and product stability. The chief advantage of acid or base additions is the rapidity of the inactivation, and storage on ice can further diminish enzymic activity and ensure product stability. For each new experimental condition, it is useful to demonstrate the efficiency of the stopping procedure, and one convenient method involves the use of radioisotopes. Here, the enzyme sample with substrates, cofactors, and buffer, as used in the assay, is subjected to the stopping protocol. Isotopically labeled substrate is then added and any conversion to product resulting from incomplete quenching may be measured. It is important to mimic the reaction conditions since substrates, cofactors, and buffers are known to affect the susceptibility of many enzymes to inactivation.

Some techniques for substrate and product separation include ion exchange, chromatography, simple extraction or adsorption, electrophoresis, derivatization, precipitation, isotope dilution, and the deliberate labilization of tritium or carbon as tritiated water or labeled CO_2. Undoubtedly, medium- and high-pressure liquid chromatography will be increasingly exploited as a high-resolution method. Indeed, poor resolution is the single most important source of high blanks in such assays, and the high sensitivity of the assay is dependent upon the minimization of the radioactivity of blank samples. The occurrence of labeled impurities in the substrate resulting from nonenzymic processes can also plague sensitive measurements. Nonetheless, the case of substrate and product resolution has been remarkably increased through the availability of DEAE-, ECTEOLA-, PEI-, and CM-cellulose papers and thin-layer plates. In addition, several types of paper-loaded polystyrene ion exchangers have been advantageously employed. The sample may be added quickly without regard to the size of the dampened spot because the ions will adsorb near the point of addition. Then, by addition of absolute ethanol or another volatile solvent, one can effect complete arrest of the reaction. With ion-exchange papers, the capacity is generally $1-10$ μeq per square centimeter; thus, it is best to consider the amount of total ions added to each spot. For complete nucleotide adsorption, the magnesium ion level should be below 1 mM. For this reason, EDTA may be required in the buffer used to develop the chromatogram.[14,15]

Sample counting can also be the source of problems, especially if the radioactive samples are counted directly on the paper support. With ^{3}H and ^{14}C the amount of quenching of the radioactivity can be sizable, and it is preferable to elute ^{3}H samples and to count them as liquid samples. In any case, the properties of the scintillation counting fluid should be considered to prevent slow release of counts from the paper support into the scintillant. It is advisable to recount one sample for a number a hours to ascertain that this is not a source of possible error. Even with dissolved compounds such as ATP, the low solubility in the scintillation fluor may require use of thixatropic agents for gel suspensions.

The radioactive substrate should be scrupulously examined for impurities that result from radiolysis. This is particularly important with highly labeled materials when ethanol or some other stabilizing agent cannot be present in the sample. The position of labeling should also be such that no primary or secondary kinetic isotope effects can lead to apparently reduced activity. Thus, isotopic substitution should be preferably away from the atoms undergoing reaction and in a position that is chemically inert. For certain cases where the position of the isotopic substitution is somewhat labile, it may be best to distill away the solvent from a small aliquot of stored radionuclide and count the distillate. In all cases, it is advisable to learn about special handling problems intrinsic to particular substances. For example, even in 50% ethanol, the stability of ^{32}P-labeled

---

[14] J. F. Morrison, *Anal. Biochem.* **24,** 106 (1968).
[15] R. M. Roberts and K. C. Tovey, *Anal. Biochem.* **34,** 582 (1970).

metabolites is not very great, and repurification just prior to use is often required. The availability of HPLC-purified compounds from commercial vendors of radioactive materials has increased the reliability of most preparations. Even so, the investigator should verify this.

Determination of Reaction Equilibrium Constants

Equilibrium constants for biochemical reactions are frequently more complex than one might suppose at first sight. Substrates and products are almost always ionic, and hydrogen ions and metal ions can substantially displace the so-called mass-action ratio. For example, consider the adenylate kinase reaction

$$ADP^{3-} + MgADP^{-} \rightleftharpoons AMP^{2-} + MgATP^{2-} \tag{7}$$

The mass-action ratio is written as

$$K = \frac{[AMP]_0[ATP]_0}{[ADP]_0^2} \tag{8}$$

where the subscripts indicate total substrate and product concentrations. Yet, even Eq. (8) is an inadequate representation of the reaction in that ATP^{4-} and $HATP^{3-}$ differ in affinity for metal ions, and the value of K will show both pH and metal ion concentration dependences. Even the buffer can affect the equilibrium constant if it interacts with metal ions, as is the case with Tris and imidazole buffers. The important point is that one must know the value of the equilibrium constant under the reaction conditions to be used in experiments, and many literature values may only serve as a rough guide. Thus, it is best to take the time and effort to obtain an accurate estimate for the equilibrium constant. This is especially important if one is interested in using Haldane relationships.

Basically, one should determine the equilibrium constant by a number of methods, and each should give a self-consistent value. Starting with substrates only (or products only) one may add a sufficient amount of enzyme to equilibrate the reaction. One may periodically sample the reaction mixture for substrate(s) and product(s) concentrations, and the system should reach a time-independent state. (The equilibrium value should also be independent of the amount of enzyme added if no competing side reactions occur.) One problem with this method is that the catalytic efficiency of the enzyme may be decreased by product accumulation, and the approach to equilibrium may be sluggish. If the enzyme is irreversibly inactivated by denaturation over a long incubation period, a false equilibrium value may be obtained. Likewise, the pH may change if protons are taken up or released as the reaction progresses, and the buffering capacity of the system must be sufficient to accommodate this change. For this reason, the pH should be carefully monitored or one may use a pH-stat to maintain the pH. It is also advisable to demonstrate that the same equilibrium constant is obtained by starting from either side of the reaction.

Another approach, which obviates some of these difficulties, involves the preparation of a group of reaction samples with different substrate(s):product(s) ratios roughly in the range of the estimated equilibrium constant. Provided one has a sensitive measure of changes in product or substrate levels, one may follow each reaction and plot the deviation in concentration attained after equilibration in the presence of the enzyme. The sample giving zero deviation must correspond to the equilibrium ratio of substrates and products. If there are several substrates and several products, [A] and [P] may be set at several constant values, and the [B]:[Q] ratio may be changed (or vice versa). The zero deviation position on the graph should be always at the same point. If it is not, careful scrutiny of the free metal ion concentration, the pH, and the enzyme's activity is indicated. This method is further illustrated in Vol. 64 [1].

Probably the best analytical measure of substrate and product levels can be achieved by use of radioisotopically labeled substrates. At isotopic equilibrium the isotope will be distributed in the substrate or product pool in strict accord with the analytical concentrations of the substrate and product. Thus, one may rigorously demonstrate the occurrence of equilibrium by double-label methods. If ^{14}C-labeled substrate and ^{3}H-labeled product were incubated with enzyme, the equilibrium condition requires that the ^{3}H:^{14}C ratio in substrate and product be identical.

Choice of Buffer Agents

With few exceptions, studies of enzyme-catalyzed systems require that a buffer agent be present in the reaction solutions. Unfortunately, the choice of the buffer substance cannot be made strictly by matching closely the pK_a value and the desired pH of the reaction medium. Too many buffers are inappropriate for enzyme studies as a result of undesired interactions with the enzyme or some reaction component. For example, few investigators recognize the oxidizing potential of arsenate and cacodylate (dimethylarsinic acid) under acidic conditions. Likewise, many phosphate esters, phosphate itself, and a variety of carboxylic acids are also natural metabolites, and they may bind to special enzyme sites that affect the catalytic activity. Borate buffers are also often unacceptable as a result of complexation with many polyols, ribonucleotides, and carbohydrates. Thus, care must be exercised in choosing from the many available buffering agents.

One solution to the difficult problem of selecting a buffer is to examine as many buffers as possible. Choose several that appear to yield the greatest activity. To determine whether the buffer interacts with the enzyme, buffer dilutions at constant ionic strength can be made (and solutions readjusted to the desired pH). Buffer dilution should not affect the activity of the enzyme provided pH and ionic strength are maintained. The use of buffers that cause activity changes should be questioned. It might also be noted that several counterions should be used to discover the best counterion. (Additional information on pH effects can be obtained in this volume [9].)

It cannot be overemphasized that control of divalent metal ion concentrations requires the correct choice of buffers, especially in nucleotide-dependent reactions.[16] The reader should consult this volume [11] and [12].

Finally, the equilibrium position of many reactions is pH dependent, and it is wise to fully consider the scope of the planned experiments. If the kinetics of a reaction are to be studied in both the forward and reverse directions, one is well advised to select a pH that permits this to be readily accomplished. For example, the mass-action ratio for the hexokinase reaction, [glucose-6-P][ADP]:[ATP][glucose], is 490 at pH 6.5 but around 4900 at pH 7.5. Measuring the reverse reaction rate at the higher pH may be quite difficult. The same is true for many dehydrogenase reactions, and a little prior consideration may eliminate considerable work.

Temperature Control

Valid enzyme assays will be obtained only when the temperature is carefully maintained. Usually, this requires a good temperature-regulated bath and circulator with a variability of less than 0.1°. Samples should be sufficiently immersed to allow full thermal equilibration, which may require several minutes with glass and plastic vessels. With spectrophotometers, the entire sample compartment is commonly thermally isolated by water circulation through the hollow walls, but the equilibration of the cuvette is subject to a rather inefficient convective heat transfer by air from the warmed or chilled walls. Thus, one may have considerably less temperature control than indicated by the temperature regulator. A more

---

[16] D. D. Perrin and B. Dempsey, "Buffers for pH and Metal Ion Control." Wiley, New York, 1974.

efficient system uses a brass block fabricated to mount directly around the cuvettes. This block should be designed to permit sufficient circulation of fluid from the regulated water bath.

The problem of adequate temperature control is most serious when reactions are monitored at more than 10° above or below ambient temperature. In such cases, sample removal from the bath for manual mixing alone may be sufficient to disturb the temperature. The sample tube and cuvette should be thermally preequilibrated, and a minimum of sample handling is desirable. It may be desirable to prewarm the enzyme solution prior to initiating the reaction. Even a 0.1 ml aliquot of ice-cold enzyme solution can perturb a 3.0 ml assay by 1 or 2 degrees, and the reaction velocity may be altered by 10 or 15%. One approach to ensure rigorous temperature control is to use a spring-loaded plunger mounted on the sample compartment lid. The end of the plunger may contain a Teflon "spoon" holding up to 0.1 ml of solution, and mixing is readily achieved by rapidly depressing the plunger several times. The "spoon" can be fashioned to have small jets that permit complete mixing in a few seconds. This method has the added advantage of permitting mixing while the photomultiplier tube is already operating. With the Cary 118C, for example, opening and reclosing of the sampling compartment leads to a 4-sec delay before the opaque safety shutter to the photomultiplier tube is reopened. With the above apparatus, mixing can be complete in 3 sec, and the spectrophotometer can give a record of the early progress of the reaction.

Reporting Initial-Rate Data

As a recommended course of action in publishing rate data, the following statements are offered. Velocity should always be expressed in terms of molarity changes per unit time. Other terms, which are proportional to molarity, are frequently reported, but the use of an intensive variable such as molarity is preferable. The conditions of the assay should be fully described, and any special treatments required for linearity should be thoroughly detailed. The specific activity and amount of enzyme used in each experiment should be stated in each figure legend, especially when these undergo change during the course of the kinetic study. Reviewers are especially grateful if the investigator provides estimates of the maximum percentage of substrate depletion during the course of rate assays. Statements regarding the number of replicate samples that were assayed and the variation in observed constants are also helpful. Unfortunately, some investigators fail to report such statistical data, and this leads to confusion for those interested in repeating the work.

Concluding Remarks

The development of a reliable initial-rate assay may appear to be an insurmountable task involving the interplay of many variables. Yet, for those interested in enzymology, this activity frequently presents the opportunity to uncover new aspects of the behavior of a particular enzyme. Many fascinating details of enzyme mechanism and metabolic regulation have evolved from such exercises, and a valid assay represents a powerful tool to probe further.

[2] Techniques in Coupled Enzyme Assays

By FREDERICK B. RUDOLPH, BENNETT W. BAUGHER, and ROBERT S. BEISSNER

The major problem in initial-rate kinetic studies is often the method of assay. If possible, an assay should be accurate, sensitive, and convenient. In addition, the ability to continuously monitor a reaction process is of great value. Unfortunately, many reactions do not produce changes in the spectral or other properties of the reactants and cannot be directly measured. To allow continuous assay of such reactions, the formation of a product can be measured by addition of an auxiliary enzyme that produces a measurable change. Such methods are sensitive and convenient, but have certain disadvantages, the most important being a lag period before the steady state is reached, which will be detailed in this chapter. Various theoretical analyses of the use of consecutive enzyme reactions for assay systems have been made and will be considered here along with practical aspects of their use, precautions to be observed, and examples of such assays.

Theory

Models and Analysis of Coupled Systems

A number of approaches describing ways to ensure valid coupled assays have appeared in recent years.[1-8] The basic problem is to deter-

---

[1] H. U. Bergmeyer, *Biochem. Z.* **324**, 408 (1953).
[2] H. Gutfreund, "An Introduction to the Study of Enzymes," pp. 302–306. Blackwell, Oxford, 1965.
[3] W. R. McClure, *Biochemistry* **8**, 2782 (1969).
[4] C. J. Barwell and B. Hess, *Hoppe-Seyler's Z. Physiol. Chem.* **351**, 1531 (1970).

mine that auxiliary enzymes will react at a rate that allows monitoring only of the steady-state concentration of the product(s) (P) of the reaction being studied, not of the rates of the auxiliary enzymes. The systems have an inherent lag time prior to the steady state that must be analyzed and minimized. The simplest example of such a system is

$$A \xrightarrow{E_1} P \xrightarrow{E_2} Q \tag{1}$$

where E_1 is the enzyme whose activity is being measured (primary enzyme), E_2 is the auxiliary enzyme, and k_1 and k_2 are the rate constants for the respective enzymes. The general approach to such assays has been to use a large excess of the auxiliary enzymes to assure steady-state conditions. The behavior of such a system was first treated quantitatively by McClure[3] using the following assumptions. (1) k_1 is the rate constant for E_1, which is assumed to be an irreversible zero-order step. To meet this criterion, all substrates for E_1 must be saturating or only a small fraction of the substrates can react. Irreversibility is assumed since P is continuously removed by E_2 during the assay. (2) The second reaction is irreversible and first order with respect to P (rate constant k_2). This necessitates that $P \ll K_p$ (the Michaelis constant for P for E_2) and that the other substrates for E_2 be nearly saturating. If the equilibrium for the reaction catalyzed by E_2 lies to the right or only a small amount of reaction occurs so that the other substrates for E_2 are not depleted, irreversibility can be assumed. Proper choice of the auxiliary enzyme will satisfy this condition. With these assumptions, it is possible to calculate the amount of E_2 required for a theoretically correct assay. For the sequence in Eq. (1)

$$dP/dt = k_1 - k_2 P \tag{2}$$

which integrates, using the limits $t = 0 \to t$ and $P = 0 \to P$, to

$$P = (k_1/k_2)(1 - e^{-k_2 t}) \tag{3}$$

as $t \to \infty$, P approaches its steady-state concentration (P_{ss}) or

$$P_{ss} = k_1/k_2 \tag{4}$$

Equation (3) can be rearranged to

$$\ln[1 - k_2 P/k_1] = -k_2 t \tag{5}$$

---

[5] B. Hess and B. Wurster, *FEBS Lett.* **9**, 73, (1970).
[6] J. Easterby, *Biochim. Biophys. Acta* **293**, 552 (1973).
[7] P. W. Kuchel, L. W. Nichol, and P. D. Jeffery, *J. Theor. Biol.* **48**, 39 (1974).
[8] A. Storer and A. Cornish-Bowden, *Biochem. J.* **141**, 205 (1974).

These equations allow calculation of the amount of E_2 required in an assay to reach the steady state for P (P_{ss}) in a given period of time. Combining Eqs. (4) and (5) results in

$$\ln(1 - P/P_{ss}) = -k_2 t \tag{6}$$

If $K_p \gg P_{ss}$ as assumed initially, then, from the Michaelis–Menten equation for E_2,

$$v_2 = \frac{V_2 P_{ss}}{K_p + P_{ss}} \simeq \frac{V_2 P_{ss}}{K_p} = k_2 P_{ss} \tag{7}$$

This expression can be substituted into Eq. (6) to give an expression for the amount of E_2 (or V_2 which is the maximal velocity for E_2) required to attain a given fraction of the steady-state phase of the coupled reaction at any given time t.

$$V_2 = \frac{-\ln(1 - P/P_{ss}) K_p}{t} \tag{8}$$

If F_p is the fraction of P_{ss} desired at time t, the equation can be expressed as

$$V_2 = \frac{-2.303 \log (1 - F_p) K_p}{t} \tag{9}$$

The time required to each F_p (t^*) can be calculated from

$$t^* = \frac{-\ln(1 - F_p)}{k_2} \tag{10}$$

To use expression (9), one decides on the F_p desired with a given lag period and calculates V_2. Only K_p has to be known to make the calculation.

Certain features of the Eqs. (7)–(9) are apparent[3]: (1) k_2 is influenced by the ratio V_2/K_p and is not just dependent on a large excess of E_2; this point is illustrated in Fig. 1. Both the enzyme concentration and K_p will influence the lag time. Either an increase in K_p or decrease in E_2 will cause an increase in the lag time. These factors should be considered if isozymes or similar enzymes are available for the same assay. (2) The time required for establishment of the steady state (P_{ss}) is independent of k_1. The system is therefore suitable for assaying any activity of the primary enzyme as long as $P_{ss} \ll K_p$. This is an important factor and suggests that a lower limit for K_p exists and should be considered when a choice of auxiliary enzymes is made. (3) P_{ss} is a function of both k_1 and k_2.

The attainment of the steady-state concentration of P is shown in Fig. 2. In this hypothetical example sufficient E_2 is present to give the steady

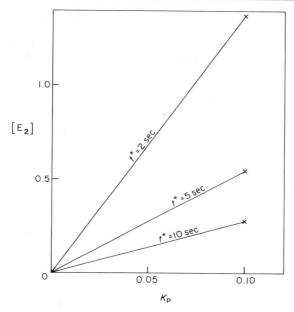

FIG. 1. A plot of amount of auxiliary enzyme, E_2, versus the K_p for various indicated lag times (t^*)

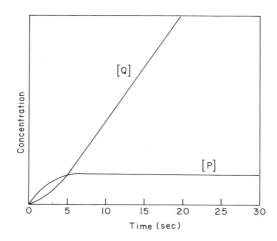

FIG. 2. A plot of the concentration of the intermediate P, product, and the final product, Q, as a function of time.

state within 5 sec, after which the formation of Q is constant ($dQ/dt = C$ and $-dP/dt = 0$) and initial velocity conditions are achieved.

Assay systems utilizing more than one auxiliary enzyme are more difficult to analyze. For the system

$$A \xrightarrow[E_1]{} P \xrightarrow[E_2]{} Q \xrightarrow[E_3]{} R \tag{11}$$

where E_1 is the primary enzyme and E_2 and E_3 are the auxiliary enzymes, McClure has applied an analysis similar to that described above for Eq. (1). An expression for [Q] analogous to Eq. (3) is

$$Q = \frac{k_1}{k_3} - \frac{k_1}{k_3 - k_2} [e^{-k_2 t} - (k_2/k_3)e^{-k_3 t}] \tag{12}$$

This equation contains the following features: (1) The first term is equal to the steady-state concentration of Q (Q_{ss}), and the negative term is the time dependence of steady-state attainment. (2) The rate of steady-state attainment is symmetrical with respect to k_2 and k_3, allowing values of k_2 and k_3 to be interchanged without affecting the lag time. (3) The time required to achieve a given fraction of Q_{ss} is not a function of k_1.

Equation (12) cannot be solved as easily for t or t^* as the two-enzyme case, but some conclusions were drawn by McClure.[3] If either k_2 or k_3 is large compared to the other rate, then the lag time is primarily dependent on the smaller value. The equation reduces to

$$t^* = \frac{-\ln(1 - F_Q)}{k_2} \quad \text{if } k_3 \gg k_2 \tag{13}$$

or

$$t^* = \frac{-\ln(1 - F_Q)}{k_3} \quad \text{if } k_2 \gg k_3 \tag{14}$$

In fact, the difference between the two constants need only be 4- to 5-fold for satisfactory prediction of t^*. When neither auxiliary enzyme is in excess, there is no single solution for t^*. McClure[3] has solved Eq. (12) numerically, and the results are presented in Fig. 3 as nomograms showing the time required to reach 99% of Q_{ss} for different values of k_2 and k_3.

A similar approach to dealing with the lag time of a coupled assay has been presented by Easterby,[6] and his approach allows calculation of some useful values. A resultant equation from his treatment of Eq. (1) is

$$[P] = v_1 (t + \tau e^{-t/\tau} - \tau) \tag{15}$$

where τ, the transient (lag) time $= K_p/V_2$ and $P_{ss} = v_1 \tau$. For this treatment the plot of Q formation versus t is shown in Fig. 4. The P_{ss} concen-

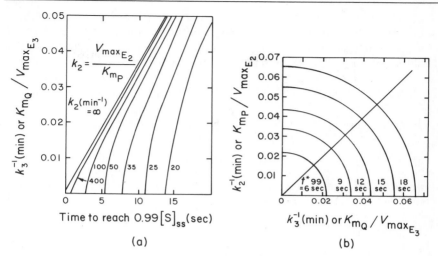

FIG. 3. Nomograms of McClure showing the time required for 99% attainment of the steady state (t^*) in a coupled assay using the auxiliary enzymes. R_2 and R_3 represent the first-order rate constants (V_{max}/K_m) for the two coupling enzymes. Adapted with permission of W. R. McClure and the American Chemical Society (copyright holder), from *Biochemistry* **8**, 2782 (1969).

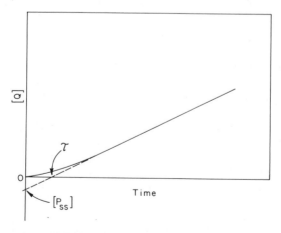

FIG. 4. A plot of the formation of the final product Q, of E_2 as a function of time. The x intercept represents the transient time (τ) as discussed in the text, and the y intercept represents the steady-state concentration of intermediate P (P_{ss}).

tration is found from the y intercept, and the transient time (τ) from the x intercept. This information allows calculation of v_1, the velocity of the reaction being studied. Determination of v_1 is done at a series of A concentrations, and the velocity dependence on substrate concentration is plotted. The transient time approximates K_p/V_2 only if $V_2/K_p \gg V_1/K_A$. The ability to determine P_{ss} allows a check on the assumptions used in McClure's derivations. If $P_{ss} \ll K_p$, then the methods of calculation of the amount of coupling enzymes required are more valid. It also allows monitoring if product inhibition is a potential problem in the assay.

Easterby[6] has also treated assays using two or more auxiliary enzymes. He found the transient time for reaching the steady state for the assay to be the sum of the transients for the individual coupling enzymes. Thus, the lag time increases as the number of auxiliary enzymes increases. Of importance is that this treatment also indicates that the transient time is independent of the initial rate-limiting enzyme. Changing the activity of E_1 does not change the lag. Easterby[6] also showed that the minimum time for accurate measurement of initial velocity (steady-state intermediate) has to be at least five times the transient time.

Storer and Cornish-Bowden[8] have considered coupled assays without assuming that the second and subsequent reactions followed first-order kinetics, as was done by McClure and Easterby. This more general treatment for the reaction sequence

$$A \xrightarrow{v_1} P \xrightarrow{v_2} Q \tag{16}$$

where v_1 and v_2 are the experimental velocities, gives an equation for the time (transient time) required for v_2 to reach any given value as

$$t^* = \phi K_p/v_1 \tag{17}$$

where ϕ is defined as

$$\phi = \frac{V_2 v_1}{(V_2 - v_1)2} \ln \left[\frac{v_1(V_2 - v_2)}{V_2(v_1 - v_2)} \right] - \frac{v_1 v_2}{(V_2 - v_2)(V_2 - v_1)} \tag{18}$$

ϕ is a dimensionless number and a function of the ratios v_2/v_1 and v_1/V_2 only. V_2 is the maximal velocity of the auxiliary enzyme. Values of ϕ can be calculated at various values of v_1/V_2 and v_2/v_1 and are given in Table I.

The value of v_2/v_1 depends on the accuracy desired. In a precise assay it is desirable that it be at least 0.99. The amount of E_2 required for a desired accuracy can be estimated for a given v_1 and the lag time calculated from Eq. (17).

The analysis of Storer and Cornish-Bowden[8] also considered multiple coupled systems. They found that, by treating each step individually, the lag time was bounded by $t_1 \ldots t_n < t_{lag} < \Sigma_{i=1}^n t_i$. That is, it was longer than any individual time but less than the sum of all. Assuming that the lag

TABLE I
ANALYSIS OF LAG TIMES FOR COUPLED ASSAYS[a,b]

	ϕ	
v_1/V_2	$v_1/v_2 = 0.95$	0.99
0	0	0
0.05	0.16	0.25
0.10	0.35	0.54
0.15	0.56	0.89
0.20	0.81	1.31
0.25	1.11	1.81
0.30	1.46	2.42
0.35	1.88	3.18
0.40	2.39	4.12
0.45	3.02	5.32
0.50	3.80	6.86
0.55	4.79	8.91
0.60	6.08	11.7
0.65	7.80	15.6
0.70	10.2	21.4
0.75	13.6	30.3
0.80	18.7	45.5
0.85	27.2	74.3
0.90	42.8	141
0.95	77.9	377

[a] Adapted from A. Storer and A. Cornish-Bowden, *Biochem. J.* **141**, 205 (1974).
[b] The time required for v_2 (for the auxiliary enzyme) to reach the specified percentage of the steady-state velocity (v_1) of the primary enzyme (E_1) is given by $t = \phi K_p/v_1$.

time is the sum gives a useful upper limit similar to the method of Easterby.[6] The experimental study by Storer and Cornish-Bowden[8] illustrated that linearity of an assay cannot be assumed unless the apparently linear portion is ten times as long as the obvious lag period.

The techniques discussed so far have assumed that the steady state reaction of the primary enzyme can be measured with accuracy. The lag period has been described only as to its duration and how to shorten it. Kuchel et al.[7] have solved the basic equations describing Eq. (1) by formulation as a set of Maclaurin polynomials. This treatment leads to the following equation

$$[P] = \frac{V_1 V_2 A_0 t^2}{2K_p (K_A + A_0)} \tag{19}$$

where A_0 is the initial A concentration.

A plot of P versus t^2 gives an initially linear plot with a slope $V_2/2K_p \cdot V_1A_0/(K_A + A_0)$, which describes a rectangular hyperbola. The slopes of a number of A_0 concentrations are plotted in double reciprocal form to give an x intercept of $-1/K_A$ and a y intercept of $2K_p/V_1V_2$. Prior analysis of the V_2 and K_p allows calculation of both V_1 and K_A. This analysis can be extended to any number of auxiliary enzymes by similar plots.[7] This treatment will apply best when the activities of the primary and auxiliary enzymes are similar. A long lag time is needed to obtain the initial linear slope of the P versus t^2 plot. If the lag is short, the data are readily analyzed by the techniques described by McClure[3], Easterby,[6] and Storer and Cornish-Bowden.[8] This technique is of use particularly if the coupling enzymes are not available in sufficient quantities to saturate the system and does avoid some difficulties inherent in coupled systems. Additional information concerning individual rate constants can be obtained from pre-steady state studies as described by Roberts.[9] These are beyond the scope and intent of this report but are available for reference.

The major conclusions of all the theoretical treatments on coupled enzyme assays are that they can be designed so that lag times are known and minimal and the measured velocity accurately represents the steady-state velocity of the primary enzyme. The calculations will usually provide the correct answers, but certain precautions and limitations must be considered and will be discussed in the following sections.

Practical Aspects

The basic question in designing a coupled enzyme assay is how much auxiliary enzyme(s) to add to ensure a short lag time and accurate measurement of the enzyme being studied. It would be convenient if one could simply say to use a 10-fold excess of the first coupling enzyme and 100-fold excess of the second enzyme if it is necessary, which has been a convention among some enzymologists. However, the systems may not always work well even with these excesses of coupling enzymes. The expense of the enzymes, contamination with other proteins, and other factors necessitate a more defined approach to the problem. The theories discussed above have the limitations that they are based on models requiring certain assumptions but they generally are experimentally applicable.

The choice of auxiliary enzymes is obviously dictated by the products formed in the reaction. A number of examples of coupled assays are listed

---

[9] D. V. Roberts, "Enzyme Kinetics." Cambridge Univ. Press, London and New York, 1977.

in Tables II and III. This listing is not an exhaustive compilation of such assays but generally represents recently developed methods. Included in Table III are a number of stopped-time assays, in which auxiliary enzymes are present during the reaction to allow formation of an identifiable product. The reactions are stopped at various time intervals, and the product is measured. The use of the coupled system requires that the assay times be done after the steady-state condition is achieved.[3] The excellent series entitled "Methods of Enzymatic Analysis" edited by H. U. Bergmeyer[10] should be the primary source for finding suitable systems for coupled enzyme assays. The series details assays for both enzymes and metabolites by using enzymic analysis. Such methods can be generally adapted to kinetic assays. Choices between systems should be made by comparing the K_m's for the measured substrates and activity of the auxiliary enzymes. Different isozymes can be more useful in coupled systems as illustrated by the fact that the H_4 isozyme of lactate dehydrogenase is much better than the M_4 isozyme on coupling systems.[11]

An interesting suggestion has been presented by Marco and Marco,[12] who proposed using an alternative substrate, 3-acetyl-NAD, for dehydrogenase coupling enzymes. This analog has a more favorable equilibrium for 3-acetyl-NADH formation, allowing assays monitoring the production of 3-acetyl-NADH to be done in the pH range of 7–8. Normal dehydrogenase assays are not as effective in this range. Also, the extinction change is higher than for NADH so the assay is even more sensitive. In general, alternative substrates can offer ways to develop better assays for a given reaction. The number of coupling enzymes should be minimized, but if the enzymes are readily available the sensitivity of the given assay should be the most important factor. Spectrophotometric assays are normally sufficiently sensitive for use if an adequate change in extinction coefficient is involved. A change of at least 0.05 to 0.1 OD unit is necessary for good analytical measurement. For NADH, a change in concentration of 8 μM will generate an OD change of 0.05 in a 1-cm cuvette. If the coupled assay contains a species that has a high absorbance, care must be taken to avoid stray light errors.[13] A single-beam spectrophotometer is not accurate at an absorbance above 2 owing to stray light effects. Beer's law should be verified for the spectrophotometer being used to ensure accurate measurements. Changes in substrate concentration or

---

[10] H. U. Bergmeyer (ed.) "Methods of Enzymatic Analysis," 2nd ed., Vols. 1–4. Academic Press, New York, 1974.
[11] M. Chang and T. Chung, *Clin. Chem.* **21**, (1975).
[12] E. J. Marco and R. Marco, *Anal. Biochem.* **62**, 472 (1974).
[13] R. L. Cavalieri and H. Z. Sable, *Anal. Biochem.* **59**, 122 (1974).

TABLE II
EXAMPLES OF COUPLED ENZYME ASSAYS

Product measured	Species monitored	Coupling enzymes	Examples
Assays for nucleotides			
ATP	Disappearance of NADH	Hexokinase, glucose-6-phosphate dehydrogenase	Adeylate kinase, creatine kinase[a]
ADP	Disappearance of NADH	Pyruvate kinase (PK), lactate dehydrogenase (LDH)	Fructokinase,[b] γ-glutamyl-cysteine synthetase,[c] mannokinase,[d] ATPase[e,f]
ADP	Disappearance of NADH	PK, LDH, 3'-nucleotidase	Adenosine-5'-phosphosulfate kinase[g]
AMP	Disappearance of NADH	Adenylate kinase	Adenosine phosphotransferase,[h] GMP synthetase[i]
GDP	Disappearance of NADH	PK, LDH	Adenylosuccinate synthetase[j]
GMP	Disappearance of NADH	GMP kinase, PK, LDH	Hypoxanthine-guanine phosphoribosyltransferase[k]
Inosine	Formation of uric acid	Nucleoside diphosphorylase, xanthine oxidase	Adenosine deaminase[l]
UDP	Disappearance of NADH	PK, LDH	Glycogen synthetase[m]
dUMP	Formation of dihydrofolate	Thymidylate synthetase	Adenosine phosphotransferase[i]
S-Adenosyl homocysteine	Disappearance of S-adenosyl methionine	Adenosine deaminase	Catechol-O-methyltransferase[n]
Assays for carbohydrates and glycolytic intermediates			
Cellobiose	O$_2$ consumption	β-1,4-Glucosidase, mutarotase, glucose oxidase, catalase	Cellulase[o]
Fructose 6-phosphate	Formation of NADH	Glucosephosphate isomerase, glucose-6-phosphate dehydrogenase	Fructokinase[b]

Galactose	Formation of NADH	Galactose dehydrogenase	Lactase[p]
β-D-Glucose	O$_2$ consumption	β-Fructofuranosidase, glucose oxidase, catalase	Aldose-1-epimerase[q]
β-D-Glucose	O$_2$ consumption	Glucose oxidase	Mutarotase[r]
Glucose 1-phosphate	Formation of NADH	Phosphoglucomutase, glucose-6-phosphate dehydrogenase, 6-phosphogluconate dehydrogenase	Galactose-1-phosphate uridylyltransferase[s]
Maltose	Formation of NADH	α-Glucosidase, hexokinase, glucose-6-phosphate dehydrogenase	α-amylase[t,u]
Mannose 6-phosphate	Formation of NADH	Mannose-6-phosphate isomerase, glucose-6-phosphate dehydrogenase	Mannokinase[d]
3-Phosphoglycerate	Disappearance of NADH	Phosphoglyceromutase, enolase, pyruvate kinase, LDH	Phosphoglycerate kinase[v]
Pyruvate	Disappearance of NADH	LDH	Neuraminidase[w]
Assays for TCA intermediates			
Acetyl-CoA	Disappearance of p-nitroaniline	Arylamine acetyltransferase	Acetyl-CoA synthetase, ATP-citrate lyase[x]
Fumarate	Formation of NADH fluorimetrically	Fumarase, malate dehydrogenase	Argininosuccinate lyase[y]
Oxaloacetate	Disappearance of NADH	Malate dehydrogenase	Phosphoenolpyruvate carboxykinase[z]
Assays for amino acids and related compounds			
Aspartate	Disappearance of NADH	Aspartate aminotransferase, malate dehydrogenase	Asparaginase[aa]
Elastin fragments	Solubilization of Congo red-elastin	Trypsin	Elastase[bb]
Histidinol	Formation of NADH	Histidinol dehydrogenase	Histidinolphosphate phosphatase[cc]

(continued)

TABLE II (continued)

Product measured	Species monitored	Coupling enzymes	Examples
		Assays for various other organic compounds	
Cystamine aminoaldehyde	Disappearance of NADH	Alcohol dehydrogenase	Diamine oxidase[dd]
Phenol	O_2 consumption	Tyrosinase	Alkaline phosphatase[ee]
R-CHO	O_2 consumption	Alcohol dehydrogenase	Glycerol dehydratase[ff]
		Assays for inorganic compounds	
H_2O_2	Formation of 2',7'-dichlorofluorescein	Peroxidase	Monoamine oxidase[gg]
H_2O_2	Oxidation of O-dianisidine	Peroxidase	Peroxisomal oxidases[hh]
NH_3	Disappearance of NADH	Glutamate dehydrogenase	D- and L-Amino acid oxidases,[ii] asparaginase[jj]
PP_i	Disappearance of NADH	Pyrophosphate: fructose-6-phosphotransferase, aldolase, triosephosphate isomerase, glycerol-3-phosphate dehydrogenase	Argininosuccinate synthetase[kk]

[a] L. T. Oliver, Biochem. J. **61**, 116 (1955).
[b] B. Sabater, J. Sebastián, and C. Asensio, Biochim. Biophys. Acta **284**, 414 (1972).
[c] W. B. Rathbun and H. D. Gilbert, Anal. Biochem. **54**, 153 (1973).
[d] B. Sabater, J. Sebastián, and C. Asensio, Biochim. Biophys. Acta **284**, 406 (1972).
[e] B. C. Monk and G. M. Kellerman, Anal. Biochem. **73**, 187 (1976).
[f] D. J. Horgan, R. K. Tume, and R. P. Newbold, Anal. Biochem. **48**, 147 (1972).

[g] J. N. Burnell and F. R. Whatley, *Anal. Biochem.* **68**, 281 (1975).
[h] W. D. Park, M. E. Tischler, R. B. Dunlop, and R. R. Fisher, *Anal. Biochem.* **54**, 495 (1973).
[i] T. Spector, R. L. Miller, J. A. Fyfe, and T. A. Krenitsky, *Biochim. Biophys. Acta* **370**, 585 (1974).
[j] T. Spector and R. L. Miller, *Biochim. Biophys. Acta* **445**, 509 (1976).
[k] A. Giacomello and C. Salerno, *Anal. Biochem.* **79**, 263 (1977).
[l] W. G. Yasmineh, K. Byrnes, C. T. Lum, and M. Abbasnezhad, *Clin. Chem.* **23**, 2024 (1977).
[m] J. S. Passonneau and D. A. Rottenburg, *Anal. Biochem.* **51**, 528 (1973).
[n] J. K. Coward and F. Y.␣␣␣Wu, *Anal. Biochem.* **55**, 406 (1973).
[o] T. R. Green, Y. W. Han, and A. W. Anderson, *Anal. Biochem.* **82**, 404 (1977).
[p] N. Asp and A. Dahlqvist, *Anal. Biochem.* **47**, 527 (1972).
[q] M. K. Weibel, *Anal. Biochem.* **70**, 489 (1976).
[r] I. Miwa, *Anal. Biochem.* **45**, 441 (1972).
[s] M. A. Pesce, S. H. Bodourian, R. C. Harris, and J. F. Nicholson, *Clin. Chem.* **23**, 1711 (1977).
[t] C. S. Wilson and M. J. Barret, *Clin. Chem.* **21**, 947 (1975).
[u] G. G. Guilbault and E. B. Rietz, *Clin. Chem.* **22**, 1702 (1976).
[v] M. Ali and Y. S. Brownstone, *Biochim. Biophys. Acta* **445**, 74 (1976).
[w] D. N. Ziegler and H. D. Hutchinson, *Appl. Microbiol.* **23**, 1060 (1972).
[x] G. Hoffman, L. Weiss, and O. H. Weiland, *Anal. Biochem.* **84**, 441 (1978).
[y] J. F. Sherwin and S. Natelson, *Clin. Chem.* **21**, 230 (1975).
[z] R. J. Hansen, H. Hinz, and H. Holzer, *Anal. Biochem.* **74**, 576 (1976).
[aa] H. N. Jayaram, P. A. Cooney, S. Jayaram, and L. Rosenblum, *Anal. Biochem.* **59**, 327 (1974).
[bb] G. Gnosspelius, *Anal. Biochem.* **81**, 315 (1977).
[cc] D. R. Brady and L. L. Houston, *Anal. Biochem.* **48**, 480 (1972).
[dd] S. Dupré, S. P. Solinas, P. Guerrieri, G. Federici, and D. Cavallini, *Anal. Biochem.* **77**, 68 (1977).
[ee] A. Kumar and G. D. Christian, *Anal. Biochem.* **48**, 1283 (1976).
[ff] M. I. Yakusheva, A. A. Malakhov, A. A. Poznanskaya, and V. A. Yakovlev, *Anal. Biochem.* **60**, 293 (1974).
[gg] H. Köchli and J. P. von Wartburg, *Anal. Biochem.* **84**, 127 (1978).
[hh] J. Duley and R. S. Holmes, *Anal. Biochem.* **69**, 164 (1975).
[ii] D. Holme and D. M. Goldberg, *Biochim. Biophys. Acta* **377**, 61 (1975).
[jj] D. A. Ferguson, *Anal. Biochem.* **62**, 81 (1974).
[kk] W. O'Brien, *Anal. Biochem.* **76**, 423 (1976).

TABLE III. MISCELLANEOUS COUPLED ASSAYS

Product measured	Species monitored	Coupling enzymes	Examples
Stop-time coupled enzyme assays			
ATP	Formation of [^{14}C]glucose 6-phosphate	Hexokinase	Pyruvate kinase[a]
NAD-^{3}H	Formation of [^{3}H]lactate	LDH	9-Hydroxyprostaglandin dehydrogenase[b]
[^{14}C]Biliverdin	Formation of [^{14}C]bilirubin	NADPH-Dependant biliverdin reductase	Heme oxygenase[c]
Octopamine	Formation of [^{14}C]synephrine	Phenylethanolamine N-methyltransferase, adenosylhomocysteinase, adenosine deaminase	Dopamine-β-hydroxylase[d]
P_i groups from ends of RNA fragments	Formation of phosphomolybdate	Alkaline phosphatase	Ribonuclease[e]
Other assays			
FMN	Disappearance of riboflavin fluorescence; binding of FMN to apoflavodoxin quenches its fluorescence	Apoflavodoxin	Flavokinase[f]
[^{3}H]S-Adenosylmethionine	Production of [^{3}H]melatonin, which can be separated from the other radioactive compounds in the assay by extraction with CHCl$_3$	Hydroxyindole-O-methyltransferase	Methionine adenosyltransferase[g]
Phosphoenol pyruvate (PEP)	Disappearance of oxaloacetate; PK converts PEP to pyruvate, which does not absorb at 280 nm, where OAA is monitored	PK	PEP carboxykinase[h]

[a] A. Preller and T. Ureta, *Anal. Biochem.* **76**, 416 (1976). [b] H. Tai and B. Yuan, *Anal. Biochem.* **78**, 410 (1977). [c] R. Tenhunen, *Anal. Biochem.* **45**, 600 (1972). [d] A. N. Karahasonoglu, P. T. Ozand, D. Diggs, and J. T. Tildon, *Anal. Biochem.* **66**, 523 (1975). [e] R. Stein and J. Wilczek, *Anal. Biochem.* **54**, 419 (1973). [f] S. G. Mayhew and J. H. Wassnik, *Biochim. Biophys. Acta* **482**, 341 (1977). [g] S. Mathysse, R. J. Baldessarini, and M. Vogt, *Anal. Biochem.* **48**, 410 (1972). [h] M. D. Hatch, *Anal. Biochem.* **52**, 280 (1973).

addition of inhibitors that absorb at the assay wavelength can cause serious errors and lead to incorrect interpretations. Fluorescence assays are 2–3 orders of magnitude more sensitive than spectrophotometric assays if a fluorescing species is involved. Self-quenching can be a problem when disappearance of fluorescense is being measured.[14] Reactions can be coupled to enzymes that release or take up protons and assayed by use of a pH stat. This method is probably not as sensitive as spectral assays but would be useful in some cases. The use of isotopes in coupled assays as illustrated in Table III can be very sensitive, limited only be separation methods and the specific radioactivity of the substrate.

Once the choice of coupling system and method of assay is made, the parameters of the auxiliary enzymes must be evaluated. The V_{max} and K_m for each auxiliary enzyme should be determined with identical buffer, ionic strength, temperature, and other conditions as will be used in the actual measurements. It may be necessary to consider the inhibitory effects of substrates or inhibitors for each enzyme on the others. For example, glucose-6-phosphate dehydrogenase is significantly inhibited by free ATP, producing a long lag time[15] and limiting its use for measurement of glucokinase and hexokinase activity in the presence of high free ATP. The measured kinetic parameters are likely to be different from manufacturer or literature values but are the proper values for the system being studied. In addition, the solution in which the auxiliary enzyme is kept must not cause inhibition of the primary enzyme. Ammonium sulfate is a particular problem requiring dialysis of the auxiliary enzymes prior to use. Even albumin, which is often added to stabilize proteins, can bind many substrates and cofactors and cause inhibition in that manner. If possible, the coupling enzymes should be in the same buffer system as the assay. Use of small molecular seive columns or rapid dialyzers will help avoid stability problems encountered during normal dialysis.

The amount of auxiliary enzymes needed can then be calculated as described either by McClure[3] or Storer and Cornish-Bowden.[8] The parameters to be specified are the ratio of v_2/v_1 or the fractional attainment (F_p) of the steady-state rate and the lag time before F_p is reached. To ensure accurate assays, F_p should usually be specified as 0.99 so that the measured rate represents 99% of the steady-state rate of the primary enzyme. Lag time will depend on the system being studied. A lag of 30 sec, or longer, may not be a problem in some assays if a long linear region is then observed. If the primary enzyme is suspected of being hysteretic,[16]

---

[14] K. Dalziel, *Biochem. J.* **80**, 440 (1961).
[15] G. Avigad, *Proc. Natl. Acad. Sci. U. S. A.* **56**, 1543 (1966).
[16] C. Frieden, *J. Biol. Chem.* **245**, 5788 (1970).

the lag time should be minimal to ensure actual measurement of the primary enzyme's behavior.

The two methods can be compared in the following examples. Storer and Cornish-Bowden[8] used the assay of glucokinase with glucose-6-phosphate dehydrogenase for illustration of their method. The K_m for glucose 6-phosphate was found to be 0.11 mM under assay conditions. They specified that the lag time prior to $v_2 = 0.99\ v_1$ would be 1 min and v_1 would not exceed 0.04 mM min^{-1}. Substitution of these values into Eq. (17), $t^* = \phi K_2/v_1$, gives $\phi = 0.36$. Using Table I and interpolating for this value of ϕ shows $v_1/V_2 = 0.08$ or $V_2 = 0.5$ mM min^{-1} or 0.5 IU/ml. Using McClure's analysis with Eq. (9) gives

$$V_2 = \frac{-2.303 \log(1 - F_p)K_p}{t} = -2.303 \log(1 - 0.99)(0.11)$$

$$= 0.51 \text{ mM min}^{-1} \text{ or } 0.51 \text{ IU/ml}$$

So, for a simple system, the two treatments will give similar answers. As v_1 decreases, the amount of E_2 added in the Storer and Cornish-Bowden[8] treatment will be somewhat less, but not proportionally since both ϕ and V_2 are related to v_1.

An example of a two-auxiliary enzyme couple was presented by McClure[3] for a generalized kinase assay using pyruvate kinase and lactate dehydrogenase. The values used were $k_1 = 0.05$ mM min^{-1}, $K_{ADP} = 0.21$ mM, and $K_{pyruvate} = 0.14$ mM. The transient time (t^*) was chosen to be 12 sec. From the nomogram in Fig. 3(b), if t^* is 12 sec then k_2 and k_3 must be larger than 1/0.043 or 23 min^{-1}. Assuming that k_2 and k_3 are approximately equal and that sufficient values are $k_1 = k_3 = 33$ min^{-1}, then $k_2 = V_2/K_{ADP}$ and $k_3 = V_3/K_{pyruvate}$ and $k_2 = 6.9$ IU/ml of pyruvate kinase and $k_3 = 4.6$ IU/ml of lactate dehydrogenase. Using equal amounts of k_2 and k_3 allows the minimum total units of enzyme to be used.[3]

Using the Storer and Cornish-Bowden[8] analysis necessitates separation of the transient time for each couple and calculation of the ϕ for each reaction as was done in the one auxiliary enzyme case. If the t_1 and t_2 are assumed to each equal 6 sec, then the ϕ for pyruvate kinase is equal to

$$\frac{(t^*)(v_1)}{K_p} = \frac{(0.1)(0.05)}{0.21} = 0.024$$

The ratio of v_1/V_2 from Table I is 0.005, so $V_2 = 10$ IU/ml. For lactate dehydrogenase v_3 is assumed $\simeq v_2$, so the calculation can be made. $\phi = (0.1)(0.05)/(0.14) = 0.036$, and from Table I $v_2/V_3 = 0.007$. Thus, the lactate dehydrogenase concentration should be 7.1 IU/ml. The two methods once again give similar results although McClures'[3] method is probably easier to handle for a two-auxiliary enzyme system. The treatment of Storer and Cornish-Bowden[8] gives high values for V_2 and V_3 because of the assumption that the lag times are independent. The actual observed lag time will be shorter than assumed in the calculations. The treatment of Storer and Cornish-Bowden[8] has the advantage that it can be adapted to 3 or more coupled enzymes.

Calculations of transient time for a given assay using defined levels of coupling enzymes can be done using the methods of McClure,[3] Easterby,[6] and Storer and Cornish-Bowden.[8] The analysis of Easterby[6] illustrated in Fig. 4 allows easy determination of the lag time and the steady-state concentration of the intermediate product. Knowledge of that concentration allows confirmation of the validity of the assumptions in the models presented in the Theory section and analysis of product inhibition effects by the intermediate product.

Making the calculation as to the amount of enzyme to be added does not free the investigator from making sure that the coupled assay really measures the velocity of the primary enzyme. Test assays should be run with levels of the coupling enzymes 2- to 10-fold higher than calculated, and the rates should all be the same. If an increase in rate is observed, further checking of the assays is required.

The calculation of required enzyme levels allows economical use of the auxiliary enzyme and will avoid many problems. McClure[3] has listed the information that should be evaluated for a coupled assay and presented as part of the experimental methods. This includes (1) the units of auxiliary enzyme added per milliliter of assay as described under the experimental conditions; (2) the apparent Michaelis constants for the measured products under the experimental conditions; (3) t^* and the F_p chosen on the basis of the calculations; and (4) the effect of the additional auxiliary substrates on the activity of the primary enzyme. Most assays currently used meet these criteria, having been determined by trial and error. The treatment described here allows a more rational approach to future design of coupled assays.

If a coupling enzyme is not available in sufficient quantity or some circumstance limits its use, the treatment of Kuchel et al.[7] will allow analysis of a system where the activities of the primary and auxiliary enzymes are similar. This technique would obviously not work with hysteretic enzymes but can be used with allosteric proteins as the primary enzyme.[7]

Precautions

Once one is sure that sufficient auxiliary enzyme(s) is present to allow accurate assays, there are a few other problems to be dealt with. The reaction mixture for the primary enzyme should be preincubated with the assay enzymes to determine whether any change occurs. Often the substrates will be contaminated with a small amount of product, such as ADP in ATP, which can react with the coupling system. The endogenous product should be exhausted prior to actual assay. Often this results in actually better kinetic experiments, since no product will be present initially in the reaction mixture. Also the system should be checked to see whether there is a reaction observed in the absence of the assay enzymes. The presence of other enzymes, particularly during purification, can give rise to a blank rate that has to be considered and accounted for.

Another problem is contamination of the auxiliary enzymes by the primary enzyme. Often commercial enzymes will have a low contamination of the activity being studied, but even a 0.1% contamination is often a serious problem. This is more critical if the studies are being done with auxiliary enzyme based on the calculations presented above. A related problem is that the auxiliary enzymes will sometimes react with one of the primary substrates. Glucose-6-phosphate dehydrogenase has a small but detectable reaction with glucose,[8] which limits the amount of enzyme that can be used to couple enzymes such as hexokinase.

The final problem deals with interpretation of inhibition experiments on the primary enzyme. If the inhibitor does not inhibit the auxiliary enzymes, no problems occur, since the activity of the primary enzyme will be accurately measured. Often, however, an inhibitor that is a structural analog of a substrate will be an inhibitor of the auxiliary enzymes and will cause an observed increase in the lag time. Generally, the advice has been simply not to use coupled enzyme assays with inhibitors. The model systems presented in the Theory section do make useful predictions regarding such effects. The inhibitor should be tested with the auxiliary enzymes and the K_i and type of inhibition be determined. If the inhibitor is a competitive inhibitor of the auxiliary enzyme with respect to the measured product, both McClure[3] and Storer and Cornish-Bowden[8] suggest that simply increasing the coupling enzyme will reduce the lag. In some cases the lag may be too long to be able to reduce to a reasonable value, but the addition of more auxiliary enzyme will usually allow the measurements to be made. In general, similar considerations can be made for a noncompetitive inhibitor. McClure's treatment suggests that an uncompetitive inhibitor will not alter the lag time since K_p and V_2 change in a constant ratio, but Storer and Cornish-Bowden[8] have shown that this is not always true. It will be approximately true only if V_2 is at least 10 times v_1.

The effects of inhibitors on the auxiliary enzymes can be readily handled using the techniques described here. The inhibitor can be determined and compensated for by adding more auxiliary enzyme, and the lag time can be analyzed on the actual assay so that the validity of the assay is assured.

Summary

The amount of auxiliary enzyme to be added to a assay system utilizing a single coupled reaction can be calculated from

$$V_2 = \frac{-2.303 \log (1 - F_p) K_p}{t} \qquad (9)$$

based on McClure's[3] analysis where V_2 is the number of units of E_2, F_p is the desired fraction of the steady-state reaction of the primary enzyme to be measured, K_p is the Michaelis constant for P for E_2, and t is the desired lag time. Alternatively the method of Storer and Cornish-Bowden[8] used Eq. (17)

$$t^* = \phi K_p / v_1 \qquad (17)$$

where t^* is the transient time, K_p is above, and v_1 is the highest velocity of the primary enzyme to be measured. A value for ϕ is calculated at a specified time, and from Table I the ratio v_1/V_2 is determined for a specified v_2/v_1 ratio.

A plot of product (Q) appearance versus time allows evaluation of the steady-state intermediate product level (P_{ss}) and the lag time as a check on the assumptions for the above equations. A check should always be done to assure that the assays are linear with a reasonable lag time.

Storer and Cornish-Bowden's[8] treatment can be applied to systems with several coupling enzymes, and the nomogram of McClure (Fig. 3) is useful for a two-auxiliary enzyme system.

A check should always be done by adding a small amount of the primary enzyme product and determining that it reacts very rapidly with the assay enzymes, ensuring that the assay system is actually working.

Concluding Remarks

The use of coupled enzyme assays affords a convenient, reliable method of measuring the steady-state activity of an enzyme. Certain precautions must be taken to assure accuracy, and the system used should be well documented as described above. Under proper conditions even inhibition studies can be done with assurance of accurate measurements. New assays can be designed with confidence avoiding trial-and-error determinations.

Acknowledgments

This work was supported by Grants CA14030 awarded by the National Cancer Institute and C-582 from the Robert A. Welch Foundation. B. W. B. and R. S. B. are Robert A. Welch Foundation Predoctoral Fellows.

[3] Summary of Kinetic Reaction Mechanisms

By HERBERT J. FROMM

Kinetic mechanisms for enzyme-catalyzed reactions are ordinarily proposed in order to explain initial-rate data. It was this very sort of attempted correlation that led Brown[1] in 1902 to suggest that the enzyme and its substrate must combine for a finite period before catalysis can occur. Since then, a large number of kinetic mechanisms have been proposed in order to explain the molecular mechanism of enzyme action.

An understanding of how the enzyme functions as a catalyst, and how it is regulated in the cell requires in many cases a knowledge of the enzyme's kinetic mechanism. This information is necessary even in order to evaluate inhibition constants for multisubstrate enzymes. Establishment of the kinetic mechanism for an enzyme may provide information on, among other things, the chemical mechanism, the nature of the transition state, the geometry of the active site, substrate specificity, inhibition and activation, acidic and basic groups associated with catalysis, allosteric properties, and the mode of regulation.

Exactly why two different enzymes that catalyze single displacement reactions should exhibit different pathways of enzyme and substrate interaction leading to similar productive central complexes is a question that remains to be answered. From the point of view of catalysis exclusively, there seems to be no clear advantage for random substrate binding to be preferred over ordered substrate binding by an enzyme, or vice versa. Bell and Koshland[2] have summarized data for approximately 60 different enzyme systems in which there is strong evidence for the participation of covalent enzyme–substrate intermediates in catalysis. They have also listed a number of ways in which such intermediates may be catalytically important, but they also state that, considering the great catalytic potential of enzymes, such intermediates are probably not essential for catalysis. On the other hand, the order of substrate binding and prod-

---

[1] A. J. Brown, *J. Chem. Soc.* **81**, 373 (1902).
[2] R. M. Bell and D. E. Koshland, Jr., *Science* **172**, 1253 (1971).

uct release may play an important role in how enzymes are regulated *in vivo*.[3] For example, substrate inhibition, as manifested through abortive ternary complex formation, may completely inhibit an enzyme that exhibits an Ordered mechanism. On the other hand, if the mechanism is Random, complete substrate inhibition requires the presence on the enzyme of an inhibitory allosteric site.

Although literally hundreds of kinetic mechanisms have been proposed for enzymes, only those pathways that are either well documented, or seem to be a logical extension of established mechanisms, will be presented in this chapter. In addition, mechanisms that have been proposed to explain hysteresis, activation, inhibition, half-site reactivity, and allostery will not be considered here.

A Note on Nomenclature[4]

Enzyme mechanism can be segregated into two classes as suggested by Alberty,[5] Sequential and Ping Pong, by inspection of double-reciprocal plots of 1/initial-velocity versus 1/substrate concentration. These in turn may be divided into Random and Ordered types. It is possible to further classify enzyme systems, using Cleland's nomenclature,[4] into one (Uni), two (Bi), three (Tri), and four (Quad) reactant systems.

Cleland's nomenclature will be used throughout this chapter and may be illustrated in the case of a one-substrate system as follows:

$$
\begin{array}{cccc}
\quad A & \quad P & \quad Q & \\
k_1 \downarrow k_2 & k_3 \uparrow k_4 & k_5 \uparrow k_6 & \quad (1) \\
\hline
E & \begin{pmatrix} EA \\ EPQ \end{pmatrix} & EQ & E
\end{array}
$$

The mechanism described in Eq. (1) is Uni Bi Ordered Sequential (one substrate and two products). The well known Michaelis–Menten equation is derived for a Uni Uni system. Substrates are designated A, B, C, and D, and products are P, Q, R, S. If the mechanism is Ordered, the substrates will add to the enzyme as A first, B second, etc., and the first product to dissociate from the enzyme will be P, followed by Q, etc. In the

---

[3] D. L. Purich and H. J. Fromm, *Curr. Top. Cell. Regul.* **6**, 131 (1972).
[4] W. W. Cleland, *Biochim. Biophys. Acta* **67**, 104 (1963).
[5] R. A. Alberty, *J. Am. Chem. Soc.* **75** 1928 (1953).

case of Random mechanisms, substrate addition and product release will not occur in any particular order.

Those enzyme forms that can break down in a unimolecular step to substrates and products, or those enzyme forms that can isomerize to these species, are referred to as transitory complexes. These complexes are of the binary, ternary, or quarternary type and will be described by the substrate molecules with which they are associated, e.g., EA, EPQ. Central complexes are those transitory complexes that can only decompose in a unimolecular step to substrates or products, or transitory complexes that isomerize to such forms. Central complexes cannot participate in bimolecular reactions and are enclosed in parentheses in Eq. (1).

Sequential Mechanisms

Sequential mechanisms exhibit initial-rate data of the kind depicted in Figs. 1 and 2 for a bireactant system. The cardinal feature of these pathways of enzyme and substrate interaction is that the family of lines in double-reciprocal plots intersect at a common point, either to the left of the $1/v$ axis as shown, or on the axis. Implicit in all Sequential mechanisms is the assumption that all substrates must be present simultaneously at the enzyme's active site before product formation can occur.

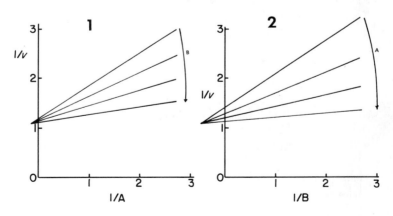

FIG. 1. Reciprocal of initial-reaction velocity (v) versus the reciprocal of the concentration of substrate A at four different fixed concentrations of substrate B.

FIG. 2. Reciprocal of initial-reaction velocity (v) versus the reciprocal of the concentration of substrate B at four different fixed concentrations of substrate A.

A. *Bireactant Sequential Mechanisms*

1. Ordered Bi Uni

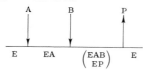

4. Ordered Bi Bi – Subsite Mechanism

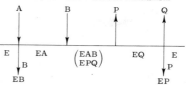

2. Random Bi Uni

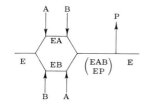

5. Random Uni Bi

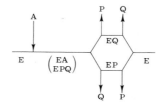

3a. Ordered Bi Bi

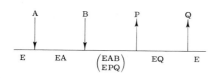

6. Random Bi Bi

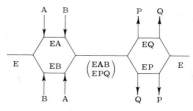

3b. Ordered Bi Bi – Theorell-Chance

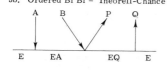

7. Modified Ping Pong

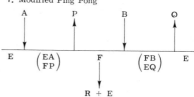

B. *Terreactant Sequential Mechanisms*

1. Ordered Bi Ter

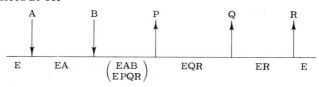

2a. Ordered Ter Ter

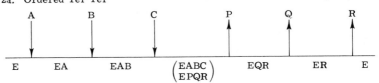

2b. Ordered Ter Ter – Theorell-Chance

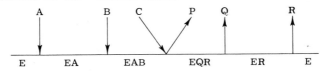

3. Random Ter Ter

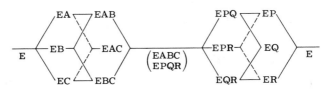

4. Random AB, Random QR Ter Ter

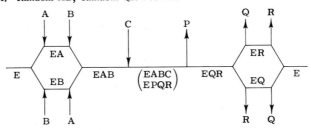

5. Random BC, Random PQ Ter Ter

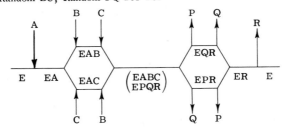

6. Random AC, Random PR Ter Ter

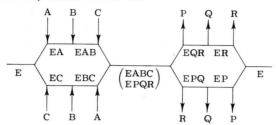

A number of additional mechanisms, which are modifications of the Sequential mechanisms presented here may be visualized. For example, Mechanism B6 can be altered so as to include an EAC complex, which may arise from interactions (EA + C) and (EC + A). The ternary complex may then add substrate B to form the productive quarternary complex. Mechanism B3 (Random Ter Ter) can also be modified so as to eliminate complexes EC, EBC, ER, and EQR.

C. *Bireactant Ping Pong Mechanisms*

Initial-rate data for a Ping Pong Bi Bi mechanism are illustrated in Figs. 3 and 4. In theory, the lines in the double-reciprocal plots are parallel as contrasted with Sequential mechanisms, which exhibit converging lines. The specific rate equations that give rise to the data of Figs. 1–4 are derived in this volume [5]. The cardinal feature of the Ping Pong mecha-

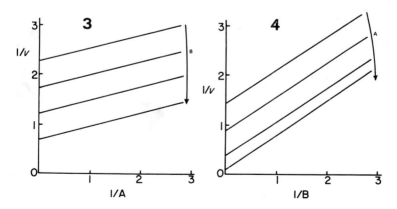

FIG. 3. Reciprocal of initial-reaction velocity (v) versus the reciprocal of the concentration of substrate A at four different fixed concentrations of substrate B.

FIG. 4. Reciprocal of initial-reaction velocity (v) versus the reciprocal of the concentration of substrate B at four different fixed concentrations of substrate A.

nism is that the enzyme reacts with one of the substrates to form a Michaelis-type complex, which then breaks down to yield a modified enzyme with dissociation of product P before the second substrate binds the enzyme. It is often suggested that the modified enzyme represents a "covalent intermediate"; however, this does not necessarily follow from the kinetic mechanism, which only indicates that a product must dissociate before the second substrate forms a Michaelis complex with the enzyme.

1. Ping Pong Bi Bi

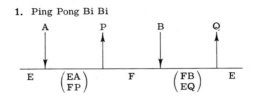

2. Two-Site Ping Pong Bi Bi

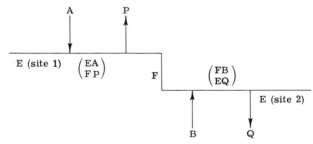

D. *Terreactant Ping Pong Mechanisms*

In the case of terreactant Ping Pong mechanisms only the Hexa Uni Ping Pong mechanism will give data of the type shown in Figs. 3 and 4 for all substrates. All Ping Pong mechanisms, however, will exhibit parallel line data in double-reciprocal plots for at least one substrate.

1. Hexa Uni Ping Pong

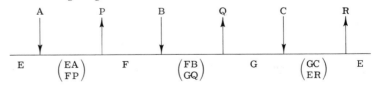

2. Ordered Bi Uni Uni Bi Ping Pong

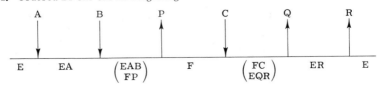

3. Ordered Uni Uni Bi Bi Ping Pong

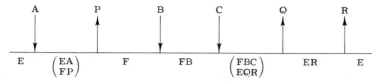

4. Random A-B, Random Q-R, Bi Uni Uni Bi Ping Pong

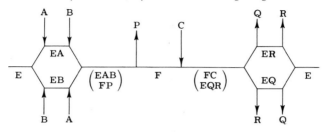

5. Random B-C, Random Q-R, Uni Uni Bi Bi Ping Pong

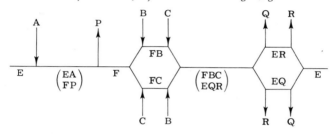

6. Random B-C, Uni Uni Bi Bi Ping Pong

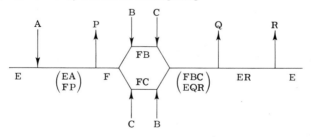

7. Random Two-Site Ping Pong

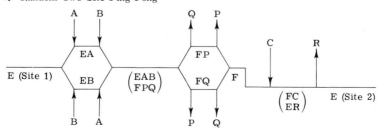

There are only three different Ordered terreactant Ping Pong mechanisms (B1, B2, and B3). The Uni Uni Bi Bi Ping Pong (B3) and the Bi Bi Uni Uni Ping Pong mechanisms are equivalent, as are the Bi Uni Uni Bi Ping Pong (B2) and the Uni Bi Bi Uni Ping Pong mechanisms.

Mechanisms B4 and B5 are random with respect to two substrates and two products. Analogous partially random terreactant Ping Pong mechanisms have been proposed.

E. *Isomerization Mechanisms*

Peller and Alberty[6] have rigorously shown that V_1/E_0 (maximal velocity/total enzyme concentration) cannot be greater than any unimolecular rate constant involved in the direction of substrate going to product. If calculation of the rate constant for a mechanism leads to negative values, or if the numerical value of a rate constant is less than V_1/E_0, or if abnormal Dalziel ϕ relationships[7] or product inhibition patterns[8] are obtained, then isomerization of one or more "stable" enzyme forms may be occurring.[9,10]

Experimental evidence has accumulated indicating that a large number of enzyme systems exhibit anomalous kinetic parameters that can best be explained by invoking the idea that certain "stable" enzyme forms isomerize. Alberty[9] and his co-workers have compiled a list of pyridine-linked anerobic dehydrogenases that very probably undergo isomerization steps in their kinetic mechanisms. It is now well recognized that isomerization of stable enzyme forms leads to alterations in rate equations when compared to analogous mechanisms in which isomerizations do not occur. On the other hand, isomerization of central complexes will not be detectable from initial-rate studies. Although it is true that isomerization steps may serve to complicate initial-rate studies, these effects, if recognized, give additional insight into the enzyme's kinetic mechanism.

All the enzyme mechanisms considered in this chapter can be rewritten in terms of stable enzyme isomerization steps; however, only two examples will be presented to illustrate "Iso" mechanisms. Depending upon the number of isomerization steps involved, the mecha-

[6] L. Peller and R. A. Alberty, *J. Am. Chem. Soc.* **81,** 5907 (1959).
[7] K. Dalziel, *Acta Chem. Scand.* **11,** 1706 (1957).
[8] H. J. Fromm, "Initial Rate Enzyme Kinetics," pp. 196, 197. Springer-Verlag, Berlin and New York, 1975.
[9] V. Bloomfield, L. Peller, and R. A. Alberty, *J. Am. Chem. Soc.* **84,** 4375 (1962).
[10] H. R. Mahler, R. H. Baker, Jr., V. J. Shiner, Jr., *Biochemistry* **1,** 47 (1962).

EXAMPLES OF ENZYME MECHANISMS

Mechanism	Enzyme	Reference[a]
	Bireactant Sequential	
Ordered Uni Bi	Adenylosuccinate lyase	11
or		
Ordered Bi Uni	Malyl coenzyme A lyase	12
Ordered Bi Bi	Alcohol dehydrogenase	13
	Carbamate kinase	14
	Lactate dehydrogenase	15
	Ribitol dehydrogenase	16
Ordered Bi Bi (Rapid Equilibrium)	Creatine kinase (pH 7)	17
Random Bi Bi	Adenylate kinase	18
	Citrate synthase	19
	Creatine kinase (pH 8)	20
	Hexokinase	21
	Phosphorylase *b*	22
	Phosphofructokinase	23
	Bireactant Ping Pong	
Ping Pong Bi Bi	Adenine phosphoribosyltransferase	24
	Coenzyme A transferase	25
	Glutamic–alanine transaminase	26
	Glucose oxidase	27
	Lipoamide dehydrogenase	28
	Nucleoside diphosphokinase	29, 30
Modified Ping Pong Bi Bi	Glucose-6-phosphatase	31
	Transglutaminase	32
Two-Site Ping Pong Bi Bi	Transcarboxylase	33
	Terreactant Sequential	
Ordered Bi Ter	Malic enzyme	34
or		
Ordered Ter Bi	Glyceraldehyde-3-phosphate dehydrogenase	35
Ordered Ter Ter	Glutamate dehydrogenase (pH 8.8)	36
Random Ter Ter	Adenylosuccinate synthetase	37, 38
	Glutamate dehydrogenase	39
	Glutamine synthetase	40
	Glutathione synthetase	41
	Formyltetrahydrofolate synthetase	42
Random AB, Random QR	Guanylate cyclase	43
Random BC, Random PQ	Citrate cleavage enzyme	44
	Galactosyltransferase	45
	γ-Glutamylcysteine synthetase	46

(*continued*)

EXAMPLES OF ENZYME MECHANISMS

Mechanism	Enzyme	Reference[a]
	Terreactant Ping Pong	
Bi Uni Uni Uni	NAD[+]-specific glyceraldehyde-3-P dehydrogenase	47
Bi Uni Uni Bi	Melilotate hydroxylase	48
	Threonyl-tRNA synthetase	49
Bi Uni Uni Bi (Random)	Leucyl-tRNA synthetase	50
Bi (Random) Uni Uni Bi (Random)	Asparagine synthetase	51
Two-Site Bi (Random) Bi (Random) Uni Uni	Pyruvate carboxylase	52
Hexa Uni	Pyruvate, phosphate dikinase	53

[a] See text footnotes 11–53.

nisms are described as "Mono-Iso," "Di-Iso," "Tri-Iso," etc.[4] Examples of the first two types are as follows:

1. Mono-Iso Ping Pong Bi Bi

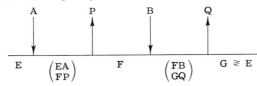

2. Di-Iso Theorell-Chance

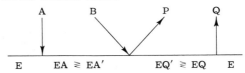

The table illustrates a few examples of some of the kinetic mechanisms described in this chapter. It should be borne in mind that the kinetic mechanisms for many enzyme systems have been proposed to date, and this listing represents only a very small fraction of the studies in the literature.[11–53]

[11] W. A. Bridger and L. H. Cohen, *J. Biol. Chem.* **243**, 644 (1968).
[12] L. B. Hersh, *J. Biol. Chem.* **249**, 5208 (1974).
[13] C. C. Wratten and W. W. Cleland, *Biochemistry* **2**, 935 (1963).

Acknowledgments

This work was supported by research grants from the National Institutes of Health (NS 10546) and the National Science Foundation (PCM 77-09018).

---

[14] M. Marshall and P. P. Cohen, *J. Biol. Chem.* **241**, 4197 (1966).
[15] M. T. Hakala, A. J. Glaid, and G. W. Schwert, *J. Biol. Chem.* **221**, 191 (1956).
[16] R. C. Nordlie and H. J. Fromm, *J. Biol. Chem.* **234**, 2523 (1959).
[17] M. I. Schimerlik and W. W. Cleland, *J. Biol. Chem.* **248**, 8418 (1973).
[18] D. G. Rhoads and J. M. Lowenstein, *J. Biol. Chem.* **243**, 3963 (1968).
[19] Y. Matsuoka and P. A. Srere, *J. Biol. Chem.* **248**, 8022 (1973).
[20] J. F. Morrison and E. James, *Biochem. J.* **97**, 37 (1965).
[21] J. Ning, D. L. Purich, and H. J. Fromm, *J. Biol. Chem.* **244**, 3840 (1969).
[22] H. D. Engers, W. A. Bridger, and N. B. Madsen, *J. Biol. Chem.* **244**, 5936 (1969).
[23] R. L. Hanson, F. B. Rudolph, and H. A. Lardy, *J. Biol. Chem.* **248**, 7852 (1973).
[24] M. Hori and J. F. Henderson, *J. Biol. Chem.* **241**, 3404 (1966).
[25] L. B. Hersh and W. P. Jencks, *J. Biol. Chem.* **242**, 3468 (1967).
[26] B. Bulos and P. Handler, *J. Biol. Chem.* **240**, 3283 (1965).
[27] Q. H. Qibson, B. E. P. Swoboda, and V. Massey, *J. Biol. Chem.* **239**, 3927 (1964).
[28] J. K. Reed, *J. Biol. Chem.* **248**, 4834 (1973).
[29] N. Mourad and R. E. Parks, Jr., *J. Biol. Chem.* **241**, 271 (1966).
[30] E. Garces and W. W. Cleland, *Biochemistry* **8**, 633 (1969).
[31] W. J. Arion and R. C. Nordlie, *J. Biol. Chem.* **239**, 2752 (1964).
[32] M. Gross and J. E. Folk, *J. Biol. Chem.* **248**, 1301 (1973).
[33] D. B. Northrop, *J. Biol. Chem.* **244**, 5808 (1969).
[34] R. Y. Hsu, H. A. Lardy, and W. W. Cleland, *J. Biol. Chem.* **242**, 5315 (1967).
[35] B. A. Orsi and W. W. Cleland, *Biochemistry* **11**, 102 (1972).
[36] E. Silverstein, *Biochemistry* **13**, 3750 (1974).
[37] F. B. Rudolph and H. J. Fromm, *J. Biol. Chem.* **244**, 3832 (1969).
[38] G. D. Markham and G. H. Reed, *Arch. Biochem. Biophys.* **184**, 24 (1977).
[39] P. C. Engel and K. Dalziel, *Biochem. J.* **118**, 409 (1970).
[40] F. C. Wedler and P. D. Boyer, *J. Biol. Chem.* **247**, 984 (1972).
[41] A. Wendel and H. Heinle, *Hoppe-Seyler's Z. Physiol. Chem.* **356**, 33 (1975).
[42] B. K. Joyce and R. H. Hines, *J. Biol. Chem.* **241**, 5716 (1966).
[43] D. L. Garbers, J. G. Hardman, and F. B. Rudolph, *Biochemistry* **13**, 4166 (1974).
[44] K. M. Plowman and W. W. Cleland, *J. Biol. Chem.* **242**, 4239 (1967).
[45] J. E. Bell, T. A. Beyer, and R. L. Hill, *J. Biol. Chem.* **251**, 3003 (1976).
[46] B. P. Yip and F. B. Rudolph, *J. Biol. Chem.* **251**, 3563 (1976).
[47] R. G. Duggleby and D. T. Dennis, *J. Biol. Chem.* **249**, 167 (1974).
[48] S. Strickland and V. Massey, *J. Biol. Chem.* **248**, 2953 (1973).
[49] C. C. Allende, H. Chaimovich, M. Gatica, and J. E. Allende, *J. Biol. Chem.* **245**, 93 (1970).
[50] C.-S. Linn, R. Irwin, and J. G. Chirikjian, *J. Biol. Chem.* **250**, 9299 (1975).
[51] H. Cedar and J. H. Schwartz, *J. Biol. Chem.* **244**, 4122 (1969).
[52] R. E. Barden, C.-H. Fung, M. F. Utter, and M. C. Scrutton, *J. Biol. Chem.* **247**, 1323 (1972).
[53] Y. Milner, G. Michaels, and H. G. Wood, *J. Biol. Chem.* **253**, 878 (1978).

[4] Derivation of Initial Velocity and Isotope Exchange Rate Equations

By CHARLES Y. HUANG

A rate equation for an enzymic reaction is a mathematical expression that depicts the process in terms of rate constants and reactant concentrations. It serves as a link between the experimentally observed kinetic behavior and a plausible model or mechanism. The characteristics of the rate equation permit tests to be designed to verify the mechanism. Conversely, the experimental observations provide clues to what the mechanism may be, hence, what form the rate expression shall take.

Derivation of rate equations is an integral part of the effective usage of kinetics as a tool. Novel mechanisms must be described by new equations, and familiar ones often need to be modified to account for minor deviations from the expected pattern. The mathematical manipulations involved in deriving initial velocity or isotope exchange-rate laws are in general quite straightforward, but can be tedious. It is the purpose of this chapter, therefore, to present the currently available methods with emphasis on the more convenient ones.

Derivation of Initial Velocity Equations

The derivation of initial velocity equations invariably entails certain assumptions. In fact, these assumptions are often conditions that must be fulfilled for the equations to be valid. Initial velocity is defined as the reaction rate at the early phase of enzymic catalysis during which the formation of product is linear with respect to time. This linear phase is achieved when the enzyme and substrate intermediates reach a steady state or quasi-equilibrium. Other assumptions basic to the derivation of initial rate equations are as follows:

1. The enzyme and the substrate form a complex.
2. The substrate concentration is much greater than the enzyme concentration, so that the free substrate concentration is equivalent to the total concentration. This condition further requires that the amount of product formed is small, such that the reverse reaction or product inhibition is negligible.
3. During the reaction, constant pH, temperature, and ionic strength are maintained.

Steady-State Treatment

During the steady state, the concentrations of various enzyme intermediates are essentially unchanged; that is, the rate of formation of a given intermediate is equal to its rate of disappearance. This assumption was first introduced to the derivation of enzyme kinetic equation by Briggs and Haldane.[1]

To derive a rate equation, the first step is to write a reaction mechanism. The nomenclature used by Fromm in this volume [3] will be adopted here with the exception that rate constants in the forward and reverse directions will be denoted by positive and negative subscripts. For example, the simplest one substrate–one product reaction can be written as:

$$E + A \underset{k_{-1}}{\overset{k_1}{\rightleftharpoons}} EA \overset{k_2}{\longrightarrow} E + P$$

or (1)

$$E \underset{k_{-1}}{\overset{k_1 A}{\rightleftharpoons}} EA \overset{k_2}{\longrightarrow}\!\!{}^P\!\!E$$

Since both the k_{-1} and k_2 steps (or branches) lead from EA to E, the two branches, as has been shown by Volkenstein and Goldstein,[2] can be combined into a single branch. This simplification procedure will be used whenever feasible.

$$E \underset{k_{-1} + k_2}{\overset{k_1 A}{\rightleftharpoons}} EA$$

The initial rate is given by

$$v = dP/dt = k_2 (EA)$$

Applying the steady-state assumption, we have

$$d(EA)/dt = k_1 A(E) - (k_{-1} + k_2)(EA) = 0 \quad (2)$$

To obtain an expression for (EA), the enzyme conservation equation

$$\text{Total enzyme} = E_0 = E + EA \quad (3)$$

is required. Substitution of $(E) = (E_0 - EA)$ into Eq. (2) yields

$$(EA) = \frac{E_0 A}{[(k_{-1} + k_2)/k_1] + A}$$

[1] G. E. Briggs and J. B. S. Haldane, *Biochem. J.* **19**, 338 (1925).
[2] M. V. Volkenstein and B. N. Goldstein, *Biochim. Biophys. Acta* **115**, 471 (1966).

and

$$v = k_2(\text{EA}) = \frac{k_2 E_0 A}{[(k_{-1} + k_2)/k_1] + A} \qquad (4)$$

$$= \frac{V_1 A}{K_m + A}$$

where V_1 is the maximum velocity in the forward direction and K_m is the Michaelis constant.

It should be noted that the validity of the steady-state method does not depend on the assumption $d(\text{EA})/dt = 0$. Without setting Eq. (2) equal to zero, one can obtain the following expression from Eqs. (2) and (3):

$$(\text{EA}) = \frac{k_1 A E_0 - d(\text{EA})/dt}{k_1 A + k_{-1} + k_2}$$

Wong[3] has pointed out that the steady-state approximation only requires that $d(\text{EA})/dt$ be small compared with $k_1 A E_0$. In the early phase of the reaction, if $A \gg E_0$, the rate of change of EA due to diminishing A will be relatively slow. It is clear that the validity of steady state is intimately tied to the condition of high substrate to enzyme ratio.

The Determinant Method

For a mechanism involving several enzyme-containing species, derivation of the rate equation can be done by solving the simultaneous algebraic equations by the determinant method. Consider the mechanism described by Eq. (1) with the addition of an EP intermediate.

$$E \underset{k_{-1}}{\overset{k_1 A}{\rightleftharpoons}} EA \underset{k_{-2}}{\overset{k_2}{\rightleftharpoons}} EP \overset{k_3}{\longrightarrow} E + P \qquad (5)$$

The three simultaneous equations are given in the following form:

$$\begin{array}{l} dE/dt = \\ dEA/dt = \\ dEP/dt = \end{array} \begin{vmatrix} E & EA & EP \\ -k_1 A & k_{-1} & k_3 \\ k_1 A & -(k_{-1} + k_2) & k_{-2} \\ 0 & k_2 & -(k_{-2} + k_3) \end{vmatrix} \begin{array}{l} = 0 \\ = 0 \\ = 0 \end{array}$$

The determinant, or distribution term, for E, for example, can be calculated from the coefficients listed above, after deleting the E column. For a mechanism of n intermediates, only $n - 1$ equations are needed. Thus, by leaving out the dEP/dt row, we can write

---

[3] J. T. Wong, "Kinetics of Enzyme Mechanisms." Academic Press, New York, 1975.

$$(E) = \begin{vmatrix} k_{-1} & k_3 \\ -(k_{-1} + k_2) & k_{-2} \end{vmatrix} = k_{-1}k_{-2} + k_3(k_{-1} + k_2)$$

If the dE/dt row is omitted instead, we have

$$(E) = \begin{vmatrix} -(k_{-1} + k_2) & k_{-2} \\ k_2 & -(k_{-2} + k_3) \end{vmatrix}$$

$$= k_{-1}(k_{-2} + k_3) + k_2(k_{-2} + k_3) - k_2 k_{-2}$$
$$= k_{-1}k_{-2} + k_3(k_{-1} + k_2)$$

Note that deletion of different equations often leads to different amounts of algebraic manipulations. Application of the same operations to EA and EP yields

$$(EA) = k_1(k_{-2} + k_3)A$$
$$(EP) = k_1 k_2 A$$

The rate equation is readily obtained as

$$\frac{v}{E_0} = \frac{k_3(EP)}{(E) + (EA) + (EP)}$$

$$= \frac{k_1 k_2 k_3 A}{k_{-1}k_{-2} + k_3(k_{-1} + k_2) + k_1(k_{-2} + k_3)A + k_1 k_2 A}$$

or

$$v = \frac{k_2 k_3 E_0 A/(k_2 + k_{-2} + k_3)}{\{[k_{-1}k_{-2} + k_3(k_{-1} + k_2)]/[k_1(k_2 + k_{-2} + k_3)]\} + A} \tag{6}$$

$$= \frac{V_1 A}{K_m + A}$$

Equation (6) is identical in form with Eq. (4). In fact, if $k_3 \gg k_2, k_{-2}$, Eq. (6) reduces to Eq. (4). Although Eq. (5) is a more realistic mechanism compared with Eq. (1), especially when the rapid-equilibrium treatment is applied to the reversible reaction, the information obtainable from initial-rate studies of such unireactant system remains nevertheless the same: V_1 and K_m. This serves to justify the simplification used by the kineticist; that is, the elimination of certain intermediates to maintain brevity of the rate equation (provided the mathematical form is unaltered). Thus, the *forward* reaction of an ordered Bi Bi mechanism is generally written as diagrammed below.

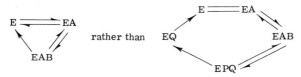

The use of the determinant method for complex enzyme mechanisms is time-consuming because of the stepwise expansion and the large number of positive and negative terms that must be canceled. It is quite useful, however, in computer-assisted derivation of rate equations (cf. Chapter [5] by Fromm, in this volume).

The King and Altman Method

King and Altman[4] developed a schematic approach for deriving steady-state rate equations, which has contributed to the advance of enzyme kinetics. The first step of this method is to draw an *enclosed* geometric figure with each enzyme form as one of the corners. Equation (5), for instance, can be rewritten as:

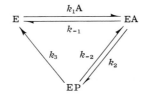

The second step is to draw all the possible patterns that connect all the enzyme species without forming a loop. For a mechanism with n enzyme species, or a figure with n corners, each pattern should contain $n - 1$ lines. The number of valid patterns for any single-loop mechanism is equal to the number of enzyme forms. Thus, there are three patterns for the triangle shown above:

The determinant for a given enzyme species is obtained as the summation of the product of the rate constants and concentration factors associated with all the branches in the patterns *leading toward* this particular enzyme species. The same patterns are used for each species, albeit the direction in which they are read will vary. However, when an irreversible step is present, e.g., the EP → E step, some patterns become invalid for certain enzyme forms.

$$(E) = \quad \text{[pattern 1]} \quad + \quad \text{[pattern 2]} \quad + \quad \text{[pattern 3]}$$

$$= k_{-1}k_3 + k_2k_3 + k_{-1}k_{-2}$$

---

[4] E. L. King and C. Altman, *J. Phys. Chem.* **60**, 1375 (1956).

[4] DERIVATION OF INITIAL VELOCITY EQUATIONS

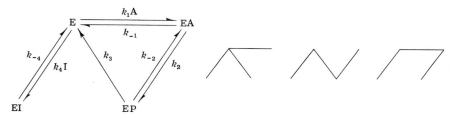

The rate equation is obtained as

$$\frac{v}{E_0} = \frac{k_3(EP)}{(E) + (EA) + (EP)}$$

where (E), (EA), and (EP) are the determinants for E, EA, and EP, respectively.

The presence of an enzyme intermediate(s) that is not part of a loop will not affect the number of King–Altman patterns. For instance, the addition of a competitive inhibitor, I, to the above system will result in the same number of patterns.

The additional E ⇌ EI branch is present in *all* the diagrams. Thus, in calculating the number of valid King–Altman patterns, only the *closed loops* need be considered. The determinants of E, EA, EAB, and EI can be obtained by the method just described:

$$(E) = k_{-4}(k_{-1}k_3 + k_2k_3 + k_{-1}k_{-2})$$
$$(EA) = k_{-4}(k_1k_3A + k_1k_{-2}A)$$
$$(EP) = k_{-4}(k_1k_2A)$$
$$(EI) = k_4I(k_{-1}k_3 + k_2k_3 + k_{-1}k_{-2})$$

It is more convenient, however, to treat this case by first considering only the loop portion (ignoring the additional E ⇌ EI step for the time being).

$$(E) = k_{-1}k_3 + k_2k_3 + k_{-1}k_{-2}$$
$$(EA) = k_1(k_{-2} + k_3)A$$
$$(EP) = k_1k_2A$$

The determinant for EI is then obtained as

$$(EI) = (E)k_4I/k_{-4}$$
$$= (k_{-1}k_3 + k_2k_3 + k_{-1}k_{-2})k_4I/k_{-4}$$
$$= (k_{-1}k_3 + k_2k_3 + k_{-1}k_{-2})I/K_i$$

where $K_i = k_{-4}/k_4$.

The King–Altman method is most convenient for single-loop mechanisms. In practice, there is no need to write down the patterns. One can use an object, say, a paper clip, to block one branch of the loop, write down the appropriate term for each enzyme species, then repeat the process until every branch in the loop has been blocked once.

For more complex mechanisms having alternative pathways that form several closed loops, the precise number of valid King–Altman patterns must be calculated to avoid omission of terms. To illustrate the various situations that may occur in such calculation, let us consider Scheme 1.

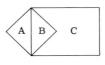

Scheme 1

The total number of patterns with $n - 1$ lines is given by the equation

$$\frac{m!}{(n-1)!(m-n+1)!}$$

where m = the number of lines in the complete geometric figure. In the above scheme, $m = 8$ and $n = 6$, and the total number of patterns with 5 lines is

$$\frac{8!}{5!\,3!} = \frac{(8 \times 7 \times 6 \times 5 \times 4 \times 3 \times 2)}{(5 \times 4 \times 3 \times 2)(3 \times 2)} = 56$$

This number, however, includes patterns that contain the following loops, which must be subtracted from the total:

The number of patterns for a loop with r lines is given by

$$\frac{(m-r)!}{(n-1-r)!(m-n+1)!}$$

According to this equation, for loops A and B, $r = 3$, we have 10 patterns each; for loops A + B and B + C, $r = 4$, we have 4 patterns each; and for loops C and A + B + C, $r = 5$ (note that 0! = 1), we have 1 pattern each. The total number of loop-containing patterns to be subtracted is 30. One of the patterns, however, occurred three times in the above calculations, but should be discarded only once. This pattern involves both loop A and loop B (solid lines indicate the loop that gives rise to this pattern).

Thus, the total of loop-containing patterns is 28, and the total number of valid patterns is 56 − 28 = 28.

The 28 5-lined patterns are shown below.

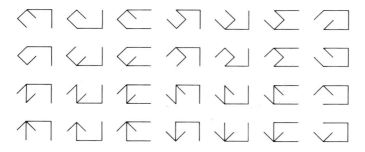

The conventional way of computing the valid King–Altman patterns is rather tedious. A set of formulas developed by the author allows the calculation of the desired number in a very short time. Each of these formulas is applicable to a particular geometric arrangement. For any figure consisting of three subfigures arrayed in sequence like the one shown in Scheme 1, the general formula for calculating the exact number of the valid King–Altman pattern, π, is

$$\pi = a \cdot b \cdot c - (l_{AB}^2 \cdot c + l_{BC}^2 \cdot a)$$

where a, b, and c = the number of lines in subfigures A, B, and C; l_{AB} and l_{BC} = the number of lines in the common boundaries between A and B, and B and C, respectively. For $a = 3$, $b = 3$, $c = 5$, $l_{AB} = 1$, and $l_{BC} = 2$ (Scheme 1), we have

$$\begin{aligned} \pi &= 3 \times 3 \times 5 - (1^2 \times 5 + 2^2 \times 3) \\ &= 45 - (5 + 12) \\ &= 28 \end{aligned}$$

In the case of two subfigures A and B sharing a common boundary as shown in Scheme 2

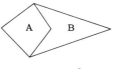

Scheme 2

the formula is given by

$$\pi = a \cdot b - l_{AB}^2$$
$$= 4 \times 4 - 2^2$$
$$= 12$$

Formulas for calculation of up to four subfigures in every possible geometric arrangement have been established.

The Method of Volkenstein and Goldstein

Volkenstein and Goldstein[2] have applied the theory of graphs to the derivation of rate equations. Their approach has three main features: the use of an auxiliary "node," the "compression" of a path into a point, and the addition of parallel branches. These can be best explained by an example (Scheme 3).

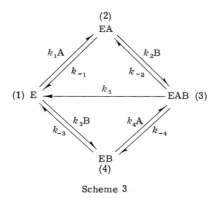

Scheme 3

Each enzyme-containing species is assigned a number and referred to as a node.

Suppose we want to calculate the determinant for EA (node 2). First, we choose another node, say node 3, as the auxiliary node (a reference starting point). The choice of the auxiliary node is arbitrary; it will not affect the outcome of the derivation, but may affect the amount of work involved. All the possible pathways (flow patterns) leading from (3) to (2) are then written (marked by solid branches).

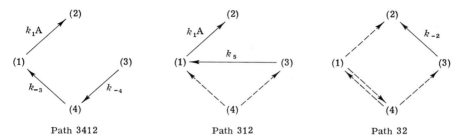

Path 3412   Path 312   Path 32

The nodes not included in the pathways retain the branches leading *away* from them (dashed branches). Since path 3412 flows through all the nodes, it is one of the terms of the determinant with a path value of $k_1 k_{-3} k_{-4} A$. Path 312 ($= k_1 k_5 A$) and path 32 ($= k_{-2}$) are now compressed into points.

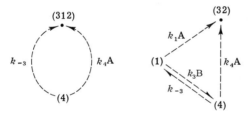

The two parallel branches leading from (4) to the compressed point (312) can be added together to yield

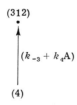

and the expression for this part is $(P312)(k_{-3} + k_4 A) = k_1 k_5 A (k_{-3} + k_4 A)$. The part containing point (32) can be treated by selecting a secondary auxiliary node, say node (4), and repeating the procedure described at the onset.

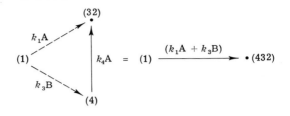

and

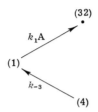

The contribution from this part is $(P432)(k_1A + k_3B) = k_{-2}k_4A(k_1A + k_3B)$ and $(P32)(k_1k_{-3}A) = k_1k_{-2}k_{-3}A$. Adding the terms together, we obtain the determinant for EA

$$(EA) = k_1k_{-3}k_{-4}A + k_1k_5A(k_{-3} + k_4A) + k_{-2}k_4A(k_1A + k_3B) \\ + k_1k_{-2}k_{-3}A \\ = k_1k_{-3}(k_{-2} + k_{-4} + k_5)A + k_{-2}k_3k_4AB + k_1k_4(k_{-2} + k_5)A^2$$

The determinants for E, EB, and EAB can be obtained in a similar fashion. The complete rate equation is given by

$$\frac{v}{E_0} = \frac{k_5(EAB)}{(E) + (EA) + (EB) + (EAB)}$$

Rate equations for more complex mechanisms can be derived by repeating the procedure described above as many times as necessary. The choice of the auxiliary point becomes important for reaction schemes containing several loops. The process is analogous to deciding which row (equation) should be omitted from the matrix in the determinant method. In general, one should choose, by inspection of the geometric structure of the mechanism, a node such that, if one removes from the figure the auxiliary node and the node whose determinant is desired, the remaining nodes do not form a closed loop. In addition, one should select a node situated in a symmetrical position with respect to the desired node. For instance, node (4) is a better choice as an auxiliary node for the calculation of the determinant for node (2). Node (3) was chosen for the sole purpose of illustrating the use of secondary auxiliary nodes.

The Systematic Approach

The systematic approach for deriving rate equations was first devised by Fromm[5] based on certain concepts advanced by Volkenstein and Goldstein.[2] Its underlying principles, however, are more akin to the graphic method of King and Altman.[4] The procedure to be described here[6] is a modified method that includes the contributions from the aforementioned workers and from Wong and Hanes.[7]

Let us use as an example the ordered Bi Bi mechanism, in which an alternative substrate, A', for the first substrate, A, is present (Scheme 4).

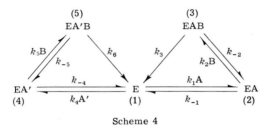

Scheme 4

Each enzyme-containing species is assigned a node number as previously described in the Volkenstein and Goldstein method. For each node, a node value is written, which is simply the summation of all branch values (rate constant and concentration factor) leading *away* from the node[5]:

$$(1) = k_1A + k_4A'$$
$$(2) = k_{-1} + k_2B$$
$$(3) = k_{-2} + k_3$$
$$(4) = k_{-4} + k_5B$$
$$(5) = k_{-5} + k_6$$

The determinant of a given enzyme species is equal to the noncyclic terms generated by multiplying together all the node values, excluding its own.[7] For example,

$$(E) = (2)(3)(4)(5)$$
$$= (k_{-1} + k_2B)(k_{-2} + k_3)(k_{-4} + k_5B)(k_{-5} + k_6)$$

and

$$(EAB) = (1)(2)(4)(5)$$
$$= (k_1A + k_4A')(k_{-1} + k_2B)(k_{-4} + k_5B)(k_{-5} + k_6)$$

[5] H. J. Fromm, *Biochem. Biophys. Res. Commun.* **40**, 692 (1970).
[6] C. Y. Huang, *Fed. Proc., Fed. Am. Soc. Exp. Biol.* **37**, 1423 (1978).
[7] J. T. Wong and C. S. Hanes, *Can. J. Biochem. Physiol.* **40**, 763 (1962).

The cyclic terms—terms that contain the products of reversible steps (underlined: $k_1A \cdot k_{-1}$, $k_2B \cdot k_{-2}$, $k_4A' \cdot k_{-4}$ and $k_5B \cdot k_{-5}$; readily identified by their subscripts differing only in positive and negative signs)—or the product of a closed loop [underscored by dashed lines; e.g., $k_4A' \cdot k_5B \cdot k_6$, constituting the (1) (4) (5) loop] are to be deleted to obtain the correct expression. The "reversible-step" terms can be eliminated during expansion:

$$(E) = (k_{-1} + k_2B)(k_{-2} + k_3)(k_{-4} + k_5B)(k_{-5} + k_6)$$
$$= [(k_{-1} + k_2B)\underset{X}{k_{-2}} + (k_{-1} + k_2B)k_3][(k_{-4} + k_5B)\underset{X}{k_{-5}}$$
$$+ (k_{-4} + k_5B)k_6]$$
$$= k_{-1}k_{-4}(k_{-2} + k_3)(k_{-5} + k_6)$$
$$+ [k_2k_3k_{-4}(k_{-5} + k_6) + k_{-1}k_5k_6(k_{-2} + k_3)]B$$
$$+ k_2k_3k_5k_6B^2$$

Note that the canceled terms are marked by an X. Also note that k_2B and k_5B, instead of k_{-2} and k_{-5}, are crossed out because the products $k_{-1}k_{-2}$ and $k_{-4}k_{-5}$ are not "cyclic."

The "loop terms" can be eliminated prior to expansion by a "branching" technique. From the expression

$$(EAB) = (1)(2)(4)(5)$$

a quick glance at Scheme 4 will reveal that (1) (4) (5) form a closed loop. Thus, the "one-branch" approach of Fromm[5] is first applied. With this approach, the determinant of an enzyme species, e.g., EAB, is obtained as the summation of the products of the nearest branch values leading to it (for EAB, there is only one nearest branch, EA → EAB or 23) and the remaining node values

$$(EAB) = (23)(1)(4)(5)$$

Similarly

$$(EA) = (12)(3)(4)(5) + (32)(1)(4)(5)$$

Note that the (1) (4) (5) loop is not eliminated by the one-branch approach. We now apply the "consecutive-branch" technique[6]

$$(EAB) = (23)(12)(4)(5)$$
$$(EA) = (12)(3)(4)(5) + (32)X$$

The procedure for using this technique is as follows: (a) Only the *loop-containing* terms require further branching; e.g., since (3) (4) (5) do not form a loop, the term (12) (3) (4) (5) remains unchanged. *Unnecessary consecutive-branching may result in omission of terms.* (b) For the loop-containing terms, the first branch(es) is followed by its nearest branches and the remaining nodes not involved in these branches; e.g., the 23 branch is followed by the 12 branch. This is done by inspection of the reaction scheme. (c) In the case of (32) (1) (4) (5), since there is no branch leading from other nodes to (3), the whole term is deleted (marked by X).

When all the loops have been removed, the resultant expression is expanded to obtain the desired determinant. Thus,

$$(EAB) = k_2B \cdot k_1A(k_{-4} + \underline{k_5B})(\underline{k_{-5}} + k_6)$$
$$= k_1k_2k_{-4}(k_{-5} + k_6)AB + k_1k_2k_5k_6AB^2$$
$$(EA) = k_1A(k_{-2} + k_3)(k_{-4} + \underline{k_5B})(\underline{k_{-5}} + k_6)$$
$$= k_1k_{-4}(k_{-2} + k_3)(k_{-5} + k_6)A + k_1k_5k_6(k_{-2} + k_3)AB$$

Note that the product of reversible steps, $k_5B \cdot k_{-5}$, is canceled in the expanding process, as has been previously described.

The determinants of EA' and $EA'B$ can be obtained by the same approach, and the complete rate equation is expressed as v/E_0. For the example given here, if the common product P is measured, we have

$$\frac{v}{E_0} = \frac{k_3(EAB) + k_6(EA'B)}{(E) + (EA) + (EAB) + (EA') + (EA'B)}$$

The first rule for using the systematic approach, broadly stated, is as follows:

Rule 1: The determinant of a given enzyme-containing species is equal to the product of the node values of the other enzyme species, minus the reversible-step terms. When the nodes form one or more closed loops, apply the one-branch approach. Apply the consecutive-branch approach to any *remaining* loop-containing terms until all loops are eliminated.

It should be noted that unnecessary application of the one-branch approach may lead to needless algebraic manipulations. Furthermore, the branching technique often results in "redundant terms" that must be searched out and deleted. As an example, let us write the determinant for E in Scheme 4 by the one-branch method:

$$(E) = (21)(3)(4)(5) + (31)(2)(4)(5) + (41)(2)(3)(5) + (51)(2)(3)(4)$$
$$= k_{-1}(k_{-2} + k_3)(k_{-4} + k_5B)(k_{-5} + k_6)$$
$$+ k_3(k_{-1} + k_2B)(k_{-4} + k_5B)(k_{-5} + k_6)$$
$$+ k_{-4}(k_{-1} + k_2B)(k_{-2} + k_3)(k_{-5} + k_6)$$
$$+ k_6(k_{-1} + k_2B)(k_{-2} + k_3)(k_{-4} + k_5B)$$

Expansion of the above equation will generate many redundant terms; e.g., $k_{-1}k_3k_{-4}k_6$ can be found in all four terms above, but only one is needed. These redundant terms can be eliminated by using Rule 2.

Rule 2: The nearest branch values cannot appear in *subsequent* node values. When the consecutive-branch approach is used, the *product* of the consecutive branches cannot appear in *subsequent* terms.

Consequently, the 21 term, k_{-1}, is crossed out from all subsequent terms; the 31 term, k_3, from all subsequent node (3) terms, but not from the node (3) term preceding it; and the 41 term, k_{-4}, from the subsequent node (4) term:

$$\begin{aligned}
(E) = &\; k_{-1}(k_{-2} + k_3)(k_{-4} + k_5B)(k_{-5} + k_6) \\
&+ k_3(k_{-1} + k_2B)(k_{-4} + k_5B)(k_{-5} + k_6) \\
& \text{X} \\
&+ k_{-4}(\underline{k_{-1}} + \underline{k_2B})(\underline{k_{-2}} + k_3)(k_{-5} + k_6) \\
&\phantom{+\;k_{-4}(\;}\text{X}\text{X} \\
&+ k_6(\underline{k_{-1}} + \underline{k_2B})(\underline{k_{-2}} + k_3)(k_{-4} + k_5B) \\
&\text{X}\text{X}\text{X}
\end{aligned} \Big\} \text{Canceled}$$

After elimination of the redundant terms, the last two terms are canceled because they all contain the $k_2B \cdot k_{-2}$ reversible-step product. The remaining terms are identical with the expression obtained from (E) = (2) (3) (4) (5), demonstrating the fact that unnecessary branching may lead to wasteful algebraic exercise.

There are situations where certain redundant terms are difficult to eliminate. The following example serves to illustrate a procedure useful for the complicated cases. Consider the hypothetical mechanism shown in Scheme 5.

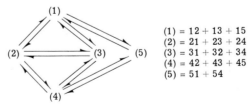

(1) = 12 + 13 + 15
(2) = 21 + 23 + 24
(3) = 31 + 32 + 34
(4) = 42 + 43 + 45
(5) = 51 + 54

Scheme 5

The numerical notation used for Scheme 5 has several advantages. (a) The tedium of writing down all the rate constants and concentration factors is avoided. The constants can be substituted into the final expression (after canceling all the loop terms, redundant terms, and reversible-step terms) to obtain the desired determinant. (b) When two or more steps are assigned the same rate constant, e.g., k_1 and k_{-1}, this method prevents the cancelation of $k_1 \cdot k_{-1}$ product formed from different steps. (c) The reversible-step terms can still be easily identified and eliminated by their numbers, e.g., 12 · 21, 35 · 53, etc.

Suppose we want to calculate the determinant for (5), we can write $D_5 = (1) (2) (3) (4)$. Applying the consecutive-branch technique, we have

$$\begin{aligned}
D_5 &= (15) (2) (3) (4) + (45) (1) (2) (3) \\
&= 15 [21 (3) (4) + 31 (2) (4)] + 45 [24 (1) (3) + 34 (1) (2)] \\
&= 15 [21 (31 + 32 + 34)(42 + 43 + 45) \\
&\quad + 31 (\underline{21} + 23 + 24)(42 + 43 + 45)] \\
&\quad + 45 [\overline{24} (12 + 13 + \underline{15})(\underline{31} + 32 + 34) \\
&\quad + 34 (12 + 13 + \underline{15})(\underline{21} + 23 + \underline{24})]
\end{aligned}$$

Note that within the same brackets, the redundant terms (underlined by solid lines) are readily removed, but the products of consecutive branch terms, 15 · 21 and 15 · 31 (underlined by dashed lines) within the second set of brackets cannot be canceled without further manipulation. One can either eliminate them during further expansion of the equation or repeat the consecutive-branch approach until *all* the consecutive branches end with a branch leading away from a *common* node. This "common-node consecutive branching" approach is always valid—even when all the loop-containing terms have been eliminated. Using this approach, we can write

$$\begin{aligned}
D_5 &= 15 [21 (3) (4) + 31 (2) (4)] + 45 [24 (1) (3) + 34 (1) (2)] \\
&= 15 \{21 (3) (4) + 31 [23 (4) + 43 \cdot 24]\} \\
&\quad + 45 \{24 (1) (3) + 34 [23 (1) + 13 \cdot 21]\}
\end{aligned}$$

In the above operation, we chose node (2) as the common node; we could have chosen node (3) and obtained the same determinant. The principle involved here is analogous to the "auxiliary node" used by Volkenstein and Goldstein. It is not routinely used because the extra operations are not needed under most circumstances.

Comparison of Different Steady-State Methods

For relatively simple mechanisms, all the diagrammatic and systematic procedures illustrated in the foregoing sections are quite convenient. The King–Altman method is best suited for single-loop mechanisms, but becomes laborious for more complex cases with five or more enzyme forms because of the work involved in the calculation and drawing of valid patterns. With multiloop reaction schemes involving four to five enzyme species, the systematic approach requires the least effort, especially when irreversible steps are present, since it does away with pattern drawing. When the number of enzyme forms reaches six or more in a mechanism with several alternate pathways, all the manual methods become tedious owing to the sheer number of terms involved.

The Rapid-Equilibrium Treatment

The first rate equation for an enzyme-catalyzed reaction was derived by Henri and by Michaelis and Menten, based on the rapid-equilibrium concept. With this treatment it is assumed that there is a slow catalytic conversion step and the combination and dissociation of enzyme and substrate are relatively fast, such that they reach a state of quasi-equilibrium or rapid equilibrium.

The derivation of initial-velocity equations for any rapid equilibrium system is quite simple. When the equilibrium relationship among various enzyme–substrate complexes are defined, the rate equation can be written simply by inspection. Consider the one-substrate system

$$E + A \underset{}{\overset{K_a}{\rightleftharpoons}} EA \xrightarrow{k_2}$$

$$(EA) = [(E)(A)/K_a, \quad v = k_2(EA)$$

We can write

$$\frac{v}{E_0} = \frac{k_2(EA)}{(E) + (EA)} = \frac{k_2(E)A/K_a}{(E) + (E)A/K_a}$$
$$= \frac{k_2(A/K_a)}{1 + A/K_a} = \frac{k_2 A}{(K_a + A)}$$

It is clear that there is no need to write down (E), and one can obtain the rate expression by replacing (E) with 1. Thus, for the Random Bi Bi mechanism shown below

$$E + A \underset{}{\overset{K_{ia}}{\rightleftharpoons}} EA \qquad EA + B \underset{}{\overset{K_b}{\rightleftharpoons}} EAB \xrightarrow{k_5} E + P$$

$$E + B \underset{}{\overset{K_{ib}}{\rightleftharpoons}} EB \qquad EB + A \underset{}{\overset{K_a}{\rightleftharpoons}} EAB \xrightarrow{k_5} E + P$$

we can quickly write

$$\frac{v}{E_0} = \frac{k_5 AB/(K_{ia} K_b)}{1 + A/K_{ia} + B/K_{ib} + AB/K_{ia} K_b}$$

Note that although two equations describe the formation of EAB, only one of them is used (because $K_{ia} K_b = K_{ib} K_a$).

In using the equilibrium treatment, one should bear in mind that the rate laws so obtained are generally different in form from those derived by the steady-state assumption for the same mechanism.

The Combined Equilibrium and Steady-State Treatment

There are a number of reasons why a rate equation should be derived by the combined equilibrium and steady-state approach. First, the experimentally observed kinetic patterns necessitate such a treatment. For example, several enzymic reactions have been proposed to proceed by the rapid-equilibrium random mechanism in one direction, but by the ordered pathway in the other. Second, steady-state treatment of complex mechanisms often results in equations that contain many higher-order terms. It is at times necessary to simplify the equation to bring it down to a manageable size and to reveal the basic kinetic properties of the mechanism.

The procedure to be described here was originally developed by Cha.[8] The basic principle of his approach is to treat the rapid-equilibrium segment as though it were a single enzyme species at steady state with the other species. Let us consider the hybrid Rapid-Equilibrium Random-Ordered Bi Bi system:

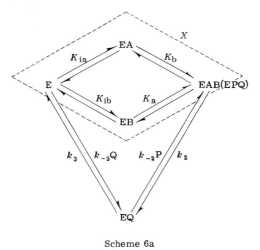

Scheme 6a

The area enclosed by dashed lines in Scheme 6a is the rapid-equilibrium random segment. The segment is called X, and Scheme 6 is reduced to a basic figure consisting of X and EQ connected by two pathways (Scheme 6b):

[8] S. Cha, *J. Biol. Chem.* **243**, 820 (1968).

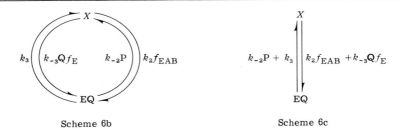

Scheme 6b   Scheme 6c

Scheme 6b can be further condensed into one line (Scheme 6c) according to the principle of "addition of parallel branches." The rate constants k_2 and $k_{-3}Q$ are multiplied by the fractional concentration of the enzyme species within the equilibrium segment that participates in that pathway, f_{EAB} and f_E.

$$f_{EAB} = \frac{EAB}{E + EA + EB + EAB} = \frac{AB/K_{ia}K_b}{1 + A/K_{ia} + B/K_{ib} + AB/K_{ia}K_b}$$

$$f_E = \frac{1}{1 + A/K_{ia} + B/K_{ib} + AB/K_{ia}K_b}$$

From Scheme 6c, we have

$$(EQ) = k_2 f_{EAB} + k_{-3}Q f_E$$
$$(X) = k_{-2}P + k_3$$

Thus, we obtain the rate equation

$$\frac{v}{E_0} = \frac{k_2 f_{EAB}(X) - k_{-2}P(EQ)}{(X) + (EQ)}$$

$$= \frac{k_2 k_3 f_{EAB} - k_{-2} k_{-3} PQ f_E}{k_{-2}P + k_3 + k_2 f_{EAB} + k_{-3} Q f_E}$$

$$= \frac{k_2 k_3 (AB/K_{ia}K_b) - k_{-2}k_{-3}PQ}{(k_{-2}P + k_3)(1 + A/K_{ia} + B/K_{ib} + AB/K_{ia}K_b) + k_2(AB/K_{ia}K_b) + k_{-3}Q}$$

This equation reveals atypical product inhibition patterns for a random mechanism: P is noncompetitive with both A and B; Q is competitive with both A and B. Whenever abnormal product inhibition patterns are observed, therefore, partial equilibrium treatment of the mechanism may be considered.

In the case of a rate-limiting step *within* a rapid equilibrium segment, it is necessary to include such a rate-limiting step in the velocity equation. Scheme 7a serves as an example for this type of mechanism:

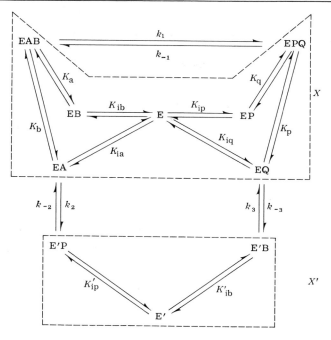

Scheme 7a

The two rapid-equilibrium segments X and X' are again indicated by areas enclosed by dashed lines. Within X, there is a rate-limiting step involving the interconversion of EAB and EPQ. By treating X and X' as though they were two enzyme species, adding parallel branches together, and multiplying the rate constants with appropriate fractional enzyme concentrations, we arrive at Scheme 7b.

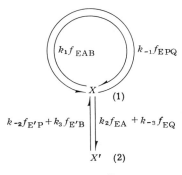

Scheme 7b

To obtain the determinants of X and X', the internal rate-limiting step in X is not included since it leads both away from and toward X. This can be readily shown by applying the systematic method described in the derivation of steady-state equations. Recall that the node values for X and X' are the summation of branch values leading *away* from them, as follows:

$$(1) = k_2 f_{EA} + k_{-3} f_{EQ}$$
$$(2) = k_{-2} f_{E'P} + k_3 f_{E'B}$$

The determinants for X and X' are defined as the product of other node values excluding their own; that is, $X = (2) = k_{-2} f_{E'P} + k_3 f_{E'B}$ and $X' = (1) = k_2 f_{EA} + k_{-3} f_{EA}$.

The fractional enzyme concentrations are obtained as their fractions in the appropriate equilibrium segments:

$$f_{EA} = \frac{A/K_{ia}}{1 + A/K_{ia} + B/K_{ib} + AB/K_{ia}K_b + P/K_{ip} + Q/K_{iq} + PQ/K_{ip}K_q}$$

$$f_{EQ} = \frac{Q/K_{iq}}{1 + A/K_{ia} + B/K_{ib} + AB/K_{ia}K_b + P/K_{ip} + Q/K_{iq} + PQ/K_{ip}K_q}$$

$$f_{EAB} = \frac{AB/K_{ia}K_b}{1 + A/K_{ia} + B/K_{ib} + AB/K_{ia}K_b + P/K_{ip} + Q/K_{iq} + PQ/K_{iq}K_q}$$

$$f_{EPQ} = \frac{PQ/K_{ip}K_q}{1 + A/K_{ia} + B/K_{ib} + AB/K_{ia}K_b + P/K_{ip} + Q/K_{iq} + PQ/K_{ip}K_q}$$

$$f_{E'B} = \frac{B/K'_{ib}}{1 + B/K'_{ib} + P/K'_{ip}}$$

$$f_{E'P} = \frac{P/K'_{ip}}{1 + B/K'_{ib} + P/K'_{ip}}$$

In writing an expression for the initial velocity for Scheme 7b, however, one must include the internal rate-limiting step. Thus, we have

$$v = (k_1 f_{EAB} - k_{-1} f_{EPQ})X + k_2 f_{EA} X - k_{-2} f_{E'P} X'$$

Note that in steady-state treatment, any of the pathways can be used to write the velocity expression. Either one of the two pathways linking X and X' (see Scheme 7a) will yield the same expression:

$$k_2 f_{EA} X - k_{-2} f_{EP} X'$$
$$= k_2 f_{EA}(k_{-2} f_{E'P} + k_3 f_{E'B}) - k_{-2} f'_{EP}(k_2 f_{EA} + k_{-3} f_{EQ})$$
$$= k_2 f_{EA} k_3 f_{E'B} - k_{-2} f_{E'P} k_{-3} f_{EQ}$$
$$k_3 f_{E'B} X' - k_{-3} f_{EQ} X$$
$$= k_3 f_{E'B}(k_2 f_{EA} + k_{-3} f_{EQ}) - k_{-3} f_{EQ}(k_{-2} f_{E'P} + k_3 f_{E'B})$$
$$= k_2 f_{EA} k_3 f_{E'B} - k_{-2} f_{E'P} k_{-3} f_{EQ}$$

The complete rate equation is given by

$$\frac{v}{E_0} = \frac{(k_1 f_{EAB} - k_{-1} f_{EPQ})X + k_2 f_{EA}X - k_{-2} f_{E'P}X'}{X + X'}$$

$$= \frac{(k_1 f_{EAB} - k_{-1} f_{EPQ})(k_{-2} f_{E'P} + k_3 f_{E'B}) + k_2 f_{EA} k_3 f_{E'B} - k_{-2} f_{E'P} k_{-3} f_{EQ}}{k_{-2} f_{E'P} + k_3 f_{E'B} + k_2 f_{EA} + k_{-3} f_{EQ}}$$

For mechanisms involving three or more rapid-equilibrium segments, once the segments are properly represented as "nodes" in a scheme, the rate equation can be obtained by the usual "systematic approach." For example, consider the case of one substrate–one product reaction in which a modifier M is in rapid equilibration with E, EA, and EP.

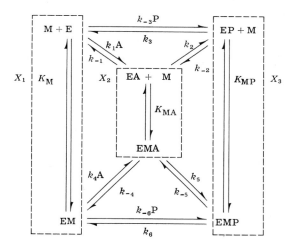

Treating X_1, X_2, and X_3 as nodes and adding parallel branches together, we obtain

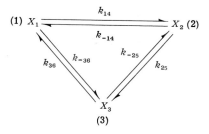

where $k_{14} = k_1 A f_E + k_4 A f_{EM}$, $k_{-14} = k_{-1} f_{EA} + k_{-4} f_{EMA}$, $k_{25} = k_2 f_{EA} + k_5 f_{EMA}$, $k_{-25} = k_{-2} f_{EP} + k_{-5} f_{EMP}$, $k_{36} = k_3 f_{EP} + k_6 f_{EMP}$, and $k_{-36} = k_{-3} P f_E + k_{-6} P f_{EM}$. Note that $v = k_{36} X_3 - k_{-36} X_1$.

Derivation of Isotope-Exchange Rate Equations

Isotope-exchange rate equations can be classified into two types: exchange at equilibrium and at steady state. The theory and technique of equilibrium exchange was pioneered by Boyer.[9] Most applications of isotope-exchange methods to enzyme systems have been of this type. Under equilibrium conditions there are no net chemical changes; whereas under steady-state conditions there is a net catalysis. Although the derivation of these two types of rate equations differs in the assumptions involved, an equation derived by the steady-state approach can be readily converted into one for equilibrium exchange. Therefore, procedures intended for steady-state exchange are equally applicable to the derivation of equilibrium-exchange rate equations.

Equations for Exchanges at Equilibrium

The derivation of equilibrium isotope-exchange equations for enzymic reactions was first formulated by Boyer.[9] Others have subsequently contributed to its development. Yagil and Hoberman[10] and Flossdorf and Kula[11] have devised generalized approaches that treat the flux of a label in a chemical reaction in a way analogous to the flow of charge in an electrical circuit. In this approach, (a) the equilibrium velocity of a reaction proceeding through n parallel steps is equal to the sum of the n individual rates; (b) when proceeding through n consecutive steps, the reciprocal of equilibrium velocity is equal to the sum of the n reciprocals. To demonstrate the use of this method, let us consider the B → P exchange in an Ordered Bi Bi mechanism (Scheme 8).

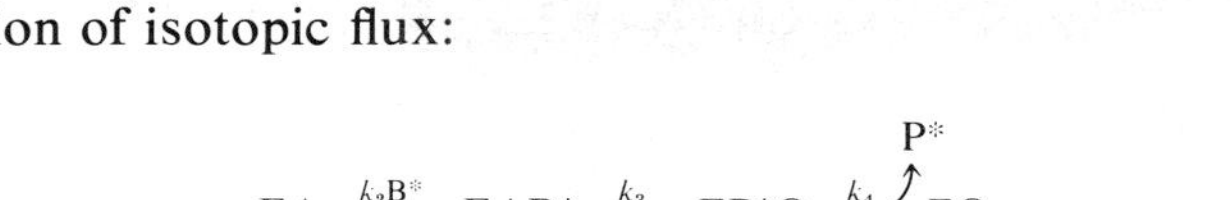

$$E \underset{k_{-1}}{\overset{k_1 A}{\rightleftarrows}} EA \underset{k_{-2}}{\overset{k_2 B^*}{\rightleftarrows}} EAB^* \underset{k_{-3}}{\overset{k_3}{\rightleftarrows}} EP^*Q \underset{k_{-4}P}{\overset{k_4 \nearrow P^*}{\rightleftarrows}} EQ \underset{k_{-5}Q}{\overset{k_5}{\rightleftarrows}} E$$

Scheme 8

The asterisks mark the labeled species. Since the system is at equilibrium and the exchange involves only three steps, we can write a new scheme in the direction of isotopic flux:

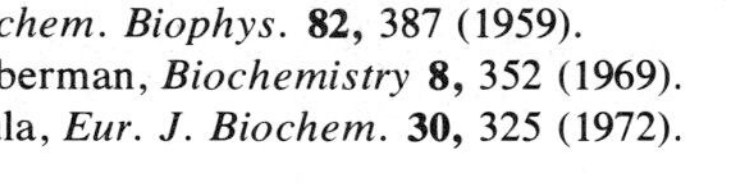

$$EA \xrightarrow{k_2 B^*} EAB^* \xrightarrow{k_3} EP^*Q \xrightarrow{k_4 \nearrow P^*} EQ$$

The reverse steps are not shown because they are included in the equilibrium relationships to be substituted into the equation later.

---

[9] P. D. Boyer, *Arch. Biochem. Biophys.* **82**, 387 (1959).
[10] G. Yagil and H. D. Hoberman, *Biochemistry* **8**, 352 (1969).
[11] J. Flossdorf and M. Kula, *Eur. J. Biochem.* **30**, 325 (1972).

Using the rule governing consecutive steps, we have

$$\frac{1}{v^*_{B\to P}} = \frac{1}{k_2 B^*(EA)} + \frac{1}{k_3(EAB^*)} + \frac{1}{k_4(EP^*Q)} \tag{7}$$

From the equilibrium relationships

$$(EAB^*) = \frac{k_2 B^*(EA)}{k_{-2}}$$

$$(EP^*Q) = \frac{k_3(EAB^*)}{k_{-3}} = \frac{k_2 k_3 B^*(EA)}{k_{-2} k_{-3}}$$

we can substitute the expressions of (EAB^*) and (EP^*Q) into Eq. (7) to obtain

$$\frac{1}{v^*_{B\to P}} = \frac{k_3 k_4 + k_{-2}(k_{-3} + k_4)}{k_2 k_3 k_4 B^*(EA)} \tag{8}$$

or

$$\frac{v^*_{B\to P}}{E_0} = \frac{k_2 k_3 k_4 B^*(EA)/E_0}{k_3 k_4 + k_{-2}(k_{-3} + k_4)} \tag{9}$$

where

$$\frac{(EA)}{E_0} = \frac{(EA)}{(E) + (EA) + (EAB) + (EPQ) + (EQ)}$$

Hence,

$$\frac{v^*_{B\to P}}{E_0} = \frac{k_1 k_2 k_3 k_4 A B^*}{k_{-1}[k_3 k_4 + k_{-2}(k_{-3} + k_4)]} \\ [1 + k_1 A/k_{-1} + k_1 k_2 AB/k_{-1}k_{-2} + k_{-5}Q/k_5 + k_{-4}k_{-5}PQ/k_4 k_5]$$

The derivation of exchange-rate equations for mechanisms with branched pathways requires the use of rules governing consecutive and parallel steps. Consider as an example the $A \to P$ exchange in the Random Bi Bi mechanism (Scheme 9):

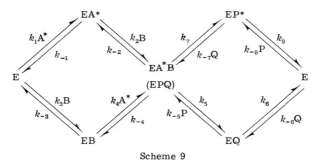

Scheme 9

Again, only the flux in one direction need be considered, and the steps not involved in the flux are ignored (Scheme 10).

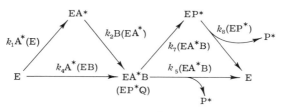

Scheme 10

The first step is to obtain the expressions for the E → EA* → EA*B and EA*B → EP* → E pathways. Let us call them k_{12} and k_{78}:

$$\frac{1}{k_{12}} = \frac{1}{k_1 A^*(E)} + \frac{1}{k_2 B(EA)}; \qquad k_{12} = \frac{k_1 A^*(E) \cdot k_2 B(EA)}{k_1 A^*(E) + k_2 B(EA)}$$

$$\frac{1}{k_{78}} = \frac{1}{k_7(EA^*B)} + \frac{1}{k_8(EP^*)}; \qquad k_{78} = \frac{k_7(EA^*B) \cdot k_8(EP^*)}{k_7(EA^*B) + k_8(EP^*)}$$

Scheme 10 can now be represented by Scheme 11.

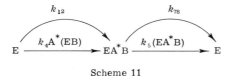

Scheme 11

Since parallel steps are additive, we have

$$E \xrightarrow{k_4 A^*(EB) + k_{12}} EA^*B \xrightarrow{k_5(EA^*B) + k_{78}} E$$

and finally

$$\frac{1}{v^*_{A \to P}} = \frac{1}{k_4 A^*(EB) + k_{12}} + \frac{1}{k_5(EA^*B) + k_{78}} \tag{10a}$$

The exchange-rate equation is obtained by substituting the expressions for k_{12}, k_{78}, and the equilibrium expressions for the enzyme intermediates into Eq. (10a).

From the following equilibrium relationships

$$(EA^*) = k_1 A^*(E)/k_{-1}$$
$$(EB) = k_3 B(E)/k_{-3}$$
$$(EA^*B) = k_4 A^* k_3 B(E)/k_{-4} k_{-3}$$
$$(EP^*) = k_7(EA^*B)/k_{-7} Q$$

and

$$k_1 A k_2 B = k_3 A k_4 B, \quad k_{-1} k_{-2} = k_{-3} k_{-4}$$

it can be shown that

$$\frac{1}{v^*_{A \to P}} = \frac{1}{k_4 A^*(EB) + [k_1 A^*(E) \cdot k_2 B(EA^*)]/[k_1 A^*(E) + k_2 B(EA^*)]}$$
$$+ \frac{1}{k_5(EA^*B) + [k_7(EA^*B) \cdot k_8(EP^*)]/[k_7(EA^*B) + k_8(EP^*)]}$$

$$= \frac{1}{(E)} \frac{[(k_{-1} + k_2 B)(k_{-7} Q + k_8)(k_{-4} + k_5)}{[k_1 A k_2 B + k_4 A k_3 B (k_{-1} + k_2 B)/k_{-3}]} \quad (10b)$$
$$\phantom{= \frac{1}{(E)}} \frac{+ k_7 k_8 (k_{-1} + k_2 B) + k_{-1} k_{-2} (k_{-7} Q + k_{-8})]}{[k_5(k_{-7} Q + k_8) + k_8 k_7]}$$

Derivation by the Steady-State Method

Britton[12] first derived isotope flux equations under steady-state rather than equilibrium conditions. To illustrate his procedure, we shall again use Scheme 8, the B → P exchange in Ordered Bi Bi mechanism, as an example, so that the results can be compared (Scheme 8a).

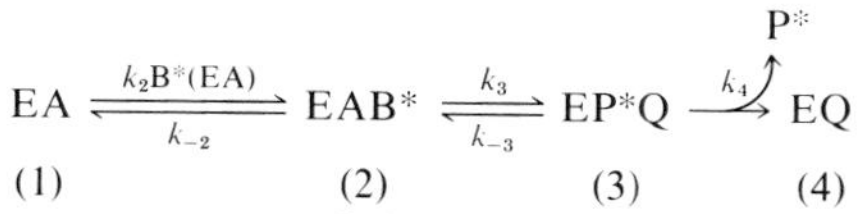

$$EA \underset{k_{-2}}{\overset{k_2 B^*(EA)}{\rightleftarrows}} EAB^* \underset{k_{-3}}{\overset{k_3}{\rightleftarrows}} EP^*Q \overset{k_7}{\underset{}{\to}} \overset{P^*}{} EQ$$
$$(1) (2) (3) (4)$$

Scheme 8a

Note that (a) the reverse steps are needed for steady-state treatment; (b) the k_{-4} step is not shown because the initial rate of exchange is being measured; and (c) only the unlabled enzyme concentration is included in the derivation because the concentration factors of labeled enzyme forms will cancel out during the derivation. Also, each enzyme form is assigned a number for reference purposes.

The procedure is to calculate the B → P or 1 → 4 isotope transfer in a stepwise manner using partition theory:

$$\text{Flux } 1 \to 2 = k_2 B^*(EA)$$

$$\text{Flux } 1 \to 3 = (1 \to 2) \cdot \frac{(2 \to 3)}{(2 \to 3) + (2 \to 1)}$$

$$= \frac{k_2 B^*(EA) k_3}{k_3 + k_{-2}}$$

---

[12] H. G. Britton, *J. Physiol. (London)* **170**, 1 (1964).

$$\text{Flux } 1 \to 4 = (1 \to 3) \cdot \frac{(3 \to 4)}{(3 \to 4) + (3 \to 1)}$$

$$= \frac{k_2 B^*(EA) \, k_3}{k_3 + k_{-2}} \cdot \frac{k_4}{k_4 + [k_{-2}k_{-3}/(k_3 + k_{-2})]}$$

$$= \frac{k_2 k_3 k_4 B^*(EA)}{k_4(k_3 + k_{-2}) + k_{-2}k_{-3}}$$

$$= v^*_{B \to P} \tag{11}$$

Note that Eq. (11) is identical with Eq. (8). More recently, Cleland[13] developed an approach that is more convenient. His procedure starts with the release of labeled product and works backward as follows:

$$\text{Flux } 3 \to 4 = k_4$$

$$\text{Flux } 2 \to 4 = (2 \to 3) \cdot \frac{(3 \to 4)}{(3 \to 4) + (3 \to 2)}$$

$$= \frac{k_3 k_4}{k_4 + k_{-3}}$$

$$\text{Flux } 1 \to 4 = (1 \to 2) \cdot \frac{(2 \to 4)}{(2 \to 4) + (2 \to 1)}$$

$$= \frac{k_2 B^*(EA) \, k_3 k_4/(k_4 + k_{-3})}{[k_3 k_4/(k_4 + k_{-3})] + k_{-2}}$$

$$= \frac{k_2 k_3 k_4 B^*(EA)}{k_3 k_4 + k_{-2}(k_4 + k_{-3})}$$

These procedures, however, are more suitable for deriving exchange-rate equations for mechanisms without branched pathways. For more complex mechanisms, the schematic method of Cleland[14] based on the approach of King and Altman is quite convenient. This method can be adapted to the systematic approach and combined with the deletion of steps linking unlabeled species previously described to further reduce the amount of work involved.[6] Consider the $A \to P$ exchange in the Random Bi Bi mechanism (cf. Scheme 9; the figure has been redrawn in a folded geometric form) shown in Scheme 12.

The pathways connecting the unlabeled enzymes species, $E \rightleftharpoons EB$ and $E \rightleftharpoons EQ$, as has been shown in Scheme 10, can be eliminated; and the $EB \rightleftharpoons EAB$ and $EAB \to EQ$ steps can be directly linked to E (Scheme 12a).

By combining the k_{-4} and k_5 branches, Scheme 12a can be further reduced to Scheme 12b.

---

[13] W. W. Cleland, *Biochemistry* **14**, 3220 (1975).
[14] W. W. Cleland, *Annu. Rev. Biochem.* **36**, 77 (1967).

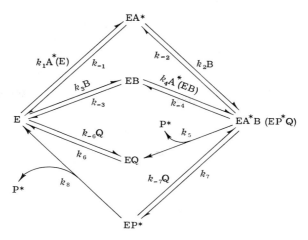

Scheme 12

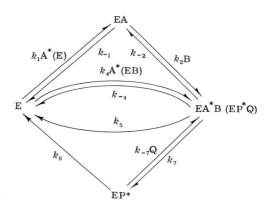

Scheme 12a

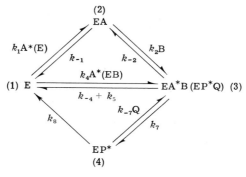

Scheme 12b

Now we apply the systematic method to Scheme 12b by assigning node numbers to the enzyme species and write down the node values:

$$(1) = k_1 A^*(E) + k_4 A^*(EB)$$
$$(2) = k_{-1} + k_2 B$$
$$(3) = k_{-2} + k_{-4} + k_5 + k_7$$
$$(4) = k_{-7} Q + k_8$$

We then use Cleland's procedure to write the expression for the initial rate of formation of labeled P from A.

$$v^* = dP^*/dt = k_5(EA^*B) + k_8(EP^*)$$
$$= \frac{k_5 N_{EA*B} + k_8 N_{EP*}}{D}$$

where N_{EA*B}, N_{EP*}, and D are the "determinants" of EA^*B, EP^*, and E in Scheme 12b. The N terms vary with the mechanism, depending on the enzyme species from which the labeled product is released, but the D term is always obtained as the "determinant" of E.

Thus, we can write

$$N_{EA*B} = (1)(2)(4) = [k_1 A^*(E) + k_4 A^*(EB)](k_{-1} + k_2 B)(k_{-7} Q + k_8)$$
$$= k_1 A^*(E)(k_2 B)(k_{-7} Q + k_8)$$
$$+ k_4 A^*(EB)(k_{-1} + k_2 B)(k_{-7} Q + k_8)$$
$$N_{EP*} = (34)(1)(2) = k_7[k_1 A^*(E) + k_4 A^*(EB)](k_{-1} + k_2 B)$$
$$= k_7 \cdot k_1 A^*(E) \cdot k_2 B + k_7 k_4 A^*(EB)(k_{-1} + k_2 B)$$
$$D = (2)(3)(4) = (k_{-1} + k_2 B)(k_{-2} + k_{-4} + k_5 + k_7)(k_{-7} Q + k_8)$$
$$= (k_{-4} + k_5)(k_{-1} + k_2 B)(k_{-7} Q + k_8) + k_{-1} k_{-2}(k_{-7} Q + k_8)$$
$$+ k_7 k_8 (k_{-1} + k_2 B)$$

The complete exchange equation is obtained as

$$\frac{v^*}{E_0} = \frac{(k_5 N_{EA*B} + k_8 N_{EP*})/D}{(E) + (EA) + (EB) + (EAB) + (EP) + (EQ)}$$

It should be noted that (E), (EA), (EB), (EAB), (EP), and (EQ) are now the *normal* determinants of E, EA, EB, EAB, EP, and EQ obtained from the steady-state treatment of the intact reaction scheme. If the equilibrium-exchange rate equation is desired, (E), (EA), (EB), (EAB), (EP), and (EQ) should be obtained from the equilibrium relationships.

The following derivations demonstrate that an isotope exchange rate equation obtained by the steady-state treatment can be converted to one of exchange at equilibrium:

$$v^* = \frac{\begin{array}{l}k_5[k_1A^*k_2B(k_{-7}Q + k_8)(E) + k_4A^*(k_{-1} + k_2B)(k_{-7}Q + k_8(EB)] \\ + k_8[k_7k_1A^*k_2B(E) + k_7k_4A^*(k_{-1} + k_2B)(EB)]\end{array}}{\begin{array}{l}(k_{-4} + k_5)(k_{-1} + k_2B)(k_{-7}Q + k_8) \\ + k_{-1}k_{-2}(k_{-7}Q + k_8) + k_7k_8(k_{-1} + k_2B)\end{array}}$$

Substitution of $(EB) = k_3B(E)/k_{-3}$ into the above expression yields

$$v^* = \frac{[k_5(k_{-7}Q + k_8) + k_8k_7][k_1A^*k_2B + k_4A^*k_3B(k_{-1} + k_2B)/k_{-3}](E)}{\begin{array}{l}(k_{-4} + k_5)(k_{-1} + k_2B)(k_{-7}Q + k_8) \\ + k_{-1}k_{-2}(k_{-7}Q + k_8) + k_7k_8(k_{-1} + k_2B)\end{array}}$$

which is identical with Eq. 10b derived by the equilibrium treatment.

The systematic method is equally convenient for the derivation of rate equations for simple mechanisms. Scheme 8, for example, can be redrawn as an enclosed figure after deleting the pathways between unlabeled enzyme forms.

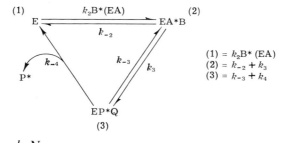

$$v^*_{B \to P} = \frac{k_4 N_{EP*Q}}{D}$$

$$N_{EP*Q} = (1)(2) = k_2B^*(EA)(k_{-2} + k_3) = k_2k_3B^*(EA)$$
$$D = (2)(3) = (k_{-2} + k_3)(k_{-3} + k_4) = k_{-2}(k_{-3} + k_4) + k_3k_4$$

Thus

$$v^* = \frac{k_2k_3k_4B^*(EA)}{k_{-2}(k_{-3} + k_4) + k_3k_4}$$

Concluding Remarks

In this chapter, the derivation of initial-velocity equations under steady-state, rapid-equilibrium, and the hybrid rapid-equilibrium and steady-state conditions has been covered. The derivation of isotope-exchange rate laws at chemical equilibrium and steady state has also been dealt with. Derivation of initial velocity equation for the quasi-equilibrium case is quite straightforward once the equilibrium relationships among various enzyme-containing species are defined. The combined rapid-equilibrium and steady-state treatment can be reduced to the steady-state method by treating the equilibrium segments as though they were enzyme intermediates. Rate equations for isotope exchange at equilibrium can be obtained by substitution of equilibrium expressions for unlabeled enzyme forms into equations derived by the steady-state procedure. Clearly, different types of rate equations are within reach when one has mastered the steady-state approach.

Albeit several methods for deriving steady-state rate equations have been described, one has to be proficient at only one of them since, for a given reaction scheme, they all lead to the same equation. The method to be recommended here is the *systematic* approach. It requires the least amount of time and work, especially when irreversible steps are present in the reaction scheme. Furthermore, no pattern drawing is needed in this approach. The initial setup is done by systematic inspection of the reaction diagram, and the risk of omission of terms (due to omission of certain patterns) is minimized.

It should be emphasized that the validity of a rate equation depends also on whether the reaction diagram is properly constructed; that is, whether the number and types of enzyme species involved, the pathways linking these species, the sequence of reaction steps, etc., are carefully considered and appropriately arranged. While the procedures presented in this chapter allow one to obtain correct algebraic expressions, the rate equations so obtained are only as valid as the assumptions made in constructing the reaction scheme.

[5] Computer-Assisted Derivation of Steady-State Rate Equations

By HERBERT J. FROMM

A number of methods are currently in vogue for deriving initial-rate equations for enzyme mechanisms in which the steady-state assumption is invoked. Algebraic methods are commonly used to derive rate equations for simple kinetic mechanisms, i.e., those in which only a few different enzyme forms are present. However, as anyone who has attempted to derive rate expressions for complex mechanisms knows, the task becomes quite formidable as the number of different enzyme forms increases. A number of procedures have been advanced in recent years in an attempt to circumvent this problem. These methods have involved schematic[1] and systematic[2,3] approaches; however, even these proce-

[1] King, E. L., and Altman, C., *J. Phys. Chem.* **60**, 1375 (1956).

dures become extremely tedious when four or more different enzyme forms are present in the kinetic mechanism.

The digital computer is now being used to solve the various differential equations obtained for a particular kinetic mechanism. A description of the method and the computer program that we have used in our laboratory since 1970 is presented here.

Program Description: ENZ-EQ

The determinant procedure described by Hurst[4] has been adapted for use with the IBM/360 computer in PL/1 language. The determinant method is based on the solution of $n - 1$ of n equations describing a steady-state enzyme system. The steady-state equations describing each enzyme form are input into the program, and a determinant is evaluated that gives an algebraic value for the concentration of each enzyme form in the total reaction under steady-state conditions. The method is, in general terms, to generate a description of each determinant, to expand the determinant while eliminating zero terms, and finally to substitute actual rate constants and concentration terms into the expanded determinant to give the final distribution expression.

Although the problem and output are algebraic in nature, the program's processing is entirely numeric except for the output of the final answers. This fact makes the program easily convertible to FORTRAN computer language. A numeric approach is possible because the nonnumeric symbols needed are limited to a given set, namely, a given number of rate constants and concentrations. Thus, these symbols are stored in two arrays, and a reference to a rate constant or a concentration is the subscript of an element in an array. An algebraic term is represented by an array containing one or more such subscripts. An algebraic expression is represented by a linear list of pairs of arrays (one for rate constants and another for concentrations), together with a coefficient and a pointer, the latter referring to the next element in the list. The end of a list is indicated by a zero pointer. The terms (elements) in each list are kept in a specific order so that algebraically similar terms will be kept next to each other and easily combined.

Since the terms from which the determinants are composed form a (2 to n, 1 to n) array, a particular determinant can be represented by specifying which terms are in it. And, since each determinant contains $n - 1$

---

[2] Volkenstein, M. V., and Goldstein, B. N., *Biochim. Biophys. Acta* **115,** 471 (1966).
[3] Fromm, H. J., *Biochem. Biophys. Res. Commun.* **40,** 692 (1970).
[4] Hurst, R. O., *Can. J. Biochem.* **45,** 2015 (1967).

columns of the array of terms and all the terms in those columns, it suffices to specify the column numbers. Furthermore, each term of an expanded determinant contains one factor from each row of the determinant (and hence of the original array) and one from each column. If the factors in a term are ordered according to row number, it will again suffice to specify only the column numbers. Thus, e.g., if the original array is designated "A," then (2,4,3) represents both

$$\begin{vmatrix} A_{22} & A_{24} & A_{23} \\ A_{32} & A_{34} & A_{33} \\ A_{42} & A_{44} & A_{43} \end{vmatrix} \quad \text{and} \quad A_{22}A_{34}A_{43}.$$

Finally, these two representations can be combined so that (2,4,3) also represents

$$A_{22} \begin{vmatrix} A_{34} & A_{33} \\ A_{44} & A_{43} \end{vmatrix} \quad \text{and} \quad A_{22}A_{34}|A_{43}|.$$

Thus, various stages in the expansion of a determinant may be represented by an array of column numbers. The expansion of the determinants is carried out by the use of minors.

To obtain the final steady-state equation, the determinants are substituted into the velocity expression for the particular enzyme mechanism under consideration. For example, in the case of the Ordered Bi Bi mechanism cited below, $v/E_0 = [k_7(EQ) - k_8(Q)(E)]/E_0$. The determinants for E and EQ are then substituted into the numerator of the velocity expression, and the determinants for E_0 (E, EA, EAB, and EQ) are placed in the denominator.

Input Deck for Program ENZ-EQ

The desired mechanism is set up in steady-state $d(EX)/dt$ form as described for the following mechanism:

$$E + A \underset{k_2}{\overset{k_1}{\rightleftharpoons}} EA$$

$$EA + B \underset{k_4}{\overset{k_3}{\rightleftharpoons}} EAB$$

$$EAB \underset{k_6}{\overset{k_5}{\rightleftharpoons}} EQ + P$$

$$EQ \underset{k_8}{\overset{k_7}{\rightleftharpoons}} E + Q$$

$$d(EA)/dt = k_1 A(E) - (k_2 + k_3 B)(EA) + k_4(EAB)$$
$$d(EAB)/dt = k_3 B(EA) - (k_4 + k_5)(EAB) + k_6 P(EQ)$$
$$d(EQ)/dt = k_5(EAB) - (k_6 P + k_7)(EQ) + k_8 Q(E)$$

The $d(E)/dt$ term need not be set up, as it is not used. The other equations completely describe the $d(E)/dt$ term.

The steady-state equations are then set up in a table of coefficients as described by Hurst.[4]

E	EA	EAB	EQ	$n+1$
1	1	1	1	1
$k_1 A$	$-(k_2 + k_3 B)$	k_4	0	0
0	$k_3 B$	$-(k_4 + k_5)$	$k_6 P$	0
$k_8 Q$	0	k_5	$-(k_6 P + k_7)$	0

The data deck is then coded from this table. Only positions containing nonzero terms are coded, and the first row is already in the program. A rate constant is represented by a $\pm XY$, where XY is the subscript and the $+$ or $-$ depends on the sign of the rate constant in the table. For example, $-k_4$ would be -04. Three spaces are required for each rate-constant description. The substrates A, B, C are represented by -01, -02, -03, respectively, and products P, Q, R are represented by $+01$, $+02$, $+03$, respectively. When no substrate or product term appears in the coefficient, 000 is used.

Each coefficient is referred to by its row and column location. $k_1 A$ is in row two, column one and is designated by 0201. The first two numbers represent the row; the second two, the column.

Each coefficient is described on a single data card as follows: The location is described in columns 1 to 4 on the card, then the complete coefficient is described in six column sets. For $k_1 A$ in row two, column one, the data card would read:

$$0\ 2\ 0\ 1 + 0\ 1 - 0\ 1$$
$$1\ 2\ 3\ 4\ 5\ 6\ 7\ 8\ 9\ 10$$

where 0201 represents the location, $+01$ represents k_1, and -01 represents A; for the other coefficients the following code would be used:

$-(k_2 + k_3 B)$ in row 2, column 2 ≡ 0202 $-$ 02000 $-$ 03 $-$ 02
k_4 in row 2, column 3 ≡ 0203 $+$ 04000
$k_3 B$ in row 3, column 2 ≡ 0302 $+$ 03 $-$ 02
$-(k_4 + k_5)$ in row 3, column 3 ≡ 0303 $-$ 04000 $-$ 05000
$k_6 P$ in row 3, column 4 ≡ 0304 $+$ 06 $+$ 01
$k_8 Q$ in row 4, column 1 ≡ 0401 $+$ 08 $+$ 02
k_5 in row 4, column 3 ≡ 0403 $+$ 05000
$-(k_6 P + k_7)$ in row 4, column 4 ≡ 0404 $-$ 06 $+$ 01 $-$ 07000

A lead card is used which has the value of N (the number of enzyme species) and NPS (the maximum number of substrate and product terms in any one term of the final determinants) in columns 1–3 and 4–6, respectively. For the mechanism above, the lead card would be 004003. NPS is a dimensioning number and may have to be determined by trial and error. It should not be made larger than necessary to conserve space.

The final complete deck, which consists of the lead card, the data cards describing the coefficients, and a blank card at the end to terminate input is placed after the //Go Cards DD* card.

Output from ENZ-EQ

The output consists of the individual determinants for each enzyme species $E(1)$—$E(n)$ in the order that they were input and the total sum of all the determinants. The terms may be either all + or all − but never both. If they are mixed, an error has been made in the input. The terms are ordered in terms of increasing rate constants. The output for the mechanism described above is as follows:

$EX(1)[(E)] = k_2k_4k_6P + k_2k_4k_7 + k_2k_5k_7 + k_3k_5k_7B$
$EX(2)[(EA)] = k_1k_4k_6AP + k_1k_4k_7A + k_1k_5k_7A + k_4k_6k_8PQ$
$EX(3)[(EAB)] = k_1k_3k_6ABP + k_1k_3k_7AB + k_2k_6k_8PQ + k_3k_6k_8BPQ$
$EX(4)[(EQ)] = k_1k_3k_5AB + k_2k_4k_8Q + k_2k_5k_8Q + k_3k_5k_8BQ$
$EX(0)$ is the sum of all the determinants and is output last.

ENZ-EQ: Flowchart Notes

1. N is the number of enzyme forms in the given system; NPS is the maximum number of concentration factors which may appear in any term of the calculations. (NPS may have to be determined by trial and error.)

5. I is zero when a blank card (which terminates input) is read.

8. If A (I, J) is not zero, then the Jth term of the Ith equation has already been read in.

9. I1 is the subscript in the array KS(50) of a rate constant; I2 is the subscript in the array PSS(−4:4) of a concentration. I2 is optional since concentration terms do not always appear.

10. UU is the subscript of the next unused element. If UU is zero, then all available storage has been used.

12. (a) If I1 is zero, then the last field on the card has been read.
    (b) If I1 is less than zero, then the term is negative.

18–23. Unused

25. I indexes the determinants to be expanded.

28. The Ith determinant consists of all but the Ith column of the A array.

30. J indexes the level of expansion of the determinants.

31. L refers to successive terms of the expansion of the Ith determinant; L2 refers to successive terms of the *next* level of expansions; L3 is a flag that indicates whether or not the signs of the terms will need correction due to the method of expansion, which changes signs of determinants of odd order.

32. I1 indexes the new terms created from one term in the present expansion.

33. Zero terms are eliminated immediately, not during the evaluation process.

34. If $J = N - 2$, then the order of the minor being expanded is two and both terms must be checked for zero values now (blocks 33 and 35) since order 1 minors are not expanded and hence not checked.

46. I indexes the expanded determinants through the evaluation process.

47. The partial sum array is a device used to equalize the lengths of the expressions added together in order to increase the efficiency of the ADD subroutine.

48. L refers to successive terms in the expansion of the Ith determinant.

49. I1 is a partial product in multiplying out the factors of the term being evaluated.

50. J indexes the remaining factors of the term.

51. I2 refers to successive terms in the value of the Jth factor.

52. L2 is the new partial product.

ENZ-EQ Flowchart Notes: ADD *Subroutine*

1. On entry A1 refers to the expression to which the expression referred to by A2 is to be added; if F1 is one, then the A2 expression is to be saved; if F1 is zero, then the A2 expression may be destroyed. On exit, A1 refers to the sum.

2. L1 refers to the next term of the A1 expression, L2 to that of the A2 expression, and L3 to the last term of the sum.

ENZ-EQ Flowchart Notes: MULT *Subroutine*

1. On entry, M1 refers to the multiplicand expression and M2 to the multiplier term (a single element). If F is one, then the multiplicand must

be saved; if F is zero, then the multiplicand may be destroyed. On exit, P refers to the product.

2. L1 refers to the next term in the multiplicand; L3 refers to the last term in the product.

12. and 22. I refers to the next factor of the current multiplicand term, J to that of the multiplier, and L to the last term of the new product term.

Acknowledgments

This work was supported by research grants from the National Institutes of Health (NS 10546) and the National Science Foundation (PCM 77-09018).

```
ENZ_EQ: PROC OPTIONS(MAIN);                                              00050
 /* THIS PROGRAM SOLVES THE RATE EQUATIONS FOR COMPLEX ENZYME SYSTEMS.   00060
    THE SOLUTIONS ARE ALGEBRAIC AND ARE IN TERMS OF RATE CONSTANTS AND   00070
    CONCENTRATIONS OF ENZYMES,PRODUCTS AND REACTANTS. NO OTHER EXPRES-   00080
    SIONS OR VALUES ARE ALLOWED. THE CONCENTRATIONS ARE REPRESENTED BY   00090
    THE PRESENCE OF THEIR SUBSCRIPTS IN LISTS OF TERMS.                  00100
 */                                                                      00110
    DCL BULK(11) CHAR (16383) STATIC EXT;                                00120
    DCL (XI,XN,XM) FIXED BIN(15);                                        00130
                                                      /*    1    */      00140
    GET FILE (CARDS) EDIT (XN,XM) (2 F(3));                              00150
    XI = 90000/(XN+XM+2);                                                00160
                                                      /*    2    */      00170
    BEGIN;                                                               00180
 /*                                                                      00190
    DECLARATIONS:                                                        00200
 */                                                                      00210
    N = XN;                                                              00220
    NPS = XM;                                                            00230
    DCL 1 EL(XI)       DEF BULK ,                                        00240
          2 (COEF,NXT) FIXED BIN(15),                                    00250
          2 K(XN) FIXED BIN(15), /* HOLDS SUBSCRIPTS OF RATE CONST.*/    00260
          2 PS(XM) FIXED BIN(15); /* HOLDS SUBSCRIPTS OF PRODUCT         00270
          AND REACTANT CONCENTRATIONS: POS FOR PROD; NEG FOR REACT. */   00280
    DCL (ERR,UU,I,J,L,I1,I2,L1,L2,PRSM(0:17),NMR,EXO) FIXED BIN(16);     00290
    DCL (L3,M1,M2,N,NPS) FIXED BIN(16);                                  00300
    DCL (NU1,NU2,NUTST) FIXED BIN(16);                                   00310
    DCL CARDS FILE STREAM INPUT;                                         00320
    DCL (NT(XN),EX(XN),SYN(0:XN),A(2:XN,XN)) FIXED BIN(16);              00330
    DCL LABEL CHARACTER(4);                                              00340
    DCL KS(50) CHAR(3) STATIC INIT('K1','K2','K3','K4','K5','K6','K7',   00350
        'K8','K9','K10','K11','K12','K13','K14','K15','K16','K17',       00360
        'K18','K19','K20','K21','K22','K23','K24','K25','K26','K27',     00370
        'K28','K29','K30','K31','K32','K33','K34','K35','K36','K37',     00380
        'K38','K39','K40','K41','K42','K43','K44','K45','K46','K47',     00390
        'K48','K49','K50') VARYING;                                      00400
 DCL PSS(-5:5) CHAR(1)STATIC INIT('M','D','C','B','A','*','P','Q','R','S'00410
    ,'I');                                                               00420
    DCL P7 CHARACTER(5);                                                 00430
 /*                                                                      00440
    INITIALIZATIONS:                                                     00450
 */                                                                      00460
                                                      /*    3    */      00470
    PUT EDIT ('THE ARRAY OF STRUCTURES EL HAS BEEN ALLOCATED AS EL(',    00480
              XI,')') (SKIP,A,F(5),A);                                   00490
    I = XI;                                                              00500
    UU=1;   /* UU IS SUBSCRIPT OF FIRST UNUSED ELEMENT. */               00510
    DO J=1 TO I-1;                                                       00520
         NXT(J)=J+1;   /*  CHAIN UNUSED ELEMENTS TOGETHER   */           00530
    END;                                                                 00540
    NXT(I)=0;                                                            00550
    COEF=0;                                                              00560
    A=0; /* A(I,J) IS SUBSCR OF 1ST TERM OF I,J ELEMENT OF DETERM. */    00570
    NT=0;   /* NT IS NUMBER OF TERMS IN THE EXPANSION A DETERMINANT */   00580
    ERR=0; J=1;                                                          00590
    DO I=0 TO N;                                                         00600
```

```
            SYN(I)=J;                                              00610
            J=-J;                                                  00620
        END;                                                       00630
  /*                                                               00640
    INPUT:                                                         00650
  */                                                               00660
IN_LOOP:                                                           00670
                                                         /*  4  */ 00680
        GET SKIP FILE(CARDS) EDIT(I,J) (2 F(2)); /* ONE CARD FOR EACH ELM*/ 00690
                                                         /*  5  */ 00700
        IF I=0 THEN   /*  SIGNAL END OF INPUT WITH BLANK CARD */   00710
        GO TO DO_IT;                                               00720
                                                         /*  6  */ 00730
        IF I<2|I>N|J<1|J>N THEN                                    00740
                                                         /*  7  */ 00750
ER: DO;                                                            00760
            ERR=1;                                                 00770
            GO TO IN_LOOP;                                         00780
        END;                                                       00790
                                                         /*  8  */ 00800
        IF A(I,J)¬=0 THEN                                          00810
        GO TO ER;                                                  00820
                                                         /*  9  */ 00830
F_L:GET FILE(CARDS) EDIT(I1,I2) (2 F(3));                          00840
                                                         /* 10  */ 00850
        IF UU=0 THEN                                               00860
        DO;                                                        00870
                                                         /* 11  */ 00880
            GO TO OVERFLOW;                                        00890
        END;                                                       00900
                                                         /* 12A */ 00910
        IF I1=0 THEN    /*  I1 IS SUBSCR OF K FACTOR */            00920
        GO TO IN_LOOP;                                             00930
                                                         /* 12B */ 00940
        IF I1<0 THEN                                               00950
                                                         /* 13  */ 00960
        DO;                                                        00970
            COEF(UU)=-1;                                           00980
            I1=-I1;                                                00990
        END; ELSE                                                  01000
                                                         /* 14  */ 01010
        COEF(UU)=1;                                                01020
                                                         /* 15  */ 01030
        IF I1 > 50|I2 < -5|I2 > 5    THEN                          01040
        GO TO ER;                                                  01050
                                                         /* 16  */ 01060
        K(UU,1)=I1;                                                01070
        K(UU,2)=0;                                                 01080
        PS(UU,1)=I2;   /* I2 IS SUBSCR OF PROD OR REACT FACTOR(IF ANY) */ 01090
        PS(UU,2)=0;                                                01100
                                                         /* 17  */ 01110
        IF A(I,J)=0 THEN                                           01120
        A(I,J)=UU; ELSE                                            01130
        NXT(L)=UU;                                                 01140
        L=UU;                                                      01150
        UU=NXT(UU);                                                01160
```

```
           NXT(L)=0;    /*  MAY BE SEVERAL TERMS ON EACH CARD.  */        01170
           GO TO F_L;                                                      01180
     /*                                                                    01190
           SUBROUTINES:                                                    01200
     */                                                                    01210
           DCL ADD ENTRY(FIXED BIN(16),FIXED BIN(16),FIXED BIN(16));       01220
                                                            /*   1   */    01230
     ADD: PROC(A1,F1,A2);   /* ADDS A2 TO A1; RESULT 'IN' A1     */        01240
           DCL (A1,F1,A2)         FIXED BIN(16);                           01250
           DCL (L1,L2,L3,L,F,C,I) FIXED BIN(16);                           01260
                                                            /*   2   */    01270
           L1=A1;                                                          01280
           L2=A2;                                                          01290
           L3=0;                                                           01300
                                                            /*   3   */    01310
           IF F1=0                                                         01320
                                                            /*   4   */    01330
           THEN A2=0;                                                      01340
     ADD_LOOP:                                                             01350
                                                            /*   5   */    01360
           IF L2=0 THEN                                                    01370
                                    DO;                                    01380
                                    GO TO A30;                             01390
                                    END;                                   01400
                                                            /*   6   */    01410
           IF L1=0 THEN                                                    01420
           DO;                                                             01430
                                                            /*   7   */    01440
                IF L3=0 THEN                                               01450
                                                            /*   8   */    01460
                     IF F1 = 0 THEN                                        01470
                     DO;                                                   01480
                                                            /*   9   */    01490
                        A1 = L2;                                           01500
                                    GO TO A30;                             01510
                     END; ELSE                                             01520
                     DO;                                                   01530
                                                            /*  10   */    01540
                        A1=0;                                              01550
                        GO TO COPY;                                        01560
                     END; ELSE                                             01570
                                                            /*  15   */    01580
                     IF F1 = 0 THEN                                        01590
                     DO;                                                   01600
                                                            /*  16   */    01610
                        NXT(L3)=L2;                                        01620
                                    GO TO A30;                             01630
                     END; ELSE                                             01640
                     DO;                                                   01650
                                                            /*  17   */    01660
                        L1 = L3;                                           01670
                        PUT EDIT ('SORRY WRONG NUMBER')  ( SKIP,A);        01680
                        GO TO COPY;                                        01690
                     END;                                                  01700
           END;                                                            01710
           DO I=1 TO N;                                                    01720
```

```
                    IF K(L1,I)¬=K(L2,I) THEN               /* 18 */    01730
                                                                       01740
                                                           /* 26 */    01750
                    IF K(L1,I)<K(L2,I) THEN                             01760
                    GO TO MOVE_L1; ELSE                                 01770
                    GO TO MOVE_L2;                                      01780
                    IF K(L1,I)=0 THEN                                   01790
                    GO TO PS_LOOP;                                      01800
            END;                                                        01810
PS_LOOP:                                                                01820
            DO I=1 TO NPS;                                              01830
                                                           /* 18 */    01840
                    IF PS(L1,I)¬=PS(L2,I) THEN                          01850
                                                           /* 26 */    01860
                    IF PS(L1,I)<PS(L2,I) THEN                           01870
                    GO TO MOVE_L1; ELSE                                 01880
                    GO TO MOVE_L2;                                      01890
                    IF PS(L1,I)=0 THEN                                  01900
                    GO TO EQUAL;                                        01910
            END;                                                        01920
EQUAL:  /* COMBINE ALGEBRAICALLY SIMILAR TERMS */                       01930
                                                           /* 19 */    01940
        C=COEF(L1)+COEF(L2);                                            01950
                                                           /* 20 */    01960
        IF F1=0 THEN    /* IF F1 IS ZERO, A2 IS NOT SAVED */            01970
                                                           /* 21 */    01980
        DO;                                                             01990
                I=NXT(L2);                                              02000
                NXT(L2)=UU;                                             02010
                UU=L2;                                                  02020
                L2=I;                                                   02030
        END; ELSE                                                       02040
                                                           /* 22 */    02050
        L2=NXT(L2);                                                     02060
                                                           /* 23 */    02070
        IF C=0 THEN                                                     02080
                                                           /* 24 */    02090
        DO;                                                             02100
                I=NXT(L1);                                              02110
                NXT(L1)=UU;                                             02120
                UU=L1;                                                  02130
                L1=I;                                                   02140
                IF L3=0 THEN                                            02150
                IF C ¬= 0 THEN                                          02160
                A1,L3=I; ELSE A1 = L1; ELSE                             02170
                NXT(L3)=I;                                              02180
        END; ELSE                                                       02190
                                                           /* 25 */    02200
        DO;                                                             02210
                COEF(L1)=C;                                             02220
                L3=L1;                                                  02230
                L1=NXT(L1);                                             02240
        END;                                                            02250
        GO TO ADD_LOOP;                                                 02260
COPY:   DO;                                                             02270
                                                           /* 11 */    02280
```

```
            DO WHILE(L2 ¬= 0);                                    02290
                                                    /* 12 */     02300
                    IF UU=0 THEN                                  02310
        DO;                                                       02320
                                                    /* 13 */     02330
                GO TO OVERFLOW;                                   02340
        END;                                                      02350
                                                    /* 14 */     02360
                    L=NXT(UU);                                    02370
                    EL(UU)=EL(L2);                                02380
                    L3=NXT(L2);                                   02390
                    NXT(UU)=0;                                    02400
                    IF A1=0 THEN                                  02410
                    A1=UU; ELSE                                   02420
                    NXT(L1) = UU;                                 02430
                    L1 = UU;                                      02440
                    L2=L3;                                        02450
                    UU=L;                                         02460
            END;                                                  02470
                            GO TO A30;                            02480
        END;                                                      02490
MOVE_L1:                                                          02500
                                                    /* 27 */     02510
    L3=L1;                                                        02520
    L1=NXT(L1);                                                   02530
    GO TO ADD_LOOP;                                               02540
MOVE_L2:  /*  MERGE TERM OF A2 INTO A1  */                        02550
                                                    /* 28 */     02560
    IF F1=0 THEN                                                  02570
                                                    /* 32 */     02580
    L=L2; ELSE                                                    02590
    DO;                                                           02600
                                                    /* 29 */     02610
            IF UU=0 THEN                                          02620
        DO;                                                       02630
                                                    /* 30 */     02640
                GO TO OVERFLOW;                                   02650
        END;                                                      02660
                                                    /* 31 */     02670
            L=UU;                                                 02680
            UU=NXT(UU);                                           02690
            EL(L)=EL(L2);                                         02700
    END;                                                          02710
                                                    /* 33 */     02720
    L2=NXT(L2);                                                   02730
    NXT(L)=L1;                                                    02740
    IF L3=0 THEN                                                  02750
    A1=L; ELSE                                                    02760
    NXT(L3)=L;                                                    02770
    L3=L;                                                         02780
    GO TO ADD_LOOP;                                               02790
                            A30: ;                                02800
END ADD;                                                          02810
    DCL MULT ENTRY(FIXED BIN(16),FIXED BIN(16),FIXED BIN(16),     02820
                    FIXED BIN(16));                               02830
                                                    /* 1 */      02840
```

```
MULT: PROC(M1,F,M2,P);   /* MULTIPIES EXPRESSION M1 BY TERM M2; */          02850
      DCL (M1,F,M2,P) FIXED BIN(16);   /*  RESULT 'IN' P */                 02860
      DCL (L1,L3,I,J,L) FIXED BIN(16);                                      02870
                                                          /*   2   */       02880
      L1=M1;                                                                02890
                                                          /*   3   */       02900
      IF F=0 THEN                                                           02910
                                                          /*   4   */       02920
      M1=0;                                                                 02930
                                                          /*   5   */       02940
      L3=0;                                                                 02950
      IF M2=0 THEN                                                          02960
      DO;                                                                   02970
                                                          /*   6   */       02980
         P=0;                                                               02990
         RETURN;                                                            03000
      END;                                                                  03010
MULT_LOOP:                                                                  03020
                                                          /*   7   */       03030
      IF L1=0 THEN                                                          03040
      DO;                                                                   03050
                                                          /*   8   */       03060
         IF L3=0 THEN                                                       03070
                                                          /*   9   */       03080
         P=0;                                                               03090
         RETURN;                                                            03100
      END;                                                                  03110
                                                          /*  10   */       03120
      IF UU=0 THEN                                                          03130
      DO;                                                                   03140
                                                          /*  11   */       03150
         GO TO OVERFLOW;                                                    03160
      END;                                                                  03170
                                                          /*  12   */       03180
      I,J=1;                                                                03190
      L=0;                                                                  03200
FAC_LOOP1:                                                                  03210
                                                          /*  13   */       03220
      IF I>N THEN                                                           03230
      GO TO X; ELSE                                                         03240
      IF K(L1,I)=0 THEN                                                     03250
                                                          /*  20   */       03260
   X: DO;                                                                   03270
         DO J=J TO N WHILE(K(M2,J)¬=0);                                     03280
            L=L+1;                                                          03290
            IF L>N THEN                                                     03300
                     DO;                                                    03310
                     PUT EDIT ('228 ') (A);                                 03320
                     GO TO ERROR;                                           03330
                     END;                                                   03340
            K(UU,L)=K(M2,J);                                                03350
         END; IF L<N THEN K(UU,L+1)=0;                                      03360
         GO TO PS_S;                                                        03370
      END;                                                                  03380
                                                          /*  14   */       03390
      IF J>N THEN                                                           03400
```

```
              GO TO Y; ELSE
              IF K(M2,J)=0 THEN
                                                                   /*   21   */
        Y: DO;
                 DO I=I TO N WHILE(K(L1,I)¬=0);
                      L=L+1;
                      IF L>N THEN
                                       DO;
                                       PUT EDIT ('242 ') (A);
                             GO TO ERROR;
                                       END;
                      K(UU,L)=K(L1,I);
                 END;
                 IF L<N THEN
                 K(UU,L+1)=0;
                 GO TO PS_S;
              END;                                                 /*   15   */
              IF L>N THEN
                                       DO;
                                       PUT EDIT ('250 ') (A);
                             GO TO ERROR;
                                       END;
                                                                   /*   16   */
              L=L+1;
                                                                   /*   17   */
              IF K(L1,I)>K(M2,J) THEN
                                                                   /*   18   */
              DO;
                   K(UU,L)=K(M2,J);
                   J=J+1;
              END; ELSE
                                                                   /*   19   */
              DO;
                   K(UU,L)=K(L1,I);
                   I=I+1;
              END;
              GO TO FAC_LOOP1;
        PS_S:
                                                                   /*   22   */
              I,J=1;
              L=0;
        FAC_LOOP2:
                                                                   /*   23   */
              IF I>NPS THEN
              GO TO Z; ELSE
              IF PS(L1,I)=0 THEN
                                                                   /*   30   */
        Z: DO;
                 DO J=J TO NPS WHILE(PS(M2,J)¬=0);
                      L=L+1;
                      IF L>NPS THEN
                                       DO;
                                       PUT EDIT ('271 ') (A);
                             GO TO ERROR;
                                       END;
```

```
                PS(UU,L)=PS(M2,J);                     03970
            END;                                       03980
            IF L<NPS THEN                              03990
            PS(UU,L+1)=0;                              04000
            GO TO PROD;                                04010
        END;                                           04020
                                            /* 24 */  04030
        IF J>NPS THEN                                  04040
        GO TO W; ELSE                                  04050
        IF PS(M2,J)=0 THEN                             04060
                                            /* 31 */  04070
    W: DO;                                             04080
            DO I=I TO NPS WHILE(PS(L1,I)¬=0);          04090
                L=L+1;                                 04100
                IF L>NPS THEN                          04110
                        DO;                            04120
                        PUT EDIT ('283 ') (A);         04130
                        GO TO ERROR;                   04140
                        END;                           04150
                PS(UU,L)=PS(L1,I);                     04160
            END;                                       04170
            IF L<NPS THEN                              04180
            PS(UU,L+1)=0;                              04190
            GO TO PROD;                                04200
        END;                                           04210
                                            /* 25 */  04220
        IF L>N THEN                                    04230
                        DO;                            04240
                        PUT EDIT ('293 ') (A);         04250
        GO TO ERROR;                                   04260
                        END;                           04270
                                            /* 26 */  04280
        L=L+1;                                         04290
                                            /* 27 */  04300
        IF PS(L1,I)>PS(M2,J) THEN                      04310
                                            /* 28 */  04320
        DO;                                            04330
            PS(UU,L)=PS(M2,J);                         04340
            J=J+1;                                     04350
        END; ELSE                                      04360
                                            /* 29 */  04370
        DO;                                            04380
            PS(UU,L)=PS(L1,I);                         04390
            I=I+1;                                     04400
        END;                                           04410
        GO TO FAC_LOOP2;                               04420
                                            /* 32 */  04430
    PROD: COEF(UU)=COEF(L1)*COEF(M2);                  04440
                                            /* 33 */  04450
        IF L3=0 THEN                                   04460
        P=UU; ELSE                                     04470
        NXT(L3)=UU;                                    04480
        L3=UU;                                         04490
        UU=NXT(UU); NXT(L3)=0;                         04500
                                            /* 34 */  04510
        IF F=0 THEN                                    04520
```

```
            DO;                                                    /*  35   */    04530
                     L=NXT(L1);                                                   04540
                     NXT(L1)=UU;                                                  04550
                     UU=L1;                                                       04560
                     L1=L;                                                        04570
            END; ELSE                                                             04580
                                                                                  04590
            L1=NXT(L1);                                            /*  36   */    04600
            GO TO MULT_LOOP;                                                      04610
   ERROR:                                                                         04620
                                                                                  04630
            PUT EDIT('TOO MANY FACTORS IN A TERM') (SKIP(2),A);    /*  37   */    04640
            STOP;                                                                 04650
   END MULT;                                                                      04660
   /*                                                                             04670
            PROCESSING:                                                           04680
   */                                                                             04690
   DO_IT:                                                                         04700
                                                                                  04710
            IF ERR = 1 THEN                                        /*  24   */    04720
            GO TO FIN;                                                            04730
                                                                                  04740
            DO I=1 TO N;  /* THIS LOOP GENERATES DESCRIPTIONS OF THE TERMS IN*/   04750
                                                                   /*  25   */    04760
                     IF UU=0 THEN      /* THE EXPANSION OF THE DETERMINANT */     04770
            DO;                                                    /*  26   */    04780
                                                                                  04790
                     GO TO OVERFLOW;                               /*  27 -  */   04800
            END;                                                                  04810
                                                                                  04820
                     EX(I)=UU;                                     /*  28   */    04830
                     NT(I)=1;                                                     04840
                     COEF(UU)=SYN(I-1);                                           04850
                     I1=0;                                                        04860
                     DO J=1 TO N-1;                                               04870
                             IF J=I THEN                                          04880
                             I1=I1+1;                                             04890
                             I1=I1+1;                                             04900
                             K (UU,J)=I1;                                         04910
                     END;                                                         04920
                                                                                  04930
                     L=UU;                                         /*  29   */    04940
                     UU=NXT(UU);                                                  04950
                     NXT(L)=0;                                                    04960
                                                                                  04970
                     DO J=1 TO N-2;                                /*  30   */    04980
                                                                                  04990
                             L=EX(I);                              /*  31   */    05000
                             L2=0;                                                05010
                             L3=SYN(N-J-2);                                       05020
                                                                                  05030
   EXPAND_LOOP:        DO I1=1 TO N-J;                             /*  32   */    05040
                                                                                  05050
                             IF A(J+1,K(L,J+I1-1))¬=0 THEN         /*  33   */    05060
                                                                                  05070
                                                                   /*  34   */    05080
```

```
                    IF  J=N-2  THEN
                                                       /*   35   */
                            IF  A(N,K(L,N-I1))¬=0  THEN
                            GO  TO  EXPAND;  ELSE;  ELSE
EXPAND:                     DO;
                                                       /*   36   */
                                    IF  UU=0  THEN
    DO;
                                                       /*   37   */
        GO  TO  OVERFLOW;
    END;
                                                       /*   38   */
                                IF  L2=0  THEN
                                EX(I)=UU;  ELSE
                                NXT(L2)=UU;              /*NEW  ELEMENT  */
                                L2=UU;                   /*LAST  ELEMENT  IN  LIST  */
                                UU=NXT(UU);
        NXT(L2)=0;
                                NT(I)=NT(I)+1;
                                                       /*   39   */
                                DO  I2=1  TO  J-1;       /*COPIES  UNCHNGD*/
                                    K  (L2,I2)=K  (L,I2);
                                END;
                                                       /*   40   */
                                COEF(L2)=COEF(L)*SYN(I1-1);
                                IF  L3=-1  THEN
                                COEF(L2)=COEF(L2)*SYN(I1);
                                                       /*   41   */
                                L1=I1+J-1;
                                DO  I2=J  TO  N-1;
                                    IF  L1>N-1  THEN     /*DOES  PERMUTING*/
                                    L1=J;
                                    K  (L2,I2)=K  (L,L1);
                                    L1=L1+1;
                                END;
                            END  EXPAND;
                        END  EXPAND_LOOP;
                                                       /*   42   */
            L1=NXT(L);                        /*PUT  ORIGINAL  ELEMENT  BACK  */
            NXT(L)=UU;
            UU=L;
            L=L1;
            NT(I)=NT(I)-1;
                                                       /*   43   */
                    IF  L¬=0  THEN
                    GO  TO  EXPAND_LOOP;
    END;
END;                                    /*END  OF  DESCRIPTION  OF  DETERMINANTS  */
                                                       /*   44   */
DO  I=1  TO  N;
    PUT  EDIT  ('EX(',I,')=')  (SKIP,A,F(1),A);
    L=EX(I);
    DO  WHILE(L  ¬=  0);
        IF  COEF(L)<0  THEN
        PUT  EDIT('-')  (SKIP,A);  ELSE
        PUT  EDIT('+')  (SKIP,A);
```

```
                PUT EDIT (('A(',J+1,',',K(L,J),')' ' DO J=1 TO N-1))
                     (A,F(1),A,F(1),A);
            L=NXT(L);
          END;
        END;
/*
    EVALUATE DETERMINANTS.
*/
                                                  /*   45   */
    IF UU = 0 THEN
    DO;
        GO TO OVERFLOW;
    END;
    M1 = UU;
    UU = NXT(UU);
    EL(M1) = 0;
    COEF(M1) = -1;
                                                  /*   46   */
    DO I=1 TO N;
                                                  /*   47   */
        PRSM=0;
        L=EX(I);
                                                  /*   48   */
        DO WHILE(L¬=0);
                                                  /*   49   */
            I1=0;
            I2=A(2,K(L,1));     /*GETS 1ST FACTOR OF TERM */
            CALL ADD(I1,1,I2);
                                                  /*   50   */
            DO J=2 TO N-1;                 /* MULT TERMS */
                                                  /*   51   */
                I2=A(J+1,K(L,J));
                L2=0;
                                                  /*   52   */
                DO WHILE(I2¬=0);           /* MULTI A TERM */
                    CALL MULT(I1,1,I2,L1);
                    CALL ADD(L2,0,L1);
                    I2=NXT(I2);
                END;
                L1=I1;
                                                  /*   53   */
END_LIST_LOOP1:     I2=NXT(L1);
                IF I2¬=0 THEN
                DO;
                    L1=I2;
                    GO TO END_LIST_LOOP1;
                END;
                NXT(L1)=UU;
                UU=I1;
                I1=L2;
            END;
                                                  /*   54   */
        IF COEF(L) < 0 THEN
                                                  /*   55   */
        CALL MULT(I1,0,M1,I1);
                                                  /*   56   */
```

```
            PRSM(0)=L1;                                             06210
            DO J=1 TO 17 WHILE(PRSM(J)¬=0);                         06220
                CALL ADD(PRSM(J),0,PRSM(J-1));                      06230
            END;                                                    06240
            PRSM(J)=PRSM(J-1);                                      06250
            PRSM(J-1)=0;                                            06260
                                                    /* 57 */        06270
                                                                    06280
            L1=NXT(L);                                              06290
            NXT(L)=UU;                                              06300
            UU=L;                                                   06310
            L=L1;                                                   06320
         END;                                                       06330
         L1=0;                                      /* 58 */        06340
                                                                    06350
         DO J=1 TO 17;                                              06360
            IF PRSM(J)¬=0 THEN                                      06370
                CALL ADD(L1,0,PRSM(J));                             06380
         END;                                                       06390
         EX(I)=L1;                                   /* 59 */       06400
                                                                    06410
         CALL PRNT('EX('||I||')=',EX(I));                           06420
      END;                                                          06430
      EXO=EX(1);                    /*DENOMINATOR */                06440
      DO I=2 TO N;                                                  06450
         CALL ADD(EXO,0,EX(I));                                     06460
      END;                                                          06470
      CALL PRNT('EXO',EXO);                                         06480
      IF P7 = 'PUNCH' THEN CALL PUNCH(EXO);                         06490
      GOTO FIN;                                                     06500
OVERFLOW:PUT EDIT ('OVER')(A);                                      06510
      GOTO FIN;                                                     06520
   DCL PRNT ENTRY (CHAR(*),FIXED BIN(16));                          06530
   PRNT: PROC(LABEL,PTR);                                           06540
   DCL LABEL CHAR(*), (J,L,PTR)FIXED BIN(16);                       06550
         PUT EDIT  (LABEL)      (SKIP(3),A,F(1),A);                 06560
      L = PTR;                                                      06570
         DO WHILE (L ¬= 0);                                         06580
            PUT EDIT(COEF(L),' ') (SKIP,X(4),F(3),A);               06590
            DO K11 = 1 TO N WHILE (K(L,K11) ¬= 0);                  06600
               PUT EDIT (KS(K(L,K11))) (X(1),A);                    06610
            END;                                                    06620
            DO K12 = 1 TO NPS WHILE(PS(L,K12) ¬= 0);                06630
               PUT EDIT (PSS(PS(L,K12)))  (X(1),A);                 06640
            END;                                                    06650
            L=NXT(L);                                               06660
         END;                                                       06670
   END PRNT;                                                        06680
   DCL PUNCH ENTRY (FIXED BIN(16));                                 06690
   PUNCH:PROC(PTR);                                                 06700
   DCL (J,L,PTR) FIXED BIN(16);                                     06710
   L = PTR;                                                         06720
   DO WHILE (L ¬= 0);                                               06730
   PUT FILE(OUT) EDIT('+') (SKIP,COL(6),A);                         06740
   DO K1=1 TO N WHILE (K(L,K1) ¬= 0);                               06750
   IF K1 > 1 THEN PUT FILE(OUT) EDIT('*') (A);                      06760
   PUT FILE(OUT) EDIT (KS(K(L,K1))) (A);
```

```
        END;                                                                    06770
        DO K2 = 1 TO NPS WHILE(PS(L,K2) ¬= 0);                                  06780
        PUT FILE(OUT)    EDIT('*',PSS(PS(L,K2)),'(I)') (A,A,A);                 06790
        END;                                                                    06800
        PUT FILE(OUT) EDIT(' + ') (A);                                          06810
        L=NXT(L);                                                               06820
        END;                                                                    06830
        END PUNCH;                                                              06840
   FIN: PUT SKIP LIST('THE END');   END ENZ_EQ;                                 06850
                                                                                  */
```

[6] Statistical Analysis of Enzyme Kinetic Data

By W. WALLACE CLELAND

Although graphical analysis is a quick and useful way to visualize enzyme kinetic data, for any serious study the data must be subjected to statistical analysis so that the precision of the derived kinetic constants can be evaluated. This is particularly important when one is trying to distinguish between possible patterns, such as competitive or noncompetitive inhibition. In this case the question is whether or not the intercepts of reciprocal plots are a function of inhibitor concentration; when such variation is small or absent, only statistical analysis can give one a quantitative estimation of the probability that the data represent one pattern or the other.

It is important to note at the start what statistics can and cannot do. Statistical analysis is a matter of calculating probabilities, and where there is no useful information present in the data, statistical analysis will not magically produce any. Likewise when the data clearly define the nature of the rate equation, statistical analysis is not needed to do this, but will provide precision estimates of the fitted kinetic constants. Finally, one always has to use common sense and not be led blindly by the results of any statistical analysis.

Least Squares Method

We will discuss only one method of fitting data to rate equations; namely, the least squares method, which is the one most commonly used and is the proper one to use where errors in the data are normally distributed. We will assume the error to be present only in the experimental parameter (velocity, or kinetic constant derived from velocities, such as V, V/K, $1/K_i$), and that the concentrations of substrates or inhibitors, or such variables as pH, are free of error, at least relative to the error in the experimental parameters.[1] This is normally a fairly good assumption, and the errors in velocity are certainly greater than those in substrate concentrations, at least if one uses care in making up reaction mixtures. Even if the *absolute* concentration of the substrate is not accurately known, the *relative* concentrations of the substrate should be accurate if a single stock solution is used to prepare the reaction mixtures, and when one is trying to determine the kinetic pattern this will be sufficient. This point should be kept in mind when comparing the absolute values of kinetic constants; the standard errors reflect experimental variation, and the absolute error is determined by how well you know the concentration of the stock solution used.

In the least squares method, one picks values for the constants in the rate equation so that the sum of the squares of the differences between experimental and calculated velocities is minimized. Mathematically this is done by writing out the appropriate expression, taking partial derivatives with respect to each constant to be determined, and setting these partial derivatives equal to zero. Simultaneous solution of the resulting equations gives the desired constants. This procedure is quite straightforward when the equation being fitted is linear in all the constants, since it results in a set of linear simultaneous equations. The rate equations one deals with in enzyme kinetic studies are not of this type, however, but are always the ratio of two expressions, with the denominator at least (and in some cases the numerator) being the sum of a number of separate terms, each containing kinetic constants. Application of the procedure outlined above produces a set of nonlinear simultaneous equations that are not readily solved, and thus a trick is needed to generate a substitute equation to which the data can be fitted that *is* linear in the constants, or in parameters that will allow estimates of the constants.

Nonlinear Least Squares

A large number of strategies have been developed for nonlinear least squares analysis, but all involve picking preliminary estimates for the constants (in the most powerful methods, one can pick any values) and then calculating better estimates in some way so that the residual least square (sum of squares of differences between experimental and calculated velocities) is minimized. The most powerful nonlinear methods succeed simply because they are cautious—they do not adjust the preliminary esti-

---

[1] For a treatment of the general case where this is not true, see G. Johansen and R. Lumry, *C. R. Trav. Lab. Carlsberg* **32**, 185 (1961).

mates very much at each cycle of iteration, and thus the fit rarely diverges and tends to converge cleanly (if slowly) to the minimum point. The disadvantages to such methods are the large number of iterations required and the resulting expense. The method that we will describe here (which is used in all the computer programs included here) requires good preliminary estimates, but converges very rapidly to the final answer, and thus is very cheap to use. When the fit fails to converge, the fault is usually with the data, but construction of a least squares surface (see below) allows one to diagnose what the problem is.

Proper Weighting

Most rate equations for enzyme-catalyzed reactions can be made linear by inversion, and thus good preliminary estimates can be obtained by normal least squares fitting to these reciprocal equations. Such fits must be properly weighted, however, since the variance of $1/v$ will be the variance of v divided by v^4, and one uses the reciprocal of the variance as the weighting factor. The variances of the velocities (or other parameters being fitted) may be determined experimentally, which is a lot of work, or one can make a reasonable assumption about them. Wilkinson[2] assumed the variance of the velocities to be constant, calling for v^4 weights in reciprocal fits, but no weights in the final iterative fits. This assumption is reasonable when the range of experimental velocities is not greater than a factor of 5 and corresponds to constant absolute errors (that is, 5 ± 0.2 and 1 ± 0.2). It is the assumption made in most of the computer programs included in this chapter. When the range of experimental velocities is a power of 10 or more, however, the assumption of constant variance is not appropriate and causes the lower velocities to be ignored in the fitting process. In fitting pH profiles for V, V/K, and $1/K_i$ values this assumption is also inappropriate, since these parameters vary a factor of 10 per pH unit above or below the pK which causes loss of activity or binding. In these cases it is better to assume that the variance of the velocities, or other parameter, is proportional to the square of the velocity. This corresponds to constant proportional error (that is, 5 ± 1 and 1 ± 0.2), and calls for v^2 weights in reciprocal plots. The final iterative fits may then be unweighted if the equation is expressed in log form, since the variance of log v is the variance of v divided by v^2, and the variance of v in this case is proportional to v^2. The programs for fitting pH profiles, and also the alternate one for simple reciprocal plots, make this assumption.

[2] G. N. Wilkinson, *Biochem. J.* **80**, 324 (1961).

Iterative Fitting by the Gauss–Newton Method

When good preliminary estimates are available, either from properly weighted fits to the equation in reciprocal form, or from graphical analysis (necessary when inversion does not generate a linear form of the equation), the Gauss–Newton method of iteration converges rapidly and allows simple calculation of standard errors of the fitted constants. The basic equation used is

$$v = F_0 + (a - a_0)(\partial F/\partial a)_0 + (b - b_0)(\partial F/\partial b)_0 + \cdots \qquad (1)$$

where there are as many terms containing partial derivatives as there are nonlinear constants in the rate equation. The constants a_0, b_0, ... are the preliminary estimates of the constants $a, b, \ldots$; F_0 is the function evaluated using these preliminary values a_0, b_0, ...

This is now a linear equation, and fitting data to it by the least squares method gives as constants (1) any linear constant in the original rate equation, and (2) correction factors such as $(a - a_0)$, which can be used to adjust the preliminary estimates of the nonlinear constants. The process is then repeated with the new preliminary estimates; after 3–5 cycles of iteration, no further change will occur. At this point, although $(a - a_0)$ has become zero, its variance is finite and is readily used to calculate the standard error of a.

The above brief description is designed to give the user some feel for what actually occurs during least squares analysis, but we will not present any further description of the mathematics here, in order to be able to include as many computer programs as possible. The theory and mathematics for the procedures used in these programs are given by Wilkinson[2] and Cleland.[3] Those interested in the full statistical complexities involved in fitting enzyme kinetic data should consult the review by Garfinkel,[4] which evaluates all these matters thoroughly. Since, however, the average biochemist will understand none of these articles, we believe it is more helpful simply to present the computer programs and show how to use them. Thus a number of programs are included at the end of this chapter and briefly described, and a list of other equations that have been fitted is included.

Least Squares Surfaces

The computer programs included in this chapter will handle most decent data with little difficulty, but in some cases where data are bad,

---

[3] W. W. Cleland, *Adv. Enzymol.* **29**, 1 (1967).
[4] L. Garfinkel, M. C. Kohn, and D. Garfinkel, *Crit. Rev. Bioeng.* **2**, 329 (1977).

where the preliminary estimates cannot be obtained by fitting in reciprocal form, or where the equation is simply a difficult one to fit, the program may fail to converge on a position of minimum residual least square. This is indicated by the failure of the 3–5 lines after the printed title to show convergence, or if they are not even printed, by an error indication such as attempted division by zero or taking of a negative square root. In these cases it is useful to examine the actual shape of the least squares surface. For an equation with only two constants, such as Eq. (2)

$$v = VA/(K + A) \qquad (2)$$

the surface is simply a contour map of residual least square (that is, sum of squares of differences between experimental v and the value of v calculated with the assumed values of K and V) as a function of V and K (see Fig. 1). The elongated diagonal shape of the contours results from V being in the numerator and K in the denominator and shows that K and V can be raised or lowered together with a much smaller effect on residual least square than if one is raised and the other lowered.

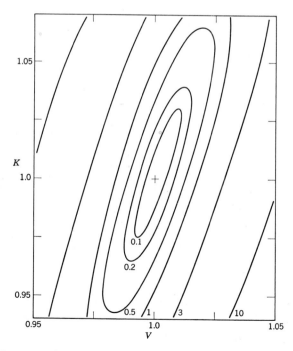

FIG. 1. Contours of equal residual least square for a set of data that fit Eq. (2).

The construction of such a least squares surface is very simple; one simply progresses through a grid of K and V values and calculates

$$\Sigma[v_i - VA_i/(K + A_i)]^2 \tag{3}$$

This is easily done with a short computer program, and if the results are printed out in grid form, one can then draw contours directly on the printed sheet.

Where there are three constants in the equation, the least squares surface consists of contours in three dimensions and may be constructed by calculating two-dimensional contour maps for two constants at various levels of the third constant. The problem is further compounded with four or more constants. What is often more useful, however, is to vary two nonlinear constants and make a least squares fit to the equation in which only the remaining constants are now the variables. For example, consider Eq. (4).

$$\log y = \log \left[\frac{c}{(1 + K_1/H)(1 + K_2/H)} \right] \tag{4}$$

which describes the drop in V or V/K at high pH when two groups ionizing independently both must be protonated for activity. To construct a least squares surface with K_1 and K_2 (or actually, pK_1 and pK_2) as the variables, we assume a grid of these, and at each set of pK values we make a least squares fit to Eq. (4) with c as the only variable. This gives:

$$\log c = \frac{\Sigma \log y_i D_i}{n} \tag{5}$$

where

$$D_i = (1 + K_1/H_i)(1 + K_2/H_i) \tag{6}$$

n is the number of data points, and y_i and H_i are the experimental data. When we substitute the value of $\log c$ from Eq. (5) into the expression for the residual least square

$$\Sigma(\log y_i - \log c + \log D_i)^2 \tag{7}$$

we get

$$\Sigma(\log y_i D_i)^2 - (\Sigma \log y_i D_i)^2/n \tag{8}$$

which is the value plotted on the contour map as a function of pK_1 and pK_2.

A more complex example is the equation for hyperbolic competitive inhibition, which can be fitted by the Gauss–Newton method only when good preliminary estimates are available:

$$v = \frac{VA}{[K(1 + I/K_{in})/(1 + I/K_{id})] + A} \tag{9}$$

Since there is no way to make this equation linear by inversion, preliminary estimates of K_{in} and K_{id} must be obtained by analysis of the slopes of reciprocal plots by graphical methods, or by using the HYPRPLT program. When the data are good and properly placed this works nicely, but in some cases divergence has been observed. In these cases, examination of the least squares surface was a useful way to tell whether any minimum really existed.

The procedure used was to vary K_{in} and K_{id} over a grid of values, and at each set of values to fit the data to Eq. (9) with K_{in} and K_{id} considered to be constants, and K and V the variables. The grid covered a power of 10 for both K_{in} and K_{id}, with the values stepped by factors of 1.26 (the tenth root of 10). The computer program read in the starting values of K_{in} and K_{id} (as well as the experimental data), and these could be altered as needed to determine different parts of the least squares surface. The program then printed out for each set of K_{in} and K_{id} values the fitted values of K and V and the residual least square. The results of this study were rather interesting. In some cases a minimum was found, and with the values of K_{in} and K_{id} from this minimum as preliminary estimates, the computer program using the Gauss–Newton method then did converge. In some of the cases, however, the minimum was reached at a negative value of K_{in} or K_{id}, suggesting that this equation was not really the proper one for the data. In other cases no minimum was found, and the contours of residual least square formed a long shallow valley with one end closed, but the other end open, and the floor showing a very gradual drop toward the open end. This shows that one constant is reasonably well determined (set by the coordinate of the floor of the valley), but the other one has only a minimum value (set by the head of the valley) and is not significantly different from infinity (which corresponds to the open end of the valley). In such cases, one will get a good fit by assuming the constant to be infinity (that is, by leaving the appropriate inhibition term out of Eq. 9).

The use of least squares surfaces is thus of great value in stubborn cases in diagnosing why the data do not seem to fit the assumed rate equation, and should be strongly encouraged. Computer programs to accomplish the analysis are very easy to write, and the author will be happy to assist anyone desiring to perform this sort of analysis.

Evaluation of Results

Once one has fitted the data to a given rate equation (or to several possible rate equations), one has to evaluate the results and draw proper conclusions. First, let us consider the case where several possible rate equations have been considered. The following criteria are used in picking the best equation.

1. Residual least square, or its square root, SIGMA. Adding extra terms to a rate equation will lower SIGMA only when the fit is really improved; thus the fit with the lowest SIGMA is usually the best.

2. Standard errors of the constants. When the standard errors are less than 25% of the values, one can consider the values to be well determined, and thus the term containing this constant definitely present. On the other hand, values such as 2 ± 5 or -1 ± 3 show complete lack of significance, and suggest that the term may be absent. If no appreciable rise in SIGMA results from leaving the term out, it should be discarded. The meaning of this is that the data do not detect the presence of the term; it is possible of course that more precise experiments, or a different spread of data points would establish the presence of the term.

3. Randomness of the residuals. The size and sign of the residuals (DIFF values in the table printed out) should be random and show no trends. Thus in a competitive inhibition experiment if the residuals for the middle line are all of one sign, and for the top line of opposite sign, one should suspect parabolic or hyperbolic inhibition (that is, nonlinearity of the slope replot). The size of the residuals can also check on proper weighting. Thus if residuals at high substrate levels are much larger than those at low levels, one should use log weighting.

Two examples of the use of these methods are shown in Tables I and II. In Table I two initial velocity patterns were run in which DPN was varied at different levels of either deuterated or nondeuterated cyclohexanol. The entire set of data were fitted to a rate equation which assumed that an isotope effect (that is, a value different from 1.00) was present on 4, 3, 2, or only 1 kinetic parameters, as shown. In this case there are only

TABLE I
ISOTOPE EFFECTS WITH CYCLOHEXANOL-1-D AND LIVER ALCOHOL DEHYDROGENASE[a]

Isotope effects on				
$V/K_{\text{cyclohexanol}}$	V	V/K_{DPN}	$K_{\text{i DPN}}$	SIGMA
3.14 ± 0.30	1.08 ± 0.09	0.90 ± 0.18	0.43 ± 0.66	0.285
3.04 ± 0.23	1.10 ± 0.09	0.85 ± 0.20	(1.00)	0.283
3.04 ± 0.24	1.06 ± 0.08	(1.00)	(1.00)	0.282
3.19 ± 0.16	(1.00)	(1.00)	(1.00)	0.280

[a] Unpublished data of Dr. P. F. Cook.

small changes in SIGMA, but it is clear from the size of the standard errors that isotope effects on $K_{i\,\text{DPN}}$ and V/K_{DPN} do not exist, and if there is an isotope effect on V it is very small.

Table II illustrates another example of computer fits to rate equations assuming isotope effects on V, V/K, or both.[5] In this case TPN was held constant at a saturating level, and only the concentration of deuterated or nondeuterated isocitrate was varied. At pH 7.45, none of the values is significantly different from 1.00, and thus there are no isotope effects on either V or $V/K_{\text{isocitrate}}$. At pH 9.5, however, the much lower SIGMA value for the fit with effects on both V and V/K, as well as the small standard errors, show clearly that small but real effects on both parameters are present.

Description of Computer Programs

These programs are written in very simple FORTRAN and should run on any computer with minimal changes in input and output statements. All are designed for card input and will accept as many sets of data as desired (a blank card on the end of the data deck stops the program). Each program makes a least squares fit to a given rate equation and prints out the kinetic constants and various combinations thereof, together with standard errors. In addition, the average residual least square (VARIANCE) and its square root (SIGMA) are printed out, so that one can compare the results of fitting the data to different rate equations.

Each program also prints out a table of residuals showing calculated as well as experimental values for each data point, and the differences. Several of the programs also examine the residuals, discard each data point

TABLE II
Isotope Effects with Isocitrate-2-D and Isocitrate Dehydrogenase[a]

pH	Isotope effects on		SIGMA
	V	$V/K_{\text{isocitrate}}$	
7.45	0.98 ± 0.04	(1.00)	0.40
	(1.00)	0.99 ± 0.09	0.42
	0.95 ± 0.06	1.07 ± 0.16	0.44
9.5	1.07 ± 0.01	(1.00)	1.48
	(1.00)	1.12 ± 0.02	1.90
	1.04 ± 0.005	1.06 ± 0.008	0.47

[a] Unpublished data of Dr. P. F. Cook.

[5] "Isotope Effects on Enzyme-Catalyzed Reactions" (W. W. Cleland, M. H. O'Leary, and D. B. Northrop, eds.), p. 261. Univ. Park Press, Baltimore, Maryland, 1977.

when the difference between the calculated and experimental value exceeds 2.6 × SIGMA, and make a revised fit to the remaining data points. This procedure is designed to throw out bad points and is based on a 99% probability that any difference greater than 2.6 × SIGMA is not the result of random error.

To use these programs, the data are placed on cards as follows. The first card is always a title card having in columns 1–3 the number of data points in I3 format (that is, a number with the decimal point assumed to lie between columns 3 and 4). Columns 4–20 are normally left blank, except that all programs except for replots and pH profiles accept reciprocal velocities as input when a number is punched in column 20. Columns 21–68 will be read verbatim from the title card and printed out as a title to identify the output.

The title card is followed by as many data cards as are indicated in columns 1–3 of the title card. There is one data card for each point. If replicates have been run, treat each as a separate data point, and do not average them first. In columns 1–10 of the data card place the velocity, or whatever experimental parameter is being fitted (V, V/K, $1/K_i$ for pH profiles, for example), using F10.5 format (a number with a decimal point but no exponent, such as 1.26, 101., or 0.013). The number can be placed anywhere in the 10 columns, and the decimal point should always be punched. Columns 11–20 are used for substrate concentration, pH, etc., depending on the program, and in some cases data will be placed in columns 21–30 and subsequent fields. All data are in F10.5 format. The individual programs included at the end of this chapter are described in more detail below.

A. Programs for single reciprocal plots

1. HYPER. This program fits Eq. (2), assuming equal variance for the velocities. Data input: velocities in columns 1–10 (or reciprocal velocities if a number is placed in column 20 of the title card), substrate concentration in columns 11–20. Express concentrations as molar, millimolar, or micromolar, so that the size of the numbers is reasonable (between 0.01 and 100); the value of K will come out in the same units. Likewise, scale velocities to some reasonable units; V will come out in these units.

2. HYPERL. This program fits data to

$$\log v = \log \left(\frac{VA}{K + A}\right) \tag{10}$$

Data input is the same as for HYPER. Because of the log fit, the units of SIGMA are in log v units, rather than in units of velocity, and thus SIGMA values for HYPER and HYPERL can not be directly compared. If one is in doubt which fit to use, look at the residuals (DIFF column in table of residuals) for the two fits. If the residuals are randomly distributed and about the same size at high and low velocities, HYPER is the appropriate fit. If the DIFF values for v are low at low velocities and high for high velocities, but the DIFF values for log v are about the same size, HYPERL is the correct fit.

3. SIGMOIL. The rate equation fitted is

$$\log v = \log \left(\frac{VA^2}{a + 2bA + A^2} \right) \tag{11}$$

Data input is similar to HYPER. The log fit is used because of the wider range of velocities commonly seen when the kinetics are of this form.

4. TWOONL. The rate equation fitted is

$$\log v = \log \left(\frac{V(dA + A^2)}{c + bA + A^2} \right) \tag{12}$$

Data cards are similar to those for HYPER. Because this equation does not become linear upon inversion, it is necessary to supply preliminary estimates from graphical analysis in double-reciprocal form. The following are placed on a single card, which follows the title card and precedes the data cards (it is not counted in the total in columns 1–3 of the title card). Columns 1–10: initial slope of reciprocal plot (when $1/A = 0$). Columns 11–20: slope of asymptote (when $1/A = \infty$). Columns 21–30: $1/v$ intercept of the reciprocal plot. Columns 31–40: $1/v$ intercept of the asymptote. All data are F format. Be careful to see that these parameters have the correct dimensions. If the fit diverges, construct a least squares surface for a grid of b and c values with d and V as the variables. Pick the approximate values of b and c to start with from the equations:

$$d = (SL_0 - SL_\infty)/(INT_\infty - INT_0) \tag{13}$$
$$b = d + SL_0/INT_0 \tag{14}$$
$$c = d \, SL_\infty/INT_0 \tag{15}$$

where SL is slope, INT is $1/v$ intercept, and the subscripts 0 and ∞ refer to the values when $1/A$ is zero or infinity. If a minimum in residual least square is found, modify the program to accept preliminary estimates of b, c, and d directly on the extra card, and now a convergent fit should be obtained.

5. SUBIN. This program fits a reciprocal plot showing linear (that is, total) substrate inhibition

$$v = \frac{VA}{K + A + A^2/K_i} \tag{16}$$

Data input same as for HYPER. It is important that the data cover both high and low concentrations of substrate so that the velocity falls at least a factor of 2 on both sides of its maximum observed value.

B. Programs for initial velocity patterns

6. SEQUEN. This program fits the rate equation for an intersecting initial velocity pattern

$$v = \frac{VAB}{K_{ia}K_b + K_aB + K_bA + AB} \tag{17}$$

Data input: velocities in columns 1–10, concentrations of A in columns 11–20, concentrations of B in columns 21–30. The units of A and B do not need to be the same; the values of K_{ia} and K_a come out in the units of A, and the values of K_b and K_{ib} (calculated from the assumed relationship $K_{ia}K_b = K_aK_{ib}$) come out in the units used for B. Although one usually runs this sort of pattern as 3 or 4 lines with variable A at different B levels (or vice versa), the data do not have to fall on single reciprocal plots so long as a grid of values of A and B is used. All data are combined and processed with one single title card [the data may of course be separately fitted to Eq. (2) for each of the separate lines, if desired]. Either substrate may be A, since the equation is symmetrical with respect to A and B.

7. PINGPONG. This program fits the parallel initial velocity pattern

$$v = \frac{VAB}{K_aB + K_bA + AB} \tag{18}$$

Data input is the same as for SEQUEN. Again, either substrate may be A.

8. EQORD. This program fits the equilibrium ordered initial velocity pattern

$$v = \frac{VAB}{K_{ia}K_b + K_bA + AB} \tag{19}$$

Data input is similar to SEQUEN, except that, unlike SEQUEN and PINGPONG, it matters which reactant is considered to be A and which is B. This is the result of the equation not being symmetrical in A and B, with $K_a = 0$, and $K_{ia}K_b/K_a = \infty$. Thus for input, velocities go in columns 1–10, A concentrations in 11–20, and B concentrations in 21–30.

C. Programs for inhibition patterns

9. COMP. This program fits data to a linear competitive inhibition pattern

$$v = \frac{VA}{K(1 + I/K_{is}) + A} \tag{20}$$

Data input: v in columns 1–10, A concentration in columns 11–20, I concentrations in columns 21–30. K comes out in the units of A, and K_{is} in the units of I, and these units may be different. Note that all data are combined and processed together with a single title card, although each reciprocal plot at a separate I level can be fitted separately to Eq. (2).

10. UNCOMP. The fit is to a linear uncompetitive inhibition pattern

$$v = \frac{VA}{K + A(1 + I/K_{ii})} \tag{21}$$

Data input is the same as for COMP.

11. NONCOMP. The fit is to a linear noncompetitive inhibition pattern

$$v = \frac{VA}{K(1 + I/K_{is}) + A(1 + I/K_{ii})} \tag{22}$$

Data input is the same as for COMP.

D. Programs for replot analysis

12. LINE. This program fits kinetic parameters such as K/V or $1/V$ to a straight line as a function of inhibitor or reciprocal substrate concentration. It is thus useful for analysis of slope or intercept replots. All possible combinations of slope, vertical or horizontal intercepts are provided plus standard errors. Data input: K/V or $1/V$ in columns 1–10, inhibitor or reciprocal substrate concentration in columns 11–20, weights if desired in columns 21–30. If no weights are provided, they are set equal to 1. Programs of types A, B, and C above provide the reciprocal of the square of the standard error of each parameter as a possible weighting factor.

13. PARA. A kinetic parameter is assumed to be a parabola

$$y = a + bX + cX^2 \tag{23}$$

where y is the kinetic parameter and X is inhibitor or reciprocal substrate concentration. Data input is the same as for LINE.

14. HYPRPLT. This program assumes hyperbolic variation of a kinetic parameter

$$y = a\frac{(1 + X/K_{in})}{(1 + X/K_{id})} \tag{24}$$

where X is again inhibitor or reciprocal substrate or activator concentration. It can be used for analysis of replots or for any set of data where y changes hyperbolically to a new value as a function of X. If $K_{id} < K_{in}$, y decreases with increasing X, while if $K_{id} > K_{in}$, it increases to the new limiting value of aK_{id}/K_{in}. Data input: y in columns 1–10, X in columns 11–20, weights if desired in columns 21–30. The first data card after the title card must be one where $X = 0$. If no such point exists, estimate the value of y when $X = 0$ from graphical analysis and include this point with a very low weighting factor (.000000001). Weights do not need to be supplied for the rest of the data; if not supplied, they are assumed equal to 1. The reason for this procedure is that the preliminary estimates of K_{in} and K_{id} are obtained by raising the horizontal axis so that the curve goes through the origin (using the value of y on the first data card), and then making a double-reciprocal plot of the data in the normal manner.

E. Programs for pH profiles

15. BELL. This program fits kinetic constants, such as V, V/K, or $1/K_i$ from a series of inhibition experiments, to a pH profile where the kinetic constant drops at both high and low pH by a factor of 10 per pH unit:

$$\log y = \log \left(\frac{c}{1 + H/K_a + K_b/H} \right) \tag{25}$$

where K_a and K_b are the dissociation constants of the groups that ionize, and c is the pH independent value of the parameter y. The fit is in the log form because of the wide range of y values involved. Data input: y in columns 1–10 (scaled so that the numbers are of a reasonable size), pH in columns 11–20, weights if desired in columns 21–30. Although the reciprocal of the square of the standard error of y can be used as a weighting factor, such weights will have considerable random variation, and in practice we have preferred fits where no weights were used. Wilkinson discusses this problem at length.[2]

When two groups ionizing independently are responsible for the form of Eq. (25), the true form of the denominator is: $1 + K_2/K_1 + H/K_1 + K_2/H$, where K_1 and K_2 are the dissociation constants of the two groups. The minimum separation between pK_a and pK_b in Eq. (25) is then $\log 4 = 0.6$ (corresponding to $pK_1 = pK_2$), and the true values for the pK's of the two groups can be calculated from the apparent ones given by the fit to Eq. 25 (see Cleland[6] for a full discussion of this fitting problem). The separation of pK_a and pK_b is very easy to establish when it is 2 or more pH units, but when the pK's are closer together than this, the curve has almost the same shape regardless of the pK's, and small experimental errors cause considerable uncertainty in the pK separation (although not in the sum $pK_a + pK_b$). It is, in fact, common in such cases for pK_a to appear higher than pK_b in Eq. (25), a mathematical possibility, but a physical impossibility. The program thus calculates the values of pK_1 and pK_2, unless $pK_b - pK_a < 0.6$, in which case it assumes that $pK_1 = pK_2$ and computes this value instead.

16. HABELL. This program assumes a drop in kinetic constant only at low pH:

---

[6] W. W. Cleland, *Adv. Enzymol.* **45**, 273 (1977).

$$\log y = \log \left(\frac{c}{1 + \text{H}/K}\right) \qquad (26)$$

Data input is similar to BELL.

17. HBBELL. The drop in parameter with pH occurs only at high pH.

$$\log y = \log \left(\frac{c}{1 + K/\text{H}}\right) \qquad (27)$$

Data input is similar to BELL.

18. WAVL. This program assumes that the parameter y decreases at high or low pH, but levels out at a new value

$$\log y = \log \left(\frac{\text{YL} + \text{YH}(K/\text{H})}{1 + K/\text{H}}\right) \qquad (28)$$

where YL is the value of y at low pH, YH the value at high pH, and K is the pK of the group whose ionization or protonation decreases activity. Data input is similar to BELL, except that a card carrying preliminary estimates from graphical analysis must be placed between the title and data cards (it is not counted in the value placed in columns 1–3 of the title card). On this card, place preliminary estimates in columns 1–10 of pK (the pH at which log y has dropped 0.3 below its higher plateau value, regardless of whether this is YL or YH), and in columns 11–20 of YL/YH.

Additional Programs

The programs included here, and those published previously, should allow statistical analysis of most kinetic data. A number of other programs have been written for less commonly encountered rate equations, and the author will make listings of these available to people who have data requiring them. Investigators should send graphs of their data and an explanation of what rate equations they think may fit their data to Dr. W. W. Cleland (Department of Biochemistry, University of Wisconsin, Madison, Wisconsin 53706). Include a phone number in case consultation is necessary. Requests for "all programs," requests from libraries, or requests unaccompanied by justifying data will not be honored.

Programs for the following rate equations are available.[7]

A. Programs for single-reciprocal plots

Programs to fit rate equations similar to Eqs. (11) and (12), except not in the log form, and Eq. (16) in the log form.

---

[7] Existing programs can be used for other equations by altering the input. For example, $1/v$ vs I plots can be fitted by using HYPER or HYPERL with v in columns 1–10 and $1/I$ in columns 11–20. K_i is then given by the reciprocal of the printed K value.

```
C     MOST PROGRAMS USE THE SAME MATRIX SOLUTION SUBROUTINE AND END
C     FOR THE PROGRAM, WHICH FOLLOWS.

C     MATRIX SOLUTION SUBROUTINE
  2   DO 3 J = 1,N2
      DO 3 K = 1,M1
  3   S(K,J) = 0
      DO 4 I = 1,NP
      GO TO (15,17), M
  13  DO 4 J = 1,N1
      DO 4 K = 1,N
  4   S(K,J) = S(K,J) + Q(K) *Q(J)
      DO 5 K = 1,N
  5   SM(K) = 1./SQRT (S(K,K))
      SM(N1) = 1.
      DO 6 J = 1,N1
      DO 6 K = 1,N
  6   S(K,J) = S(K,J)*SM(K)*SM(J)
      SS(1,N2) = 1.
      S(1,N2) = 1.
      DO 7 K = 1,N
  7   SS(K) = S(K,1)
      DO 8 J = 1,N1
  8   S(K,J) = S(K+1,J+1) = SS(K+1)*S(1,J+1)/SS(1)
      DO 9 K = 1,N
  9   S(K,1) = S(K,1)*SM(K)
      GO TO (16,18), M
  36  FORMAT(23H PROGRAM COMPLETED FOR I4, 6H LINES  )
  99  PRINT 36, JJ
      STOP
      END

C     REPLOT AND PH PROFILE PROGRAMS WHICH USE WEIGHTS SUBSTITUTE THE
C     FOLLOWING FOR STATEMENT 4.

  4   S(K,J) = S(K,J) + Q(K) * Q(J) * W(I)

C     HYPER
      DIMENSION V(100), A(100), S(3,4), Q(3),         SM(3), SS(3)
      PRINT 100
 100  FORMAT(33H FIT TO HYPERBOLA  V = VMAX/(K+A) ///)
  11  FORMAT(I3,I17,48H ANYTHING HERE WILL BE PRINTED DURING OUTPUT  )
  1   FORMAT (5F10.5, F10.7)
      JJ = 0
  14  READ 11, NP, NO
      IF (NP) 99, 99, 12
  12  M = 1
      II = 0
      GO TO 2
  15  READ 1, V(I), A(I)
      IF (NO) 19, 19, 32
  32  V(I) = 1. /V(I)
  19  Q(1) = V(I)**2/A(I)
      Q(2) = V(I)**2
      Q(3) = V(I)
      GO TO 13
  16  CK = S(1,1)/S(2,1)
      JJ = JJ + 1
      PRINT 11, JJ, NP
      PRINT 1, CK
      NT = 0
      M = 2
      GO TO 2
  17  D = CK + A(I)
      Q(1) = A(I)/D
      Q(2) = A(I)/D**2
      Q(3) = V(I)
      GO TO 13
  18  CK = CK - S(2,1)/S(1,1)
      PRINT 1, CK
      NT = NT + 1
      IF (NT-3) 2,87,87
  87  S2 = 0
      PRINT 71
  71  FORMAT (60H         CONC       1/CONC       EXPTL V     1/EXPTL V     CALC V      DI
 1FF  /)
      DO 26 I = 1,NP
      X2 = 1./A(I)
      X3 = 1./V(I)
      X1 = S(1,1)/(1. + CK/A(I))
      DX1 = V(I) - X1
      PRINT 1, A(I), X2, V(I), X3, X1, DX1
  26  S2 = S2 + (V(I) - S(1,1)*A(I))/CK + A(I)))**2
      LP = NP - 2
      S2 = S2/P
      S1=SQRT(S2)
      SL = CK/S(1,1)
      VK = 1./SL
      VINT = 1./S(1,1)
      DO 10 J = 2,3
  10  DO 10 K = 1,2
      S(K,J) = S(K,J)*SM(K)*SM(J+1)
      SEVK=S1*SQRT(S(1,2))
      SECK=S1*SQRT(S(2,3))/S(1,1)
      SEVI = SEV*S(1,3)**2
      S(1,3) = S1*SQRT (CK**2*S(1,2)+ S(2,3) + 2.*CK*S(1,3))
      SESL = S(1,3)/S(1,1)**2
      SEVK = S(1,3)/CK**2
      WCK = 1./SECK**2
      WV = 1./SEVK**2
      WSL = 1./SESL**2
      WVI = 1./SEVI**2
      WVK = 1./SEVK**2
```

118

```
      GRAPH = SL*(1./A(NP)) + VINT
      PRINT 34, CK, SECK, WCK
      CVV = SEV/S(1,1)
      WLV = 1./CVV**2
      PRINT 121, CVV, WLV
  121 FORMAT(11H C.V.(V) = F11.6,15H    WL = 1/CV = E14.5)
      PRINT 37, SL, SESL, WSL
      PRINT 38, VINT, SEVI, WVI
      PRINT 39, VK, SEVK, WVK
      CVVK = SEVK/VK
      WLVK = 1./CVVK**2
      PRINT 122, CVVK, WLVK
  122 FORMAT(13H C.V.(V/K) = F11.6,15H    WL = 1/CV = E14.5)
      PRINT 41, A(NP), GRAPH
      PRINT 98, S1
      IF(II) 30,30,14
   30 S1 = 2.6*S1
      DO 24 I=1,NP
      IF(ABS(V(I)-S(I,1)*A(I)/(CK+A(I)))-S1)25,20,20
   20 FORMAT(35H POINT DEVIATES MORE THAN 2.6*S1  2F10.5/)
      GO TO 24
   25 II = II + 1
      V(II) = V(I)
      A(II) = A(I)
   24 CONTINUE
      IF(NP-II)14,14,22
   22 NP = II
      NT = 0
      PRINT 23
   23 FORMAT(14H  REVISED FIT//)
      GO TO 2
   34 FORMAT(9H      K = F12.6,15H   S.E.(K) =   F11.6,7H    W = F14.5)
   35 FORMAT(9H      V = F12.6,15H   S.E.(V) =   F11.6,7H    W = F14.5)
   37 FORMAT(9H    K/V = F12.6,15H   S.E.(K/V) = F11.6,7H    W = F14.5)
   38 FORMAT(9H    1/V = F12.6,15H   S.E.(1/V) = F11.6,7H    W = F14.5)
   39 FORMAT(9H    V/K = F12.6,15H   S.E.(V/K) = F11.6,7H    W = F14.5)
   41 FORMAT(12H 1/V AT A = F12.6,4H IS F12.6)
   98 FORMAT(9H  SIGMA = F13.9)
   40 FORMAT(12H VARIANCE = E14.5,5H FOR I4,20H DEGREES OF FREEDOM //)
C     MATRIX SOLUTION SUBROUTINE

C     HYPER1
      DIMENSION V(100), A(100), S(3,4), Q(3)    , SM(3), SS(3)
      PRINT 100
  100 FORMAT(32H FIT TO   LOG V = LOG(VMA/(K+A))///)
   11 FORMAT(13,J17,48H ANYTHING HERE WILL BE PRINTED DURING OUTPUT       )
    1 FORMAT(3F10.5,F10.7,2F12.5,F10.7)
      JJ = 0
   14 READ 11, NP, NO
      IF (NP) 99, 99, 12
   12 M = 1
      N = 2
      P = NP = N
      N1 = N + 1
      N2 = N + 2
      GO TO 2
   15 READ 1, V(I), A(I)
      IF (NO) 19, 19, 32
   32 V(I) = 1./V(I)
   19 Q(1) = V(I) / A(I)
      Q(2) = V(I)
      Q(3) = 1.
      GO TO 13
   16 CK = S(1,1)/S(2,1)
      JJ = JJ + 1
      PRINT 11, JJ, NP
      PRINT 1, CK
      NT = 2
      M = 2
      GO TO 2
   17 Q(1) = 1.
      Q(2) = 1./(CK + A(I))
      Q(3) = LOG (V(I)*(1. + CK/A(I)))
      GO TO 13
   18 CK = CK - S(2,1)
      PRINT 1, CK
      NT = NT + 1
      IF (NT-3) 2,21,21
   21 CV = EXP (S(1,1))
      S2 = 0
      PRINT R1
   81 FORMAT(71H  A CONC    EXPTL V      DIFF   /)
    1 LC LOG V   DIFF   /)
      DO 22 I = 1, NP
      X1 = CV/(1. + CK/A(I))
      DX1 = V(I) - X1
      X2 = LOG (V(I))
      X3 = LOG (X1)
      DX2 = X2 - X3
      PRINT 1, A(I), V(I), X1, DX1, X2, X3, DX2
   22 S2 = S2 + DX2**2
      S2 = S2/P
      S1 = SQRT (S2)
      SL = CK/CV
      VK = 1./SL
      VINT = 1./CV
      DO 10 J = 2,N1
   10 S(K,J) = S(K,J)*SM(K)*SM(J-1)
      SEV = S1*SQRT (S(1,2))*CV
      DX1 = V(I) - X1
      SECK = S1*SQRT (S(2,3))
      SEVI = SEV/CV**2
      SESL = S1*SQRT (S(2,3)+CK**2*S(1,2)+2.*S(1,3)*CK)/CV
```

```
       SEVK = SESL/SL**2
       PRINT 34, CK, SECK
       PRINT 35, CV, SEV
       CVV = SEVK/CV
       WLV = 1./CVV**2
       PRINT 121, CVV, MLV
  121  FORMAT(11H C.V.(V) = F11.6,15H     WL = 1/CV = E14.5)
       PRINT 37, SL, SESL
       PRINT 38, VINT, SEVI
       PRINT 39, VK, SEVK
       CVVK = SEVK/VK
       WLVK = 1./CVVK**2
       PRINT 122, CVVK, WLVK
  122  FORMAT(13H C.V.(V/K) = F11.6,15H    WL = 1/CV = E14.5)
       PRINT 33, S2, S1
  34   FORMAT(9H      K = F12.6,15H   S.E.(K)   = F11.6)
  35   FORMAT(9H      V = F12.6,15H   S.E.(V)   = F11.6)
  37   FORMAT(9H    K/V = F12.6,15H   S.E.(K/V) = F11.6)
  38   FORMAT(9H    1/V = F12.6,15H   S.E.(1/V) = F11.6)
  39   FORMAT(9H    V/K = F12.6,15H   S.E.(V/K) = F11.6)
  33   FORMAT(12H VARIANCE = E14.5,  9H SIGMA = F12.7//)
       GO TO 14
C      MATRIX SOLUTION SUBROUTINE
C
       SIGMOIL
       DIMENSION V(100), A(100), S(4,5), Q(4)       , SM(4), SS(4)
       PRINT 100
  100  FORMAT(40H FIT TO LOG(V)=LOG(V*X**2/(A+2B*X+X**2))///)
  11   FORMAT(13,I17,48H   ANYTHING HERE WILL BE PRINTED DURING OUTPUT   )
  1    FORMAT(8F10.5)
       JJ = 0
  14   READ 11, NP, NO
       IF (NP) 99,99,12
  12   M = 1
       P = NP - 3
       N = 3
       N1 = N + 1
       N2 = N + 2
       GO TO 2
  15   READ 1, V(I), A(I)
       IF(NO)19,19,32
  32   V(I) = 1./V(I)
  19   Q(1) = V(I)         /A(I)**2
       Q(2) = V(I)         /A(I)
       Q(3) = V(I)
       Q(4) = 1
       GO TO 13
  16   CA = S(1,1)/S(3,1)
       CB = .5*S(2,1)/S(3,1)
       JJ = JJ + 1
       PPINT 11, JJ, NP
       PRINT 1, CA, CB, S(3,1)
       NT = 0
       M = 2
       GO TO 2
  17   D = CA + 2.*CB*A(I) + A(I)**2
       Q(1) = 1.
       Q(2) = 1./D
       Q(3) = Q(2)=2.*A(I)
       Q(4) = LOG(V(I)*D/A(I)**2)
       GO TO 13
  18   CA = CA - S(2,1)
       CB = CB - S(3,1)/ 2.
       NT = NT + 1
       PRINT 1, CA, CB, S(1,1)
       IF (NT - 5) 2, 87, 87
  87   S2 = 0
       CV2 = EXP(S(1,1))
       C1V = 1./CV2
       PRINT 200
  200  FORMAT(80H   LOG V   CALC LOG V    A CONC       EXPTL V        CALC V         DIFF           DIFF   CALC 1/V  EXPTL
      1LOG V  CALC LOG V   DIFF             /)
       DO 82 I=1,NP
       X=CV2*A(I)**2/(CA+2.*CB*A(I)+A(I)**2)
       DXSV(I)=X
       RX=1./X
       X1 = LOG(X)
       X2 = LOG(V(I))
       DX1 = X2 - X1
       PRINT 1, A(I),V(I),X,DX,RX ,X2, X1, DX1
  82   S2=S2+DX1**2
       S2 = S2/P
       S1 = SQRT (S2)
       DO 10 J = 2,4
       DO 10 K = 1,3
  10   S(K,J) = S(K,J)*SM(K)*SM(J-1)
       SEA = S1*SQRT (S(2,3))
       SER = .5*S1*SQRT (S(3,4))
       SEV = S1*SQRT (S(1,2))    *CV2
       SFIV = SEV/CV2**2
       WV = 1./SEV**2
       W1V = 1./SEIV**2
       AR = CA/CB
       SEAB = S1*SQRT (S(2,3)+AB**2*S(3,4)/4.-AB*S(2,4))/CR
       PRINT 37, CB, SEA
       PRINT 38, CB, SER
       PRINT 49, AR, SEAR
       PRINT 39, CV2, SEV, WV
       PRINT 40, C1V, SEIV, W1V
  37   FORMAT(6H   A = F12.6,15H      S.E.(A)   = F11.6)
  38   FORMAT(6H   B = F12.6,15H      S.E.(B)   = F11.6)
  39   FORMAT(6H   V = F12.6,15H      S.E.(V)   = F11.6)
  40   FORMAT(7H  1/V = F12.6,10H   S.E.(1/V) = F11.6,7H   W = E14.5)
  49   FORMAT(7H  A/B = F12.6,10H   S.E.(A/B) = F11.6,7H   W = E14.5)
       X = CR**2 + CA
       IF (X) 83,R4,R5
```

```
   83 PRINT 41
   41 FORMAT(25H K CALCULATION IMPOSSIBLE    )
      GO TO 88
   84 PRINT 42
   42 FORMAT(6H K = B)
      GO TO 88
   85 CK = CB + SQRT (X)
      Z = S(2,3) + CK**2*S(3,4) + 2.*CK*S(2,4)
      SEK = S1*SQRT (Z/X)/2.
      WK = 1./SEK**2
      PRINT 43, CK, SEK, WK
   43 FORMAT(6H  K = F12.6,15H       S.E.(K) = F11.6,7H    W = E14.5)
   98 FORMAT(9H SIGMA = F13.9)
   88 PRINT 44, S2
   44 FORMAT(12H VARIANCE = E14.5//)
      GO TO 14
C     MATRIX SOLUTION SUBROUTINE
C
C
      TWOONL
      DIMENSION V(100),A(100),         S(5,6),Q(5)       ,SM(5),SS(5)
      PRINT 100
  100 FORMAT(42H LOG(Y)=LOG(V(A**2 + D*A)/(A**2 + B*A +C))      )
      PRINT 101
  101 FORMAT(40H  K = B/2 - D + SQRT(B/2 - D)**2 + C)  ///)
   11 FORMAT(13,117,48H ANYTHING HERE WILL BE PRINTED DURING OUTPUT   )
    1 FORMAT(8F10.5)
      JJ = 0
   14 READ 11, NP,NO
      IF (NP) 99, 99, 12
   12 READ 1, CSL1, CSL2, CINT1, CINT2
      CD = (CSL1 - CSL2)/(CINT2 - CINT1)
      CB = CSL1/CINT1 + CD
      CC = CSL2*CD/CINT1
      JJ = JJ + 1
      PRINT 11, JJ  ,NP
      PRINT 1, CINT1, CB, CC, CD
      M = 1
      P = NP - N
      N = 4
      N1 = N + 1
      N2 = N + 2
      NT = 0
      GO TO 2
   15 V(I), V(I), A(I)
   32 V(I) = 1./V(I)
      IF (NO) 17, 17, 32
   17 DN = A(I) + CD
      D = A(I) + CB + CC/A(I)
      Q(1) = 1.
      Q(2) = 1./D
      Q(3) = Q(2)/A(I)
      Q(4) = 1./DN
      Q(5) = LOG(V(I))*D/DN)
      GO TO 13
   16 M = 2
   18 NT = NT + 1
      CB = CB - S(2,1)
      CC = CC - S(3,1)
      CD = CD + S(4,1)
      CV = EXP(S(1,1))
      PRINT 98, CV,CB, CC, CD
      IF (NT - 5) 2, 87, 87
   87 S2 = 0
      PRINT 81
   81 FORMAT(83H  A  CONC    EXPTL V    CALC V    DIFF    1/V CALC    LO
     1G V  LOG CALC V  DIFF LOG/)
      DO 82 I = 1, NP
      X = CV   *(A(I)+CD)/((A(I)+CB+CC/A(I))
      RX = 1./X
      X2 = LOG(V(I))
      X3 = LOG(X)
      DX2 = X2 - X3
      PRINT 1, A(I),V(I),X,DX    ,RX,X2, X3, DX2
   82 S2 = S2 + DX2**2
      S2 = S2/P
      S1 = SQRT (S2)
      DO 10 J = 2,N1
      DO 10 K = 1,N
   10 S(K,J) = SV(K,J)*SM(K)*SM(J-1)
      CIV = 1./CV
      SEV = S1*SQRT (S(1,2))
      SEIV = SEV/CV**2
      WV = 1./SEV**2
      WIV = 1./SEIV**2
      SER = S1*SQRT (S(2,3))
      SEC = S1*SQRT (S(3,4))
      SED = S1*SQRT (S(4,5))
      PRINT 39,CV   ,SEV, WIV
   39 FORMAT(6H  V = F12.6,15H     S.E.(V)   = F11.6,7H   W = E14.5)
   40 FORMAT(7H 1/V = F12.6,15H     S.E.(1/V) = F11.6,7H   W = E14.5)
      PRINT 38, CB, SER
      PRINT 37, CC, SEC
      PRINT 42, CD, SED
   38 FORMAT(6H  B = F12.6,15H    S.E.(B)  = F11.6)
   37 FORMAT(6H  C = F12.6,15H    S.E.(C)  = F11.6)
   42 FORMAT(6H  D = F12.6,15H    S.E.(D)  = F11.6)
      COR = CC/CB
      SECB = S1*SQRT (COB**2*S2,3) + S(3,4) - 2.*COB*S(2,4))/CB
      ROD = CB/CD
      CARD = C1V*CR/CD
      SEROD = S1*SQRT (ROD**2*S(4,5)+S(2,3)+2.*ROD*S(2,5))/CD
      SEABOD = C1V*CR/CD
      SEAROD = S1*SQRT (CR**2*CD**2*S(1,2) + CD**2*S(2,3) + CR**2*S(4,5)
     1 + 2.*CR*CD**2*S(1,3) + 2.*CB**2*CD*S(1,5) + 2.*CR*CD*S(2,5))/CD
```

```
            2**2/CV
45    FORMAT(7H C/A = F12.6,15H            S.E.(C/B) = F11.6)
46    FORMAT(7H B/D = F12.6,15H            S.E.(A/D) = F11.6)
47    FORMAT(8H 1/V* = F12.6,15H           S.E.(1/V*)= F11.6)
      PRINT 45, COB, SECOB
      PRINT 46, ROD, SEROD
      PRINT 47, CABOD, SEABOD
      X = CR**2 - 4.*CC
      TF(X) 201,201,202
201   PRINT 203
203   FORMAT(30H SEPARATE KM VALUES IMPOSSIBLE/)
      GO TO 206
202   X = SQRT(X)
      CK1 = (CB+X)/2.
      CK2 = (CB-X)/2.
      SEK1 = S1*SQRT(CK1**2*S(2,3)+S(3,4)-2.*CK1*S(2,4))/X
      SEK2 = S1*SQRT(CK2**2*S(2,3)+S(3,4)-2.*CK2*S(2,4))/X
      PRINT 204, CK1, SEK1
      PRINT 205, CK2, SEK2
204   FORMAT(9H   KM1 = F12.6,15H          S.E.(KM1) = F11.6)
205   FORMAT(9H   KM2 = F12.6,15H          S.E.(KM2) = F11.6)
      CV1 = CV  *(CD-CK1)/X
      CV2 = CV  *(CK2-CD)/X
      MM = 1
      PS1 = CD - CK1
      PS2 = (CR+CD - 2.*CC)/X**2
      PS3 =-(2.*CD-CR)/X**2
302   XX = S1*SQRT(PS1**2*S(1,2)+PS2**2*S(2,3)+PS3**2*S(3,4)+S(4,5)+
     1  2.*PS1*PS2*S(1,3)+2.*PS1*PS3*S(1,4)+2.*PS1*S(1,5)+2.*PS2*PS3*S(2,
     2  4)+2.*PS2*S(2,5)+2.*PS3*S(3,5))/X
      GO TO (300,301), MM
300   SEV1 = XX       *CV
      MM = 2
      PS1 = CD-CK2
      GO TO 302
301   SEV2 = XX       *CV
      PRINT 304, CV1, SEV1
      PRINT 305, CV2, SEV2
304   FORMAT(9H   VM1 = F12.6,15H          S.E.(VM1) = F11.6)
305   FORMAT(9H   VM2 = F12.6,15H          S.E.(VM2) = F11.6)
206   CONTINUE
      X = CB/2. - CD
      XX = X**2 + CC
      IF (XX) 83, 83, 85
83    PRINT 41
41    FORMAT(25H K CALCULATION IMPOSSIBLE              )
      GO TO 88
85    CK = X + SQRT (XX)
      Z = CK**2*S(2,3)/4. + S(3,4)/4. + CK**2*S(4,5) + CK*S(2,4)/2. +
     1CK**2*S(2,5) + CK*S(3,5)
      SEK = S1*SQRT (Z/XX)
      PRINT 43, CK, SEK
43    FORMAT(6H   K = F12.6,15H           S.E.(K)   = F11.6)
      PRINT 98, S1
```

```
98    FORMAT(9H SIGMA = F13.9)
88    PRINT 44, S2
44    FORMAT(12H VARIANCE = E14.5/)
      GO TO 14
C     MATRIX SOLUTION SUBROUTINE

C
      SURIN
      DIMENSION V(100), A(100), S(4,5),Q(4),         SM(4), SS(4)
      PRINT 100
100   FORMAT(34H FIT TO V = VMA*(K + A + A**2/KI)///)
11    FORMAT(I3,I17,48H ANYTHING HERE WILL BE PRINTED DURING OUTPUT  )
1     FORMAT(6F10.5)
      JJ = 0
14    READ 11, NP, NO
      IF (NP) 99, 99, 12
12    M = 1
      N = 3
      P = NP - N
      LP = N + 1
      N1 = N + 1
      N2 = N + 2
      GO TO 2
15    READ 1, V(I), A(I)
      IF (NO) 19, 19, 32
32    V(I) = 1./V(I)
19    Q(1)  = V(I)**2*2/A(I)
      Q(2)  = V(I)**2
      Q(3)  = V(I)**2*2*A(I)
      Q(4)  = V(I)
      GO TO 13
16    CK = S(1,1)/S(2,1)
      B = S(3,1)/S(2,1)
      JJ = JJ + 1
      PRINT 11,JJ,NP
      PRINT 1, CK, B, S(2,1)
      NT = 0
      GO TO 2
17    D = 1. + CK/A(I) + B*A(I)
      Q(1)  = 1./D
      Q(2)  = 1./D**2*2/A(I)
      Q(3)  = A(I)/D**2
      Q(4)  = V(I)
      GO TO 13
18    CK = CK - S(2,1)/S(1,1)
      B = B - S(3,1)/S(1,1)
      PRINT 1, CK, B, S(1,1)
      NT = NT + 1
      IF (NT-5) 2,87,87
87    S2 = 0
      PRINT 81
81    FORMAT (60H   A CON        V EXPTL        V CALC        DIFF 1/VEXPTL 1/VC
```

122

```
1ALC                   /)
      DO 82 I = 1,NP
      X = S(1,1)/(1. + CK/A(I) + B*A(I))
      DX = V(I) - X
      X1 = 1./V(I)
      X2 = 1./X
      PRINT 1, A(I),V(I),X,DX              ,X1,X2
82    S2 = S2 + DX**2
      S2 = S2/P
      S1 = SQRT (S2)
      SL = CK/S(1,1)
      VINT = 1./S(1,1)
      CKI = 1./B
      DO 10 J = 2,N1
10    S(K,J) = S(K,J)*SM(K)*SM(J-1)
      SFV = S1*SQRT (S(1,2))
      SECK = S1*SQRT (S(2,3))/S(1,1)
      SER = S1*SQRT (S(3,4))/S(1,1)
      SEVI = SFV/S(1,1)**2
      S(1,3) = S1*SQRT (CK**2*S(1,2) + 2.*CK*S(1,3))
      SESL = S(1,3)/S(1,1)**2
      SEKI = SER/B**2
      WCK = 1./SECK**2
      WV = 1./SEVA**2
      WSL = 1./SESL**2
      WVI = 1./SEVI**2
      PRINT 34, SECK, WCK
      PRINT 35, S(1,1), SEV, WV
      PRINT 37, SL, SESL, WSL
      PRINT 38, VINT, SEVI, WVI
      PRINT 39, CKI, SEKI
      PRINT 41, B, SER
      PRINT 98, S1
      PRINT 40, S2, LP
34    FORMAT(9H   K    = F12.6,15H    S.E.(K)   = F11.6,7H    W = E14.5)
35    FORMAT(9H   V    = F12.6,15H    S.E.(V)   = F11.6,7H    W = E14.5)
37    FORMAT(9H   K/V  = F12.6,15H    S.E.(K/V) = F11.6,7H    W = E14.5)
38    FORMAT(9H   1/V  = F12.6,15H    S.E.(1/V) = F11.6,7H    W = E14.5)
39    FORMAT(9H   KI   = F12.6,15H    S.E.(KI)  = F11.6)
41    FORMAT(9H   1/KI = F12.6,15H    S.E.(1/KI)= F11.6)
98    FORMAT(9H SIGMA  = F13.9)
40    FORMAT(12H VARIANCE = E14.5,5H FOR I4,20H DEGREES OF FREEDOM //)
      GO TO 14
C     MATRIX SOLUTION SUBROUTINE
C
      SEQUEN
      DIMENSION V(100),A(100),B(100),S(5,6),Q(5),    SM(5),SS(5)
      PRINT 100
100   FORMAT(43H FIT TO Y = V*A*B/(K*A*B+KB*A+KI*AB + KI*A*KB) ///)
11    FORMAT(I3,I17,48H ANYTHING HERE WILL BE PRINTED DURING OUTPUT )
1     FORMAT (8F10.5, F10.7)
      JJ = 0
14    READ 11, NP, NO
      IF (NP) 99,99,12
12    M = 1
      N1 = N + 1
      N2 = N + 2
      II = 0
      GO TO 2
15    READ 1, V(I), A(I), B(I)
      IF (NO) 19, 19, 32
32    V(I) = 1./V(I)
19    Q(1) = V(I)**2/B(I)
      Q(2) = V(I)**2/A(I)
      Q(3) = V(I)**2
      Q(4) = V(I)**2/A(I)/B(I)
      Q(5) = V(I)
      GO TO 13
16    CKA = S(2,1)/S(5,1)
      CKB = S(1,1)/S(3,1)
      CKIA = S(4,1)/S(1,1)
      JJ = JJ + 1
      PRINT 11, JJ, NP
      NT = 0
      M = 2
      PRINT 1,CV,CKA,CKB,CK1A
      GO TO 2
17    D = CKA/A(I) + CKB/B(I) + 1. + CKIA*CKB/A(I)/B(I)
      Q(1) = 1./D
      Q(2) = 1./A(I)/D**2
      Q(3) = (1. + CKIA/A(I))/B(I)/D**2
      Q(4) = (1./A(I)/B(I))/D**2
      Q(5) = V(I)
      GO TO 13
18    CV = S(1,1)
      CKA = CKA - S(2,1)/S(1,1)
      CKB = CKB - S(3,1)/S(1,1)
      CKTA = CKIA - S(4,1)/S(1,1)/CKB
      PRINT 1, CV, CKA, CKB, CKIA
      NT = NT + 1
      IF (NT - 5) 2, 87, 87
87    S2 = 0
      PRINT 71
71    FORMAT (90H    CONC A     1/A     CONC B     1/B    EXPTL V     1/EX
     1PTL V   CALC V   1/CALC V    V DIFF //)
      DO 82 I = 1,NP
      X2 = 1./A(I)
      X3 = 1./V(I)
      X1 = CV/(CKA/A(I) + CKB/B(I) + CKIA*CKB/A(I)/B(I) + 1. + CK1A*CKB*A(I)/B(I))
      X5 = 1.
      DX1 = V(I) - X1
      PRINT 1, A(I), X2, B(I), X4, V(I), X3, X1, X5, DX1
82    S2=S2+(DX1)**2
```

```
      P = NP - N
      S2 = S2/P
      S1 = SQRT (S2)
      DO 10 J = 2,N1
      DO 10 K = 1,N
   10 S(K,J) = S(K,J)+SM(K)*SM(J-1)
      CKIB = CKIA*CKB/CKA
      HAL = CKIA*CKA/CV
      R = CV/CKA
      VKA = CV/CKA
      VKA = CV/CKB
      VCPT = (1. - CKA/CKIA)
      RKIA = 1./CKIA
      SEV =S1*SQRT (S(1,2))
      SEKA =S1*SQRT (S(2,3))/S(1,1)
      SEKB =S1*SQRT (S(3,4))/S(1,1)
      SEKIA =S1*SQRT (S(4,5))/CKB/S(1,1)
      SEKIB = S1*SQRT (CKIB**2*S(2,3) + CKIA**2*S(3,4) + S(4,5)
     1-2.*CKIACKCB*S(2,4) -2.*CKIB*S(2,5) + 2.*CKIA*S(3,5))/CKA/S(1,1)
      SER = S1*SQRT (R**2*S(2,3) + S(4,5)/CKB**2 -2.*R*S(2,5)/CKB)/CKA/S
     1(1,1)
      SEHAL = S1*SQRT (CKIA**2*CKB**2*S(1,2) + CKIA**2*S(3,4) + S(4,5)
     1 + 2.*CKIA**2*CKRAS(1,4) + 2.*CKIA*CKB*S(1,5) + 2.*CKIA*S(3,5))/
     2 S(1,1)**2
      SEVKA = S1*SQRT(S(1,2) + S(2,3)/CKA**2 + 2.* S(1,3)/CKA)/CKA
      SEVKB = S1*SQRT(S(1,2) + S(3,4)/CKB**2 + 2. * S(1,4)/CKB)/CKB
      SEKTA = S1*SEKIA/CKIA**2
      SEVCPT = S1*SQRT(VCPT*VCPT**2*S(1,2)+RKIA**2*S(2,3)+RKIA**2*S(4,5)/CKIB
     1**2-2.*VCPT*RKIA*S(1,3)+2.*VCPT*RKIA*S(1,5)/CKIB-2.*RKIA**2*S(2,5)
     2/CKIB)/S(1,1)**2
      VCPT = VCPT/S(1,1)
      HKA = 1./SEKA**2
      HKB = 1./SEKB**2
      HV = 1./SEV**2
      HKIA = 1./SEKIA**2
      HKIB = 1./SEKIB**2
      HR = 1./SER**2
      WHAL = 1./SEHAL**2
      WRVKA = 1./SEVKA**2
      WRVKB = 1./SEVKB**2
      WRKIA = 1./SERKIA**2
      WVCPT = 1./SEVCPT**2
      PRINT 39, Cv, SEV, HV
      CVV = SEV/CV
      MLV = 1./CVV**2
  121 FORMAT(11H C,v,(v) = F11.6,15H    WL = 1/CV = E14.5)
      PRINT 37, CKA,SEKA, WKA
      PRINT 38, CKB, SEKB, WKB
      PRINT 40, CKIA, SEKIA, WKIA
      PRINT 41, CKIB, SEKIB, WKIB
      PRINT 42, R, SER, WR
      PRINT 43, HAL, SEHAL, WHAL
      PRINT 47, VKA, SEVKA, WRVKA
      
      CVVKA = SEVKA/VKA
      WLVKA = 1./CVVKA**2
      PRINT 123,CVVKA, WLVKA
  123 FORMAT(1H C,v,(V/KA) = F11.6,15H    WL = 1/CV = E14.5)
      PRINT 48, VKB, SEVKB, WRVKB
      CVVKB = SEVKB/VKB
      WLVKB = 1./CVVKB**2
      PRINT 124,CVVKB, WLVKB
  124 FORMAT(1H C,v,(V/KB) = F11.6,15H    WL = 1/CV = E14.5)
      PRINT 45, RKIA, SERKIA, WRKIA
      PRINT 46, VCPT, SEVCPT, WVCPT
   37 FORMAT(12H     KA = F12.6,20H    S.E.(KA)     = F12.6,7H    W =
     1E14.5)
   38 FORMAT(12H     KB = F12.6,20H    S.E.(KB)     = F12.6,7H    W =
     1E14.5)
   39 FORMAT(12H     V  = F12.6,20H    S.E.(V)      = F12.6,7H    W =
     1E14.5)
   40 FORMAT(12H    KIA = F12.6,20H    S.E.(KIA)    = F12.6,7H    W =
     1E14.5)
   41 FORMAT(12H    KIB = F12.6,20H    S.E.(KIB)    = F12.6,7H    W =
     1E14.5)
   42 FORMAT(12H  KIA/KA = F12.6,20H    S.E.(KIA/KA) = F12.6,7H    W =
     1E14.5)
   43 FORMAT(12H KIA*KB/V = F12.6,20H    S.E.(KIA*KB/V)= F12.6,7H    W =
     1E14.5)
   47 FORMAT ( 12H    V/KA = F12.6,20H    S.E.(V/KA)   = F12.6,7H    W
     1= E14.5)
   48 FORMAT ( 12H    V/KB = F12.6,20H    S.E.(V/KB)   = F12.6,7H    W =
     1= F14.5)
   45 FORMAT(12H   1/KIA = F12.6,20H    S.E.(1/KIA)  = F12.6,7H    W =
     1 E14.5)
   46 FORMAT (17H 1/V(1-KA/KIA) = F12.6,23H S.E.(1/V(1-KA/KIA)) = F12.6,
     2 6H M = F14.5)
   98 FORMAT(9H SIGMA = F13.9)
      PRINT 98, S1
      PRINT 44, S2
   44 FORMAT(12H VARIANCE = E14.5/)
      IF (II) 30, 30, 14
   30 S1 = 2.6*S1
      DO 24 I = 1, NP
      IF (ABS (V(I) - CV/(CKA/A(I) + CKB/B(I) + 1.+CKIA*CKB/A(I)/B(I)))
     C-S1) 25, 20, 20
   20 PRINT 21, V(I), A(I), B(I)
   21 FORMAT(33H POINT DEVIATES MORE THAN 2.6*S1 3F10.5/)
      GO TO 24
   25 II = II + 1
      V(II) = V(I)
      A(II) = A(I)
      B(II) = B(I)
   24 CONTINUE
      IF (NP-II) 14, 14, 22
   22 NP = II
      NT = 0
      PRINT 23
```

```
   23 FORMAT(14H   REVISED FIT//)
      GO TO 2
    C MATRIX SOLUTION SUBROUTINE

    C PINGPONG
      DIMENSION V(100),A(100),B(100),S(4,5),Q(4)      ,SM(4),SS(4)
      PRINT 100
  100 FORMAT(35H FIT TO Y = V*A*B/(KA*B+KB*A+A*B ) ///)
   11 FORMAT (8F10.5, F10.7)
      FORMAT(I3,I17,48H  ANYTHING HERE WILL BE PRINTED DURING OUTPUT   )
      JJ = 0
   14 READ 11, NP, NO
      IF (NP) 99,99,12
   12 P = NP - 3
      M = 1
      N = 3
      N1 = N + 1
      N2 = N + 2
      GO TO 2
   15 READ 1, V(I), A(I), B(I)
      IF (NO) 19, 19, 32
   32 V(I) = 1./V(I)
   19 Q(1) = V(I)**2/B(I)
      Q(2) = V(I)**2/A(I)
      Q(3) = V(I)**2
      Q(4) = V(I)
      GO TO 13
   16 CKA = S(2,1)/S(3,1)
      CKB = S(1,1)/S(3,1)
      JJ = JJ + 1
      PRINT 11, JJ,NP
      NT = 0
      M = 2
      GO TO 2
   17 D = CKA/A(I) + CKB/B(I) + 1.
      Q(1) = 1./D
      Q(2) = 1./A(I)/D**2
      Q(3) = 1./B(I)/D**2
      Q(4) = V(I)
      GO TO 13
   18 CV = S(1,1)
      CKA = CKA - S(2,1)/S(1,1)
      CKB = CKR - S(3,1)/S(1,1)
      CKAKB = CKA*CKB
      NT = NT + 1
      PRINT 1, CV,CKA,CKB
      IF (NT - 5) 2, 87, 87
   87 S2 = 0
   71 FORMAT (90H CALC V    CONC A        1/A        CONC B        1/B
     1PTL V    CALC V    1/CALC V   V DIFF  ///                EXPTL V     1/EX
      DO 82 I = 1,NP
         X2 = 1./A(I)
         X4 = 1./B(I)
         X3 = 1./V(I)
         X1 = CV/(CKA/A(I) + CKB/B(I) +1.)
         X5 = 1./X1
         DX1 = V(I) - X1
         PRINT 1, A(I), X2, B(I), X4, V(I), X3, X1, X5, DX1
   A2 S2=S2+DX1**2
      S2 = S2/P
      S1 = SQRT (S2)
      DO 10 J = 2,N1
      DO 10 K = 1,N
   10 S(K,J) = S(K,J)*SM(K)*SM(J-1)
      SEKA =S1*SQRT (S(2,3))/S(1,1)
      SEKB =S1*SQRT (S(3,4))/S(1,1)
      SEV =S1*SQRT (S(1,2))
      OSEKAR = S1*SQRT (CKB**2*S(2,3)+CKA**2*S(3,4)+2.*CKA*CKB*S(2,4))/S(
     11,1)
      WKA = 1./SEKA**2
      WKR = 1./SEKR**2
      WV = 1./SEV**2
      WKAKB = 1./SEKAB**2
   80 PRINT 37, CKA,SEKA, WKA
      PRINT 38, CKR, SEKR, WKR
      PRINT 39, CV, SEV, WV
      CVV = SEV/CV
      WLV = 1./CVV**2
      PRINT 121, CVV, WLV
  121 FORMAT(11H  C.V.(V) = F11.6,15H    WL = 1/CV = E14.5)
      PRINT 41, CKAKB, SEKAB, WKAKB
   37 FORMAT(6H KA = F12.6,15H   S.E.(KA)   = F11.6,7H    W = E14.5)
   38 FORMAT(6H KB = F12.6,15H   S.E.(KB)   = F11.6,7H    W = E14.5)
   39 FORMAT(6H  V = F12.6,15H   S.E.(V)    = F11.6,7H    W = E14.5)
   41 FORMAT(8H KA*KB = F12.6,17H   S.E.(KA*KB) = F11.6,7H   W = E14.5)
      VKA = CV/CKA
      VKB = CV/CKB
      SEVKA = S1*SQRT(S(1,2) + S(2,3)/CKA**2 + 2.* S(1,3)/CKA)/CKA
      SEVKR = S1*SQRT(S(1,2) + S(3,4)/CKB**2 + 2.* S(1,4)/CKB)/CKB
      WRVKA = 1./SEVKA**2
      WRVKB = 1./SEVKB**2
      PRINT 47, VKA, SEVKA, WRVKA
      CVVKA = SEVKA/VKA
      WLVKA = 1./CVVKA**2
      PRINT 123, CVVKA, WLVKA
  123 FORMAT(14H C.V.(V/KA) = F11.6,15H   WL = 1/CV = E14.5)
      PRINT 48, VKB, SEVKB, WRVKB
      CVVKB = SEVKB/VKB
      WLVKB = 1./CVVKB**2
      PRINT 124,CVVKB, WLVKB
  124 FORMAT( 12H   V/KA     V/KB = F12.6,20H     S.E.(V/KA)   = F12.6,7H
     1= E14.5)
   47 FORMAT( 12H   V/KB = F12.6,20H     S.E.(V/KB)   = F12.6,7H
   48 FORMAT( 
     1= E14.5)
```

```
         PRINT 98, S1
      98 FORMAT(9H SIGMA = F13.9)
         PRINT 44, S2
      44 FORMAT(12H VARIANCE = E14.5//)
         GO TO 14
       C MATRIX SOLUTION SUBROUTINE

       C
         EQORD
         DIMENSION V(100),A(100),B(100),S(4,5),Q(4),SM(4),SS(4)
         PRINT 100
     100 FORMAT(43H FIT TO Y = V*A*B/(    KB*A+A*B + KIA*KB)   ///)
      11 FORMAT(I3,I17,48H  ANYTHING HERE WILL BE PRINTED DURING OUTPUT  )
       1 FORMAT(5F10.5,F10.6)
         JJ = 0
      14 READ 11, NP, NO
         IF (NP) 99,99,12
      12 M = 1
         N = 3
         P = NP - N
         N1 = N + 1
         N2 = N + 2
         GO TO 10
      15 READ 1, V(I), A(I), B(I)
      32 IF (NO) 19, 19, 32
         V(1) = 1./V(I)
      19 Q(1) = V(I)**2/B(I)
         Q(2) = V(I)**2*A(I)/B(I)
         Q(3) = (1. + CKIA*A(I))/B(I)/D**2
         Q(4) = V(I)
         GO TO 13
      18 CV = S(1,1)
         CKR = CKR - S(3,1)/S(1,1)
         CKIA = CKIA - S(2,1)/S(1,1)/CKB
         NT = NT + 1
         PRINT 1, S(1,1),            CKB, CKIA
      34 IF (NT - 5) 2, 87, 87
      87 S2 = 0
         PRINT 88
      88 FORMAT(56H A CONC    B CONC    EXPTL V     CALC V    CALC 1/V    DIFF
```

```
         DO 82 I = 1,NP
         X=CV/(CKB/B(I) +1. +CKIA*CKB*A(I)/B(I))
         DX = V(I)-X
         RX = 1./X
         PRINT 1, A(I),   B(I),   V(I), X, RX, DX
      82 S2 = S2 +DX**2
         S2 = S2/P
         S1 = SQRT (S2)
         DO 10 J = 2,N1
      10 S(K,J) = S(K,J)*SM(K)*SM(J-1)
         SEKB =S1*SQRT (S(3,4))/S(1,1)
         SEV =S1*SQRT (S(1,2))
         SEKIA =S1*SQRT (S(2,3))/CKB/S(1,1)
         CKAB = CKIA*CKB
         SEKAB = S1*SQRT(S(2,3)+CKIA**2*S(3,4)+2.*CKIA*S(2,4))/S(1,1)
         VKR = S(1,1)/CKB
         SEVKB = S1*SQRT(S(1,2)+S(3,4)/CKB**2+2.*S(1,4)/CKB)/CKB
         PRINT 39, CV, SEV
         PRINT 40, CKIA, SEKIA
         PRINT 38, CKB, SEKB
         PRINT 41,CKAB,SEKAB
         PRINT 50, VKR, SEVKR
         PRINT 98, S1
      38 FORMAT(6H KB  = F12.6,15H     S.E.(KB)  = F11.6)
      39 FORMAT(6H V   = F12.6,15H     S.E.(V)   = F11.6)
      40 FORMAT(10H KIA*KB  = F12.6,18H   S.E.(KIA*KB) = F11.6)
      41 FORMAT(6H KIA = F12.6,15H     S.E.(KIA) = F11.6)
      50 FORMAT(7H V/KR = F12.6,15H   S.E.(V/KR) = F11.6)
      98 FORMAT(9H SIGMA = F13.9)
         PRINT 44, S2
      44 FORMAT(12H VARIANCE = E14.5//)
         GO TO 14
       C MATRIX SOLUTION SUBROUTINE

       C
         COMP
         DIMENSION V(100),A(100),CI(100),S(5,6),Q(5)      ,SM(5),SS(5)
         PRINT 100
     100 FORMAT(33H FIT TO    Y = V*A/(K(1+I/KI) + A)  ///)
      11 FORMAT(I3,I17,48H  ANYTHING HERE WILL BE PRINTED DURING OUTPUT  )
       1 FORMAT (8F 10.5)
         JJ = 0
      14 READ 11, NP, NO
         IF (NP) 99,99,12
      12 M = 1
         N = 3
         N1 = N + 1
         N2 = N + 2
         II = 0
         GO TO 10
      15 READ 1, V(I),A(I),CI(I)
```

```
       IF (NO) 19, 19, 32
   32  V(I) = 1./V(I)
   19  Q(1) = V(I)**2/A(I)
       Q(2) = V(I)**2*CI(I)/A(I)
       Q(3) = V(I)**2
       Q(4) = V(I)
       GO TO 13
   16  CV = 1./S(3,1)
       CK = S(1,1)/S(3,1)
       CKIS = S(1,1)/S(2,1)
       JJ = JJ + 1
       PRINT 11, JJ, CK, CKIS,CV
       NT = 0
       M = 2
       GO TO 2
   17  Q(1) = (1.+CI(I)/CKIS)*CK/A(I) + 1.)
       Q(2) = (1.+CI(I)/CKIS)/A(I))/D**2
       Q(3) = CI(I)/A(I)/D**2
       Q(4) = V(I)
       GO TO 13
   18  CV = S(1,1)
       CK = CK - S(2,1)/S(1,1)
       CKIS = CKIS*(1. + S(3,1)*CKIS/S(1,1)/CK)
       PRINT 1, CK, CKIS,CV
       NT = NT + 1
       IF (NT - 5) 2, 87, 87
   87  S2 = 0
   88  FORMAT ( 80H     A CONC       I CONC       V EXP       V CAL       DIFF     1/V
      1CALC  1/V EXPTL       1/A      /)
       DO 82 I = 1,NP
       X=S(1,1)/CK/A(I)*(1.+CI(I)/CKIS)*(1.+CI(I)/CKIS)+1.)
       DX=V(I)-X
       RX=1./X
       X1 = 1./V(I)
       X2 = 1./A(I)
       PRINT 1, A(I), CI(I), V(I), X, DX, RX, X1, X2
   82  S2=S2+DX**2
       P = NP - N
       S2 = S2/P
       S1 = SQRT (S2)
       DO 10 J = 2,N1
   10  S(K,J) = S(K,J)*SM(K)*SM(J-1)
       SEV =S1*SQRT (S(1,2))
       SEK =S1*SQRT (S(2,3))/S(1,1)
       SEKIS =S1*CKIS**2*SQRT (S(3,4))/S(1,1)/CK
       WV = 1./SEV**2
       WK = 1./SEK**2
       WKIS = 1./SEKIS**2
       PRINT 43, CK, SEK, WK
       PRINT 39, CV, SEV, WV
       PRINT 37, CKIS, SEKIS, WKIS
       RKIS = 1./CKIS
       CVKIS = SEKIS/CKIS
       WRKIS = 1./CVKIS**2
       PRINT 110, RKIS, CVKIS, WRKIS
  110  FORMAT(9H 1/KIS = F12,6,15H    C.V.(KIS) = F11.6,21H    WL = 1/CVKIS
      1**2 = E14.5)
       PRINT 98, S1
   43  FORMAT(6H    K = F12.6,15H    S.E.(K)   = F11.6,7H    W = E14.5)
   39  FORMAT(6H    V = F12.6,15H    S.E.(V)   = F11.6,7H    W = E14.5)
   37  FORMAT(6H  KIS = F12.6,15H    S.E.(KIS) = F11.6,7H    W = E14.5)
   98  FORMAT(9H SIGMA = F13.9)
       PRINT 44, S2
   44  FORMAT(12H VARIANCE = E14.5//)
       IF(II)30,30,14
   30  S1 = 2.6*S1
       DO 24 I = 1,NP
       IF(ABS (V(I)-CV*A(I)/(CK*(1.+CI(I)/CKIS)+A(I))) - S1)25,20,20
   20  PRINT 21, V(I), A(I), CI(I)
   21  FORMAT(35H POINT DEVIATES MORE THAN 2.6*S1   3F10.5/)
   25  II = II + 1
       V(II) = V(I)
       A(II) = A(I)
       CI(II) = CI(I)
   24  CONTINUE
   22  NP = II
       NT = 0
       PRINT 23
   23  FORMAT(14H    REVISED FIT/)
       GO TO 2
C      MATRIX SOLUTION SUBROUTINE
C
C
       UNCOMP
       DIMENSION V(100),A(100),CI(100),S(5,6),Q(5),SM(5),SS(5)
       PRINT 100
  100  FORMAT(33H FIT TO  Y = V*A/(K + A(1+I/KI))  ANYTHING HERE WILL BE PRINTED DURING OUTPUT   )
    1  FORMAT (8F 10.5)
       JJ = 0
       II = 0
   14  READ 11, NP, NO
       IF (NP) 99,99,12
   12  M = 1
       N = 3
       P = NP - N
       N1 = N + 1
       N2 = N + 2
       GO TO 2
   15  READ 1, V(I), A(I), CI(I)
       IF (NO) 19, 19, 32
```

```
   32 V(I) = 1./V(I)
   19 Q(1) = V(I)**2/A(I)
      Q(2) = V(I)**2
      Q(3) = V(I)**2*CI(I)
      Q(4) = V(I)
      GO TO 13
   16 CV = 1./S(2,1)
      CK = S(1,1)/S(2,1)
      CKII = S(2,1)/S(3,1)
      JJ = JJ + 1
      PRINT 11, JJ, NP
      NT = 2
      M = 2
      GO TO 2
   17 D = CK/A(I) + 1. + CI(I)/CKII
      Q(1) = 1./D
      Q(2) = 1./A(I)/D**2
      Q(3) = CI(I)/D**2
      Q(4) = V(I)
      GO TO 13
   18 CV = S(1,1)
      CK = CK - S(2,1)/S(1,1)
      CKII = CKII*(1. + S(3,1)*CKII/S(1,1))
      NT = NT + 1
      PRINT 1, CV,CK,CKII
      IF (NT - 5) 2, 87, 87
   87 S2 = 0
      PRINT 88
   88 FORMAT ( ROW  A CONC   I CONC     V EXP       V CAL        DIFF  1/V
     1CALC  1/V EXPTL    1/A        /)
      DO 82 I = 1,NP
      X=CV/(CK/A(I)+1.+CI(I)/CKII)
      DX=V(I)-X
      RX=1./X
      X1 = 1./V(I)
      X2 = 1./A(I)
      PRINT 1, A(I),CI(I), V(I), X, DX, RX,X1,X2
   82 S2=S2+DX**2
      S2 = S2/P
      S1 = SQRT(S2)
      DO 10 J = 2,N1
      DO 10 K = 1,N
   10 S(K,J) = S(K,J)+SM(K)*SM(J-1)
      SEV = S1*SQRT(S(1,2))
      SEK = S1*SQRT(S(2,3))/S(1,1)
      SEKII = S1*CKII**2*SQRT(S(3,4))/S(1,1)
      WV = 1./SEV**2
      WK = 1./SEK**2
      WKII = 1./SEKII**2
      PRINT 43, CV, CK, SEK, WK
      PRINT 39, CV, SEV, WV
      PRINT 38, CKII, SEKII, WKII
      RKII = 1./CKII
      CVKII = SEKII/CKII
```
```
      WRKII = 1./CVKII**2
      PRINT 111, RKII, CVKII, WRKII
  111 FORMAT(9H 1/KII = F12.6,15H   C.V.(KII) = F11.6,21H   WL = 1./CVKII
     1**2 = E14.5)
      PRINT 98, S1
   43 FORMAT(6H   K = F12.6,15H      S.E.(K)   = F11.6,7H   W = E14.5)
   39 FORMAT(6H   V = F12.6,15H      S.E.(V)   = F11.6,7H   W = E14.5)
   38 FORMAT(6H KII =F12.6,15H      S.E.(KII) = F11.6,7H   W = E14.5)
   98 FORMAT(9H SIGMA = F13.9)
   44 FORMAT(12H VARIANCE = E14.5//)
      IF(II)30,30,14
   30 S1 = 2.*S1
      DO 24 I = 1,NP
      IF(ABS(V(I)-CV/(CK/A(I)+1.+CI(I)/CKII))-S1)25,20,20
   20 PRINT 21, V(I), A(I), CI(I)
   21 FORMAT(35H POINT DEVIATES MORE THAN 2.*S1    3F10.5/)
      GO TO 24
   25 II = II + 1
      V(II) = V(I)
      A(II) = A(I)
      CI(II) = CI(I)
   24 CONTINUE
      IF(NP - II) 14,14,22
   22 NP = II
      NT = 0
      PRINT 23
   23 FORMAT(14H REVISED FIT//)
      GO TO 2
C     MATRIX SOLUTION SUBROUTINE

C
      NONCOMP
      DIMENSION V(100),A(100),CI(100),S(5,6),Q(5)     ,SM(5),SS(5)
      PRINT 100
  100 FORMAT(44H FIT TO   Y = V*A/(K(1+I/KIS) + A(1+I/KII)) /)
      PRINT 101
  101 FORMAT(20H       CPT = K*KII/KIS///)
   11 FORMAT(I3,I17,48H   ANYTHING HERE WILL BE PRINTED DURING OUTPUT   )
    1 FORMAT (8F 10.5)
      JJ = 0
      II = 0
   14 READ 11, NP, NO
      IF (NP) 99,99,12
   12 M = 1
      N = 4
      N1 = N + 1
      N2 = N + 2
      GO TO 2
   15 READ 1, V(I), A(I), CI(I)
      IF (NO) 19, 19, 32
   32 V(I) = 1./V(I)
   19 Q(I) = V(I)**2/A(I)
```

```
      Q(2) = V(I)**2*CI(I)/A(I)
      Q(3) = V(I)**2
      Q(4) = V(I)**2*CI(I)
      Q(5) = V(I)
      GO TO 13
   16 CV = 1./S(5,1)
      CK = S(1,1)/S(3,1)
      CKIS = S(1,1)/S(2,1)
      CKII = S(3,1)/S(4,1)
      JJ = JJ + 1
      PRINT 11, JJ, NP
      NT = 0
      M = 2
      GO TO 2
   17 D = (1.+CI(I)/CKIS)*CK/A(I) + 1. + CI(I)/CKII
      Q(1) = (1.+CI(I)/CKIS)/A(I)/D**2
      Q(3) = CI(I)/A(I)/D**2
      Q(4) = CI(I)/D**2
      Q(5) = V(I)
      GO TO 13
   1A CV = S(1,1)
      CK = CK - S(2,1)/S(1,1)
      CKIS = CKIS*(1. + S(3,1)*CKIS/S(1,1)/CK)
      CKII = CKII*(1. + S(4,1)CKII/S(1,1))
      PRINT 1, CV, CK, CKIS, CKII
      NT = NT + 1
      IF (NT - 5) 2, 87, 87
   87 S2 = 0
      PRINT 88
   88 FORMAT ( 80H    A CONC     I CONC      V EXP       V CAL      DIFF    1/V
     1CALC 1/V EXPTL   1/A       ,NP
      DO A2 I = 1,NP
      X=S(1,1)/(((CK/A(I)*(1.+CI(I)/CKIS))+1.+CI(I)/CKII)
      DX=V(I)-X
      RX=1./X
      X1 = 1./V(I)
      X2 = 1./A(I)
      PRINT 1, A(I), CI(I), V(I), X, DX, RX,X1,X2
   A2 S2=S2+DX**2
      P = NP - N
      S2 = S2/P
      S1 = SQRT (S2)
      DO 10 J = 2,N1
      DO 10 K = 1,N
   10 S(K,J) = S(K,J)*SM(K)*SM(J-1)
      SEV =S1*SQRT (S(1,2))
      SFK =S1*SQRT (S(2,3))/S(1,1)
      SEKIS=S1*CKIS**2*SQRT (S(3,4))/S(1,1)/CK
      SEKII=S1*CKII**2*SQRT (S(4,5))/S(1,1)
      WV = 1./SEV**2
      WK = 1./SFK**2
      WKIS = 1./SEKIS**2
      WKII = 1./SEKII**2
      CPT = CK*CKII/CKIS
      SECPT = S1*CPT*SQRT (S(2,3) + CKIS**2*S(3,4) + CKII**2*CK**2*S(4,5
     1)  - 2.*CKIS*S(2,4) - 2.*CKII*CK*S(2,5) - 2.*CKIS*CKII*CK*S(3,5))/S(
     21,1)/CK
      WCPT = 1./SECPT**2
      PRINT 43, CK, SFK, WK
      PRINT 39, CV, SEV, WV
      PRINT 37, CKIS, SEKIS, WKIS
      PRINT 3A, CKII, SEKII, WKII
      RKIS = 1./CKIS
      CVKIS = SEKIS/CKIS
      WRKIS = 1./CVKIS**2
      PRINT 110, RKIS, CVKIS, WRKIS
  110 FORMAT(9H 1/KIS = F12.6,15H C.V.(KIS) = F11.6,21H WL = 1/CVKIS
     1**2 = E14.5)
      RKII = 1./CKII
      CVKII = SEKII/CKII
      WRKII = 1./CVKTI**2
      PRINT 111, RKII, CVKII, WRKII
  111 FORMAT(9H 1/KII = F12.6,15H C.V.(KII) = F11.6,21H WL = 1/CVKII
     1**2 = E14.5)
      PRINT 40, CPT, SECPT, WCPT
   9A FORMAT(9H SIGMA = F13.9)
      PRINT A4, S2
   43 FORMAT(6H K = F12.6,15H S.E.(K) = F11.6,7H W = E14.5)
   39 FORMAT(6H V = F12.6,15H S.E.(V) = F11.6,7H W = E14.5)
   37 FORMAT(6H KIS = F12.6,15H S.E.(KIS) = F11.6,7H W = E14.5)
   38 FORMAT(6H KII = F12.6,15H S.E.(KII) = F11.6,7H W = E14.5)
   40 FORMAT(6H CPT = F12.6,15H S.E.(CPT) = F11.6,7H W = E14.5)
   44 FORMAT(12H VARIANCE = E14.5//)
      IF(II)30,30,14
   30 S1 = 2.6*S1
      DO 24 I = 1,NP
      IF(ABS(V(I)-CV*A(I)/(CK*(1.+(CI(I)/CKIS))+A(I)*(1.+CI(I)/CKII)))-
     1S1)25,20,20
   20 PRINT 21, V(I), A(I), CI(I)
   21 FORMAT(35H POINT DEVIATES MORE THAN 2.6*S1    3F10.5/)
      GO TO 24
   25 II = II + 1
      A(II) = A(I)
      V(II) = V(I)
      CI(II) = CI(I)
   24 CONTINUE
      IF(NP - II) 14,14,22
   22 NP = II
      NT = 0
      PRINT 23
   23 FORMAT(14H REVISED FIT//)
      GO TO 2
C     MATRIX SOLUTION SUBROUTINE
```

```
C      LINE
       DIMENSION V(100), A(100), W(100)
       PRINT 100
100    FORMAT(28H FIT TO LINE  Y = A*X + B   ///)
11     FORMAT(I3,I17,48H ANYTHING HERE WILL BE PRINTED DURING OUTPUT   )
1      FORMAT(F10.5, F10.5)
       JJ = 0
14     READ 11, NP, NO
       IF (NP) 99,99,12
12     S1 = 0
       S2 = 0
       S3 = 0
       S4 = 0
       PP = 0
       DO 2   I = 1, NP
       READ 1, V(I), A(I), W(I)
17     IF (W(I)) 15,15,16
15     W(T) = 1.
16     S1 = S1 + W(I)*A(I)
       S2 = S2 + W(I)*A(I)**2
       S3 = S3 + W(I)*V(I)*A(I)
       S4 = S4 + W(I)*V(I)
2      PP = PP + W(I)
       DF = PP*S2 - S1**2
       AL = PP*S3 - S1*S4
       BE = S2*S4 - S1*S3
       AA = AL/DF
       B = BF/DF
       VH = B/AA
       RR = 1./B
       AR = AA/B
       S = 0
       DO 3   I = 1, NP
3      S = S + (V(I) - A*A(I) - B)**2*W(I)
       P = NP - 2
       S = S/P
       SS = SQRT (S/DF)
       VA = SS*SQRT (PP)
       VR = SS*SQRT (S2)
       VRA = VR/B**2
       VBA = SS*SQRT (S2 + VH**2*PP + 2.*VH*S1)/AA
       VAR = VBA*AB**2
       JJ = JJ + 1
40     FORMAT(F12.6, 13H SLOPE        A =    F12.6,  13H  S.E.(A) =  F12.7)
41     FORMAT(23H VERT.INTERCEPT      B =    F12.6,  13H  S.E.(B) =  F12.7)
42     FORMAT(23H HOR.INTERCEPT     B/A =    F12.6,  15H  S.E.(B/A) =  F12.7)
43     FORMAT(23H                   1/B =    F12.6,  15H  S.E.(1/B) =  F12.7)
44     FORMAT(23H                   A/B =    F12.6,  15H  S.E.(A/B) =  F12.7)
98     FORMAT(10H SIGMA =   F13.9)
45     FORMAT(12H VARIANCE =  E14.6///)
19     PRINT 11, JJ, NP
       PRINT 40, AA, VA
       PRINT 41, B, VB
       PRINT 42, VH, VBA
       PRINT 43, RB, VRB
       PRINT 44, AB, VAR
       PRINT 98, SS
       PRINT 45, S
       GO TO 14
36     FORMAT(23H PROGRAM COMPLETED FOR I4, 6H LINES)
99     PRINT 36, JJ
       STOP
       END
C      PARA
       DIMENSION V(50), A(50), S(4,5), Q(4),    SM(4) , SS(4) , W(50)
       PRINT 100
100    FORMAT(41H FIT TO PARABOLA   Y = A + A*X + C*X**2  ///)
11     FORMAT(I3,I17,48H ANYTHING HERE WILL BE PRINTED DURING OUTPUT   )
1      FORMAT (F10.5, F10.5,F10.5)
       JJ = 0
14     READ 11, NP, NO
       IF (NP) 99,99,12
12     DO 3 J = 1,5
       DO 3 K = 1,4
3      S(K,J) = 0
       DO 4 I = 1,NP
       READ 1, V(I), A(I), W(I)
17     IF (W(I)) 15,15,16
15     W(I) = 1.
16     Q(1) = 1.
       Q(2) = A(I)
       Q(3) = A(I)**2
       Q(4) = V(I)
       DO 4 J = 1,4
       DO 4 K = 1,3
4      S(K,J) = S(K,J) + Q(K,I)*Q(J,I)*W(I)
5      SM(K) = 1./SQRT (S(K,K))
       SW(4) = 1.
       DO 6 K = 1,3
6      S(K,J) = S(K,J)*SM(K)*SM(J)
       SS(4) = 1.
       S(1,5) = 1.
       DO 80 L = 1,3
80     SS(K) = S(K,1)
       DO 8 J = 1,4
       DO 8 K = 1,3
8      S(K+1,J+1) = S(K+1,J+1) - SS(K+1)*S(1,J+1)/SS(1)
       DO 9 K = 1,3
9      S(K,1) = SS(K)*SM(K)
       DO 10 K = 1,3
       DO 10 J = 2,4
```

```
       HYPRPLT
       DIMENSION V(50),A(50),W(50),S(4,5),Q(4)
       PRINT 100
100    FORMAT(43H FIT TO HYPERBOLA  Y = A(1+X/KT2)/(1+X/KTD)///)
 11    FORMAT(I3,I17,48H ANYTHING HERE WILL BE PRINTED DURING OUTPUT
   1   FORMAT(6F10.5)
 14    READ 11, NP, NO
       IF (NP) 99, 99, 12
 12    READ 1, V(I), A(I), W(I)
       JJ = JJ + 1
 10    S(K,J) = S(K,J)+SM(K)*SM(J-1)
       AB = S(1,1)/S(2,1)
       BC = S(2,1)/S(3,1)
       AC = S(1,1)/S(3,1)
       C1A = 1./S(1,1)
       JJ = JJ + 1
       PRINT 11, JJ, NP
 19    S2 = 0
       DO 20 I = 1, NP
 20    S2=S2+(V(I)-S(1,1)*A(I)-S(3,1)*A(I)**2)**2*W(I)
       P = NP - 3
       S2 = S2/P
       S1 = SQRT (S2)
       SEAB = S1*SQRT (AB**2*S(2,3)+S(1,2)-2.*AB*S(1,3))/S(2,1)
       SEBC = S1*SQRT (BC**2*S(3,4)+S(2,3)-2.*BC*S(2,4))/S(3,1)
       SEAC = S1*SQRT (AC**2*S(3,4)+S(1,2)-2.*AC*S(1,4))/S(3,1)
       DO 21 K = 1, 3
 21    S(K,K+1) = S1*SQRT (S(K,K+1))
       SE1A = S(1,2)/S(1,1)**2
       PRINT 40, S(1,1), S(1,2)
 41    PRINT 41, S(2,1), S(2,3)
       PRINT 42, S(3,1), S(3,4)
       PRINT 43, C1A, SE1A
       PRINT 46, AB, SEAB
       PRINT 47, AC, SEAC
       PRINT 45, S2
 40    FORMAT(8H       A  =  F12.6,13H   S.E.(A)  =  F12.7)
 41    FORMAT(8H       B  =  F12.6,13H   S.E.(B)  =  F12.7)
 42    FORMAT(8H       C  =  F12.6,13H   S.E.(C)  =  F12.7)
 43    FORMAT(8H     1/A  =  F12.6,15H   S.E.(1/A) =  F12.7)
 44    FORMAT(8H     A/B  =  F12.6,15H   S.E.(A/B) =  F12.7,8H    (KI1))
 46    FORMAT(8H     A/C  =  F12.6,15H   S.E.(A/C) =  F12.7,8H    (KI2))
 47    FORMAT(8H     A/C  =  F12.6,15H   S.E.(A/C) =  F12.7,12H (KI1*KI2))
 45    FORMAT(12H VARIANCE =  F12.4//)
       GO TO 14
 36    FORMAT(23H PROGRAM COMPLETED FOR I4, 6H LINES)
 99    PRINT 36, JJ
       STOP
       END

       PRINT 11, JJ, NP
       IF (A(I)) 90, 91, 90
 90    PRINT 92
 92    FORMAT(28H X ON FIRST CARD IS NOT ZERO///)
       DO 93 I = 2, NP
 93    READ 1, V(I), A(I), W(I)
       GO TO 14
 91    N = 1
       N1 = N + 1
       N2 = N + 2
       NS = 2
       GO TO 2
 15    READ 1, V(I), A(I), W(I)
       IF (W(I)) 94, 94, 95
 94    W(I) = 1.
 95    Q(1) = (V(I) - V(I))**2/A(I)
       Q(2) = (V(I) - V(I))**2
       Q(3) = V(I) - V(I)
       GO TO 13
 16    C = S(2,1)/S(1,1)
       B = C + 1./S(1,1)*V(I))
       PRINT 1, B, C
       N = 3
       N1 = 4
       N2 = 5
       NS = 1
       NT = 2
       IF (W(I)) 97, 97, 2
 97    W(I) = 1.
       GO TO 2
 17    DD = 1. + B*A(I)
       D = 1. + C*A(I)
       Q(1) = DD/D
       Q(2) = DD*A(I)/D**2
       Q(3) = A(I)/D
       Q(4) = V(I)
       GO TO 13
 18    B = B + S(3,1)/S(1,1)
       C = C - S(2,1)/S(1,1)
       PRINT 1, B, C
       NT = NT + 1
       IF (NT - 5) 2, 87, 87
 87    S2 = 0
 81    FORMAT(55H     X CONC    EXPTL Y    CALC Y     DIFF    CALC 1/Y    W )
       DO 82 I = 1, NP
       Y = S(1,1)+A(I)*B)/(1.+A(I)*C)
       RY = 1./Y
       DY = V(I) - Y
       PRINT 1, A(I), V(I), Y, DY, RY, W(I)
 82    S2 = S2 + DY**2*W(I)
       P = NP - 3
```

131

```
       BELL
       DIMENSION S(4,5), Y(100), A(100), W(100), Q(4), SM(4), SS(4)
       PRINT 100
100    FORMAT(41H FIT TO LOG V = LOG( C/(1 + H/KA + KB/H))///)
       PRINT 1
 1     FORMAT(13,17,4H ANYTHING HERE WILL BE PRINTED DURING OUTPUT
       FORMAT (7F10.5  )
30     FORMAT (2E14.5)
       JJ = 0
14     READ 11, NP, NO
       M = 1
       NT = 0
       N = 3
       N1 = 4
       N2 = 5
       IF (NP) 99,99,12
12     DO 3 J = 1,N2
       DO 3 K = 1,N1
3      S(K,J) = 0
       DO 4 I = 1,NP
       GO TO (13,16,71),M
13     READ 1, Y(I), A(I), W(I)
       A(I) = 1./EXP (2.3026*A(I))
       IF (W(I)) 15,15,17
15     W(I) = 1.
17     Q(1) = Y(I)
       Q(2) = A(I)*Q(1)
       Q(3) = Q(1)/A(I)
       Q(4) = 1.
       GO TO 27
16     D =1. + A(I)/CKA + CKA/A(I)
       R(2) = A(I)/D
       Q(2) = A(I)/D
       Q(3) = 1./A(I)/D
       Q(4) = LOG (Y(I)/D)
       GO TO 27
71     D = 1. + A(I) /CKA+ CKA/A, /A(I)
       Q(1) = (.25*A(I) - A(I)/CKA**2)/D
       Q(2) = LOG(Y(I))*D
       Q(3) = LOG(Y(I))*D
27     DO 4 J = 1,N1
       DO 4 K = 1,N
4      S(K,J) = S(K,J) + Q(K)*Q(J)*W(I)
       DO 5 K = 1,N
5      SM(K) = 1./SQRT (S(K,K))
       DO 6 J = 1,N1
       DO 6 K = 1,N
6      S(K,J) = S(K,J)*SM(K)*SM(J)
       SS(N1) = -1.
       S(1,N2) = 1.
       DO 8 L = 1,N
       DO 7 K = 1,N
7      SS(K) = S(K,1)
       DO 8 J = 1,N1
```

```
       S2 = S2/P
       S1 = SQRT (S2)
       DO 10 J = 2,N1
       DO 10 K = 1,N
10     S(K,J) = S(K,J)*SM(K)*SM(J-1)
       RA = 1./S(1,1)
       RB = 1./A
       RC = B/C
       SEA = S1*SQRT (S(1,2))
       PRINT 40, S(1,1), SEA
       SFRA = SFA/S(1,1)**2
       PRINT 41, RA, SERA
       SERR = S1*SQRT (S(3,4))/S(1,1)/B**2
       PRINT 42, RB, SERR
       SERC = S1*SQRT (S(2,3))/S(1,1)/C**2
       PRINT 43, RC, SERC
       SERC = S1*SQRT (RC**2*S(2,3) + S(3,4) + 2.*RC*S(2,4))/C/S(1,1)
       PRINT 45, RC, SFRC
       SER = S1*SQRT(S(3,4))/S(1,1)
       PRINT 46, B, SER
       SEC = S1*SQRT(S(2,3))/S(1,1)
       PRINT 47, C, SEC
       YINF = S(1,1)*RC
       HYF = 1./YINF
       SEYF = S1*SQRT(R**2*S(1,2) + RC**2*S(2,3) + 2.*B*RC*S(1,3
      1) + 2.*R*S(1,4) + 2.*RC*S(2,4))/C
       SERYF = SF*YF/YINF**2
       PRINT 48, YINF, SEYF
       PRINT 49, HYF, SERYF
       YY = S(1,1)*(RC-1.)
       SFYY = S1*SQRT((B-C)**2*S(1,2)+RC**2*S(2,3)+S(3,4)+2.*(B-C)*BC*S(1
      1,3)+2.*(H-C)*S(1,4)+2.*RC*S(2,4))/C
       PRINT 50, YY, SFYY
       PRINT 9R, S1
       PRINT 44, S2
42     FORMAT(11H KTNIM    = F12.6,15H  S.E. (KIN)    = F11.6)
43     FORMAT(11H KIDENOM  = F12.6,15H  S.E. (KID)    = F11.6)
45     FORMAT(11H KIN/KID   = F12.6,19H  S.E.(KIN/KIN)  = F11.6)
46     FORMAT(9H 1/KIN    = F12.6,17H  S.E.(1/KIN)   = F11.6)
41     FORMAT(11H A         = F12.6,15H  S.E. (A)      = F11.6)
47     FORMAT(9H 1/KID 1/KID = F12.6,17H  S.E.(1/KID)  = F11.6)
48     FORMAT(13H A*KID/KIN  = F12.6,21H  S.E.(A*KID/KIN)  = F11.6)
49     FORMAT(13H KIN/KID/A  = F12.6,21H  S.E.(KIN/KID/A)  = F11.6)
50     FORMAT(16H A(KID/KIN-1)  = F12.6,24H  S.E.(A(KID/KIN-1)) = F11.6)
98     FORMAT(9H SIGMA = F13.9)
44     FORMAT(12H VARIANCE = E14.5//)
       GO TO 14
C      MATRIX SOLUTION SUBROUTINE EXCEPT THAT THE OUTER LIMIT ON DO
C      LOOP 4 IS ALTERED AS FOLLOWS (FOURTH CARD)

       DO 4 I = NS, NP
```

```
      DO 8 K = 1,N
    8 S(K,J) = S(K+1,J+1) - SS(K+1)*S(1,J+1)/SS(1)
      DO 9 K = 1,N
    9 S(K,1) = S(K,1)*SM(K)
      GO TO (18,19,72),M
   18 CKA = S(3,1)/S(1,1)
      CKA = S(1,1)/ S(2,1)
      JJ = JJ + 1
      PRINT 11, JJ, NP
      SCKA = CKA
      SCKB = CKA
      M = 2
      GO TO 12
   19 CKA = CKA + S(2,1)*CKA**2
      CKA = CKA - S(3,1)
      PRINT 30, CKA, CKB
      IF (CKA) 50, 50, 51
   50 CKA = SCKA
   51 IF (CKB) 52, 52, 53
   52 CKB = SCKB
   53 CV = EXP (S(1,1))
      NT = NT + 1
      IF (NT-5) 12,12,28
   72 CKA = CKA - S(2,1)
      PRINT 30, CKA
      NT = NT + 1
      IF(NT-5) 12,12,78
   78 CKB = EXP(S(1,1))
      CKB = CKA/4.
   28 DO 10 J = 2,N1
      DO 10 K = 1,N
   10 S(K,JJ) = S(K,J) *SM(K)*SM(J-1)
      S2 = 0
      PRINT 101
  101 FORMAT(72H                PH         EXPTL V    CALC V       DIFF               EXPTL LOG V
     1CALC LOG V            DIFF/)
      DO 20 I = 1,NP
      X = CV/C1, + A(I)/CKA + CKB/A(I))
      PH = -.43429*LOG(A(I))
      X2 = LOG10 (Y(I))
      X3 = LOG10 (X)
      DX1 = Y2 - X3
      DX = Y(I)-X
      PRINT 1,PH,Y(I),X,DX,X2,X3,DX1
   20 S2 = S2 + W(I)*DY1**2
      P = NP - N
      S2 = S2/P
      S1 = SQRT (S2)
      SEV = S1*SQRT (S(1,2)) * CV
      CR = 1./CV
      SECR = SEV/CV**2
      PRINT 40, CV, SEV
      PRINT 45, CR, SECR

      GO TO(80,80,81),M
   80 IF (CKA) 21, 21, 22
   21 PRINT 44
   44 FORMAT(12H KA NEGATIVE)
      GO TO 23
   22 PKA = -.43429*LOG (CKA)
      SEPKA = S1*.43429*CKA*SQRT (S(2,3))
      PRINT 41, PKA, SEPKA
   23 IF (CKB) 24, 24, 25
   24 PRINT 46
   46 FORMAT(12H KB NEGATIVE)
      GO TO 26
   25 PKB = -.43429*LOG(CKB)
      SEPKB = S1*.43429*SQRT (S(3,4))/CKB
   26 PRINT 42, PKB, SEPKB
      PRINT 98, S1
      PRINT 43, S2
      GO TO (70,70,14),M
   40 FORMAT(9H     C   = F12.6,15H    S.E.(C)   = F12.7)
   41 FORMAT(9H    PKA  = F12.6,15H    S.E.(PKA) = F12.7)
   42 FORMAT(9H    PKB  = F12.6,15H    S.E.(PKB) = F12.7)
   98 FORMAT(9H    SIG  = F13.9)
   45 FORMAT(9H   1/C   = F12.6,15H   S.E.(1/C)  = F12.7)
   43 FORMAT(12H VARIANCE = E14.5/)
   70 X = CKA**2-4.*CKA*CKB
      IF(X)82,82,84
   84 X = SQRT(X)
      CKR = (CKA-X)/2.
      CKA = (CKA+X)/2.
      PKA = -.43429*LOG(CKA)
      PKR = -.43429*LOG(CKB)
      PRINT 85
   85 FORMAT(45H IF TWO GROUPS ARE INVOLVED, TRUE PKS ARE      )
      PRINT 41, PKA
      PRINT 42, PKB
      GO TO 14
   82 PRINT 83
   83 FORMAT(101H PKS TOO CLOSE,  ASSUME MIN SEPARATION(0.6PH UNITS) AND
     1 THUS IF 2 GROUPS INVOLVED, TRUE PKS IDENTICAL            )
      M = 3
      N = 2
      N1 = 3
      N2 = 4
      NT = 0
      GO TO 12
   81 PKA = -.43429*LOG(CKA)
      PKB = PKA +.60206
      SEPKA = S1*.43429     *SQRT(S(2,3))            /CKA
      SEPKB = SEPKA
      PRINT 41, PKA, SEPKA
      PRINT 42, PKB, SEPKB
      GO TO 26
   36 FORMAT(23H PROGRAM COMPLETED FOR I4, 6H LINES)
   99 PRINT 36, JJ
```

```
      STOP
      END

C     HAPELL
      DIMENSION S(3,4), Y(100), A(100), W(100), Q(3), SM(3), SS(3)
      PRINT 100
100   FORMAT(29H FIT TO LOG V = LOGC/(1+H/KA)///)
11    FORMAT(I3,17X,48H ANYTHING HERE WILL BE PRINTED DURING OUTPUT
1     FORMAT (7F10.5   )
30    FORMAT (F14.5)
      JJ = 0
14    READ 11, NP
      M = 1
      NT = 0
      IF (NP) 99,99,12
12    DO 3 J = 1,4
3     S(K,J) = 0
      DO 4 I = 1,NP
      GO TO (13,16),M
13    READ 1, Y(I), A(I), W(I)
      A(I) = 1./FEXP (2.3026*A(I))
      IF (W(I)) 15,15,17
15    W(I) = 1.
17    Q(1) = Y(I)
      Q(2) = A(I)*Q(1)
      Q(3) = 1.
      GO TO 27
16    D=1. + A(I)/CKA
      Q(1) = 1.
      Q(2) = A(I)/D
      Q(3) = LOG (Y(I)*D)
27    DO 4 J = 1,3
      DO 4 K = 1,2
4     S(K,J) = S(K,J) + Q(K)*Q(J)*W(I)
      DO 5 K = 1,2
5     SM(K) = 1./SQRT (S(K,K))
      SM(3) = 1.
      DO 6 J = 1,3
      DO 6 K = 1,2
6     S(K,J) = S(K,J)*SM(K)*SM(J)
      SS(3) = -1.
      SS(1,4) = 1.
      DO 8 L = 1,2
      DO 7 K = 1,2
7     SS(K) = S(K,1)
      DO 8 J = 1,3
8     S(K,J) = S(K+1,J+1) - SS(K+1)*S(1,J+1)/SS(1)
      DO 9 K = 1,2
9     S(K,1) = S(K,1)*SM(K)
      GO TO (18,19),M

1R    CKA = S(1,1)/S(2,1)
      PRINT 30, CKA
      SCKA = CKA
      M = 2
      GO TO 12
19    CKA = CKA + S(2,1)*CKA**2
      PRINT 30, CKA
      IF (CKA) 50, 50, 53
50    CKA = SCKA
53    CV = EXP (S(1,1))
      NT = NT + 1
      IF (NT-5) 12,12,28
28    DO 10 J = 2,3
      DO 10 K = 1,2
10    S(K,J) = S(K,J) *SM(K)*SM(J-1)
      JJ = JJ + 1
      PRINT 11, JJ
      S2 = 0
      PRINT 101
101   FORMAT (72H                                                    EXPTL         PH        EXPTL V     CALC V       DIFF         EXPTL LOG V
     1CALC LOG V    DIFF/)
      DO 20 I = 1, NP
      X = CV/(1. + A(I)/CKA)
      PH = -LOG10(A(I))
      X2 = LOG10 (Y(I))
      X3 = LOG10 (X)
      DX1 = X2 - X3
      PRINT 1,PH,Y(I),X,DX,X2,X3,DX1
20    S2 = S2 + W(I)*DX1**2
      P = NP-2
      S2 = S2/P
      S1 = SQRT (S2)
      SEV = S1*SQRT (S(1,2)) * CV
      CR = 1./CV
      SECR = SEV/CV**2
      PRINT 40, CV, SEV
      PRINT 45, CR, SECR
      IF (CKA) 21, 21, 22
21    PRINT 44
44    FORMAT(12H KA NEGATIVE)
      GO TO 23
22    PKA = -.43429*LOG (CKA)
      SEPKA = S1*.43429*CKA*SQRT (S(2,3))
      PRINT 41, PKA, SEPKA
23    PRINT 9R, S1
      PRINT 43, S2
40    FORMAT(9H       C = F12.6,15H        S.E.(C)   = F12.7)
41    FORMAT(9H     PKA = F12.6,15H        S.E.(PKA) = F12.7)
98    FORMAT(9H   SIGMA = F13.9)
43    FORMAT(12H VARIANCE = E14.5/)
45    FORMAT(9H     1/C = F12.6,15H        S.E.(1/C) = F12.7)
      GO TO 14
36    FORMAT(23H PROGRAM COMPLETED FOR I4, 6H LINES)
```

```
99 PRINT 56, JJ
   STOP
   END

C      HBRELL
       DIMENSION S(3,4), Y(100), A(100), W(100), Q(3), SM(3), SS(5)
       PRINT 100
100 FORMAT(29H FIT TO LOG V = LOGC/(1+KB/H)//)
 11 FORMAT(13,17X,48H ANYTHING HERE WILL BE PRINTED DURING OUTPUT
  1 FORMAT (7F10.5
 30 FORMAT (E14.5)
    JJ = 0
 14 READ 11, NP
    M = 1
    NT = 0
    IF (NP) 99,99,12
 12 DO 3 J = 1,4
  3 S(K,J) = 0
    DO 4 I = 1,NP
    GO TO (13,16),M
 13 READ 1, Y(I), A(I), W(I)
    A(I) = 1./FXP (2.3026*A(I))
    IF (W(I)) 15,15,17
 15 W(I) = 1.
 17 Q(1) = Y(I)
    Q(2) = Q(1)/A(I)
    Q(3) = 1.
    GO TO 27
 16 D = 1. + CKB/A(I)
    Q(1) = 1.
    Q(2) = 1./A(I)/D
    Q(3) = LOG (Y(I)*D)
 27 DO 4 J = 1,3
  4 S(K,J) = S(K,J) + Q(K)*Q(J)*W(I)
  5 SM(K) = 1./SQRT (S(K,K))
    SM(3) = 1.
    DO 6 J = 1,2
  6 S(K,J) = S(K,J)*SM(K)*SM(J)
    SS(3) = -1.
    S(1,4) = 1.
    DO 8 L = 1,2
    DO 7 K = 1,2
  7 SS(K) = S(K,1)
  8 S(K,J+1) = S(K+1,J+1) - SS(K+1)*S(1,J+1)/SS(1)
    DO 9 K = 1,2
  9 S(K,1) = S(K,1)*SM(K)
    GO TO (18,19),M
 18 CKB = S(2,1)/S(1,1)
    PRINT 30, CKB
    SCKB = CKB
    M = 2
    GO TO 12
 19 CKB = CKB - S(2,1)
    PRINT 30, CKB
 51 IF (CKB) 52, 52, 53
 52 CKB = SCKB
 53 CV = EXP(S(1,1))
    NT = NT + 1
    IF (NT-5) 12,12,28
 28 DO 10 J = 2,3
    DO 10 K = 1,2
 10 S(K,J) = S(K,J) *SM(K)*SM(J-1)
    JJ = JJ + 1
    PRINT 11, JJ
    S2 = 0
101 FORMAT (72H            PH        EXPTL V        CALC V       DIFF         EXPTL LOG V
   1CALC LOG V        DIFF//)
    PRINT 101
    DO 20 I = 1,NP
    X = CV/(1. + CKB/A(I))
    PH = -LOG10 (A(I))
    X2 = LOG10 (Y(I))
    X3 = LOG10 (X)
    DX1 = X2 - X3
    DX = Y(I) - X
    PRINT 1, PH, Y(I), X, DX, X2, X3, DX1
 20 S2 = S2 + W(I)*DX1**2
    P = NP-2
    S2 = S2/P
    S1 = SQRT (S2)
    SEV = S1*SQRT (S(1,2)) * CV
    CR = 1./CV
    SECR = SEV/CV**2
    PRINT 40, CV, SEV
    PRINT 45, CR, SECR
    IF (CKB) 24,24,25
 24 PRINT 46
 46 FORMAT(12H KB NEGATIVE)
    GO TO 23
 25 PKB = -.43429*LOG (CKB)
    SEPKB = S1*.43429*SQRT (S(2,3))/CKB
    PRINT 42, PKB, SEPKB
 23 PRINT 98, S1
    PRINT 43, S2
 40 FORMAT(9H        C  = F12.6,15H  S.E.(C)    = F12.7)
 42 FORMAT(9H      PKB  = F12.6,15H  S.E.(PKB)  = F12.7)
 43 FORMAT(12H VARIANCE = E14.5//)
 45 FORMAT(9H      1/C  = F12.6,15H  S.F.(1/C)  = F12.7)
 98 FORMAT(9H    SIGMA = F13.9)
    GO TO 14
```

```
        WAVL
        DIMENSION S(4,5), Y(100), H(100), W(100), Q(4), SM(4), SS(4)
        PRINT 100
  100   FORMAT(36H FIT TO LOG((YL+YH*K/H)/(1+K/H))       ///)
   11   FORMAT(I3,17X,08H          TITLE       HERE
    1   FORMAT(7F10.5,E14.5)
        JJ = 0
   14   READ 11, NP
        IF (NP) 99,99,12
   12   READ 1, PK     , R
        CK = 1./EXP(2.3026*PK)
        JJ = JJ + 1
        PRINT 11, JJ
        N = 3
        P = NP - N
        N1 = N + 1
        N2 = N + 2
        NT = 0
        M = 0
        GO TO 2
   15   READ 1, Y(I), H(I), W(I)
        H(I) = 1./EXP(2.3026**H(I))
        IF (W(I)) 115,115,17
  115   W(I) = 1.
   17   D = 1. + CK/H(I)
        DD = R + CK/H(I)
        Q(1) = 1./DD
        Q(2) = 1.,00
        Q(3) = Q(2)*(1.-R)/D/H(I)
        Q(4) = LOG(Y(I)*D/DD)
        GO TO 13
   16   M = 2
   18   NT = NT + 1
        CK = CK + S(3,1)
        R = R + S(2,1)
        PRINT 1, S(1,1) , S(2,1), S(3,1), S(1,2), S(2,3), S(3,4),R,CK
        IF (NT-5) 2,87,87
   87   S2 = 0
        YH = EXP(S(1,1)  )
        YL = R*YH
        PRINT 88
   88   FORMAT(69H  PH       EXP Y     CALC Y      DIFF      EXP LOG Y    CALC
       1 LOG Y  DIFF/)
        DO 82 I = 1,NP
        PH = -LOG10(H(I))
        X=YH*(R+CK/H(I))/(1.+CK/H(I))
        DX1= Y(I) - X
```

```
        X1 = LOG10(Y(I))
        X2 = LOG10(X)
        DX = X1 -X2
        PRINT 1, PH, Y(I), X, DX1, X1, X2, DX
   A2   S2 = S2 + DX**2*W(I)
        S2 = S2/P
        S1 = SQRT(S2)
        DO 10 J = 2,N1
        DO 10 K = 1,N
   10   S(K,J) = S(K,J)*SM(K)*SM(J-1)
        PK = -LOG10(CK)
        SEYH = S1*SQRT(S(1,2))*YH
        SEYL = S1*SQRT(S(1,2)*R**2+S(2,3)*2.*R*S(1,3))*YH
        SEPK = S1*SQRT(S(3,4))/CK                          *.43429
        PRINT 40, YL,     SEYL
        PRINT 41, YH,     SEYH
        SER = S1*SQRT(S(2,3))
        PRINT 43, R, SER
   43   FORMAT(5H R = F12.6,13H    S.E.(R) = F11.6)
        PRINT 42, PK, SEPK
   40   FORMAT( 6H YL = F12.6,14H   S.E.(YL) = F11.6)
   41   FORMAT( 6H YH = F12.6,14H   S.E.(YH) = F11.6)
   42   FORMAT( 6H PK = F12.6,14H   S.E.(PK) = F11.6)
        PRINT 44, S1, S2
   44   FORMAT(9H SIGMA = F11.6, 12H VARIANCE = E14.5//)
        GO TO 14
C       MATRIX SOLUTION SUBROUTINE
```

```
   36   FORMAT(23H PROGRAM COMPLETED FOR I4, 6H LINES)
   99   PRINT 36, JJ
        STOP
        END
```

B. Programs for initial velocity patterns, including ones showing substrate inhibition

Programs to fit Eqs. (17), (18), and (19) in the log form

Equation (17) with factored denominator [i.e.: $(K_a + A)(K_b + B)$]

Parallel pattern, competitive substrate inhibition by B

Parallel pattern, competitive substrate inhibition by both A and B

Intersecting pattern, competitive, noncompetitive, or uncompetitive substrate inhibition by B

$$v = \frac{VABC}{\text{const} + (\text{Coef A})A + (\text{Coef B})B + (\text{Coef C})C + K_aBC + K_bAC + K_cAB + ABC} \quad (29)$$

Equation (29) with B term or B and C terms missing

Equation (17) in log form with either AB, or B and AB terms missing from the denominator

$$\log v = \log\left(\frac{VA}{K + A} + PA\right) \quad (30)$$

$$\log v = \log\left(\frac{(V/K_b)A^2B}{K_1 + K_2A + A^2}\right) \quad (31)$$

Equation (31) with an AB term in denominator

C. Programs for inhibition patterns

Equations (20), (21), (22) in the log form; S-parabolic I-linear noncompetitive; S-linear I-parabolic noncompetitive; parabolic noncompetitive (both slope and intercept replots are parabolic); hyperbolic uncompetitive; S-linear I-hyperbolic uncompetitive; hyperbolic competitive; parabolic competitive; S-hyperbolic I-hyperbolic noncompetitive (either intersecting where K_{id} is the same for slopes and intercepts, or not intersecting); Eq. (11) (not log form) with a and b being linear functions of I; induced substrate inhibition in an ordered mechanism (vary B and I, with A constant).

$$v = \frac{VA}{[K/(1 + I/K_{isa})] + [A(1 + I/K_{in})/(1 + I/K_{id})]} \quad (32)$$

$$v = \frac{V}{1 + [I/K_i(1 + A/K_a)]} \quad (33)$$

D. Programs for replot analysis

$$y = (a + X/K_i)/(1 + X/K_i) \quad (34)$$

E. Programs for pH profiles

$$\log y = \log\left(\frac{c}{H(1 + H/K)}\right) \quad (35)$$

$$\log y = \log \left(\frac{c\mathrm{H}}{(1 + K/\mathrm{H})}\right) \qquad (36)$$

$$\log y = \log \left(\frac{c}{1 + K_1/\mathrm{H} + K_1 K_2/\mathrm{H}^2}\right) \qquad (37)$$

$$\log y = \log \left(\frac{c}{1 + \mathrm{H}/K_1 + \mathrm{H}^2/(K_1 K_2)}\right) \qquad (38)$$

F. Programs for determining isotope effects

Programs that assume an isotope effect on V or V/K, or on both when reciprocal plots have been run with deuterated and nondeuterated substrates are published.[5] The programs used in Table I, which accept entire intersecting initial velocity patterns assuming isotope effects on the parameters shown, are available.

Concluding Comments

Use of the computer programs reproduced here will make possible statistical analysis of most enzyme kinetic data in a manner satisfactory for all but the most dedicated statistician. Emphasis should thus be placed on improving precision and repeating experiments a number of times, rather than on more sophisticated techniques for analysis of bad data. It is important that concentrations be properly picked to cover appropriate ranges of the variable substrates or inhibitors; that pH, ionic strength, and temperature be properly controlled; that the methods for determining rates be sensitive enough and the amount of enzyme used be suitable for precise measurement of initial rates. And above all, it is important to use your head in making evaluations based on statistical analysis; all the mathematics in the world is no substitute for a reasonable amount of common sense.

[7] Plotting Methods for Analyzing Enzyme Rate Data

By FREDERICK B. RUDOLPH and HERBERT J. FROMM

Visual inspection and numerical analysis of initial-rate data often permit an evaluation of both kinetic parameters and the binding sequence or mechanism for enzyme-catalyzed reactions. Segal *et al.*[1] were among the first to suggest that kinetic mechanisms could be differentiated by

---

[1] H. L. Segal, J. F. Kachmar, and P. D. Boyer, *Enzymologia* **15**, 187 (1952).

inspection of initial rate plots. Alberty[2] demonstrated in 1953 how sequential and Ping Pong mechanisms could be distinguished. (The nomenclatures of Cleland[3] and Dalziel[4] will be used through this chapter. The majority of the mechanisms are described in this volume [3].) A large number of techniques were subsequently developed that allow considerable insight to be gained into the binding order of substrates to a multisubstrate enzyme from simple kinetic experiments. A distinction can usually be made as to the highest-order enzyme–substrate complex present during the course of catalysis, and very often the actual sequence of substrate addition can be deduced. A variety of approaches to analysis of kinetic data have been developed and will be considered here with applications to both two- and three-substrate systems.

There are two basic procedures used in making decisions concerning kinetic mechanisms from initial-rate data. The first of these involves determination of whether a ternary enzyme–substrate complex is formed in a two-substrate mechanism and if either ternary or quaternary complexes are formed in a three-substrate system. This segregation will give rise to two classes of mechanism, sequential and Ping Pong, which can be separated graphically or by numerical analysis of the initial-rate date.

The second procedure involves either determining numerical values for certain combinations of kinetic constants, or the relationships between different groups of kinetic constants. Different mechanisms have different terms present in their rate equations and combinations of kinetic constants, and in most cases these factors can be evaluated. Frequently, these two approaches will totally define a kinetic binding mechanism, and how this is done will be detailed in this chapter.

Definition of Initial-Rate Studies

The first type of experiment one does in evaluating an enzyme mechanism involves varying the levels of one substrate at fixed levels of other substrates (if the enzyme is a multisubstrate enzyme) and measuring the initial reaction velocity. A plot of product formation with time for an enzyme reaction is shown in Fig. 1. The product formation is linear with time in the indicated initial velocity period prior to product accumulation, substrate depletion, or other factors that cause the reaction to slow down. During this initial velocity portion of the reaction progress curve, the concentration of the enzyme form [EP] that gives rise to the product P is in

---

[2] R. A. Alberty, *J. Am. Chem. Soc.* **75**, 1928 (1953).
[3] W. W. Cleland, *Biochim. Biophys. Acta* **67**, 104 (1963).
[4] K. Dalziel, *Acta Chem. Scand.* **11**, 1706 (1957).

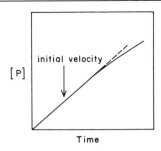

FIG. 1. A plot of product concentration [p] versus time for an enzyme-catalyzed reaction. The linear portion is the initial reaction velocity.

the steady state where $d[\text{EP}]/dt = 0$. Only with initial-velocity measurements are the steady-state assumptions used in derivation of kinetic equations valid (see this volume [4]).

The rate equations describing the steady-state kinetics of an enzyme reaction have the form

$$v = \frac{dp}{dt} = \frac{a(\text{A,B---})}{b(\text{A,B---})} \tag{1}$$

where the velocity is the ratio of the two linear polynomials $a(\text{A,B---})$ and $b(\text{A,B---})$. The equations can be integrated to describe the time course of the reaction, but such treatment does not easily allow distinctions to be made among kinetic mechanisms. Therefore the differential (dp/dt) is actually measured as a function of substrate concentration.

In certain cases it is possible to analyze velocities as a function of substrate concentration from progress curves of the reaction and determine the $V_{\max}$ and Michaelis constants. This technique requires a high equilibrium constant and a knowledge of the mode of inhibition by the products. If products do not inhibit the reaction, a single experiment can allow determination of $V_{\max}$ and K_m. Generally, however, the products do inhibit and an analysis of the inhibition is required to evaluate the kinetic parameters. Such experiments are most useful for study of the effects of pH, temperature, or other parameters, such as ionic strength on the $V_{\max}$ and K_m's of a particular enzyme.

Treatment of Initial Rate Data

The dependence of reaction velocity on substrate concentration has to be carefully determined. The interpretation of which mechanism fits an experimental result will very often be a function of the accuracy of the data. Factors such as substrate concentrations, pH, temperature, and ionic strength need to be controlled as discussed in this volume [1]. In addition, a general discussion of experimental protocol is available.[5] The method of assay should afford accurate data for the analysis (discussed in this volume [1] and [2]) and metal–ligand complexes should be treated properly (this volume [12]).

The data are then plotted to evaluate the kinetic mechanism and determine kinetic parameters. The basic Michaelis–Menten equation has the form

$$v = VA/(K_a + A) \tag{2}$$

where v, V, A, and K_a represent initial velocity, maximal velocity, substrate A, and Michaelis constant for A, respectively. A plot of v versus A is hyperbolic and difficult to use for analysis. To avoid this problem, Eq. (2) has been transformed into various linear forms. The most common transformation of the equation is the double-reciprocal or Lineweaver–Burk[6] plot.

$$\frac{1}{v} = \frac{1}{V} + \frac{K_a}{V}\left(\frac{1}{A}\right) \tag{3}$$

Another transformation, suggested by Hanes,[7] is

$$\frac{A}{v} = \frac{K_a}{V} + \frac{A}{V} \tag{4}$$

Hofstee et al.,[8] on the other hand, proposed the form

$$v = V - \frac{K_a v}{A} \tag{5}$$

More recent methods of plotting include the direct linear plot suggested by Eisenthal and Cornish-Bowden,[9] which is based on the equation

$$\frac{V}{v} - \frac{K_a}{A} = 1 \tag{6}$$

The plots described by Eqs. (3)–(6) are shown in Fig. 2.

The method of plotting described in this chapter will be the double-reciprocal, or Lineweaver–Burk, plot, as it is the plot most often used by

---

[5] H. J. Fromm, "Initial Rate Enzyme Kinetics." Springer-Verlag, Berlin and New York, 1975.
[6] H. Lineweaver and D. Burk, *J. Am. Chem. Soc.* **56**, 658 (1934).
[7] C. S. Hanes, *Biochem. J.* **26**, 1406 (1932).
[8] B. H. Hofstee, M. Dixon, and E. C. Webb, *Nature (London)* **184**, 1296 (1959).
[9] R. Eisenthal and A. Cornish-Bowden, *Biochem. J.* **139**, 715 (1974).

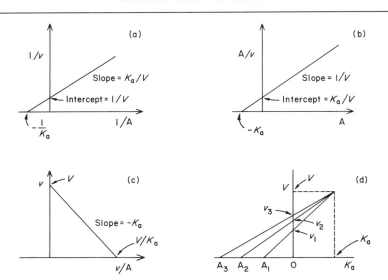

FIG. 2. Plots of $1/v$ versus $1/A$ (a), A/v versus A (b), v versus v/A (c), and V versus K_a (d) for the Michaelis–Menten equation based on Eqs. (3)–(6), respectively.

enzymologists. Although it has been suggested that this plot may not be as statistically valid as some others,[5,7] proper analysis using data weighting and computer methods obviates this problem.[10]

Graphical Analysis

The initial rate data for an enzyme-catalyzed reaction can be analyzed graphically using a variety of procedures. Although it is desirable to vary the substrate concentration over as great a range as possible, this is usually not technically feasible. Thus, the range of substrate concentrations should be around the K_m to provide a maximal velocity change. The experiments discussed in this chapter are all done in the absence of products or inhibitors, and the interpretation depends entirely on the effect of changing substrate concentrations. It also should be noted that this analysis is concerned only with the substrate side of the reaction. The number of products released is not important kinetically, and the order of their release is not evaluated unless both the forward and reverse reactions are studied. For example, Uni Bi and Bi Ter reactions are simply treated as Uni, Bi, or Ter reactions.

Two- and three-substrate mechanisms will be considered separately.

---

[10] D. B. Siano, J. W. Zyskind, and H. J. Fromm, *Arch. Biochem. Biophys.* **170**, 587 (1975).

Bireactant Mechanisms

The most general form of the rate equation for bireactant mechanisms using the nomenclature of Cleland[3] is

$$v = \frac{VAB}{K_{ia}K_b + K_aB + K_bA + AB} \quad (7)$$

It can be expressed in reciprocal form in terms of ϕ's as suggested by Dalziel[4] as

$$\frac{1}{v} = \phi_0 + \frac{\phi_A}{A} + \frac{\phi_B}{B} + \frac{\phi_{AB}}{AB} \quad (8)$$

The values of the ϕ's for each bisubstrate mechanism are listed in Table I. Graphical analysis simply involves determining whether certain ϕ values are defined, i.e., not equal to zero. A plot of the reciprocal of the concentration of A versus the reciprocal of the velocity for Eq. (8) is shown in Fig. 3. The slopes have the value of $\phi_A + \phi_{AB/B}$, and the intercepts are $\phi_0 + \phi_{B/B}$. Secondary replots of the slopes and intercepts versus the reciprocal of the concentration of B can be made as shown in Fig. 4. This method permits the evaluation of the ϕ values and consequently the kinetic parameters. The slope and intercept of the secondary intercept plot are ϕ_B and ϕ_0 respectively, while the slope and intercept from the secondary slope plot are ϕ_{AB} and ϕ_A, respectively.

TABLE I
ϕ Values for Two-Substrate Mechanisms[a]

Mechanism	ϕ_0	ϕ_A	ϕ_B	ϕ_{AB}
Ordered Bi Bi	$\dfrac{k_5 + k_7}{k_5 k_7}$	$\dfrac{1}{k_1}$	$\dfrac{k_4 + k_5}{k_3 k_5}$	$\dfrac{k_2(k_4 + k_5)}{k_1 k_3 k_5}$
Theorell–Chance Bi Bi	$\dfrac{1}{k_5}$	$\dfrac{1}{k_1}$	$\dfrac{1}{k_3}$	$\dfrac{k_2}{k_1 k_3}$
Ping Pong Bi Bi	$\dfrac{k_3 + k_7}{k_3 k_7}$	$\dfrac{k_2 + k_3}{k_1 k_3}$	$\dfrac{k_6 + k_7}{k_5 k_7}$	0
Rapid Equilibrium Random Bi Bi	$\dfrac{1}{k_1}$	$\dfrac{K_A}{k_1}$	$\dfrac{K_B}{k_1}$	$\dfrac{K_{AB}}{k_1}$
Equilibrium Ordered Bi Bi	$\dfrac{1}{k_3}$	0	$\dfrac{k_2 + k_3}{k_1 k_3}$	$\dfrac{k_2 + k_3}{k_1 k_3 K_A}$

[a] Absence of a particular ϕ term is indicated by a 0.

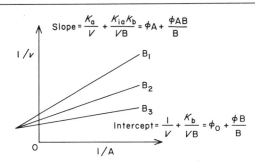

FIG. 3. A plot of $1/v$ versus $1/A$ for a two-substrate sequential enzyme at different fixed levels of B. The parameters describing the slope and intercepts are described in terms of K's as suggested by W. W. Cleland (*Biochim. Biophys. Acta* **67**, 104 (1963) and ϕ values as suggested by K. Dalziel (*Acta Chem. Scand.* **11**, 1706 (1957).

The ϕ values are related to the K values of Cleland[3] as follows:

$K_a = \phi_A/\phi_0 =$ Michaelis constant for A
$K_b = \phi_B/\phi_0 =$ Michaelis constant for B
$K_{ia} = \phi_{AB}/\phi_B =$ dissociation constant for EA
$V = E_0/\phi_0 =$ maximum velocity

where E_0 is the enzyme concentration in normality. The same information can be derived from the primary plot of $1/v$ versus $1/B$ (based on the same data as the $1/A$ plot), and the answers should be identical. The absence of certain ϕ values causes the reciprocal plots to exhibit different patterns. The most common two-substrate patterns are as shown in Fig. 3, where the lines at the different fixed substrate concentrations intersect to the left of the $1/v$ axis. Values exist for all the ϕ's in Eq. (8). This finding is diagnostic of a sequential mechanism in which a ternary complex is formed prior to product release, but does not specify the order of substrate addition. The Ordered, Rapid Equilibrium Random, and

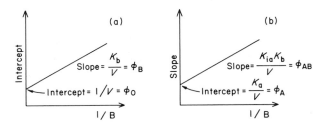

FIG. 4. Secondary plots of intercepts (a) and slopes (b) versus the reciprocal of the concentration of B based on the data from Fig. 3. The kinetic constants are evaluated from the slopes and intercepts of these plots.

Theorell–Chance mechanisms will all give this type of pattern. Some distinctions can be made within this group based on the intersection point of the lines, as will be discussed below.

A second type of initial rate pattern with intersecting plots is diagnostic of the Equilibrium Ordered mechanism. In this mechanism the addition of the first substrate A is at equilibrium and the intercepts of a 1/B reciprocal plot are not affected by changes in the concentration of A. The equation describing this mechanism is

$$\frac{1}{v} = \phi_0 + \frac{\phi_B}{B} + \frac{\phi_{AB}}{AB} \tag{9}$$

The lines on the 1/B versus 1/v plot intersect on the 1/v axis indicating a value of 0 for ϕ_A. The 1/A plot intersects to the left of the 1/v axis above the horizontal axis with a 1/v coordinate equal to 1/V. Both the 1/A and 1/B plots, or analysis of the replots, must be done to distinguish this mechanism from the usual sequential mechanisms.

The other type of bisubstrate initial rate pattern is one in which the lines at different fixed substrate levels are parallel. This is descriptive of the Ping Pong mechanism and is due to the absence of a ternary complex in the mechanism. The equation for this mechanism is

$$\frac{1}{v} = \phi_0 + \frac{\phi_A}{A} + \frac{\phi_B}{B} \tag{10}$$

If the last term in Eq. (8) is small relative to the other terms in the equation, the lines may appear parallel even though a ternary complex exists during the reaction. Plotting of the initial rate data in the form suggested by Hanes,[7] S/v versus S [Eq. (4)], transposes the slope and intercept effects so that the parallel Lineweaver–Burk plot will intersect on the S/v axis in a Hanes plot. It is much easier to draw conclusions about an intersection point than about parallel lines. For a sequential mechanism, the relationship $(\phi_0 + \phi_A/A + \phi_B/B) \gg \phi_{AB}/AB$, will probably not apply for both directions of the mechanism, and so both the forward and reverse reactions should be analyzed if possible. The pattern in one direction should be clearly intersecting if the mechanism is sequential, not Ping Pong. Alternative substrates can also give a clearer intersecting pattern than the normal substrate. Substitution of fructose for glucose in the mammalian hexokinase reaction gives a very clear intersecting pattern, whereas there is some question as to whether the lines converge with glucose as the substrate. The effect can be explained as follows: In a Ping Pong mechanism the slope = ϕ_A or ϕ_B and is independent of the nature of the other substrate. The slope for a sequential mechanism is ϕ_A or ϕ_B +

$\phi_{AB/A \text{ or } B}$, which is a function of the second substrate, and rarely will the dependence be negligible with an alternative substrate.

This effect can be used to differentiate between sequential and Ping Pong mechanisms. For hexokinase, for example, a plot of $1/v$ vs $1/\text{ATP}$ can be made at subsaturating fixed levels of both fructose and glucose, and the slopes of the lines be determined. If the mechanism is Ping Pong, the slopes of both plots will be identical, but should be different for sequential mechanisms due to the absence or the presence of the $\phi AB/AB$ term. Similar results are obtained when alternative substrates are used for the other reactant.

An additional plotting technique for confirmation of Ping Pong mechanisms utilizes variation of substrates A and B in a constant ratio. When A and B are maintained at a constant ratio, $A = a(B)$ where a is a constant. Substitution of this relationship into Eq. (10) gives

$$\frac{1}{v} = \phi_0 + \frac{\phi_A}{A} + \frac{a\phi_B}{A} \qquad (11)$$

When $1/v$ is plotted versus $1/A$, the plot will be linear with an intercept = ϕ_0. In the case of a sequential mechanism, the rate equation will have the form

$$\frac{1}{v} = \phi_0 + \frac{\phi_A}{A} + \frac{a\phi_B}{A} + \frac{a\phi_{AB}}{(A)^2} \qquad (12)$$

The substrate squared term makes the plot parabolic-up with a minimum in the second quadrant to the left of the $1/v$ axis. If ϕ_{AB} is very small compared to the other terms in Eq. (12), it will resemble Eq. (11) and the nonlinearity may not be apparent. Fromm[5] has presented a discussion of how this may be tested using the relative amounts of A and B so that the a factor in Eq. (12) becomes significant. Levels of the substrates used in the constant-ratio experiments must be below the concentration where substrate inhibition occurs.

It is also possible, using alternative substrates, to make a choice between the Ordered Bi Bi and Rapid Equilibrium Random Bi Bi mechanisms. In the case of the random pathway, the $K_{ia}K_b$ term in Eq. (7) can be replaced by $K_{ib}K_a$, but this is not true for the Ordered Mechanism. Thus, in the Random mechanism, an alternative substrate for substrate A will not alter the dissociation constant for B, K_{ib}, nor will an alternative substrate for B alter K_{ia}. These two relationships will not hold for the Ordered mechanism, and only K_{ia} will remain unchanged with an alternative substrate for B.

The Point of Convergence of the Double-Reciprocal Plots as a Criterion of Sequential Mechanisms

The intersection points of the double-reciprocal plots for a sequential mechanism such as illustrated in Fig. 2 have defined kinetic relationships that are diagnostic of particular mechanisms. The x and y coordinates of the intersection points for the 1/A and 1/B reciprocal plots are $-\phi_B/\phi_{AB}$, $\phi_0 - \phi_A\phi_B/\phi_{AB}$ and $-\phi_A/\phi_{AB}$, $\phi_0 - \phi_A\phi_B/\phi_{AB}$, respectively. The identical $1/v$ coordinates is a condition that must be satisfied for all bireactant sequential mechanisms. The same constraint applies to the intersection points in the reverse reaction and a mechanism-related relationship exists between the intersection points in the forward and reverse directions. This constraint is met in the equilibrium ordered mechanisms since all $1/v$ intercepts are at $1/V_{max}$.

The $1/v$ intersection point can be above ($K_a < K_{ia}$), on ($K_a = K_{ia}$), or below ($K_{ia} < K_a$) the horizontal axis. The intersection on the axis implies that the Michaelis constant equals the dissociation constant for A but is not indicative of a particular mechanism. This can occur with either ordered or random mechanisms. The intersection will be above the x axis when the dissociation constant is higher than the K_m and below the axis when the K_m is higher than the dissociation constant. Lueck et al.[11] have shown that a choice of mechanism listed in Table II could be made based on evaluation of the points of intersection of the double-reciprocal plots in the forward and reverse direction. The Ordered and Theorell–Chance mechanisms have certain constraints placed on their intersection points as indicated in Table II, but the Rapid Equilibrium Random mechanism has none, since only the equilibrium constant relates the forward and reverse reactions. For the Theorell–Chance mechanisms, the intersection points must be either forward above and reverse below, or vice versa, or both on the axis. The Ordered mechanisms can have any relationship if the forward intersection point is above the axis. On the other hand, if the intersection is on or below the abscissa in one direction, it can only be on or below the axis, respectively, in the reverse direction. Since the Random mechanism allows any relationship, the technique allows one to exclude the Theorell–Chance and Ordered possibilities only if the intersections do not fit the predicted patterns.

Janson and Cleland[12] have presented a similar approach in which the sum of the vertical coordinates for the points of convergence in the for-

---

[11] J. D. Lueck, W. R. Ellison, and H. J. Fromm, *FEBS Lett.* **30**, 321 (1973).
[12] C. A. Janson and W. W. Cleland, *J. Biol. Chem.* **249**, 2562 (1974).

TABLE II
Predicted Points of Convergence of Double-Reciprocal Plots for Sequential Bireactant Mechanisms[a]

Mechanism	Forward direction	Reverse direction		
		Above	On	Below
Ordered Bi Bi or Iso-Ordered Bi Bi	Above +	+	+	+
	On +	+	F[b]	F
	Below +	+	F	F
Random Bi Bi	Above +	+	+	+
	On +	+	+	+
	Below +	+	+	+
Theorell–Chance or Iso-Theorell–Chance	Above +	F	F	+
	On +	F	+[c]	F
	Below +	+	F	F

[a] Based on the table described by Lueck et al.[11]
[b] F indicates that intersection on that position is not possible for that particular mechanism.
[c] The maximal velocities in both directions must be equal under this condition.

ward and reverse directions is divided by the sum of the reciprocals of the V_{max} in both directions. This results in a dimensionless number with a value between zero and one. Both the forward and reverse reactions have to be corrected to identical enzyme concentrations, and the measurements should be done under similar conditions. A value of zero for the ratio corresponds to a Theorell–Chance mechanism, and a value of one corresponds to an Equilibrium Ordered mechanism. Analysis of this ratio is based on the same relationships as described above for the intersection point analysis. The Ordered and Random mechanisms give values of the ratio between zero and one. The value of R gives the proportion of enzyme present in the central complexes when both substrates are saturating and will have some value in analyzing rate-determining steps. The ratio method will exclude the Theorell–Chance and Equilibrium Ordered mechanisms, but not any others.

Terreactant Mechanisms

Initial-rate studies for three-substrate enzymes allow segregation of kinetic mechanisms into sequential and Ping Pong classes as in bisubstrate mechanisms and, in addition, can often define the specific mechanism. Several methods of variation of substrate concentration to evaluate velocity dependence have been developed. The ϕ values for various three-substrate mechanisms are shown in Table III. Multiple combinations of ϕ values are present, and the missing terms can be detected fairly easily.

The most useful method involves varying one substrate while holding the other two substrates in a fixed ratio at different levels in the range of their Michaelis constants as illustrated in Fig. 5. Three such experiments are required. These studies allow differentiation of kinetic mechanisms into sequential and Ping Pong types as well as considerable discrimination among these two classes of mechanisms. Ping Pong mechanisms give one or more parallel lines in double reciprocal plots from initial-rate data, and therefore are easily distinguished from sequential mechanisms, where all reciprocal plots exhibit intersecting lines. Owing to the absence of various terms in the rate equation for the different mechanisms, differentiation can be made from initial-rate double-reciprocal plots alone. The initial velocity patterns for the indicated three-substrate mechanisms are shown in Table IV. The Random AB mechanism and the Equilibrium Ordered mechanism both have 1/C reciprocal plots which intercept at the origin. Thus, these two mechanisms can be distinguished from the others from primary plots of the initial-rate data. The Hexa Uni Uni Ping Pong mechanism exhibits unique behavior, while the other four Ping Pong mechanisms are unique as a class but need additional analysis. As was pointed out in the case of the two-substrate mechanisms, observation of parallel reciprocal plots should be interpreted with caution and reliance should be placed on other supportive evidence.

Construction of secondary plots from the primary double-reciprocal data is a powerful technique for three-substrate systems. Listed in Table V are the predicted slope and intercept replots from experiments illustrated in Fig. 5. The basic rate equation for a three-substrate system

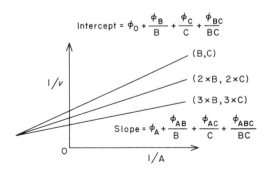

FIG. 5. A plot of $1/v$ versus $1/A$ for a three-substrate reaction. The concentrations of B and C are held in a constant ratio in the range of their Michaelis constants.

TABLE III. ϕ VALUES FOR THREE-SUBSTRATE MECHANISMS[a]

Mechanisms	ϕ_0	ϕ_A	ϕ_B	ϕ_C	ϕ_{AB}	ϕ_{AC}	ϕ_{BC}	ϕ_{ABC}
I Random Ter	$1/k_1$	K_{12}/k_1	K_{11}/k_1	K_{10}/k_1	K_7K_{11}/k_1	K_9K_{12}/k_1	K_6K_{11}/k_1	$K_1K_8K_{11}/k_1$
II(a) Ordered Ter	$1/k_7 + 1/k_9 + 1/k_{11}$	$1/k_1$	$1/k_3$	$(k_6+k_7)/k_5k_7$	k_2/k_1k_3	0	$k_4(k_6+k_7)/k_3k_5k_7$	$k_2k_4(k_6+k_7)/k_1k_3k_5k_7$
II(a') Ordered Theorell–Chance Ter	$1/k_7 + 1/k_9$	$1/k_1$	$1/k_3$	$1/k_5$	k_2/k_1k_3	0	k_4/k_3k_5	$k_2k_4/k_1k_3k_5$
II(b) Random AC (All ternary complexes present) Ter	$1/k_1$	K_8/k_1	K_9/k_1	K_7/k_1	K_6K_9/k_1	0	K_3K_7/k_1	$K_1K_3K_7/k_1$
III Random AC (Ordered B) Ter	$1/k_1$	K_6/k_1	0	K_5/k_1	K_4K_6/k_1	0	K_3K_5/k_1	$K_1K_3K_5/k_1$
IV Random A Ter	$1/k_1$	K_6/k_1	0	K_5/k_1	0	K_4K_6/k_1	K_3K_5/k_1	$K_1K_3K_5/k_1$
V Random (No EAC) Ter	$1/k_1$	K_8/k_1	0	K_7/k_1	K_6K_8/k_1	K_5K_8/k_1	K_4K_7/k_1	$K_1K_4K_7/k_1$
VI Random AB Ter	$1/k_1$	0	0	K_5/k_1	0	K_4K_5/k_1	K_3K_5/k_1	$K_1K_3K_5/k_1$
VII Random BC (Equilibrium A) Ter	$1/k_1$	0	K_5/k_1	K_4/k_1	0	0	K_2K_4/k_1	$K_1K_3K_5/k_1$
VIII Random BC (Steady State A) Ter	$1/k_3$	$1/k_1$	K_5/k_3	K_4/k_3	0	0	K_2K_4/k_3	$k_2K_3K_5/k_1k_3$
IX Equilibrium Ordered Ter	$1/k_1$	0	0	K_3/k_1	0	0	K_2K_3/k_1	$K_1K_2K_3/k_1$
X Hexa Uni Uni Ping Pong	$1/k_3 + 1/k_5 + 1/k_7$	$(k_2+k_3)/k_1k_3$	$(k_6+k_7)/k_5k_7$	$(k_{10}+k_{11})/k_9k_{11}$	0	0	0	0
XI Ordered Bi Uni Uni Bi Ping Pong	$1/k_5 + 1/k_9 + 1/k_{11}$	$1/k_1$	$(k_4+k_3)/k_3k_5$	$(k_8+k_9)/k_7k_9$	$k_2(k_4+k_5)/k_1k_3k_5$	0	$k_6(k_8+k_9)/k_5k_7k_9$	0
XII Ordered Uni Uni Bi Bi Ping Pong	$1/k_3 + 1/k_9 + 1/k_{11}$	$(k_2+k_3)/k_1k_3$	$1/k_5$	$(k_8+k_9)/k_7k_9$	0	0	0	0
XIII Random Bi Uni Uni Bi Ping Pong	$1/k_5 + 1/k_9 + 1/k_{11}$	K_4'/k_5	K_3/k_5	$(k_8+k_9)/k_7k_9$	K_1K_3/k_9	0	0	0
XIV Random Uni Uni Bi Bi Ping Pong	$1/k_3 + 1/k_5 + 1/k_7$	$(k_2+k_3)/k_1k_3$	K_4/k_5	K_3/k_5	0	0	K_1K_3/k_5	0

TABLE IV
INITIAL-RATE PATTERNS FOR THREE-SUBSTRATE MECHANISMS[a]

Class of mechanism[b]	Plot		
	1/A	1/B	1/C
I	I[c]	I	I
II	I	I	I
III	I	I	I
IV	I	I	I
V	I	I	I
VI	I	I	0[d]
VII	I	I	I
VIII	I	I	I
IX	I	I	0
X	P[e]	P	P
XI	I	I	P
XII	P	I	I
XIII	I	I	P
XIV	P	I	I

[a] Based on the experimental protocol described by H. J. Fromm [*Biochim. Biophys. Acta* **139**, 221 (1967)] in which the two substrates not being plotted are varied in a constant ratio at concentrations in the range of their Michaelis constants.
[b] The mechanisms are listed in Table III.
[c] I refers to a double-reciprocal plot in which the lines intersect to the left of the $1/v$ axis.
[d] 0 refers to a double-reciprocal plot in which the lines intersect on the $1/v$ axis.
[e] P refers to a double-reciprocal plot in which the lines are parallel.

varying B and C in a constant ratio where $B = aC$ is illustrated by that for a Rapid Equilibrium Random mechanism.

$$\frac{1}{v} = \phi_0 + \frac{\phi_A}{A} + \frac{\phi_B}{B} + \frac{\phi_C}{C} + \frac{\phi_{AB}}{aAC} + \frac{\phi_{BC}}{aC^2} + \frac{\phi_{AC}}{AC} + \frac{\phi_{ABC}}{aAC^2} \quad (13)$$

The replots of slopes and intercepts versus 1/C (or 1/B) will be parabolic-up. The extrapolated portion of the curve at infinite C allows evaluation of the kinetic constants as was done for two-substrate systems. The linearity and intercept values of the slope and intercept replots depend on the terms present in the rate equation describing a particular mechanism. The mechanisms of Classes III through X exhibit unique replot behavior. All except VIII have at least one replot that intersects the

---

Footnote to TABLE III:[a] Absence of a particular ϕ term is indicated by a zero. For many of the ϕ values, more than one combination of dissociation constants is possible. These are simply the product of the different paths for formation of the quaternary complex from any given enzyme form. The majority of the mechanisms are listed in this volume [3]; the others are indicated by their name and were discussed by F. B. Rudolph and F. C. Clayton [*Abstr. Southwestern Am. Chem. Soc. Meet.*, A116 (1974)].

TABLE V
Graphical Method for Differentiating between Various Three-Substrate Mechanisms[a]

Class of mechanisms[b]	Substrate A		Substrate B		Substrate C	
	Slope	Intercept	Slope	Intercept	Slope	Intercept
I	N[c]	N	N	N	N	N
II	N	N	N	N	N	N
III	N	N	NO[d]	L[e]	N	N
IV	N	N	NO	N	N	L
V	N	N	NO	N	N	N
VI	NO	N	NO	N	N	A[f]
VII	NO	N	N	L	N	L
VIII	N	N	N	L	N	L
IX	NO	N	NO	L	N	A
X	L	L	L	L	L	L
XI	L	L	L	L	L	N
XII	L	N	L	L	L	L
XIII	L	L	L	L	L	N
XIV	L	N	L	L	L	L

[a] The double-reciprocal plots for each substrate are made by the procedure described by H. J. Fromm [*Biochim. Biophys. Acta* **139**, 221 (1967)] as in Table IV and the slopes and intercepts for the different fixed concentrations are determined and then plotted against the reciprocal of one of the fixed concentrations.
[b] The numbers refer to the mechanisms listed in Table III.
[c] N refers to nonlinear replots with nonzero intercepts on the y axis.
[d] NO refers to nonlinear replots that intersect the origin.
[e] L refers to nonlinear replots with nonzero intercepts on the y axis.
[f] A refers to a case in which the reciprocal plot intersects on the axis.

origin. This is frequently a good discriminating test. In some cases, the nonlinearity of various replots may not be conclusive if the ranges of substrate concentrations are too small. This problem is similar to that encountered in evaluation of the ϕ constants.

Another approach has been suggested by Dalziel[13] where the concentrations of all the substrates are varied and plots are made of the varied substrate at different fixed nonsaturating levels of the other substrates. Secondary plots of the slopes and intercepts versus the concentration of a fixed substrate are made and the slopes and intercepts of the secondary plots are then plotted against the other fixed substrate in a tertiary plot to evaluate each ϕ value listed in Table III. For details of the procedure, the reader is referred to the original reference.[13]

[13] K. Dalziel, *Biochem. J.* **114**, 547 (1969).

If any of the ϕ values are zero in a particular mechanism, this will be indicated in the secondary and tertiary plots. One problem with this type of analysis is that it requires a considerable amount of extremely accurate data. Errors may be a serious problem in the tertiary plots, particularly in evaluating whether a particular ϕ value is truly zero.

A third experimental protocol that may be used in initial-rate studies involves determination of the effect of substrate saturation on the kinetic behavior. As shown in Table IV, all initial-rate reciprocal plots for the sequential mechanisms are intersecting. If, however, one of the substrates is raised to a saturating level (100 × K_m) so that all terms in the rate equation containing that substrate are effectively zero, parallel patterns may be observed. This technique, which was first suggested by Frieden,[14] may be useful in some cases, but with many enzymes such high substrate levels are impossible to achieve owing to low solubility or inhibitory effects. The effect of saturation of the enzyme by various substrates on each class of mechanism is listed in Table VI. As is indicated in the table, the absence of some terms in the rate equations for Classes IV, VI, VII, and VIII mechanisms cause the rate of reaction of the enzyme to become independent of the level of certain of the substrates. Additional parallel patterns are generated in the partial Ping Pong mechanisms (XI–XIV), but no additional differentiation is possible with this procedure. This technique may be of use for particular systems, but the point at which saturation truly occurs may be difficult to achieve. Another problem with this approach is possible changes in ionic strength with charged substrates at high concentrations.

It is apparent from initial rate studies alone that mechanism Classes III–X may be either confirmed or eliminated. Classes I and II give the same initial rate and replot patterns, but evaluation of the ϕ constants or effect of saturation by substrates will allow distinction between these two classes. The mechanisms in Class II are the Ordered, Theorell–Chance, and a Random AC mechanism in which all ternary complexes are present. Only formation of the EB complex does not occur. The existence of this mechanism, which has identical initial rate behavior as the Ordered and Theorell–Chance mechanism, requires other experiments, such as inhibition or exchange studies, to confirm an ordered binding sequence.

Evaluation of ϕ and Haldane Relationships

ϕ *Relationships.* Dalziel[4,13] has described certain relationships between kinetic constants for forward and reverse reactions that must be obeyed for initial rate studies for two- and three-substrate enzymes. These relationships are termed ϕ relations and are useful in distinguishing

[14] C. Frieden, *J. Biol. Chem.* **234**, 2891 (1959).

TABLE VI
Effect of Saturation by a Substrate on Reciprocal Plots for Three-Substrate Enzymes[a]

Class of mechanism[b]	Saturating substrate	Plot[c]		
		1/A	1/B	1/C
I	A	—	I[d]	I
	B	I	—	I
	C	I	I	—
II	A	—	I	I
	B	P[e]	—	P
	C	I	I	—
III	A	—	I	IA[f]
	B	P	—	P
	C	IA	I	—
IV	A	—	I	IA
	B	I	—	I
	C	O[g]	O	—
V	A	—	I	IA
	B	I	—	I
	C	IA	I	—
VI	A	—	I	IA
	B	I	—	IA
	C	O	O	—
VII	A	—	I	I
	B	O	—	O
	C	O	O	—
VIII	A	—	I	I
	B	P	—	P
	C	P	P	—
IX	A	—	I	IA
	B	O	—	O
	C	O	O	—
X	All still parallel—no diagnostic value			
XI	A	—	P	P
	B	P	—	P
	C	I	I	—
XII	A	—	I	I
	B	P	—	P
	C	P	P	—
XIII	A	—	P	P
	B	P	—	P
	C	I	I	—

(continued)

TABLE VI (continued)

Class of mechanism[b]	Saturating substrate	Plot[c] 1/A	1/B	1/C
XIV	A	—	I	I
	B	P	—	P
	C	P	P	—

[a] The experimental protocol involves raising one substrate to a very high constant level while varying the other two substrates in the range of their K_m's.
[b] The mechanisms are as listed in Table III.
[c] Refers to a double-reciprocal plot of reciprocal of initial velocity versus the indicated reciprocal substrate concentration.
[d] I indicates a family of intersecting lines.
[e] P indicates a family of parallel lines.
[f] IA indicates a family of lines that intercept on the $1/v$ axis.
[g] O indicates that the rate equation predicts that velocity is independent of concentration of one or more substrates. This condition is experimentally impossible and merely indicates that the indicated set of conditions will not afford any useful diagnostic information.

between kinetic mechanisms. However, the rapid equilibrium Random mechanism can have any ϕ relationship, and thus the technique can serve only to rule out ordered-type mechanisms. The ϕ relationships for various mechanisms are listed in several references[4,5,13] and will not be treated in detail here.

Isomerization mechanisms as discussed in this volume [3] give rise to altered rate equations. If the central enzyme substrate complexes isomerize, the isomerization cannot be detected from initial-rate studies. If the isomerization occurs with other enzyme forms, the evaluation of the ϕ relationships will provide insight into the mechanism. A table of such mechanisms and their ϕ relationships has been reported.[5]

Haldane Relationships. Haldane[15] has shown that a relationship exists between certain kinetic parameters for a kinetic mechanism and the apparent equilibrium constant, K_{eq}. For the simple Uni Uni mechanism, $K_{eq} = k_1 k_3 / k_2 k_4 = V_1 K_p / V_2 K_a = K_{ip}/K_{ia}$. Alberty[16] has shown that such relationships can be used to make a choice between the Theorell–Chance, Ordered Bi Bi and the rapid equilibrium Random Bi Bi mechanisms. Cleland[3] has presented a general equation relating K_{eq} to the kinetic parameters,

$$K_{eq} = \frac{(V_1)^n K_{(p)} K_{(q)} K_{(r[}}{(V_2)^n K_{(a)} K_{(b)} K_{(c)}} \qquad (14)$$

[15] J. B. S. Haldane, "Enzymes." Longmans, Green and Co., London, 1930.
[16] R. A. Alberty, *J. Am. Chem. Soc.* **75**, 1928 (1953).

The $K_{(a)}$ may be either K_a or K_{ia} and so forth. There is always at least one Haldane relationship where $n = 1$, but, depending on the mechanism, n may have several values.

Experimentally, the forward and reverse reactions are studied and the kinetic parameters are calculated. The data for a proposed mechanism should give the K_{eq}, and all the different Haldanes for the same mechanism should give similar numbers for the mechanism to be consistent with the data. The Haldane relationships for a number of mechanisms are listed elsewhere[5] and will not be tabulated here. One obvious limitation to using the Haldane relationship is that precise kinetic and equilibrium constant data are required. In addition, only those enzymes that exhibit kinetic reversibility can be studied. Similar shortcomings arise when attempts are made to employ the Dalziel ϕ relationships.

Examples of Initial-Rate Studies

A large number of initial-rate studies have been carried out on multisubstrate enzymes, particularly bireactant systems. We will not try to catalog them, but a few examples of where initial-rate gave insight into the mechanism will be presented.

Rabbit muscle phosphofractokinase is an example of an enzyme with a K_{ia} smaller than its K_a so that initial-rate patterns appear nearly parallel if substrate levels are studied above 0.5 K_m. In the study by Hanson et al.,[17] the substrate concentrations were maintained quite low and clearly intersecting lines were observed, confirming a sequential mechanism. In general, whether a mechanism is Ping Pong or sequential is an important factor to be obtained from initial rate studies with bireactant enzymes. The analysis of points of convergence, ϕ relationships, and Haldane values will confirm or exclude mechanisms, but other types of experiments as described in this volume are usually necessary to really differentiate between mechanisms.

With three-substrate systems, a good deal of information can usually be obtained from initial-rate data alone, depending on the nature of the kinetic mechanism. Examples of three-substrate enzymes that have kinetic behavior fitting many of the listed sequential mechanisms listed in this chapter are available from the literature (see this volume [3]). No examples of Class III, V, or IX mechanisms have yet been found, but these all appear to be reasonable mechanisms and are consistent with chemical theory.

Several enzymes have been suggested to have totally random binding

---

[17] R. L. Hanson, F. B. Rudolph, and H. A. Lardy, *J. Biol. Chem.* **248**, 7852 (1973).

mechanisms (Class I). These include *Escherichia coli* glutamine synthetase,[18] adenylosuccinate synthetase,[19] glutathione synthetase,[20] tetrahydrofolate synthetase,[21] and glutamate dehydrogenase.[22] Glutamate dehydrogenase was originally thought to have an Ordered (Class II) binding mechanism based on apparent saturation of the enzyme by ammonia with resultant nearly parallel reciprocal plots, as discussed in the text. It has more recently been shown, using inhibitor techniques, that the binding sequence is random. For the various synthetases, the mechanisms seem well established from studies of inhibition or isotope exchange. Initial rate studies were of particular value in the study on adenylosuccinate synthetase. Replots of the initial rate data allowed exclusion of partially random mechanisms.

The existence of the partially random mechanism in Class II (IIb) requires that more experiments than simple initial-velocity experiments be done to differentiate within this class. This mechanism will exhibit the same initial-rate behavior as the ordered mechanisms [II(a), (a')] but can easily be distinguished by inhibition studies. An example of this problem is with 6-phosphogluconate dehydrogenase.[23] From evaluation of the ϕ values alone, it was concluded that the mechanism was ordered. The kinetic behavior was consistent with a Class II mechanism, but exchange or inhibition experiments will be required to prove which specific mechanism is involved. A similar problem exists with saccharopine dehydrogenase,[24] where again an ordered mechanism was suggested for initial-rate studies alone. It should be noted that only in these subclasses of the Class II mechanisms can distinctions be made, since both steady-state and equilibrium mechanisms are described by the same initial-rate equation. The mechanism of formylglycinamide ribonucleotide (FGAR) amidotransferase[25] has been suggested to be Random BC (Class VIII, steady state A). This assumption was based upon the use of constant saturating levels of one substrate and observing the kinetic behavior with respect to the other substrates. The Random BC mechanism (VIII) should exhibit two sets of parallel lines in reciprocal plots when substrates B and C are each saturating. For the FGAR amidotransferase, two plots were "nearly" parallel. The inhibition and initial rate data are, however,

---

[18] F. C. Wedler and P. D. Boyer, *J. Biol. Chem.* **247**, 984 (1972).
[19] F. B. Rudolph and H. J. Fromm, *J. Biol. Chem.* **244**, 3832 (1969).
[20] A. Wendel, and H. Heinke, *Hoppe-Seyler's Z. Physiol. Chem.* **356**, 33 (1975).
[21] B. K. Joyce, and R. H. Hines, *J. Biol. Chem.* **241**, 5716 (1966).
[22] P. C. Engel and K. Dalziel, *Biochem. J.* **118**, 409 (1970).
[23] R. H. Villet and K. Dalziel, *Eur. J. Biochem.* **24**, 244 (1972).
[24] M. Fujioka and Y. Nakatani, *J. Biol. Chem.* **249**, 6886 (1974).
[25] H. C. Li and J. M. Buchanan, *J. Biol. Chem.* **246**, 4720 (1971).

also totally consistent with an ordered addition of substrates [Class II(a)]. More data would probably be required to distinguish between these two possibilities.

A Random AB mechanism (VI) has been postulated from initial-rate behavior for sea urchin sperm guanylate cyclase when Mn^{2+} is a reactant, and multiple nucleotide sites are considered.[26]

Interestingly, the Random BC mechanisms (VII or VIII) have a number of examples. *E. coli* CoA synthetase,[27] citrate cleavage enzyme[28] (Class VIII), γ-glutamylcysteine synthetase[29] (Class VII), and acetate:CoA ligase[30] have all been suggested to bind substrates in that order.

Limitations

The techniques described in this chapter do not necessarily prove certain mechanisms, but at least they allow separation of mechanisms into groups that can be analyzed by other techniques described in this volume. In the case of three-substrate mechanisms, initial-rate techniques can very often completely define the kinetic mechanism. Freiden[31] has presented a rapid equilibrium Ordered Bi Bi subsite mechanism (this volume [3] A4) that cannot be differentiated from the Rapid Equilibrium Random Bi Bi mechanism by initial rate studies. Analogous mechanisms can be suggested for terreactant systems. Techniques that will afford differentiation between these types of mechanisms are discussed in this volume [18].

Concluding Remarks

Initial rate studies are the most useful single method for analyzing kinetic data. They allow determination of the types of complexes involved in an enzyme reaction and can afford insight into the order of addition of substrates, particularly with in terreactant systems. We have attempted to point out the advantages and limitations of each technique discussed in this chapter. With careful experimentation, critical analysis, and use of other techniques described in this volume, the evaluation of kinetic mechanisms should be readily accomplished.

[26] D. L. Garbers, J. G. Hardman, and F. B. Rudolph, *Biochemistry* **13**, 4166 (1974).
[27] F. J. Moffet and W. A. Bridger, *Can. J. Biochem.* **51**, 44 (1973).
[28] K. M. Plowman and W. W. Cleland, *J. Biol. Chem.* **242**, 4239 (1967).
[29] B. P. Yip and F. B. Rudolph, *J. Biol. Chem.* **251**, 3563 (1976).
[30] W. W. Farror and K. M. Plowman, *Fed. Proc., Fed. Am. Soc. Exp. Biol.* **29**, 425 (1970).
[31] C. Frieden, *Biochem. Biophys. Res. Commun.* **72**, 55 (1976).

Acknowledgments

This study was supported by Grants NS 10546 and CA 14030 from the National Institutes of Health, PCM 77-09018 from the National Science Foundation, and C-582 from the Robert A. Welch Foundation.

[8] Kinetic Analysis of Progress Curves

By BRUNO A. ORSI and KEITH F. TIPTON

The analysis of the progress curves of enzyme-catalyzed reactions has often been regarded as being an attractive alternative to the initial-rate methods for the determination of kinetic parameters. The attraction lies in the use of the entire progress curve rather than just a small part of it, the initial rate, which together with the final part, is the most difficult portion to measure accurately. The method has the further advantage that it should be readily amenable to direct on-line computer analysis of the results by essentially simple procedures. Another often quoted advantage of this approach is that it should allow the kinetic parameters to be determined from the results of a single experiment, but this is in fact not the case since any valid interpretation must involve data obtained from a number of different progress curves. Despite the attractions of this method it has not been used a great deal in practice, although it has provided a relatively fertile field for theoretical studies. The reason for this is that there are several possible causes of a decrease in the velocity of an enzyme-catalyzed reaction with increasing time and not all of these are amenable to kinetic analysis. In this chapter we will develop the equations that can be used for the analysis of progress curves and examine their usefulness.

Theory

The progress curve of an enzyme-catalyzed reaction typically takes the form shown in Fig. 1, starting off linear (the initial-rate phase) but falling off with increasing time. If it is assumed that the only cause of this fall-off is the depletion of substrate, it should be possible to determine the rates of the reaction at different concentrations of the substrate remaining by drawing a series of tangents to the curve, but it is not easy to do this with any accuracy without employing some simplifying assumptions that are not statistically valid.[1] An alternative and preferable procedure is to

---

[1] A. Cornish-Bowden, "Principles of Enzyme Kinetics," pp. 150–152. Butterworth, London, 1976.

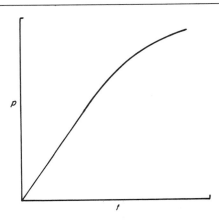

FIG. 1. The concentration of product, p, as a function of time for an enzyme reaction.

integrate the initial-rate equation to obtain an equation that describes the entire progress curve. The form of the integrated-rate equation obtained will depend upon the kinetic mechanism, and in this section we will examine the behavior of mechanisms of increasing complexity.

The initial-rate equation for a single-substrate reaction can be written as

$$v = \frac{dp}{dt} = \frac{-ds}{dt} = \frac{V}{1 + (K_m/s)} \tag{1}$$

where V is the maximum velocity and K_m the Michaelis constant; if the rate of the reaction is decreasing with increasing time only because the substrate concentration is falling, the rate at any point on the progress curve will be given by:

$$\frac{dp}{dt} = \frac{V}{1 + [K_m/(s_0 - p)]} \tag{2}$$

where s_0 is the initial substrate concentration and p the product concentration. This equation may be rearranged and integrated in the following way:

$$Vt = \int_0^p \left[1 + \frac{K_m}{s_0 - p}\right] dp + \text{constant} \tag{3}$$

where the constant can be determined from the conditions $p = 0$ when $t = 0$ giving

$$Vt = p + K_m \ln \frac{s_0}{s_0 - p} \tag{4}$$

An equation of this form was first deduced by Henri,[2] and it is sometimes referred to as the Henri equation. Since the initial substrate concentration (s_0) can be expressed in terms of the substrate remaining at any given time (s) and the product concentration

$$s_0 = s + p \tag{5}$$

Eq. (4) can also be written as

$$Vt = (s_0 - s) + K_m \ln \frac{s_0}{s} \tag{6}$$

Equation (6) is often the more useful form because it avoids any possible confusion that might arise in experiments where the progress curve is determined with a fixed amount of the product initially present. This equation can be rearranged to give

$$\frac{s_0 - s}{t} = V - K_m \frac{1}{t} \ln \frac{s_0}{s} \tag{7}$$

and thus a graph of $(s_0 - s)/t$ against $1/t \ln (s_0/s)$ or $1/t \cdot 2.303 \log (s_0/s)$ should give a straight line with a slope of $-K_m$ and intercepts on the vertical and horizontal axes of V and V/K_m; respectively, as shown in Fig. 2. A similar plot [Eq. (8)] was probably first used by Walker and Schmidt[3] in

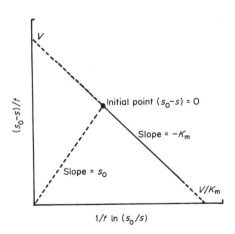

FIG. 2. Plot of the integrated rate equation [Eq. (7)] for a single-substrate enzyme reaction. The dashed line drawn through the origin with a slope of s_0 intersects the experimentally determined line at the initial point [R. J. Foster and C. Niemann, *Proc. Natl. Acad. Sci. U.S.A.* **39**, 999 (1953)].

[2] V. Henri, *C. R. Acad. Sci.* **135**, 916 (1902).
[3] A. C. Walker and C. L. A. Schmidt, *Arch. Biochem. Biophys.* **5**, 445 (1944).

studies on histidine ammonia-lyase (EC 4.3.1.3). If Eq. (7) is obeyed, the same line will be obtained if the data from a number of progress curves with different initial substrate concentrations are used. Foster and Niemann[4] in 1953 pointed out that a straight line drawn through the origin of this graph with a slope of s_0 will cut the experimentally determined line at a point corresponding to the initial point ($p = 0$) (see Fig. 2).

Equation (6) can be rearranged in other ways that will give rise to linear plots from which the kinetic parameters may be determined:

$$\frac{1}{t} \ln \frac{s_0}{s} = -\frac{1}{K_m} \frac{s_0 - s}{t} + \frac{V}{K_m} \tag{8}$$

$$\frac{t}{\ln (s_0/s)} = \frac{1}{V} \frac{s_0 - s}{\ln (s_0/s)} + \frac{K_m}{V} \tag{9}$$

$$\frac{t}{s_0 - s} = \frac{K_m}{V} \frac{\ln (s_0/s)}{s_0 - s} + \frac{1}{V} \tag{10}$$

These linear transformations are analogous to those that can be used with the Michaelis–Menten equation. Equation (7) is analogous to the plot of v against v/s, and the rearrangement to give Eq. (8) corresponds to a plot of v/s against v and will give a slope of $-1/K_m$ and a vertical axis intercept of V/K_m. Equation (9) resembles the plot of s/v against s and will give a slope of $1/V$ and a vertical axis intercept of K_m/V, and Eq. (10) corresponds to the $1/v$ against $1/s$ plot giving a slope of K_m/V and a vertical axis intercept of $1/V$. In the following sections the transformation corresponding to that in Eq. (7) will be used.

Under conditions where $s_0 \gg K_m$ and the assay time is relatively short so that there is no significant change in this relationship, s may be regarded as being constant and Eq. (4) simplifies to

$$Vt = p \tag{11}$$

which defines a linear (zero-order) portion of the progress curve. At the other extreme, where s_0, and hence p, is very much less than K_m, Eq. (2) simplifies to the first-order equation

$$dp/dt = (V/K_m)(s_0 - p) \tag{12}$$

which can be integrated to give

$$\ln (s_0/s) = Vt/K_m \tag{13}$$

and thus a graph of $\ln (s_0/s)$ against t should give a straight line that passes through the origin and has a slope of V/K_m. Under these conditions it will not be possible to determine V and K_m separately.

[4] R. J. Foster and C. Niemann, *Proc. Natl. Acad. Sci. U.S.A.* **39**, 999 (1953).

Effects of Reversible Inhibitors

The type of inhibition and the inhibitor constants may be determined from a series of progress curves determined in the presence of different initial concentrations of the inhibitor.

Competitive Inhibition

For a competitive inhibitor the Michaelis–Menten equation may be written as

$$\frac{dp}{dt} = \frac{V}{1 + [K_m/(s_0 - p)][1 + (i/K_i)]} \tag{14}$$

where i is the inhibitor concentration and K_i the inhibitor constant. This equation may be integrated and rearranged using the relationship shown in Eq. (5) to give

$$\frac{s_0 - s}{t} = V - K_m [1 + (i/K_i)] \frac{1}{t} \ln \frac{s_0}{s} \tag{15}$$

Thus the presence of a competitive inhibitor will affect the slopes but not the vertical axis intercepts of plots of $(s_0 - s)/t$ against $1/t \ln (s_0/s)$ as shown in Fig. 3a. The slope of each line will be $-K_m (1 + i/K_i)$ and therefore a graph of $-$slope against i will intersect the vertical axis at a value corresponding to K_m, and if extended it will cut the horizontal axis at a value of $-K_i$ (Fig. 3b).

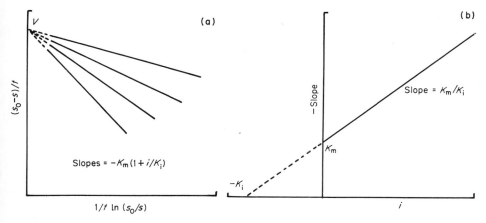

FIG. 3. (a) Plot of the integrated rate equation [Eq. (15)] for a single-substrate enzyme reaction in the presence of a competitive inhibitor. (b) Replot of the slopes against the inhibitor concentration.

Uncompetitive Inhibition

The equation for uncompetitive inhibition

$$\frac{dp}{dt} = \frac{V}{[1 + (i/K_i)] + [K_m/(s_0 - p)]} \quad (16)$$

can be integrated and rearranged to give

$$\frac{s_0 - s}{t} = \frac{V}{1 + (i/K_i)} - \frac{K_m}{1 + (i/K_i)} \frac{1}{t} \ln \frac{s_0}{s} \quad (17)$$

and thus the presence of an uncompetitive inhibitor will affect both the slopes and the vertical axis intercepts, but not the intercept on the horizontal axis of graphs of $(s_0 - s)/t$ against $1/t \ln (s_0/s)$ as shown in Fig. 4a. Replots of $-(i/\text{slope})$ against i will allow K_m and K_i to be determined (Fig. 4b), and replots of the $-(1/\text{vertical intercept})$ against i will give V and K_i (Fig. 4c).

Mixed and Noncompetitive Inhibition

The equation for mixed inhibition

$$\frac{dp}{dt} = \frac{V}{[1 + (i/K_i')] + \{[K_m/s_0 - p)][1 + (i/K_i)]\}} \quad (18)$$

can be integrated to give

$$\frac{s_0 - s}{t} = \frac{V}{1 + (i/K_i')} - \frac{K_m[1 + (i/K_i)]}{1 + (i/K_i')} \frac{1}{t} \ln \frac{s_0}{s} \quad (19)$$

and thus the presence of a mixed inhibitor will affect the slopes and the vertical and horizontal axis intercepts as shown in Figs. 5a and 5b. Replots of the reciprocals of intercepts on the vertical and horizontal axes against i will allow all the constants to be determined as shown in Figs. 5c and 5d.

In the special case of true noncompetitive inhibition, which will be given when $K_i = K_i'$ [Eq. (18)], the integrated Eq. (19) becomes

$$\frac{s_0 - s}{t} = \frac{V}{1 + (i/K_i)} - K_m \frac{1}{t} \ln \frac{s_0}{s} \quad (20)$$

Thus graphs of $(s_0 - s)/t$ against $1/t \ln (s_0/s)$ at different inhibitor concentrations will give a family of parallel lines (slope $= -K_m$) as shown in Fig. 6a, and the values of K_i and V can be determined from replots of the reciprocal of the vertical axis intercept against i (Fig. 6b).

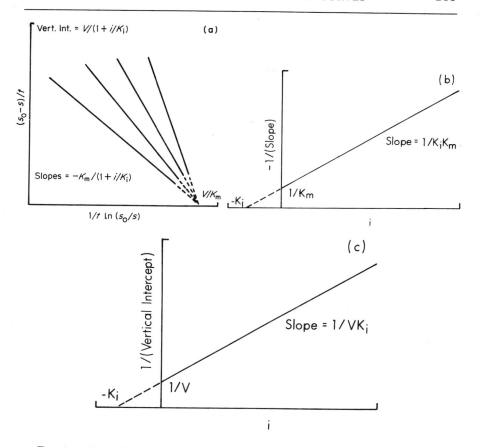

FIG. 4. (a) Plot of the integrated rate equation [Eq. (17)] for a single-substrate enzyme reaction in the presence of an uncompetitive inhibitor. Replots of the reciprocals of slopes and vertical intercepts against the inhibitor concentration are shown in (b) and (c), respectively.

Substrate Inhibition

In the cases where high concentrations of the substrate cause inhibition, the equation is

$$\frac{dp}{dt} = \frac{V}{1 + [K_m/(s_0 - p)] + [(s_0 - p)/K_i]} \tag{21}$$

which can be integrated and rearranged to give

$$\frac{s_0 - s}{t} = V - \frac{(s_0^2 - s^2)}{2K_i}\left(\frac{1}{t}\right) - K_m\left(\frac{1}{t}\right)\ln\frac{s_0}{s} \tag{22}$$

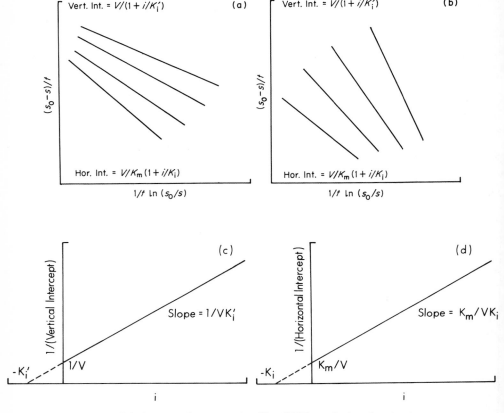

FIG. 5. Plots of the integrated rate equation [Eq. (19)] for a single-substrate enzyme reaction in the presence of a mixed inhibitor; (a) $K_i < K_i'$ and (b) $K_i > K_i'$. Reports of reciprocals of vertical and horizontal intercepts against the inhibitor concentration are shown in (c) and (d), respectively.

which results in a nonlinear plot.[5] In cases such as this, if s_0 is high enough there is an increase in rate as the reaction proceeds, after which the reaction decreases as expected and the point at which such an apparent autocatalytic reaction occurs corresponds to $s_0 > (K_m K_i)^{1/2}$.

Product Inhibition

If the product of the reaction is a reversible inhibitor, its formation will contribute to the falloff in the progress curve and the form of the integrated rate equation will depend on the type of inhibition.

[5] K. J. Laidler and P. S. Bunting, "The Chemical Kinetics of Enzyme Action." Oxford Univ. Press, London and New York, 1973.

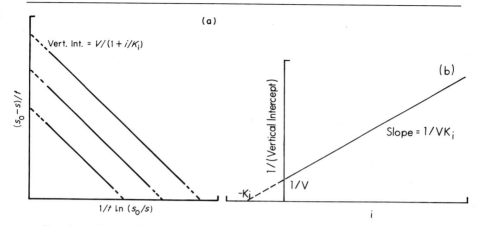

FIG. 6. (a) Plot of the integrated rate equation [Eq. (20)] for a single-substrate enzyme reaction in the presence of a true noncompetitive inhibitor. (b) Replots of reciprocal of the vertical intercepts against the inhibitor concentration.

COMPETITIVE PRODUCT INHIBITION

The rate equation will be given by substituting p for i in Eq. (14), and integration of this gives

$$Vt = \left(1 - \frac{K_m}{K_i}\right) p + K_m \left(1 + \frac{s_0}{K_i}\right) \ln \frac{s_0}{s_0 - p} \qquad (23)$$

which may be rearranged using the relationship shown in Eq. (5) to give

$$\frac{s_0 - s}{t} = \frac{V}{1 - (K_m/K_i)} - K_m \frac{K_i + s_0}{K_i - K_m} \frac{1}{t} \ln \frac{s_0}{s} \qquad (24)$$

and thus a graph of $(s_0 - s)/t$ against $1/t \ln (s_0/s)$ will give a straight line of slope $-K_m [(K_i + s_0)/(K_i - K_m)]$ and an intercept on the vertical axis of $V/(1 - K_m/K_i)$. The slope of the graph will be positive and the intercept negative if $K_i < K_m$; if $K_i > K_m$, the opposite effects will be observed; and if $K_i = K_m$, a vertical line (infinite slope and intercept) will be obtained (see Fig. 7). Thus this case can be indistinguishable from the simple case represented by Eq. (7) on the basis of the analysis of a single progress curve. However, if a series of plots are obtained from progress curves determined with different initial substrate concentrations a family of lines will be obtained that will all intersect on the vertical axis at a value of $\pm V/(1 - K_m/K_i)$ provided that K_m and K_i are different, as shown in Fig. 8a for the case in which $K_i > K_m$. The slopes of these lines will be given by

$$\pm \text{slope} = K_m \frac{K_i + s_0}{K_i - K_m} = \frac{K_m s_0}{K_i - K_m} + \frac{K_m K_i}{K_i - K_m} \qquad (25)$$

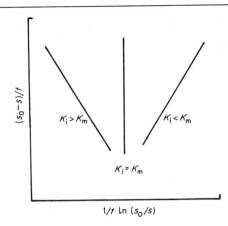

FIG. 7. Plots of the integrated rate equation [Eq. (24)] for a single-substrate enzyme reaction where the single product causes competitive inhibition.

Thus a graph of ± slope against s_0 will extend to cut the baseline at a value of $-K_i$ (Fig. 8b), and this value can be used to calculate K_m and V from the slopes and intercepts of the lines shown in Fig. 8a.

If an initial concentration of product p_0 is present at the start of the reaction, Eq. (24) becomes

$$\frac{s_0 - s}{t} = \frac{V}{1 - (K_m/K_i)} - \frac{K_m(K_i + s_0 + p_0)}{K_i - K_m} \frac{1}{t} \ln \frac{s_0}{s} \qquad (26)$$

and only the slope of the line is affected.

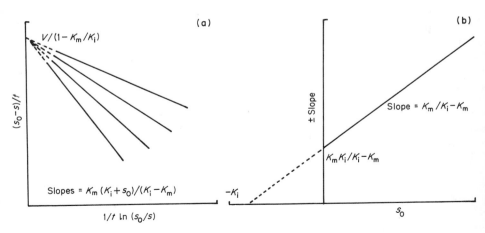

FIG. 8. (a) Plots of the integrated rate equation [Eq. (24)] for competitive product inhibition at different initial substrate concentrations. (b) Replot of the slopes against s_0.

An alternative method for analyzing the results is that of Niemann and his co-workers.[4,6,7] As in the case of an enzyme obeying Eq. (7), a line drawn through the origin with a slope of s_0 will cut the experimental line at a value corresponding to the initial conditions ($p = 0$). Thus if a series of these intersection points are determined for the lines obtained from progress curves at different values of s_0, a line joining these initial points will correspond to that which would be obtained in the absence of product inhibition. Such a plot is shown in Fig. 9 for the example where $K_i < K_m$. This type of analysis will not give the value of K_i, but this value may be determined by a similar technique if the initial points are determined for several families of progress curves, and this is repeated at a series of fixed initial concentrations of the product and the data are then analyzed in the way described in previously.

If two products are produced, both of which are competitive, with inhibitor constants of K^p and K_i^q, similar equations can be used in which $(1/K_i^p + 1/K_i^q)$ is substituted for $1/K_i$.

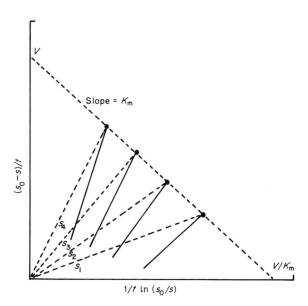

FIG. 9. The determination of the kinetic constants of an enzyme subject to competitive product inhibition by the method of Foster and Niemann.[4] $S_1 - S_4$ represent different initial substrate concentrations.

[6] R. J. Foster and C. Niemann, *J. Am. Chem. Soc.* **77**, 1886 (1955).
[7] R. R. Jennings and C. Niemann, *J. Am. Chem. Soc.* **77**, 5432 (1955).

Uncompetitive Product Inhibition

The rate equation will be given by substituting p for i in Eq. (16), and this may be integrated to give

$$Vt = p + \frac{p^2}{2K_i} + K_m \ln \frac{s_0}{s_0 - p} \quad (27)$$

and the appearance of the p^2 term in this equation precludes simple rearrangement to produce a linear graph. Rearrangement using the relationship given in Eq. (5) gives

$$\frac{s_0 - s}{t} = \frac{V}{1 + [(s_0 - s)/2K_i]} - \frac{K_m}{1 + [(s_0 - s)/2K_i]} \left(\frac{1}{t}\right) \ln \frac{s_0}{s} \quad (28)$$

The values of K_m and V may be determined from a series of plots determined at different values of s_0 by the method of Niemann et al.[4] and determining the initial points by the use of lines with slopes of s_0 drawn through the origin. The K_i value can be determined by experiments carried out with fixed amounts of the product p_0 initially present. In this case Eq. (28) becomes

$$\frac{s_0 - s}{t} = \frac{V}{1 + (p_0/K_i) + [(s_0 - s)/2K_i]}$$
$$- \frac{K_m}{1 + (p_0/K_i) + [(s_0 - s)/2K_i]} \frac{1}{t} \ln \frac{s_0}{s} \quad (29)$$

Mixed and Noncompetitive Product Inhibition

The rate equation for the mixed case will be given by substituting p for i in Eq. (18), which integrates to give

$$Vt = (1 - K_m/K_i)p + \frac{p^2}{2K_i'} + K_m \left(1 + \frac{s_0}{K_i}\right) \ln \frac{s_0}{s_0 - p} \quad (30)$$

This equation represents the general equation for all types of product inhibition. If $K_i' = \infty$, it simplifies to the form of the competitive Eq. (23); if $K_i = \infty$ the uncompetitive Eq. (27) is given, and the equation for true noncompetitive will be given when $K_i = K_i'$. Equation (30) may be rearranged, using the relationship given in Eq. (5), to give

$$\frac{s_0 - s}{t} = \frac{V}{1 + [(s_0 - s)/2K_i'] - (K_m/K_i)}$$
$$- \frac{K_m(K_i + s_0)}{K_i + [K_i(s_0 - s)/2K_i'] - K_m} \frac{1}{t} \ln \frac{s_0}{s} \quad (31)$$

This equation does not predict a linear relationship between $(s_0 - s)/t$

and $1/t \ln(s_0/s)$, since the slope and intercept will change with $(s_0 - s)$. The degree of curvature, however, will depend on the relative values of K_i and K_i' (being greatest when $K_i > K_i'$), and it may be difficult to detect in practice. As in the previous cases the values of K_m and V can be determined if the method of Niemann et al.[4] is applied to the results from a series of progress curves. The K_i values and type of inhibition can then be determined from a series of such plots obtained from progress curves in which fixed amounts of the product p_0 are originally present. In this case Eq. (31) becomes

$$\frac{s_0 - s}{t} = \frac{V}{1 + (p_0/K_i') + [(s_0 - s)/2K_i'] - (K_m/K_i)} - \frac{K_m(K_i + s_0 + p_0)}{K_i + (p_0 K_i/K_i') + [(s_0 - s)K_i/2K_i'] - K_m} \frac{1}{t} \ln \frac{s_0}{s} \quad (32)$$

More Complicated Cases

The equations derived so far have represented an essentially simple irreversible enzyme reaction involving only one substrate and one product (Uni Uni). Extension of the equations to more complicated, and perhaps more realistic, situations results in integrated rate equations of considerably greater complexity.

The Formation of More Than One Inhibitory Product

The situation in which there are two products of the reaction, both of which are competitive inhibitors of the reaction, has been mentioned in the section on competitive product inhibition; similar types of treatment can be applied if the type of inhibition given by the two products is different, but the integrated rate equations become very complicated in these cases. The results can be most conveniently analyzed by determining K_m and V by finding the initial points (product concentrations = 0) on each of a family of progress curves using the construction of Niemann et al.[4] and then determining the inhibition types and K_i values from these initial points in experiments where fixed amounts of each of the products are present initially, as described previously. Such an approach can also be used in cases where inhibition is nonlinear.

Reversible Reactions

If the reaction is reversible, the approach to equilibrium, where the net rate of the forward reaction is balanced by the rate of the backward reaction, will contribute to the observed falloff in the progress curve. For a simple single-substrate, single-product reversible reaction, the rate equation can be written as

$$\frac{dp}{dt} = \frac{V^f s}{K_m^f[1 + (p/K_m^r)] + s}$$

$$- \frac{V^r p}{K_m^r[1 + (s/K_m^f)] + p} = \frac{V^f K_m^r s - V^r K_m^f p}{K_m^r s + K_m^f p + K_m^f K_m^r} \quad (33)$$

where the superscripts f and r designate the kinetic constants for the forward and reverse reactions, respectively. If the concentration of the product at equilibrium is designated p_{eq}, the Haldane relationship will be given by

$$\frac{p_{eq}}{s_0 - p_{eq}} = \frac{V^f K_m^r}{V^r K_m^f} = K_{eq} \quad (34)$$

and Eq. (33) can then be written as

$$\frac{dp}{dt} = \frac{V^f K_m^r s_0 - V^f K_m^r s_0 p/p_{eq}}{K_m^r K_m^f + K_m^r s_0 + (K_m^f - K_m^r)p} \quad (35)$$

which can be rearranged to give

$$\frac{dp}{dt} = \frac{V^f K_m^r s_0 (p_{eq} - p)}{(K_m^r K_m^f + K_m^r s_0)p_{eq} + (K_m^f - K_m^r)p_{eq}^2 + (K_m^f - K_m^r)p_{eq}(p_{eq} - p)} \quad (36)$$

which can be integrated and rearranged to give

$$\left(\frac{V^f}{K_m^f} + \frac{V^r}{K_m^r}\right)t = \left(\frac{1}{K_m^f} - \frac{1}{K_m^r}\right)p$$
$$- \ln\left(1 - \frac{p}{p_{eq}}\right)\left\{1 + \frac{s_0}{K_m^r} + \frac{[(1/K_m^f) - (1/K_m^r)]V^r s_0}{[(V^f/K_m^f) + (V^r/K_m^r)]K_m^r}\right\} \quad (37)$$

This fearsome-looking equation can be simplified by using the Haldane relationship [Eq. (34)] to give[8]

$$\frac{V^f}{K_m^f}\left(1 + \frac{1}{K_{eq}}\right)t = \left(\frac{1}{K_m^f} - \frac{1}{K_m^r}\right)p$$
$$- \ln\left(1 - \frac{p}{p_{eq}}\right)\left\{1 + \frac{s_0}{K_m^r} + \frac{[(1/K_m^f) - (1/K_m^r)]s_0}{1 + K_{eq}}\right\} \quad (38)$$

which can be rearranged, using Eq. (5) to give

---

[8] R. A. Alberty, in "The Enzymes" (P. D. Boyer, H. Lardy, and K. Myrbäck, eds.), 2nd ed., Vol. 1, p. 151. Academic Press, New York, 1959.

$$\frac{s_0 - s}{t} = \frac{V^f[1 + (1/K_{eq})]}{1 - (K_m^f/K_m^r)}$$
$$+ \left[\frac{K_m^f K_m^r}{K_m^r - K_m^f} + \frac{(K_m^r + K_m^f K_{eq})s_0}{(K_m^r - K_m^f)(1 + K_{eq})}\right]\frac{1}{t}\ln\left(1 - \frac{s_0 - s}{p_{eq}}\right) \quad (39)$$

which will give rise to a linear graph of $(s_0 - s)/t$ against $1/t \ln[1 - (s_0 - s/p_{eq})]$. If K_{eq} and hence p_{eq} is known, it is thus possible to determine the values of all the other constants from a series of progress curves determined at different values of s_0. Alternatively $(s_0 - s)/t$ can be plotted against $1/t \ln(s_0/s)$, and the initial points on each of the curves ($p = 0$) can be determined by the procedure of Niemann et al.[4] to allow K_m^f and V^f to be determined. A similar procedure can be used for the reverse reaction, allowing K_m^r, V^r, and hence K_{eq} to be calculated. In practice the complexity of equations for reversible reactions makes it desirable, where possible, to simplify the system by using conditions where the reaction is essentially irreversible. This can sometimes be achieved by altering the reaction pH or by including a system that traps one of the products.

Reactions with More Than One Substrate

Although integrated rate equations have been derived for reactions involving two different substrates and two different products,[9,10] the equations generally are too complex for practical use, particularly where reversibility or product inhibition is taken into account. Recently, however, Duggleby and Morrison[11] have presented a method for analyzing the entire progress curve for a reversible Bi Bi reaction where one of the products is removed and recycled. If irreversibility and the absence of product inhibition is assumed, then Eq. (40) gives the Michaelis–Menten equation for a bisubstrate reaction following a sequential or Rapid Equilibrium Random mechanism.

$$\frac{dp}{dt} = \frac{V}{1 + [K_m^A/(a_0 - p)] + [K_m^B/(b_0 - p)] + [K_s^A K_m^B/(a_0 - p)(b_0 - p)]} \quad (40)$$

where K_m^A, K_m^B are the Michaelis constants for the two substrates A and B, K_s^A is the dissociation constant of A, and a_0 and b_0 are the initial concen-

---
[9] I. G. Darvey and J. F. Williams, *Biochim. Biophys. Acta* **85**, 1 (1964).
[10] C. Walter, *Arch. Biochem. Biophys.* **102**, 14 (1963).
[11] R. G. Duggleby and J. F. Morrison, *Biochim. Biophys. Acta* **526**, 398 (1978).

trations of the two substrates. This equation can be integrated to give:

$$Vt = p + K_m^A \ln \frac{a_0}{a_0 - p} + K_m^B \ln \frac{b_0}{b_0 - p} + \frac{K_s^A K_m^B}{b_0 - a_0} \ln \frac{a_0(b_0 - p)}{b_0(a_0 - p)} \quad (41)$$

This can be rearranged using Eq. (5) to give

$$\frac{a_0 - a}{t} = V - K_m^A \frac{1}{t} \ln \frac{a_0}{a} - K_m^B \frac{1}{t} \ln \frac{b_0}{b} - \frac{K_s^A K_m^B}{b_0 - a_0} \frac{1}{t} \ln \frac{a_0 b}{b_0 a} \quad (42)$$

This is a nonlinear equation; however, it can be linearized if one of the substrates is saturating at all stages; in this case, an equation identical to Eq. (7) is obtained in the nonsaturating substrate. This would enable V, K_m^A, and K_m^B to be obtained from experiments where each substrate in turn was saturating, but not K_s^A. In the case of a Ping Pong mechanism, the last term on the right of Eq. (42) is missing and all the initial-rate kinetic constants can be obtained. Saturation with one of the substrates may not be possible owing to limitations of substrate solubility or inhibition by high substrate concentrations. In such cases the K_m and V values can be determined at a series of fixed concentrations of the other substrate and the true values can be found by extrapolation to infinite concentration.[12,13]

Substrate Activation

A situation could arise where the binding of a second molecule of substrate is essential for reaction; the equation for such a system is

$$\frac{dp}{dt} = \frac{V}{[K_s K_m/(s_0 - p)^2] + [K_m/(s_0 - p)] + 1} \quad (43)$$

This equation can be integrated and rearranged to give

$$\frac{s_0 - s}{t} = \frac{V}{[1 + (K_s K_m/s_0 s)]} - K_m \frac{1}{t} \ln \frac{s_0}{s} \quad (44)$$

This is a nonlinear equation, but it is of interest that it is one of the simplest mechanisms capable of giving a sigmoid $v - s_0$ curve, characteristic of many cooperative enzymes. If maximum cooperativity is assumed for a mechanism described by Eq. (43), then this equation is modified to give

$$\frac{dp}{dt} = \frac{V}{[K/(s_0 - p)^2] + 1} \quad (45)$$

and this can be integrated and rearranged to give

---

[12] G. W. Schwert, *J. Biol. Chem.* **244**, 1278 (1969).
[13] R. G. Duggleby and J. F. Morrison, *Biochim. Biophys. Acta* **481**, 297 (1977).

$$\frac{s_0 - s}{t} = V - K\frac{1}{t}\left(\frac{1}{s} - \frac{1}{s_0}\right) \qquad (46)$$

In the case of the general Hill equation for an n site enzyme

$$\frac{dp}{dt} = \frac{V}{[K/(s_0 - p)^n] + 1} \qquad (47)$$

This can be integrated and rearranged to give

$$\frac{s_0 - s}{t} = V - \frac{K}{n-1}\frac{1}{t}\left(\frac{1}{s^{n-1}} - \frac{1}{s_0^{n-1}}\right) \qquad (48)$$

Estimation of the Amount of Enzyme Present

Equation (4) shows that Vt will be constant for any fixed values of s_0 and p; since V is a measure of enzyme concentration, the reciprocal of the time taken for a fixed amount of product to be formed will provide a measure of the amount of enzyme present if the system obeys the integrated-rate law.

If during the reaction the enzyme is losing activity in a first-order process, then the integrated-rate equation is given by

$$\frac{V}{k_0}(1 - e^{-k_0 t}) = p + K_m \ln \frac{s_0}{s_0 - p} \qquad (49)$$

where k_0 is the decay constant for the enzyme inactivation, determined from a separate experiment. As k_0 approaches zero so $(1 - e^{-k_0 t})$ approaches $k_0 t$, and thus Eq. (49) becomes the same as Eq. (4).[5] If loss of activity is not a first-order process, the analysis becomes very difficult.

Practical Methods

The application of integrated rate equations to enzyme-catalyzed reactions assumes that there is no change in the properties of the system during the reaction. Such changes could occur if the pH, ionic strength, or temperature is not controlled during the reaction, and it is important to check for drift in these values during the progress of the reaction. Changes in pH and temperature can usually be avoided by adequate buffering and thermostatting of the reaction mixture, but changes in ionic strength may be more difficult to avoid. A common approach is to work at such high ionic strengths that any changes caused by the enzyme reaction are negligible; alternatively controls should be used in order to find conditions under which the system is unaffected by any unavoidable changes that do occur. The relatively long periods over which the reaction is followed makes it sensitive to errors due to instrumental drift, and it is thus important to check the stability of the detection apparatus used. Provided that any drift is constant, it can be corrected for by subtraction. If the substrate is unstable and is slowly broken down in the absence of the enzyme, the analysis is more complicated than in initial-rate measurements, where a separate blank rate can be determined and subtracted for each substrate concentration. Newman et al.[14] have considered the case in which the substrate breaks down to give product in a first-order reaction governed by the rate constant k_0. In this case the rate equation becomes

$$\frac{dp}{dt} = \frac{V}{1 + [K_m/s_0 - p)]} + k_0(s_0 - p) \tag{50}$$

which integrates to give

$$k_0 t (V + k_0 K_m) = k_0 K_m \ln\left(1 - \frac{p}{s_0}\right)$$

$$- V \ln\left[1 - \frac{k_0 p}{V + k_0(K_m + s_0)}\right] \tag{51}$$

This equation can be treated easily only if the value of k_0 is known, and this can be calculated from the results of separate experiments in which the rate of nonenzymic breakdown is studied as a function of s_0. Newman et al.[14] have considered the effects of such a nonenzymic breakdown on the results obtained if the data are analyzed without making allowance for it [e.g., by using Eq. (7)] and have shown that this can lead to relatively large errors in K_m and V, which are greatest at high values of s_0.

Systematic errors in the estimation of s_0, t, or p can also lead to errors in the results obtained and are usually very difficult to detect. Newman et al.[14] have shown that such errors can be minimized by treating s_0 as an unknown parameter rather than as a constant, although such an approach increases the standard errors due to random error in the estimation of p. Another possible source of error of this type is uncertainty about the origin of the progress curve. Normally it should not be too difficult to estimate the zero-time point, although it is important to ensure rapid and efficient initial mixing of the components, but there may in some cases be less certainty in the estimation of the initial product concentration. In such cases Atkins and Nimmo[15] have shown that a reliable estimate of this quantity can be achieved by fitting the data to a polynomial of the form:

$$t = \alpha + \beta p + \gamma p^2 + \delta p^3 \cdots \tag{52}$$

and equating the initial value to the zero-order term.

[14] P. F. J. Newman, G. L. Atkins, and I. A. Nimmo, *Biochem. J.* **143**, 779 (1974).
[15] G. L. Atkins and I. A. Nimmo, *Biochem. J.* **135**, 779 (1973).

As discussed earlier, the analysis of progress curves is greatly simplified if the reaction can be treated as an irreversible single-substrate reaction in which there is no product inhibition or, at worst, only one product is inhibitory. This can often be achieved by chemically trapping one of the products or by removing it enzymically, perhaps as part of a coupled assay (see this volume [2]). If such an approach is used, it is important to ensure that the system used to remove the product is so efficient that the product concentration always remains very low in relation to its K_i value. If a coupled assay is used, it is important to ensure that the coupling system does not become rate-limiting at any stage of the progress curve. A decline in the activity of the coupling enzyme could occur owing to its own instability, to the instability of one of its substrates, or to its inhibition by its products; thus it is essential to carry out sufficient controls to establish that the enzyme under study always remains rate-limiting. As discussed previously, multisubstrate reactions can be simplified to obey the equations describing the single-substrate situation if it is possible to work at saturating concentrations of all but one of the substrates.[13]

Data Analysis and Plotting Methods

The amount of information that can be obtained from a progress curve is much greater than that from an initial-rate determination, and the use of the entire progress curve should result in greater accuracy. It is important, however, to use a wide range of the curve if the information is to be successfully extracted. A high initial substrate concentration (e.g., 2–10 K_m) will allow estimates to be made in the range where product inhibition is slight, and the reaction should be followed until the substrate concentration is considerably below the K_m. A limit to the upper concentration may be set by the solubility of the substrate, by a nonlinear response of the detection apparatus at high concentrations, or by the occurrence of high substrate inhibition, and changes at the lower end may be difficult to measure accurately if the detection system lacks sufficient sensitivity. The effects of substrate concentration range upon the precision of the results has been considered in detail by Atkins and Nimmo.[15]

Of the plotting methods considered earlier, the linear plots of the integrated-rate equation suffer from the disadvantages that it is difficult to analyze the results by any valid statistical method and that the plots are a relatively insensitive way of determining the type of product inhibition. The extrapolation procedure of Foster and Niemann[4] is essentially an initial-rate method and can be used to determine the type of product inhibition if the experiments are carried out with a series of fixed product concentrations present initially. Although this method appears to be a rather cumbersome way of determining the initial rate it has the advantage that it uses much more of the progress curve than the normal method of drawing a tangent, and it should therefore result in a considerable increase in accuracy. If a wide portion of the progress curve is used, the extrapolations required to determine the initial points should be short and easy to carry out accurately even when the curves are nonlinear.

Another method that can be used to estimate the initial velocity, which makes no assumptions about the mechanism involved, is to fit the progress curve either to a polynomial in time

$$p = \alpha + \beta t + \gamma t^2 + \delta t^3 \cdots \quad (53)$$

or to a polynomial in product concentration:

$$t = \alpha + \beta p + \gamma p^2 + \delta p^3 \cdots \quad (54)$$

In the former case α might be a measure of the contamination of substrate with the product or of an initial burst (positive value) or lag (negative value) before the steady-state velocity is established (see Tipton[16] for a detailed discussion of the possible causes of such effects). The initial rate is given by β. The second case is the one that we have already considered for determining the initial product concentration in cases where there is uncertainty about its origin [Eq. (52)]. Philo and Selwyn[17] have shown that fitting the data from a progress curve as a polynomial in p should give a better estimate of the initial rate than the alternative procedure of using a polynomial in t [Eq. (53)]. In either case there will be little gain in the accuracy by continuing the expansion to include higher powers than the cubed term. Although these methods yield considerably more precise estimates of the initial rates than the normal method of drawing tangents,[17,18] attempts to relate the coefficients of the polynomial shown in Eq. (54) to a polynomial expansion of the integrated-rate equation have not resulted in a reliable estimates of the kinetic parameters.[19]

A statistical method for fitting Eq. (4) in which K_m and V are calculated from a weighted least-squares regression of $(s_0 - s)/t$ against $1/t \ln (s_0/s)$ has been considered by Atkins and Nimmo,[15] who found that quite reliable estimates of K_m and V could be obtained despite the fact that such a statistical analysis is not strictly valid because both variables are functions of the error-containing variable $(s_0 - s)$. An alternative and statistically more satisfactory method of fitting the data directly to Eq. (4) by an iterative procedure was described by Fernley,[20] and this has been shown

---

[16] K. F. Tipton, in "Techniques in Biochemistry" (H. L. Kornberg, J. C. Metcalfe, D. H. Northcote, C. I. Pogson, and K. F. Tipton, eds.), Vol. 1. Elsevier, Amsterdam, in press.
[17] R. D. Philo and M. J. Selwyn, Biochem. J. **135**, 525 (1973).
[18] C. Walter and M. J. Barrett, Enzymologia **38**, 147 (1970).
[19] R. A. Alberty and B. M. Koerber, J. Am. Chem. Soc. **79**, 6379 (1957).
[20] N. H. Fernley, Eur. J. Biochem. **43**, 377 (1974).

to be preferable to the least-squares procedure under most conditions.[21] The use of nonlinear curve-fitting procedures has been extended to cover more complicated rate equations in which product inhibition occurs.[13,22] In the procedure of Nimmo and Atkins[22] an equation of the form of Eq. (14) is written in a form in which the product concentration ($s_0 - s$) is written as a function of each of the other parameters

$$p = f(t, s_0, K_m, K_i, V) \tag{55}$$

If provisional estimates of K_m, K_i, and V (which can be obtained by graphical analysis or by application of the Newton–Raphson procedure—see Duggleby and Morrison[13] for a detailed discussion of this method), are designated K'_m, K'_i, and V', Taylor's theorem (see Cleland[23]) can be used to show that

$$p \approx f(t, s_0, K'_m, K'_i, V') + \Delta K_m \frac{\delta f}{\delta K_m} + \Delta K_i \frac{\delta f}{\delta K_i} + \Delta V \frac{\delta f}{\delta V} \tag{56}$$

where ΔK_m, ΔK_i, and ΔV are the corrections that have to be added to the estimated values in order to obtain improved estimates. If the value of p that can be derived from the initial estimates is designated $\hat{p}$ [$\hat{p} = f(t, s_0, K'_m, K'_i, V')$] then Eq. (56) can be written as

$$p - \hat{p} \approx \Delta K_m \frac{\delta f}{\delta K_m} + \Delta K_i \frac{\delta f}{\delta K_i} + \Delta V \frac{\delta f}{\delta V} \tag{57}$$

where

$$\frac{\delta f}{\delta K_m} = \frac{(p/K'_i) + [1 + (s_0/K'_i)] \ln [1 - (p/s_0)]}{D} \tag{58}$$

$$\frac{\delta f}{\delta K_i} = \frac{-[K'_m/(K'_i)^2]\{p + s_0 \ln [1 - (p/s_0)]\}}{D} \tag{59}$$

$$\frac{\delta f}{\delta V} = \frac{t}{D} \tag{60}$$

and

$$D = \left(1 - \frac{K'_m}{K'_i}\right) + \left[\frac{K'_m}{(1 - (p/s_0))}\right]\left(\frac{1}{s_0} + \frac{1}{K'_i}\right) \tag{61}$$

The correction values ΔK_m, ΔK_i, and ΔV can be found from multiple linear regression of ($p - \hat{p}$) on the three partial derivatives, and this procedure can be repeated until the corrections necessary become negligible.

---

[21] I. A. Nimmo and G. L. Atkins, *Biochem. J.* **141**, 913 (1974).
[22] I. A. Nimmo and G. L. Atkins, *Biochem. J.* **157**, 489 (1976).
[23] W. W. Cleland, *Adv. Enzymol.* **29**, 1 (1967).

This procedure can also be extended to cover progress curves where the origin is unknown by including a displacement term to be minimized in the procedure.[13,14] Simulation studies have shown that this method gives reliable estimates of the kinetic parameters in cases where Eq. (14) is obeyed, provided that $K_i > K_m$. When this is not so the analysis gives rather poor results, but in this case the procedure of Philo and Selwyn,[17] which involves direct computer fitting to a polynomial of the form

$$\frac{s_0 - s}{s_0} = \frac{Vt}{K_m + s_0} - \frac{K_m}{s_0 + K_m}\left(1 + \frac{s_0}{K_i}\right)\left[\frac{V^2 t^2}{2(s_0 + K_m)^2} + \frac{V^3 t^3}{3(s_0 - K_m)^3} \cdots \right] \quad (62)$$

has been shown to yield satisfactory results.

Examples and Applications

Despite the extensive theoretical treatments of the integrated-rate method, examples of its use are relatively sparse, but it is to be hoped that the development of nonlinear fitting procedures may lead to greater use of this approach. The examples mentioned below are in no way exhaustive but have been chosen to illustrate the use of the approach with different types of systems.

The simple integrated-rate equation was probably first used by Walker and Schmidt[3] in studies of histidine ammonia lyase; they also showed that the data from progress curves determined at different values of s_0 all fell on a single line when the data were plotted in terms of the integrated equation (see also Dixon and Webb[24]). Acid phosphatase (EC 3.1.3.2) has been shown to obey the integrated-rate equation in which there is product inhibition by the inorganic phosphate produced,[25] and similar results have recently been obtained with this enzyme from a different source.[13] Other examples of enzymes that have been shown to fit Eq. (15) include prephenate dehydratase, which is competitively inhibited by its product phenyl pyruvate,[13] and chymotrypsin (EC 3.4.21.1) where the hydrolysis of acetyl-L-tyrosylhydroxamate, which was inhibited by the product acetyl-L-tyrosine, was analyzed by using the procedure of extrapolation to zero conditions.[4] The integrated rate equations for reversible reactions were applied to the reaction catalyzed by fumarase (EC 4.2.1.2) by Alberty and Koerber.[19] One of the most detailed studies that has been carried

---

[24] M. Dixon and E. C. Webb, "The Enzymes," p. 115. Longmans, Green, New York, 1964.
[25] F. Schønheyder, *Biochem. J.* **50**, 378 (1952).

out is that of Duggleby and Morrison,[13] who calculated the kinetic constants for lactate hydrogenase (EC 1.1.1.27) under conditions where the pyruvate concentration was saturating. Under these conditions K_m and V were calculated with NADH as the variable substrate, and K_i values were calculated for NADH as a competitive inhibitor and lactate as a mixed inhibitor.

Very recently Duggleby and Morrison[11] have examined the entire progress curve for aspartate aminotransferase (EC 2.6.1.1), in which the product α-ketoglutarate was recycled to glutamate. The values of the kinetic constants obtained agreed well with the values obtained from steady-state velocity studies.

Sorensen and Schack[26] have suggested a differential analysis of the entire progress curve by fitting the transformed data to a fourth-degree polynomial and testing the significance of the parameters. These parameters can be assigned to complex collections of all the kinetic constants found in the numerator and denominator of the original steady-state rate equation. By obtaining progress curves in the absence of products, in the presence of products (all possible combinations), in both the forward and reverse directions all kinetic constants can be obtained. For details, the reader is referred to the original paper, where a number of Bi Bi reactions are discussed.

Limitations

The use of the integrated-rate equation assumes that the only causes of the falloff in the reaction rate are the depletion of substrate and the accumulation of products that act as reversible inhibitors. There are, however, several other possible causes of such a falloff (see, e.g., Tipton[16]), and these could lead to erroneous results. If the enzyme is unstable under the assay conditions, this would contribute to the curvature of the progress curve and prevent the application of the equations considered previously. A simple test for determining whether the enzyme is stable over the assay period has been devised by Selwyn,[27] who pointed out that, since Vt will be a constant for any defined value of s_0 and p, the rate of product formation can be expressed as a function of the product concentration at a constant value of s_0,

$$dp/dt = ef(p) \tag{63}$$

---

[26] T. S. Sorensen and P. Schack, in "Analysis and Simulation of Biochemical Systems" (H. C. Hemker and B. Hess, eds.), p. 169. North-Holland, Amsterdam, 1972.
[27] M. J. Selwyn, *Biochim. Biophys. Acta* **105**, 103 (1965).

where e represents the enzyme concentration. This equation can be integrated to give

$$et = f(p) \qquad (64)$$

showing that the product concentration should depend only upon the product of the enzyme concentration and time. If a series of progress curves are determined at a series of different initial enzyme concentrations and the results are plotted as a graph of p against et, Eq. (64) shows that they should fall on a smooth curve as shown in Fig. 10a. If the enzyme is unstable and loses activity during the time of the assay, Eq. (64) will no longer be obeyed, since the enzyme concentration will itself be time-dependent; in this case a family of curves will be obtained as shown in Fig. 10b.

A test such as this is an essential preliminary to any attempt to apply the integrated-rate equations to progress curves; unless conditions can be found under which the enzyme is stable, analysis by this method will be inapplicable. The method of Selwyn[27] will also not yield a single curve if there is an appreciable nonenzymic reaction, if one of the products is a slowly acting inhibitor of the enzyme, if there is an appreciable lag period before the steady state is established, or if some other component of the assay, such as a coupling enzyme, becomes rate-limiting. Reversible changes in the activity of the enzyme due to changes in the reaction mix-

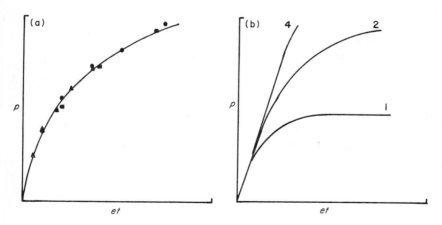

FIG. 10. (a) The concentration of product (p) as a function of time (t) multiplied by the enzyme concentration (e): ▲, ●, ■ represent different amounts of enzyme in the ratio 1:2:4, respectively. (b) An identical plot of calculated curves assuming that a unimolecular denaturation is responsible for the decrease in reaction rate. Redrawn from M. J. Selwyn, *Biochim. Biophys. Acta* **105**, 103 (1965).

ture (e.g., changes in pH) will not, however, be detected by this procedure, and care must be taken to avoid this possibility.

Acknowledgments

We are grateful to Dr. R. G. Duggleby for kindly showing us his paper on progress curve analysis in two substrate reactions prior to its publication and also to Miss Marie Gleeson for her assistance in the preparation of this manuscript.

[9] Effects of pH on Enzymes*

By KEITH F. TIPTON and HENRY B. F. DIXON

The effects of variations of the hydrogen ion concentration on the activity of enzymes have close similarities to the effects of activators and inhibitors, and the same kinetic methods and theory can be applied to both types of system. These close similarities are often obscured, however, by the fact that a logarithmic scale (pH) is usually used for hydrogen ion concentration. Treatment of the effects of hydrogen ion concentration in the same way as other effectors can yield valuable information on the nature of the kinetic mechanism obeyed by the enzyme; in addition, the characteristic ionization constants of amino acid side-chain groups has led to the use of such studies in attempts to identify specific groups as playing a role in the reaction. This latter approach has frequently been regarded as being the most important function of these studies. These two aspects of the subject, however, are complementary, and attempts to identify groups from pH studies without considering the kinetic aspects not only will miss a great deal of valuable information, but also can frequently lead to erroneous conclusions. In this chapter we will consider the effects of hydrogen ion concentration on the activity of enzymes both in terms of kinetic analysis with the hydrogen ion concentration as the variable and in terms of the use of the effects of pH to identify specific ionizing groups.

Theory

Although most enzymes contain many ionizing groups, the variation of initial velocity with pH often gives a "bell-shaped" curve like that shown in Fig. 1. Michaelis and Davidsohn[1] first interpreted this in terms

---

* Dedicated to the memory of Dr. J. Ieuan Harris.
[1] L. Michaelis and H. Davidsohn, *Biochem. Z.* **35**, 386 (1911).

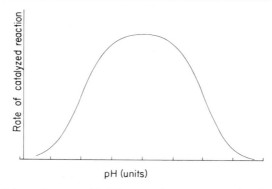

Fig. 1. Typical dependence on pH of the rate of an enzyme-catalyzed reaction. Here and in Figs. 2, 3, 8, 9, and 10 the marks on the horizontal scale indicate pH units.

of a model in which the enzyme contained only two ionizable groups that were essential for activity. At first sight this interpretation may appear oversimplistic, but it merely indicates that, of the many ionizable groups that may be needed in a particular ionic form for activity, it is only the first to protonate or deprotonate as the pH is moved up or down from the optimum pH, whose ionization can be detected by measurement of enzyme activity. The ionizations of other groups at pH values farther from the pH optimum will be undetectable because they will affect only the equilibrium between inactive forms of the enzyme. If the simple model is correct, the enzyme can be considered as being a dibasic acid, and we therefore start by giving the equations that describe the ionization of such compounds.

The Ionization of Dibasic Acids[2]

For the scheme [Eq. (1)]

$$
\begin{array}{c}
 E{<}^{X^-}_{YH} \\
K_x \nearrow x \searrow K'_y \\
E{<}^{XH}_{YH} E{<}^{X^-}_{Y^-} \\
w K_y \searrow \nearrow K'_x z \\
 E{<}^{XH}_{Y^-} \\
y
\end{array}
$$

(1)

[2] E. Q. Adams, *J. Am. Chem. Soc.* **38**, 1503 (1916).

where the concentrations of the different species are represented by the symbols w, x, y, and z, these can be related in terms of the dissociation constants shown in Eq. (1) by the equations

$$x = wK_x/[H^+] \tag{2}$$
$$y = wK_y/[H^+] \tag{3}$$
$$z = xK'_y/[H^+] \tag{4}$$
$$z = yK'_x/[H^+] \tag{5}$$

where $[H^+]$ represents the hydrogen ion concentration. Only three of the dissociation constants of Eq. (1) are required to define the concentrations, because Eqs. (2) and (4) give

$$z = wK_xK'_y/[H^+]^2 \tag{6}$$

and Eqs. (3) and (5) give

$$z = wK'_xK_y/[H^+]^2 \tag{7}$$

so that from Eqs. (6) and (7)

$$K_xK'_y = K'_xK_y \tag{8}$$

and thus fixing any three of these constants defines the fourth.

Although K_x and K'_x both refer to dissociation of a proton from the same group XH, they are not in general identical. Usually K_x will be larger than K'_x because a negative charge on the group Y^- can help to hold the proton on group XH. Nevertheless K'_x can occasionally exceed K_x: loss of a proton from YH may facilitate loss from XH by, for example, a consequent conformational change or by permitting binding of a multivalent metal ion whose higher positive charge may repel the proton on XH. Similar arguments apply to K_y and K'_y, since from Eq. (8) $K_x/K'_x = K_y/K'_y$.

The total concentration of the dibasic acid, e, will be given by

$$e = w + x + y + z \tag{9}$$

and the concentrations of each ionic form can be written as

$$w = \frac{e}{1 + [(K_x + K_y)/[H^+]] + K_xK'_y/[H^+]^2} \tag{10}$$

$$x = \frac{eK_x/[H^+]}{1 + [(K_x + K_y)/[H^+]] + K_xK'_y/[H^+]^2} \tag{11}$$

$$y = \frac{eK_y/[H^+]}{1 + [(K_x + K_y)/[H^+]] + K_xK'_y/[H^+]^2} \tag{12}$$

$$z = \frac{eK_xK'_y/[H^+]^2}{1 + [(K_x + K_y)/[H^+]] + K_xK'_y/[H^+]^2} \tag{13}$$

It can be seen from Eqs. 11 and 12 that the ratio of the concentrations of the two singly protonated species is given by

$$x/y = K_x/K_y \qquad (14)$$

and that this ratio is independent of [H$^+$]; thus any change in the concentration of one of these species with pH will be accompanied by a proportional change in the concentration of the other. Because of this there is no way of telling how much of a given effect is due to one of these two species in isolation. It is therefore more convenient to treat them as a single species whose concentration is given by

$$x + y = \frac{e}{1 + [H^+]/K_A + K_B/[H^+]} \qquad (15)$$

where

$$K_A = K_x + K_y = [H^+](x + y)/w \qquad (16)$$

and

$$K_B = K_x K'_y/(K_x + K_y) = [H^+]z/(x + y) \qquad (17)$$

(thus from Eqs. (8), (16), and (17) $K_A K_B = K_x K'_y = K'_x K_y$).

The constants K_A and K_B are termed *molecular* dissociation constants to distinguish them from the *group* dissociation constants shown in Eq. 1. Only the molecular constants can be measured experimentally, although evidence can be obtained by argument from analogy on the magnitude of group constants[3]. Any effect that is due to the singly protonated species, to whatever degree each of them contributes to it, increases as the sum of the two and obeys equations whose only parameters are the molecular constants. The constant K_A is the dissociation constant for the first proton to dissociate, regardless of the fractions of it that are derived from groups XH and YH, and K_B is that for the second proton. Because it is only molecular constants that can be determined experimentally, it is useful to express the concentrations of the other two species shown in Eq. (1) in terms of them [Eqs. (18) and (19)].

$$w = \frac{e}{1 + K_A/[H^+] + K_A K_B/[H^+]^2} \qquad (18)$$

$$z = \frac{e}{1 + [H^+]/K_B + [H^+]^2/K_A K_B} \qquad (19)$$

A theoretical curve showing the variations in the concentrations of each of the species is given in Fig. 2. If the dibasic acid under consider-

---

[3] H. B. F. Dixon, *Biochem. J.* **153**, 627 (1976).

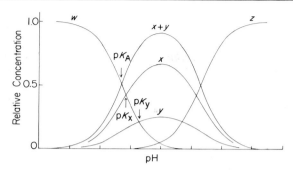

FIG. 2. pH dependence of the concentrations of the forms of a dibasic acid. The species are labeled in Eq. (1). Note that $w = x$ when $pH = pK_x$ and that $w = x + y$ when $pH = pK_A$.

ation is an enzyme that is active in one or more ionic forms and is converted into inactive forms by protonation and deprotonation, then it follows from Eq. (15) that the initial rate of the catalyzed reaction, v, varies with $[H^+]$ according to the equation

$$v = \frac{ke}{1 + [H^+]/K_A + K_B/[H^+]} \tag{20}$$

When K_A greatly exceeds K_B, so that the terms $[H^+]/K_A$ and $K_B/[H^+]$ cannot both be appreciable at any value of $[H^+]$, then K_A and K_B represent the values of $[H^+]$ that give half the maximum velocity. Since rates are often plotted against pH rather than against $[H^+]$, it is convenient to use pK values ($-\log K$) rather than K_A and K_B themselves. Thus if $pK_B \gg pK_A$, pK_A and pK_B are the pH values at which the activity is half maximal. Figure 3 shows curves of activity (i.e., of the concentration of the monoprotonated species of a dibasic acid) against pH. It can be seen that, as pK_B approaches and then falls below pK_A, the pH values at which half-maximal velocity is reached no longer approximate to the true pK values. The maximum of the curve is always at a pH equal to $(pK_A + pK_B)/2$, and this value will also be given by the mean of the two pH values that give half the maximal rate.

In analyzing the effects of pH on enzyme activity it is important to remember the two points developed above: (a) the pH values that give half-maximal activity correspond to pK values only if they are far apart (over about 3.5 units); (b) the pK values obtained will be the molecular constants derived from Eqs. (16) and (17) rather than group constants.

A simple method of estimating whether the two pH values that give half-maximal activity are far enough apart to give pK values is to measure

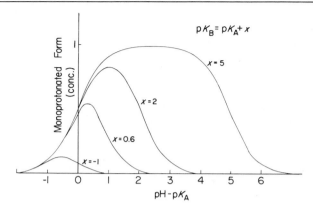

FIG. 3. pH dependence of the concentration of the monoprotonated form of a dibasic acid. Curves are shown for various values of $pK_B - pK_A$. The curve for $pK_B = pK_A + 0.6$ is the narrowest that can be achieved without positively cooperative proton binding. When $pK_B - pK_A$ is less than 0.6, there is very little of the monoprotonated form, because binding of one proton facilitates binding of the second.

the difference between them and to derive the pK difference from this by the relationship shown in Fig. 4.

Although it is molecular constants that are obtained, often they can be interpreted as group constants. Groups so closely linked (by proximity or indirect interactions) that both affect activity are unlikely not to affect each other's ionization; hence we cannot assume that $K_x = K'_x$. Nevertheless K_x and K_y may often differ appreciably, say that $K_x \gg K_y$. Then $x \gg y$ and K_A will approximate to K_x and K_B to K'_y. A typical α-amino acid

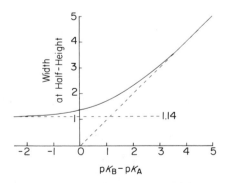

FIG. 4. Dependence of the width of a bell-shaped curve at half its height on the pK difference that determines it. If the pH values at which the concentration of the monoprotonated form of a dibasic acid is half its maximal value are called $pK^* \pm \log q$, then the molecular pK values pK_B and pK_A are given by $pK^* \pm \log(q - 4 + 1/q)$.

exemplifies this [Eq. (21)].

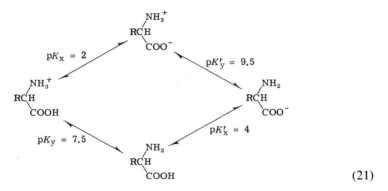

(21)

Much evidence shows that the zwitterion greatly predominates over the uncharged form. Hence the observed pK_A of 2 for the first proton to dissociate can be identified with pK_x for the dissociation of the carboxyl group in the form with a protonated amino group. Likewise the pK_B of 9.5 for the second proton can be identified with pK'_y for the amino group in the form with a dissociated carboxyl group. By analogy with other acids, we can assign to pK'_x a value of about 4. This fixes pK_y at about 7.5, which is reasonable for an amino group in view of the electron attraction of a neighboring undissociated carboxyl group.

Two points should be noted. One is that the possibility of identifying the molecular constants with group constants depends on the difference between pK_x and pK_y; it does not depend on whether or not pK_A and pK_B differ. If pK_x and pK_y were similar, then both groups would contribute to pK_A, and it would be close to both pK_x and pK_y. The second point is that pK values of groups in minor forms (here pK'_x and pK_y) are unimportant, because even if an effect such as enzyme activity depends on a minor form, its concentration depends on pH as governed by the parameters pK_A and pK_B (i.e., largely pK_x and pK'_y), not pK_y and pK'_x, as derived above.

Determination of the pK values that govern the variation of the initial rate of an enzyme-catalyzed reaction does little in itself to show the roles of specific groupings in the function of the enzyme. Even in the simplest analysis, changes in the pH may affect either the Michaelis constant (K_m) or the maximum velocity (V) of the reaction. Where both of these are affected by pH, the apparent pK values seen at an arbitrary substrate concentration are unlikely to correspond to any individual ionization. Thus, in order to obtain useful information, it is necessary to investigate the effects of pH on the kinetic parameters of the reaction catalyzed.

A Simplified System

The effects of pH on enzyme kinetic parameters are often interpreted in terms of a highly simplified kinetic mechanism. The theory will be extended later to cover more complicated, but perhaps more realistic, systems, but even in such extensions a number of assumptions that may not be easy to justify will remain, so that uncertainties are introduced into the interpretations obtained. We will first consider a simple system to illustrate the general approach that is usually applied.

Consider the system shown in Eq. (22)

$$
\begin{array}{ccc}
\mathrm{EH_2} & & \mathrm{EH_2S} \\
\updownarrow K_A^E & & \updownarrow K_A^{ES} \\
\mathrm{EH^-} \underset{k_{-1}}{\overset{k_{+1}}{\rightleftarrows}} \mathrm{EHS^-} & \xrightarrow{k_{+2}} & \mathrm{EH^-} + \mathrm{P} \\
\updownarrow K_B^E & & \updownarrow K_B^{ES} \\
\mathrm{E^{2-}} & & \mathrm{ES^{2-}}
\end{array}
\tag{22}
$$

where K_A^E and K_B^E represent the molecular dissociation constants for the free enzyme, and K_A^{ES} and K_B^{ES} represent those for the enzyme—substrate complex. The constants K_A^E and K_A^{ES} may differ either because substrate binding changes the dissociation constant of the relevant group or because the constants refer to different ionizations, since it need not be the protonation of the same group that inhibits substrate binding by the free enzyme and reaction of the enzyme—substrate complex once formed. Steady-state treatment of this mechanism yields the initial velocity equation.[4,5]

$$
\begin{aligned}
v &= \frac{\tilde{V}}{(1 + [\mathrm{H^+}]/K_A^{ES} + K_B^{ES}/[\mathrm{H^+}]) + \tilde{K}_m/s(1 + [\mathrm{H^+}]/K_A^E + K_B^E/[\mathrm{H^+}])} \\
&= \frac{\tilde{V}/(1 + [\mathrm{H^+}]/K_A^{ES} + K_B^{ES}/[\mathrm{H^+}])}{1 + (\tilde{K}_m/s)[(1 + [\mathrm{H^+}]/K_A^E + K_B^E/[\mathrm{H^+}])/(1 + [\mathrm{H^+}]/K_A^{ES} + K_B^{ES}/[\mathrm{H^+}])]}
\end{aligned}
\tag{23}
$$

where $\tilde{V} = k_{+2}[\mathrm{E}]_t$, $\tilde{K}_m = (k_{-1} + k_{+2})/k_{+1}$, $[\mathrm{E}]_t$ represents the total enzyme concentration, and s represents the concentration of unbound substrate, which can usually be taken to be the total substrate concentration.

[4] R. A. Alberty and V. Massey, *Biochim. Biophys. Acta* **13**, 347 (1955).
[5] S. G. Waley, *Biochim. Biophys. Acta* **10**, 27 (1953).

Thus $\tilde{V}$ and $\tilde{K}_m$ are pH-corrected parameters: they are the limits to which V and K_m tend at pH values between the relevant pK_A and pK_B values if these pK values are sufficiently far apart.

Comparison of Eq. (23) with the simple Michaelis equation [Eq. (24)]

$$v = \frac{Vs}{s + K_m} = \frac{V}{1 + K_m/s} \tag{24}$$

shows that the hydrogen ion concentration will affect both the apparent K_m and V values; thus hydrogen ions can be regarded as mixed (or noncompetitive) effectors of the enzyme. The apparent maximum velocity will be given by

$$V = \frac{V}{1 + [H^+]/K_A^{ES} + K_B^{ES}/[H^+]} \tag{25}$$

It will therefore depend only on ionizations of the enzyme–substrate complex [Eq. (22)]. The apparent K_m value will be given by

$$K_m = \tilde{K}_m \frac{1 + [H^+]/K_A^E + K_B^E/[H^+]}{1 + [H^+]/K_A^{ES} + K_B^{ES}/[H^+]} \tag{26}$$

and will therefore be affected by ionizations of both the free enzyme and the enzyme–substrate complex. Combination of Eqs. (25) and (26) gives

$$\frac{V}{K_m} = \frac{\tilde{V}/\tilde{K}_m}{1 + [H^+]/K_A^E + K_B^E/[H^+]} \tag{27}$$

Equation (24) shows that V/K_m is the limit approached by v/s at low substrate concentrations when $s \ll K_m$. Hence v varies with pH under these conditions in a way that depends only on ionizations of the free enzyme. Methods of finding the various ionization constants will now be given.

Methods of Obtaining Ionization Constants

Double-Reciprocal Plots, H^+ as Effector

If K_A and K_B for both E and ES differ enough, there will be pH values where terms containing one of them can be neglected and terms containing the other are appreciable. Thus at low concentrations of H^+ ($[H^+] \ll K_A^E$ and $[H^+] \ll K_A^{ES}$) it will act as a kinetically mixed essential activator of the enzyme according to Eq. (28) [derived from Eq. (23)].

$$v = \frac{\tilde{V}}{1 + K_B^{ES}/[H^+] + (\tilde{K}_m/s)(1 + K_B^E/[H^+])}$$

$$= \frac{V}{1 + \tilde{K}_m/s + K_B^{ES}/[H^+] + K_B^E \tilde{K}_m/[H^+]s} \tag{28}$$

Figure 5 shows the type of double-reciprocal plots to be expected for such a system. At higher concentrations of hydrogen ions the inhibition terms ($[H^+]/K_A^E$ and $[H^+]/K_A^{ES}$) will no longer be negligible, and this results in the inhibition at high $[H^+]$ shown in Fig. 5a and in the deviation of the lines from the simple intersecting pattern in Fig. 5b.

The situation at high concentrations of hydrogen ions where $K_B^E/[H^+]$ and $K_B^{ES}/[H^+]$ are negligible is described by Eq. (29).

$$v = \frac{\tilde{V}/(1 + [H^+]/K_A^{ES})}{1 + (\tilde{K}_m/s)[(1 + [H^+]/K_A^E)/(1 + [H^+]/K_A^{ES})]} \quad (29)$$

which shows that the hydrogen ion will function as a mixed inhibitor of the enzyme, the situation becoming more complicated at lower concentrations, where the terms $K_B^E/[H^+]$ and $K_B^{ES}/[H^+]$ are no longer negligible, as shown in Fig. 6.

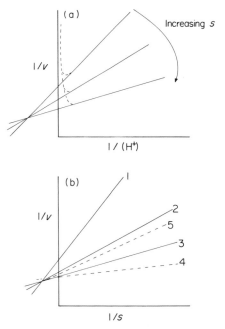

FIG. 5. Reciprocal plots of the effects of hydrogen ions on enzyme activity for a mechanism obeying Eq. (23). The solid lines correspond to the situation where $[H^+] \ll K_A^E$ and $[H^+] \ll K_A^{ES}$, and thus Eq. (28) is obeyed. The dashed lines describe the situation where these inequalities no longer hold. The solid lines intersect at a point to the left of the $1/v$ axis, which will be above the horizontal axis (as shown here) if $K_B^E > K_B^{ES}$, on that axis if $K_B^E = K_B^{ES}$, and below it if $K_B^E < K_B^{ES}$. In Fig 5b the hydrogen ion concentration increases in the order 1–5.

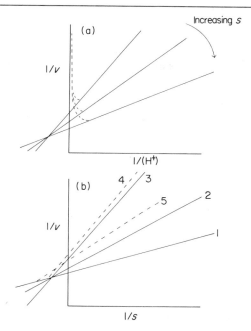

FIG. 6. Reciprocal plots of the effects of hydrogen ions on enzyme activity for a mechanism obeying Eq. (23) when $[H^+] \gg K_B^E$ and $[H^+] \gg K_B^{ES}$ (solid lines) and when this inequality breaks down (dashed lines). The solid lines intersect at a point to the left of the $1/v$ axis, which will be above the horizontal axis if $K_A^E < K_A^{ES}$, on that axis if $K_A^E = K_A^{ES}$, and below it if $K_A^E > K_A^{ES}$. In Fig. 5b the hydrogen ion concentration increases in the order 1–5.

Just as with other activators and inhibitors (this volume [15], [20]), the values of the individual ionization constants can be obtained from replots of the slopes and intercepts of plots of $1/v$ against $1/s$, i.e., of K_m/V and of $1/V$, against 1/[activator] or against [inhibitor]. Thus the slopes and intercepts of Fig. 5b are plotted against $1/[H^+]$ in Figs. 7a and 7b, and those of Fig. 6b against $[H^+]$ in Figs. 7c and 7d. The values of the intercepts on the horizontal axes of these replots allow the values of the ionization constants to be read off as shown.

Inspection of Eq. (23) and the mechanism of Eq. (22) shows that if hydrogen ions bind only to the free enzyme ($K_A^{ES} \to \infty$, $K_B^{ES} \to 0$) the kinetic patterns will be competitive (the slope K_m/V is affected, but not V). Conversely, if only the enzyme–substrate complex can ionize ($K_A^E \to \infty$, $K_B^E \to 0$) the effects will appear uncompetitive with V and K_m changed in the same proportion and the slopes of plots of $1/v$ against $1/s$ unchanged. If substrate binding does not affect the ionization constants, so that $K_A^{ES} = K_A^E$ and $K_B^{ES} = K_B^E$, the effect will be truly noncompetitive, with V and the

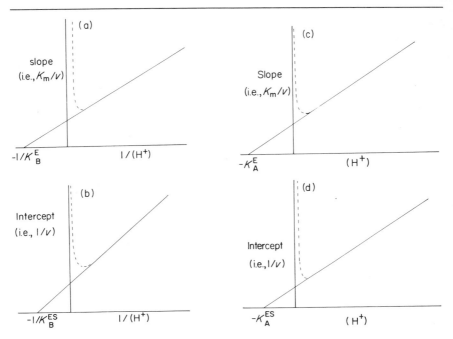

FIG. 7. Determination of ionization constants from secondary plots of the data shown in Figs. 5 and 6. Replots of the slope and vertical axis intercepts of the lines in Fig. 5b are shown in (a) and (b), respectively; those for Fig. 6b are shown in (c) and (d). The dashed lines correspond to deviations due to the breakdown of the simplifying inequalities described in the legends to Figs. 5 and 6.

slope changed but K_m unaffected, so that either kind of replot can be used to derive the ionization constants.

Plots of Kinetic Constants against pH

An alternative method of finding pK values is to plot appropriate parameter against pH, V to give the pK values of the enzyme–substrate complex and V/K_m to give those of the free enzyme. If pK_B is far above pK_A the bell-shaped curve will have a flat top, and the pK values can be read off from the pH values at which half the maximal velocity is reached (Fig. 3). If the pK values are close together or reversed, then the relationship given in Fig. 4 can be used to obtain the pK values. This is derived from the treatment of Alberty and Massey,[4] who showed that

$$K_A = [H^+]_A + [H^+]_B - 4[H^+]_{opt} \tag{30}$$

where $[H^+]_A$ and $[H^+]_B$ are the hydrogen ion concentrations that give half the maximal velocity, and $[H^+]_{opt}$ is the concentration that gives the maximal velocity. From this it follows that if pH_B and pH_A are called $pH_{opt} \pm \log q$ (note that pH_{opt} is the average of pH_A and pH_B, and also of pK_A and pK_B), then pK_A and pK_B are given by $pH_{opt} \pm \log(q - 4 + 1/q)$.[5a] The values thus found for pK_A and pK_B can be checked from the fit of the results over the whole of the bell-shaped curve by using its simplest expression $(\alpha + 1)/(\alpha + \cosh x)$ where $x = 2.303 \ (pH - pH_{opt})$ and $\alpha = (K_A/4K_B)^{1/2}$ (so that $pK_B - pK_A = 2 \log 2\alpha$).[6]

The accuracy of determination of pK_A and pK_B becomes very low when pK_B falls below pK_A (i.e., when there is strongly cooperative proton binding). This is because the width of the curve at half its maximal height becomes insensitive to changes in $pK_B - pK_A$ when the latter is small or negative (Fig. 4). Thus this width changes only from 1.53 pH units when $pK_B - pK_A = 0.6$ (the lowest it can be without positively cooperative proton binding) to 1.14 pH units as $pK_B - pK_A$ approaches minus infinity. In terms of Eq. 30, K_A becomes a small difference between large numbers as $[H^+]_A$ and $[H^+]_B$ approach $(2 \pm \sqrt{3}) \ [H^+]_{opt}$.

At the time Eq. (3) was derived it was not generally realized that it applies only to molecular ionization constants. If a group of high pK needs to be deprotonated, this does *not* cause pK_B to fall below pK_A, since these are the values for the second and first protons to dissociate, from whatever groups these come. It merely represents the situation of Eq. (1) in which $x \gg y$ and activity depends on y. Kinetics can tell us that the monoprotonated form is needed, but cannot tell us where the proton must be. A fall of pK_B below pK_A represents positively cooperative proton binding, so that the second proton to dissociate does so more readily than the first. Positive cooperativity at the group level is required, i.e., K'_x of Eq. (1) must exceed K_x, if K_B is to exceed K_A, since $K_A/K_B = (2 + K_x/K_y + K_y/K_x)K_x/K'_x$.[7]

Although K_m varies with pH in a manner determined by the K values of both free enzyme and enzyme–substrate complex [Eq. (26)], a plot of K_m against pH gives only the pK values of the enzyme–substrate complex as the pH values of the midpoints of sigmoid waves in the curve. The pK values from the numerator of Eq. (26) determine the heights of these waves (as a consequence of determining where they start) but not the positions of their centers (Brocklehurst and Dixon,[8] Fig. 3). Hence the pK values of the enzyme–substrate complex can be determined from this plot

---

[5a] H. B. F. Dixon, *Biochem. J.* (in press).
[6] H. B. F. Dixon, *Biochem. J.* **137**, 443 (1974).
[7] H. B. F. Dixon and K. F. Tipton, *Biochem. J.* **133**, 837 (1973).
[8] K. Brocklehurst and H. B. F. Dixon, *Biochem. J.* **155**, 61 (1976).

if they are well separated. If $1/K_m$ is plotted against pH, the pK values represented by the midpoints of the sigmoid waves will be those of the free enzyme,[9] since these terms will be in the denominator of the reciprocal form of Eq. (26), whereas the pK values of the enzyme–substrate complex will determine the heights of the waves.

Plots of Logarithms of Kinetic Constants against pH

The most convenient and widely used method of treating the effects of pH on enzyme activity is that of Malcolm Dixon[10]; it which involves plotting the logarithm of the kinetic constant against the pH (Fig. 8). For the maximum velocity (Fig. 8a), taking the logarithm of Eq. (25) gives

$$\log V = \log \tilde{V} - \log (1 + [H^+]/K_A^{ES} + K_B^{ES}/[H^+]) \tag{31}$$

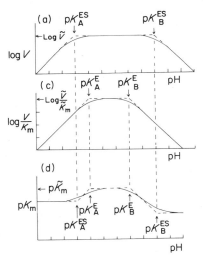
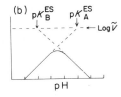

FIG. 8. Plots of logarithms of kinetic constants against pH for a system obeying Eq. (22). (a) Plot of log V (when pK_B^{ES} > pK_A^{ES}); (b) plot of log V (when pK_B^{ES} < pK_A^{ES} and proton binding is therefore positively cooperative); (c) plot of log (V/K_m); (d) plot of pK_m. The alignment shows how pK values in ES and in E (or S) appear as changes of slope of 1 unit (see text footnote 10 references). In the example given, the form of the enzyme that exists between pK_A^E and pK_B^E has slightly more affinity for the substrate than the forms that exist outside this range. Hence, between pK_B^E and pK_B^{ES} the substrate promotes protonation of the enzyme by binding mainly the protonated form. But above pK_B^{ES} the predominance of the deprotonated form of the free enzyme overcomes its lesser affinity for the substrate, and it is the main form ligated. After M. Dixon, *Biochem. J.* **55**, 161 (1953).

[9] A. R. Fersht, "Enzyme Structure and Mechanism." Freeman, Reading, 1977.
[10] M. Dixon, *Biochem. J.* **55**, 161 (1953); see also M. Dixon and E. C. Webb, "Enzymes" 2nd ed., p. 116. Longmans, Green, New York, 1974.

Since log $\tilde{V}$ is independent of pH, the dependence will be determined by the second term. At low pH values where $[H^+] \gg K_A^{ES}$ and $[H^+] \gg K_B^{ES}$, the equation will simplify to

$$\log V = \log \tilde{V} - \log \frac{[H^+]}{K_A^{ES}} = \log \tilde{V} + (pH - pK_A^{ES}) \tag{32}$$

and thus the slope of the graph will be $+1$. As the pH is raised there will be a range over which two terms contribute to the slope, and provided that $K_A^{ES} \gg K_B^{ES}$ there will then be a range where $K_A^{ES} \gg [H^+] \gg K_B^{ES}$. Over this range Eq. (31) simplifies to

$$\log V = \log \tilde{V} \tag{33}$$

so the graph will have a slope of zero. Further increase in pH will give a region in which the dependence is governed by $\log (1 + K_B^{ES}/[H^+])$, followed by one where $K_B^{ES}/[H^+]$ is dominant and Eq. (31) simplifies to

$$\log V = \log \tilde{V} - \log \frac{K_B^{ES}}{[H^+]} = \log \tilde{V} + (pK_B^{ES} - pH) \tag{34}$$

which will give a slope of -1.

Thus, provided that $pK_B \gg pK_A$, a graph of log V against pH will be composed of linear regions of slopes $+1$, 0, and -1, which will be joined by curved sections where two of the terms in the second part of Eq. 31 contribute appreciably to the slope. From Eqs. (32)–(34) it can be seen that if the linear portions are extended they will intersect at the pK values. The curved portions will pass 0.3 unit (log 2) below the intersections (Fig. 8a). If, however, pK_B is below pK_A, then the horizontal portion is missing and the intersection is not at the pK values, but at their mean (Fig. 8b).

Similar behavior should be seen if $\log (V/K_m)$ is plotted against pH [see Eq. (27)], but in this case the intersection points will give the pK values for ionization of the free enzyme rather than of the enzyme–substrate complex (Fig. 8c). The behavior of graphs involving K_m is more complicated because this constant is affected by ionizations of both the free enzyme and the enzyme–substrate complex [Eq. (26)]. Malcolm Dixon[10] has recommended that pK_m (i.e., $-\log K_m$) be plotted against pH. Since

$$pK_m = \log (V/K_m) - \log V \tag{35}$$

it follows that this plot represents a combination of the preceding two. The graph obtained should be composed of a number of sections with slopes of $+1$, 0, and -1, and each change of slope indicates a pK. Downward bends will correspond to ionizations of the free enzyme, since they will be given by the log V/K_m portion of Eq. (35), whereas upward bends will correspond to ionizations of the enzyme–substrate complex, since they will be given by the $-\log V$ portion. They are upward (i.e., they increase the slope by one unit) because $-\log V$ is here involved rather than downward for $\log V$ (Fig. 8c). A plot of pK_m is shown in Fig. 8d.

Since the plot of pK_m against pH contains all the information that can be obtained from the separate plots of $\log V$ and of $\log (V/K_m)$, one should be able to obtain all the pK values from it alone. In practice, however, pK values can be determined accurately only if they are well separated, so the presence of four pK values in the pK_m plot, compared with two in each of the others, may make it less satisfactory to use. As we will indicate later, however, there are cases in which the use of the pK_m plot in addition to the other two may help to resolve possible ambiguities.

The simple model being considered can be somewhat extended by supposing that two protonations or deprotonations are required to convert the form predominant over a given pH range into the active form. Then slopes of $+2$ or -2 may appear in the plots, although accurate determination of these requires sensitive assays capable of measuring activities that are small fractions of that at the pH optimum. This extension is just one example of the versatility of the logarithmic plot and the simplicity of the rules for interpreting it.[10]

Comparison of Graphical Methods

Three methods have been described above for obtaining ionization constants: (i) plots of $1/V$ and of K_m/V (derived, if desired, from the intercepts and slopes of plots of $1/v$ against $1/s$) against $[H^+]$ or $1/[H^+]$; (ii) plots of V, V/K_m, and K_m (or its reciprocal) against pH; (iii) plots of the logarithms of these parameters against pH. All these methods are satisfactory if the ionization constants are well separated. The method of Malcolm Dixon (iii) is the most used and gives the clearest presentation of the results. Plots of kinetic constants against pH (method ii), especially for V and V/K_m which should show only two ionization constants each, have the advantage of the ease of applying the Alberty–Massey equation [Eq. (30)] and of fitting theoretical curves. It thus copes best when pK_B is close to pK_A or below it. The method (i) of plots against $[H^+]$ or $1/[H^+]$ concentrates on a relatively narrow range of $[H^+]$ around the ionization constants (there is little useful information outside the range $0.1-10\ K$[11]). It has the advantage that it allows a more complete analysis of hydrogen ions as activators and inhibitors, using equations similar to those used in other aspects of kinetic studies (see this volume [15], [20]), especially by indicating the onset of complicating factors such as cooperativity or partial effects by curvature of the plots.

[11] S. Ainsworth, "Steady-State Enzyme Kinetics," p. 160. Macmillan, New York, 1977.

Complications

The mechanism shown in Eq. (22) oversimplifies real situations in several ways. In this section we shall look at the assumptions it makes and examine the effects of dispensing with some of them.

The Presence of More Than One Enzyme–Substrate Intermediate

The system in which two complexes of enzyme and substrate are envisaged [Eqs. (36)]

$$
\begin{array}{ccccccc}
EH_2 & & EH_2S & & EH_2P & & \\
\updownarrow K_A^E & & \updownarrow K_A^{ES} & & \updownarrow K_A^{EP} & & \\
EH^- & \underset{k_{-1}}{\overset{k_1 S}{\rightleftarrows}} & EHS^- & \underset{k_{-2}}{\overset{k_2}{\rightleftarrows}} & EHP^- & \overset{k_3}{\longrightarrow} & EH^- + P \\
\updownarrow K_B^E & & \updownarrow K_B^{ES} & & \updownarrow K_B^{EP} & & \\
E^{2-} & & ES^{2-} & & EP^{2-} & &
\end{array}
\quad (36)
$$

gives a steady-state rate equation of the form shown in Eq. (24) where

$$V = \frac{k_2 k_3 e}{(k_{-2} + k_3)(1 + [H^+]/K_A^{ES} + K_B^{ES}/[H^+]) + k_2(1 + [H^+]/K_A^{EP} + K_B^{EP}/[H^+])} \quad (37)$$

and

$$V/K_m = \frac{k_1 k_2 k_3 e}{(k_{-1}k_{-2} + k_{-1}k_3 + k_2 k_3)(1 + [H^+]/K_A^E + K_B^E/[H^+])} \quad (38)$$

Equation (38) thus shows the usual pH dependence for V/K_m, which gives the ionization constants of the free enzyme. In fact Eq. (37) is also simple in form, and the overall values it gives for $1/K_A$ and for K_B are simply $(k_{-2} + k_3)/(k_2 + k_{-2} + k_3)$ of the value of each of these for ES plus $k_2/(k_2 + k_{-2} + k_3)$ of the value of each for EP. It is thus a mean for the two complexes, weighted in the ratio of $(k_{-2} + k_3)$ to k_2 in favor of these for ES—i.e., exactly in the ratio of the steady-state concentrations of EHS$^-$ and EHP$^-$. It therefore does not matter how many enzyme–substrate complexes there are; the constants obtained from the pH dependence of V are average values weighted in favor of the predominant complex. If, further, equilibrium obtains between the complexes, then the predominance is governed by the equilibrium constant k_{-2}/k_2 instead of the steady-state ratio $(k_{-2} + k_3)/k_2$, since k_3 is then negligible in comparison with k_{-2}.

The scheme of Eq. (36) is much more plausible than that of Eq. (22), which is reasonable only when equilibrium is assumed between EH^- and EHS^-; otherwise Eq. (22) assumes that S dissociates slowly from EH^- but that P, which may be chemically similar to S, dissociates infinitely rapidly. It is therefore important that Eq. (36) gives similar kinetic relationships, although the makeup of the kinetic constants in terms of individual rate constants is changed.

More Than One Form of the Enzyme Can Bind Substrate

Elaboration of the scheme of Eq. (22) into the mechanism of Eq. (39)

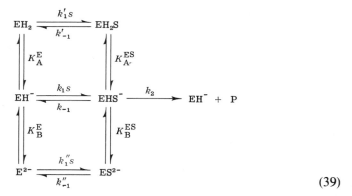

(39)

by allowing differently protonated forms of the enzyme to combine with substrate leads to several alternative pathways by which the catalytically productive EHS^- complex can be formed. If it is assumed that the rate of breakdown of the EHS^- complex to yield products is slow relative to the dissociation steps, so that the latter remain at thermodynamic equilibrium, the situation does not differ from Eq. (22), because addition of pathways cannot affect an equilibrium, and thus Eq. (23) will apply to the mechanism shown in Eq. (39). If, however, steady-state conditions are assumed to apply to the mechanism shown in Eq. (39), the resulting kinetic equation is extremely complicated[12-18] and contains terms in which the concentrations of the substrate and hydrogen ions are squared and cubed,

---

[12] K. J. Laidler, *Trans. Faraday Soc.* **51**, 528 (1955).
[13] P. Ottolenghi, *Biochem. J.* **123**, 445 (1971).
[14] L. Peller and R. A. Alberty, *J. Am. Chem. Soc.* **81**, 5709 (1963).
[15] R. M. Krupka and K. J. Laidler, *Trans. Faraday Soc.* **56**, 1467 (1960).
[16] R. A. Alberty and V. Bloomfield, *J. Biol. Chem.* **238**, 2804 (1963).
[17] J. A. Stewart and H. S. Lee, *J. Phys. Chem.* **71**, 3888 (1967).
[18] H. Kaplan and K. J. Laidler, *Can. J. Biochem.* **45**, 539 (1967).

respectively. This may lead to complex dependence of the initial velocity on the substrate and hydrogen ion concentrations, and generally the kinetic equation is too complicated to be of practical use. A number of simplifying conditions can result in equations similar to Eq. (23).[12] One assumption, which was proposed by Alberty and his co-workers,[4,14,16] is that the proton-transfer steps occur rapidly in relation to all the chemical steps, so that the proton-binding steps remain in thermodynamic equilibrium. Under these conditions the initial-rate equation becomes:

$$v = \frac{k_2 e/(1 + [H^+]/K_A^{ES} + K_B^{ES}/[H^+])}{1 + \{[k_2 + k_{-1}(1 + [H^+]/K_A^E + K_B^E/[H^+])]/k_1(1 + [H^+]/K_A^{ES} + K_B^{ES}/[H^+])\}1/s} \quad (40)$$

Because this equation differs from Eq. (23) only in the term representing K_m, V still depends solely on ionizations of the enzyme–substrate complex, so Eq. (25) still applies. The expression for V/K_m, however, shows four pK values and the plot of its logarithm against pH will show four downward turns and also two upward ones, which do not correspond to pK values. Two of the pK values are those of the free enzyme, but the other two are distorted by rate constants. It may be possible to tell which there are because the distorted ones are farther apart than the compensating upward turns by the same amount, namely $\log[(k_2 + k_{-1})/k_{-1}]$.[8] Hence if all six bends can be located the true pK values can be found, but the chances that all will be separated and within the range where the enzyme is both stable and appreciably active may not be great. As k_2 falls in comparison with k_{-1} the distorted pK values approach the compensating upward bends and therefore disappear, leaving Eq. (27) (Fig. 8c) to apply.

Change of Rate-Determining Step with pH

The above mechanism can show a change of rate-determining step with pH. The consequences of such a change are first examined in simple systems before returning to that of Eq. (39).

In a reaction of two consecutive steps, it is the ratio of the forward rate constant for the second reaction to the backward constant for the first reaction that determines which of the two is rate-limiting. Thus a reaction of the type

$$\begin{array}{ccc} AH^+ & & XH^+ \xrightarrow{k_2} \text{Product} \\ \updownarrow K & & \updownarrow K' \\ A & \underset{k_{-1}}{\overset{k_1}{\rightleftarrows}} & X \end{array} \quad (41)$$

will change its rate-determining step with pH. At low pH the reactant is in its inert form AH^+ and the intermediate in its reactive form XH^+. Thus $k_2[XH^+]$ will exceed $k_{-1}[X]$ so the first step will be rate-limiting. At high pH the reverse will hold, all four species AH^+, A, X, and XH^+ will come to equilibrium, and the second step will be rate-determining. Reactions of this type, for which imine formation provides a nonenzymic model,[19] show a typical bell-shaped pH dependence. One of the two pK values that characterize the curve is the pK of the reactant, but the other is a "mirage"[19] and is the same pK distorted by rate constants.[20] The value of pK_B [compare Eq. (20)] cannot fall below pK_A + 0.6; i.e., the width of the bell cannot be as narrow as that given by positively cooperative proton binding.

In demonstrating these statements, it was assumed[20] that the total reactant concentration, i.e., $[AH^+] + [A]$, was independent of pH, which is likely if it greatly exceeds the concentration of intermediate. In an enzymic reaction, however, no such assumption can be made. If only V is to be determined, then only species with bound substrate need be considered [Eq. (42)].

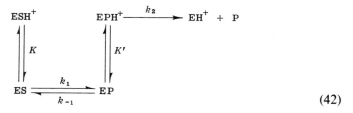

(42)

Steady-state treatment, considering $[ESH^+] + [ES] + [EP] + [EPH^+]$ constant, still gives an equation of the form of Eq. (25), with a bell-shaped curve of V against pH, but neither of the pK values that characterize it is undistorted by rate constants. It remains true, however, that $K_A \geq 4 K_B$; i.e., p$K_B \geq$ pK_A + 0.6.

The mechanism given in Eq. (39) can show a change in rate-determining step with pH[8,21] because (provided $k_2 > k_{-1}$) the term k_2 [EHS$^-$] may exceed the rate of the reverse of the first step, i.e., k'_{-1} [EH$_2$S] + k_{-1}[EHS$^-$] + k''_{-1}[ES^{2-}], at some pH values and be less than it at others. Renard and Fersht[22] have analyzed a change of rate-determining step with pH. Demonstration of its occurrence normally re-

---

[19] W. P. Jencks, "Catalysis in Chemistry and Enzymology." McGraw-Hill, New York, 1969.
[20] H. B. F. Dixon, *Biochem. J.* **131**, 149 (1973).
[21] K. Brocklehurst and H. B. F. Dixon, *Biochem. J.* **167**, 859 (1977).
[22] M. Renard and A. R. Fersht, *Biochemistry* **12**, 4713 (1973).

quires a method for observing how the concentration of intermediates varies with pH and not merely the steady-state rate of reaction.

More Than One Form of Enzyme–Substrate Complex Can Yield Products

So far we have assumed that the only ionization reactions that affect the breakdown of enzyme–substrate complex to products completely prevent this reaction. It is, however, possible that loss or gain of a proton near the active site may change the rate of the reaction. So we must modify the scheme of Eq. (39) by drawing another route to the product [Eq. (42a)].

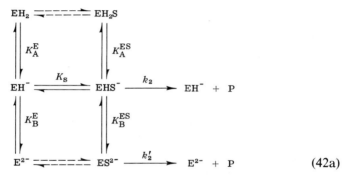

(42a)

We will deal only with rapid equilibrium of substrate and hydrogen-ion binding, because the full steady-state equations are complex. Hence we have assigned an equilibrium constant for binding of substrate to one form of the enzyme; whether or not the routes shown by dashed lines occur, their equilibrium constants are determined by K_s and the acid dissociation constants are marked. The resultant rate equation is

$$v = \frac{\{[k_2 + k_2'(K_B^{ES}/[H^+])]e\}/(1 + [H^+]K_A^{ES} + K_B^{ES}/[H^+])}{1 + (\tilde{K}_m/s)[(1 + [H^+]/K_A^E + K_B^E/[H^+])/(1 + [H^+]/K_A^{ES} + K_B^{ES}/[H^+])]} \tag{43}$$

This predicts a simple dependence of K_m on pH [Eq. (26)], in accord with the simple assumption that binding is at equilibrium, and the dependence can give all four pK values. The pH dependence of V and of V/K_m is more complicated. Putting k_2' to zero simplifies the scheme to that of Eq. (22) and simplifies Eq. (43) to Eq. (23). The possible types of pH dependence of V are shown in Fig. 9. It is possible that ES^{2-} may lose a further proton to render it completely incapable of reaction, as shown by the dashed por-

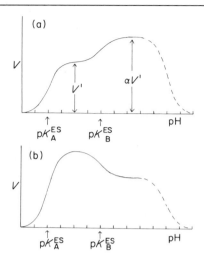

FIG. 9. pH Dependence of V when two protonic states of ES can yield products. Deprotonation of ES with a pK of pK_B^{ES} (a) increases and (b) diminishes the rate of the catalyzed reaction. Dashed lines indicate that a further deprotonation may prevent the reaction.

tions of Fig. 9, but this may not occur within the pH range under study. Provided that the pK values are adequately separated, they can be found from the curves.

For the range over which $[H^+]/K_A^E$ and $[H^+]/K_A^{ES}$ are negligible, and likewise any further inactivating dissociation of ES^{2-} is negligible—that is, the range between the two plateaus of Fig. 9—hydrogen ions will appear to be nonessential activators if $k_2' < k_2$ or partial inhibitors if $k_2' > k_2$. Thus secondary plots of $1/V$ and of K_m/V (intercepts and slopes of plots of $1/v$ against $1/s$) against $[H^+]$ or $1/[H^+]$ will be hyperbolic. Linear secondary plots can be obtained by plotting different functions, as follows. Suppose that the plateau at higher pH is the higher one (Fig. 9a), i.e., that $k_2' > k_2$, and that the value of V on the lower plateau is V', i.e., $V' = k_2 e$. Likewise let the height of the higher plateau be $\alpha V'$, so that $\alpha = k_2'/k_2$. From Eq. (43), considering only the range between the plateaus,

$$\frac{1}{V} = \frac{1 + K_B^{ES}/[H^+]}{V'(1 + \alpha K_B^{ES}/[H^+])} = \frac{[H^+] + K_B^{ES}}{V'([H^+] + \alpha K_B^{ES})} \quad (44)$$

Hence

$$\frac{1}{V} - \frac{1}{V'} = \frac{[H^+] + K_B^{ES} - [H^+] - \alpha K_B^{ES}}{V'([H^+] + \alpha K_B^{ES})} = \frac{K_B^{ES}(1 - \alpha)}{V'([H^+] + \alpha K_B^{ES})} \quad (45)$$

Taking reciprocals gives

$$1/(1/V - 1/V') = [H^+]V'/K_B^{ES}(1 - \alpha) + V'\alpha/(1 - \alpha) \quad (46)$$

Hence a plot of $1/(1/V - 1/V')$ against $[H^+]$ will be linear and will have a slope of $V'/K_B^{ES}(1 - \alpha)$, an intercept on the vertical axis of $V' \alpha/(1 - \alpha)$, and an intercept on the $[H^+]$ axis of $-K_B^{ES}$. Since α and V' are known from the heights of the plateaus, K_B^{ES} is determined from the results over the transition between them. Similar types of rearrangement can be used to obtain linear plots when $k_2' < k_2$ and also for plotting derivatives of V/K_m against $[H^+]$ or $1/[H^+]$. In plots of log V or log (V/K_m) against pH, this mechanism gives an extra bend as the velocity levels off to the finite value of V at high pH values. This bend does not correspond to a pK and will not be seen in a plot of pK_m against pH, and thus the use of this plot in addition to the other two may help to avoid a mistaken assignment of a pK value.

Clearly this mechanism can be extended to activity for more than two protonic states of the enzyme–substrate complex. Provided the pK values are separated enough so that plateaus exist, their values can be determined as above.

Abortive Complex Formation

An enzyme may be able to bind a substrate in unproductive ways. Fastrez and Fersht[23] analyzed this in connection with the hydrolysis by chymotrypsin of Ac-Tyr-NH-Ph in which the phenyl group of the aniline residue bore various substituents. If the substrate binds with this phenyl ring in the site that should bind that of the tyrosine residue, no hydrolysis can occur. In the simplest case, the same protonic state of the enzyme binds substrate to form a productive complex ES and an unproductive one ES' but cannot bind substrate in both ways at once, and both forms of binding are in equilibrium. The pK values of V/K_m remain those of the free enzyme, but those of V prove[23] to be weighted average of all forms of enzyme substrate complex whether productive or not. If abortive binding predominates, then the pK values seen in V are those of the abortive complex ES'. More complex situations involving abortive binding can easily be envisaged.

Comparison of Assumptions

In systems containing alternative pathways [see Eq. (39)] it was necessary to assume, in order to obtain manageable equations, that all protonation steps were at thermodynamic equilibrium. This assumption has been made in most studies of the effects of pH on enzyme activity, but has been

[23] J. Fastrez and A. R. Fersht, *Biochemistry* **12**, 1067 (1973).

questioned,[12,13] and in a recent detailed criticism Knowles[24] has queried its validity, pointing out that dissociation of protons may be as slow as 10^4 sec^{-1} whereas other steps in the enzymic reaction may be faster. In some branches of the mechanism shown in Eq. (39) substrate binds to the enzyme immediately after a protonation step; thus, for equilibrium conditions to apply to these proton-binding steps alone, it is necessary for their dissociation to be much faster than the subsequent substrate-binding steps (see, e.g., Dalziel[25]). Enzymes may, however, combine with their substrates at up to 5×10^8 M^{-1} sec^{-1}, so that rates of combination at natural substrate concentrations may be up to 10^6 sec^{-1}, suggesting that equilibrium conditions for protonation may not always occur when the interaction with substrate is in steady state. Indeed Knowles[24] cited evidence that increasing the buffer concentration and thus accelerating equilibration of protons increases the activity of carbonic anhydrase.[26] In nonenzymic models, e.g., catalysis of amide hydrolysis by a carboxyl group,[27] transfer of a proton can be rate limiting. Cornish-Bowden[28] also concluded that protonation reactions are unlikely to be slower than 10^4 sec^{-1} (unless compulsorily accompanied by conformational changes). He argued that the rates of individual steps in enzyme-catalyzed reactions are unlikely to be much faster than this, so that the equilibrium assumption may often be valid. We cannot assume, however, that this will always be so. Although an increase of rate on increasing buffer concentration gives evidence that a proton transfer is rate-limiting, absence of such an effect does not rule out a rate-limiting proton transfer catalyzed by groups in the enzyme, possibly at a site inaccessible to buffer components.

Cleland[29] has called attention to the fact that bound substrate can hinder proton equilibration by an enzyme; thus the proton removed by fumarate hydratase from malate only exchanges relatively slowly with water while fumarate is on the enzyme.[30] He has therefore considered the situation in which proton transfer steps in enzyme–substrate complexes are slow, even though those in the free enzyme are considered to be in equilibrium. The rate equations given by such systems contain terms in which the hydrogen ion concentration is squared, and these can lead to waves in Dixon plots as well as to displacement of pK values.

Much more general cases have been considered by many authors,[12–18]

[24] J. R. Knowles, *Crit. Rev. Biochem.* **4**, 165 (1976).
[25] K. Dalziel, *Biochem. J.* **114**, 547 (1969).
[26] C. K. Tu and D. N. Silverman, *J. Am. Chem. Soc.* **97**, 5935 (1975).
[27] M. F. Aldersley, A. J. Kirby, and P. W. Lancaster, *Chem. Commun.*, p. 570 (1972).
[28] A. Cornish-Bowden, *Biochem. J.* **153**, 445 (1976).
[29] W. W. Cleland, *Adv. Enzymol.* **45**, 427 (1977).
[30] J. N. Hansen, E. C. Dinovo, and P. D. Boyer, *J. Biol. Chem.* **244**, 6270 (1969).

who have developed complex initial-rate equations to describe them and have considered how assumptions can simplify these equations. We have shown how even fairly probable departures from the simplest system represented in Eq. (22) lead to kinetic pK values that differ from the molecular values. Although we have indicated ways for finding molecular pK values in some specific cases, they apply when it is known that a particular complication is operative, and this may not be easy to discover.

Interpretation of the Results of pH Experiments

As pointed out already, the results of studies on the effects of pH on enzyme activity are difficult to interpret with any certainty. Even in the simplest system the pK values obtained are relatively complex molecular constants [Eqs. (16) and (17)] rather than simple group constants. It is often assumed without any arguments in favor that a simple mechanism is operative, although quite simple features often found in enzyme mechanisms may shift the kinetic pK values, and those observed may bear little relation to the molecular values in more complex mechanisms if the binding steps do not approach equilibrium. Despite these reservations, many such studies have yielded results that have been shown by other methods to be substantially correct in identifying the pK values of groups involved. Even if pK values are shifted, an approximate value for a pK may give an indication of the type of group involved in a process, and so can usefully supplement other methods.

Possibly too much emphasis has been placed on the use of studies of pH dependence to identify specific ionizing groups. Other features of the responses of enzymes to pH changes may be informative about their mechanisms. Changes in the reaction mechanism of an enzyme as the pH is changed might yield valuable information on the role of ionizing groups in the reaction, and variation of the kinetic parameter V/K_m with the nature of the substrate may provide further information on the nature of the substrate-binding site in the enzyme.[31]

pH-Independence of K_m

Haldane[32] pointed out that the constancy of K_m over a pH range in which V showed a typical bell-shaped curve, observed for yeast invertase, might be explained by equality of K_m and K_s. Thus Eq. (40), given by

---

[31] J. S. Shindler and K. F. Tipton, *Biochem. J.* **167**, 479 (1977).
[32] J. B. S. Haldane, "Enzymes." Longmans, London, 1930 (reprinted 1965 by M.I.T. Press, Cambridge, Massachusetts).

the mechanism of Eq. (39), simplifies to give pH independence of K_m only if $k_2 \ll k_{-1}$, $K_A^E = K_A^{ES}$, and $K_B^E = K_B^{ES}$, and then K_m simplifies to k_{-1}/k_1, that is, K_s. Cornish-Bowden[28,33] has extended the analysis and also concluded that the constancy of K_m required that $K_A^E = K_A^{ES}$ and $K_B^E = K_B^{ES}$ and that the rate constants for binding and dissociation of substrate and product must be unaffected by the protonation and deprotonation that inactivate the enzyme–substrate complex, so that the reaction can be represented as shown in Eqs. (47)

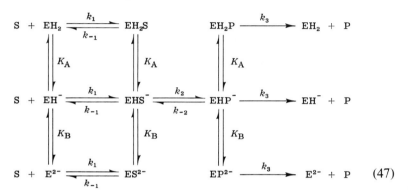

Cornish-Bowden showed, however, that this mechanism gives constancy of K_m even if substrate binding does not approach equilibrium, i.e., when the condition $k_2 \ll k_{-1}$ does not hold, provided that then $k_{-1} = k_3$, i.e., that substrate and product dissociate equally rapidly, which may be quite likely if they are chemically similar. If this is the cause of constancy of K_m, it again proves to simplify to k_{-1}/k_1, so the pH independence gives the valuable information that $K_m = K_s$. It is only when $k_{-1} \gg k_2$, however, that the pK values derived from the pH dependence of V are the true values of pK_A and pK_B; if K_m is independent of pH because of equality of k_{-1} and k_3, then the observed pK values are shifted further apart, each moved from pK_A or pK_B by $\log[(k_2 + k_{-2} + k_3)/k_3]$.

The assumptions $K_A^E = K_A^{ES}$ and $K_B^E = K_B^{ES}$ can give constancy of K_m without equality of K_m and K_s in a simpler system, such as that represented by Eqs. (22) and (23). Such systems are, however, somewhat unrealistic with these assumptions, because it would be surprising for substrate binding to have no effect on protonation equilibria and yet to be completely inhibited by these protonations and deprotonations. Hence constancy of K_m fairly reliably establishes that $K_m = K_s$, although, in

---

[33] A. Cornish-Bowden, "Principles of Enzyme Kinetics," p. 101. Butterworth, London, 1976.

view of the restrictions needed to give it, it is likely to be a much rarer phenomenon than equality of K_m and K_s.

Identification of Amino Acid Residues from Their Ionizations

pK Values

The table shows the normal ranges of pK values that have been found for the ionizing groups of amino acid residues in proteins. Comparison of the values with those obtained from studies of the effects of pH on enzyme activity may suggest the groups involved in different stages of the enzymic reaction. Such suggestions are only tentative for two reasons. The first is that, as already noted, the kinetic pK values may not be true pK values. The second is that the pK values of groups may be greatly affected by their environments, whereas the values given in the table are those for groups in an aqueous environment. A hydrophobic environment will stabilize the un-ionized forms of groups. This will be the protonated form for a "neutral acid group" (e.g., —COOH) so that its pK is raised, and the deprotonated form of a "cationic acid group" (one that bears a positive charge when protonated, e.g., —NH_3^+), whose pK will be lowered. Similarly, proximity to another charged group can greatly affect a pK, a positive charge lowering it and a negative charge increasing it. An extension of the effect of charge is the possibility of hydrogen bonding; a proton may be lost more easily because its binding site may hydrogen-bond to another bound proton and be thereby stabilized, or hydrogen bonding to a neighboring group in addition to binding to its primary site of

TYPICAL pK VALUES FOR SIDE-CHAIN IONIZING GROUPS[a]

Residue	pK	ΔH of dissociation	
		kJ/mol	kcal/mol
Asp, Glu	4	0	0
His	6	30	7
Cys	9	30	7
Lys	10	45	12
Tyr	10	30	7

[a] The pK of a terminal carboxyl is about 3.5 and that of a terminal ammonium is about 8; these values are lower than those of the corresponding groups in side chains because of the electron-withdrawing effect of the adjacent peptide bond. Other displacing factors are discussed in the text; because they include nearby charges, the values shown above do not apply to the free amino acids.

attachment may help to retain a proton. It is therefore not surprising that several groups in proteins are known to have pK values far outside the ranges shown. Perhaps the most extreme is the lysine residue of acetoacetate decarboxylase that forms an imine with the substrate. There is good evidence[34] that it has a pK of 5.9. The pK of 8.8 of a terminal amino group of δ-chymotrypsin proves to be made up of a pK of 7.9 in the conformation favored at high pH, in which the group is exposed to the medium (i.e., a normal value), and a pK of 10.0 in the conformation favored at low pH, in which it is adjacent to a negative charge inside the protein molecule.[35] Further, a carboxyl group in lysozyme is displaced in pK from 5.2 to 6.0 by ionization of a neighboring aspartic residue and to above 8 by combination with a long-chain substrate.[36] Assumptions had to be made in all these assignments, and they are discussed later.

Heats of Ionization

Examination of the table shows that even without displacement of pK values the identification of groups may be difficult because the ranges of values for the different groups overlap. Determination of the heat of dissociation of a group from the temperature dependence of its pK may allow them to be distinguished. For a dissociation reaction [Eq. (48)]

$$\Delta H = RT^2 d(\ln K)/dT = -R\, d(\ln K)/d(1/T)$$
$$= 2.303R\, d(pK)/d(1/T) \quad (48)$$

where ΔH is the enthalpy change on dissociation, R the gas constant, and T the absolute temperature. Thus a graph of pK against $1/T$ should have a slope of $\Delta H/(2.303R)$. Values of typical enthalpies of dissociation are shown in the table. Although they may be helpful when pK values are similar, they are not reliable for distinguishing groups whose pK values have been displaced by their environment, because this displacement may change the ΔH.[24]

As an approximation we may consider that a pK is displaced by a change in its ΔH of ionization. The change in ΔG^0 is $2.303RT$, i.e., 6 kJ/mol, for each unit the pK is displaced, so we might expect ΔH to be changed by the same amount. Displacement of the pK of a typical amino group from 10 to 7 would therefore change its ΔH from 45 to 27 kJ/mol; similarly, displacement of the pK of a typical carboxyl group from 4 to 7 would change its ΔH of dissociation from 0 to 18 kJ/mol. The difference in ΔH that remains is rather small on which to attribute a pK of 7 to one of these groups rather than the other. Further, it can only be an approxi-

[34] D. E. Schmidt and F. H. Westheimer, *Biochemistry* **10**, 1249 (1971).
[35] A. R. Fersht, *Cold Spring Harbor Symp. Quant. Biol.* **36**, 71 (1971).
[36] S. M. Parsons and M. A. Raftery, *Biochemistry* **11**, 1623 (1972).

mation that the changes in ΔG and ΔH are equal. An ionization may be accompanied by a change in the ordering of water molecules. Factors that displace the pK may easily alter the degree of this change. At biological temperatures (i.e., not far from the freezing point of water), this will have little effect on ΔG, but a large one on ΔH (compensated in ΔG by an opposite one in $-T\Delta S$).

Effect of Dielectric Constant

An approach that allows neutral acid and cationic acid groups to be distinguished is the effect of alteration of the dielectric constant of the medium on the observed pK. A difficulty is how to define the pH of a medium other than water, and two ways round this have been found.

The more rigorous method[37] was used for studying the pK values of pancreatic ribonuclease. The pH is defined as the pH that the buffer used would give in water, i.e., as pK_{buffer} + log([buffer base]/[buffer acid]). This may be determined by measuring the pH of the solution before addition of organic solvent.[29] Then the effect of addition of dioxane or formamide is considered. A cationic acid group ($-GH^+$ and $-G$) would react with a neutral acid buffer (HB and B^-) as follows:

$$-GH^+ + B^- \rightleftharpoons -G + HB \qquad (49)$$

Since lowering the dielectric constant favors uncharged species it will displace this reaction to the right. Hence at a fixed pH, i.e., at a fixed ratio $[B^-]/[HB]$, a greater fraction of the group will be deprotonated, so that its pK is diminished. In cationic acid buffers, however, the reaction becomes

$$-GH^+ + B \rightleftharpoons -G + HB^+ \qquad (50)$$

so the total charge on both sides is the same, and the equilibrium is unaffected by addition of organic solvent. Likewise the pK of a neutral acid group is raised in cationic acid buffers and unchanged in neutral acid buffers. Hence a study of the effect of addition of organic solvent on the pK of the group in the two types of buffer assigns the group to its class.

A more rapid method, which appears to work just as satisfactorily, is to calibrate a glass electrode in water and use it directly in the mixture of water and organic solvent. At least when dioxane is the solvent added, the electrode appears to measure correctly the activity of hydrogen ions. The two reactions to be considered, for cationic acid groups and neutral acid groups, respectively, are

$$-GH^+ \rightleftharpoons -G + H^+ \qquad (51)$$
$$-GH \rightleftharpoons -G^- + H^+ \qquad (52)$$

[37] D. Findlay, A. P. Mathias, and B. R. Rabin, *Biochem. J.* **85**, 139 (1962).

The charge is unchanged in the first reaction, and organic solvent proves to have little effect on the measured pK. In the second reaction, however, organic solvent inhibits dissociation, since this reaction produces charged species; thus addition of dioxane to water to 80% concentration progressively raises the pK of formic acid from 3.72 to 6.11.[38]

Two warnings should be given on the use of this method. One is that the organic solvent may change the conformation and ionizing properties of the enzyme under study. The other is that the molecular pK observed may have groups of different types contributing to it. Thus the carboxylate–imidazole system of chymotrypsin has a pK of 7 for protonation of form (I) to form (II) [Eq. (53)].

$$\text{(I)} \rightarrow \text{(IIa) or (IIb)} \tag{53}$$

Since form (II) has an infrared spectrum like that of an undissociated carboxylic acid,[39] and the 2-proton of the imidazole has a nuclear magnetic resonance spectrum like that of unprotonated imidazole,[40] there is strong evidence (with some qualification[3]) that form (IIa) contributes to form (II) much more than does form (IIb). In both these forms the proton is added to the side of the histidine that is accessible to water, and the net charge of the complex is that of a neutral acid. Nevertheless, the test with organic solvent gives the answer of a cationic acid.[38] Clearly the system is very different from that of Eqs. (51) and (52), where all the charged and uncharged forms are surrounded by water, since the carboxyl group and one side of the imidazole ring are buried.

Direct Titration of Groups

The identification of a specific ionizing group from its pK can be greatly strengthened if it can be shown that the group actually titrates with

[38] T. Inagami and J. M. Sturtevant, *Biochim. Biophys. Acta* **38**, 64 (1960).
[39] R. E. Koeppe II, and R. M. Stroud, *Biochemistry* **15**, 3450 (1976).
[40] M. W. Hunkapiller, S. H. Smallcombe, D. H. Whitaker, and J. H. Richards, *Biochemistry* **12**, 4732 (1973).

the determined pK. The complete titration curve of a protein with acid or alkali is usually too complex to be interpreted, because many groups are present and their titration curves overlap. The pK values of a few specific groups may be found from the differences in proton binding or release between the native enzyme and enzyme in which a specific residue has been chemically modified to prevent its ionization. Such a procedure was used to determine the pK of one aspartic residue (Asp-52) in lysozyme by comparing titration curves before and after its esterification with Et$_3$O$^+$ ion.[36] Alternatively, all groups of a certain type except the one under study may be blocked, when the specific group is protected either by its natural inertness or by specific protection (e.g., with substrate), so that the titration of this group may be studied directly (see, e.g., Fersht and Sperling[41]). Sometimes these two methods are combined.

The titration of some specific classes of residues in proteins can be followed by measuring a property that changes when the residues ionize. Perhaps the best known of these approaches uses the change in ultraviolet absorbance of tyrosine residues when they ionize, and a similar method can be used for cysteine residues (see, e.g., references in footnotes 42 and 43). The ionization of carboxyl groups can be followed by the accompanying shift in an infrared absorption band from about 1710 cm^{-1} to about 1570 cm^{-1}.[39,44,45] Because water absorbs strongly near these bands, it is necessary to carry out these studies in ^{2}H$_2$O. Proton magnetic resonance in ^{2}H$_2$O may be used to observe the titration of histidine residues, because the resonance of the proton on C-2 of the imidazole ring is sensitive to the ionization state of the ring and is shifted out of the envelope of the numerous aliphatic protons of the protein.[40,46,46a] Ionization of tyrosine residues can be similarly followed.[47] ^{13}C-Nuclear magnetic resonance has been used to titrate lysine and aspartic residues.

These methods are not quite as direct as might appear. A difference titration between a protein with one group blocked and the unmodified protein can be interpreted simply only if blocking the group does not affect the titration of any other ionizing groups in the protein. Parsons and Raftery[36] dealt ingeniously with one change in another group (see the section Examples of pH Studies, below). Similarly, the blocking of all groups ex-

[41] A. R. Fersht and J. Sperling *J. Mol. Biol.* **74** 137. (1973).
[42] S. N. Timasheff, *in* "The Enzymes" (P. Boyer, ed.), 3rd ed., Vol. 2, p. 371. Academic Press, New York, 1970.
[43] M. J. Gorbunoff, *Biochemistry* **10,** 250 (1971).
[44] H. Susi, T. Zell, and S. N. Timasheff, *Arch. Biochem. Biophys.* **85,** 437 (1959).
[45] S. N. Timasheff and J. A. Rupley, *Arch. Biochem. Biophys.* **150,** 318 (1972).
[46] G. Robillard and R. G. Schulman, *J. Mol. Biol.* **71,** 507 (1972).
[46a] J. L. Markley, *Acc. Chem. Res.* **8,** 70 (1975).
[47] S. Karplus, G. H. Snyder, and B. D. Sykes, *Biochemistry* **12,** 1323 (1973).

cept the one under study assumes that the modification does not affect this group. The methods based on specific properties of groups will give precise molecular pK values and an estimate of the contribution of the group to a molecular pK. Often all that is wanted is such an estimate, as the group may contribute overwhelmingly [$K_x \gg K_y$ in Eq. (1)]. But the precise contribution cannot be determined without assuming that the change of property on ionization is completely independent of environment.[3] Thus, in the procedures just cited for showing that form (IIa) contributes more than (IIb) in Eq. (53), the properties of carboxyl and carboxylate cannot be assumed to be exactly the same with N or HN groups of imidazole as close as they are in model compounds; the interaction that so much affects their pK is likely to affect their other properties.

Effects of pH on Enzyme Inhibition

A simple competitive inhibitor will bind to an enzyme in a rapid equilibrium, and thus the inhibitor constant will be the simple dissociation constant for the EI complex and some of the possible complications involved in consideration of the effects of substrate can be neglected. For example, for the system

$$\begin{array}{ccccc}
EH_2I & \xrightleftharpoons{K_i'} & EH_2 & & \\
\updownarrow K_A^{EI} & & \updownarrow K_A^{E} & & \\
EHI^- & \xrightleftharpoons[I]{K_i} & EH^- & \xrightleftharpoons{S} & EHS^- \longrightarrow \text{etc.}
\end{array} \quad (54)$$

the pH dependence of K_i will be given by:

$$K_i = \tilde{K}_i \frac{1 + [H^+]/K_A^E}{1 + [H^+]/K_A^{EI}} \tag{55}$$

and thus it should be possible to determine the pK values from a graph of pK_i against pH.[10]

The situation with simple uncompetitive inhibitors should also be relatively easy to analyze, giving the ionization constants for the ES and ESI complexes, but mixed and noncompetitive inhibition will be more complicated because of the alternative forms of the enzyme to which the inhibitor can bind.

Effects of pH on Individual Steps of an Enzyme Reaction

If it is possible to study the effects of pH on one step in an enzyme reaction by using rapid reaction techniques it should be possible accurately to determine the pK values of the groups controlling this step. An example of such an approach concerns the hydrolysis of 4-methyl umbelliferyl phosphate by alkaline phosphatase, where the reaction proceeds by way of a phosphoenzyme intermediate. Stopped-flow studies on the effects of pH on the reaction showed that the rate constant for the formation of the phosphoenzyme was relatively insensitive to pH in the range 5.3–7.5, whereas that for breakdown of the intermediate to yield phosphate increased with increasing pH, resulting in a change in the rate-limiting step of the reaction at about pH 6.5.[48]

Effects of pH on Chemical Modification of an Enzyme

This method is designed to identify the pK value of a specific group at the active site of an enzyme by measuring the effect of pH on the rate at which it reacts with a highly specific irreversible inhibitor.[34] If the inhibitor (X) reacts with the enzyme in a single irreversible step

$$\begin{array}{c} \text{EH} \\ \| K_A \\ \text{E}^- \xrightarrow{\tilde{k}_o x} \text{E}^- - \text{X} \end{array} \qquad (56)$$

the observed rate constant (k_0) will be given by

$$k_0 = \frac{\tilde{k}_o x}{1 + [\text{H}^+]/K_A} \qquad (57)$$

where x represents the concentration of X. Thus the value of pK_A can be found by determining the effect of pH on k_0. The scheme shown in Eq. (56) can easily be expanded to allow for two ionizations of the free enzyme.

If x is much greater than the initial enzyme concentration ($[\text{E}]_0$) *pseudo* first-order conditions apply, and thus

$$[\text{E}] = [\text{E}]_0 - [\text{E} - \text{X}] = [\text{E}]_0\, e^{-k_0' t} \qquad (58)$$

where [E] is enzyme concentration at any given time, t, and the *pseudo*

---

[48] S. E. Halford and M. J. Schlesinger, *Biochem. J.* **141**, 845 (1974).

first-order rate constant (k_0') is given by

$$k_0' = k_0 x \tag{59}$$

Equation (58) can be written as

$$\ln \frac{[\mathrm{E}]_0}{[\mathrm{E}]} = k_0' t \tag{60}$$

and thus k_0' can be determined from a plot of log $[\mathrm{E}]_0/[\mathrm{E}]$ against time.

If the inhibitor first forms a reversible complex with the enzyme according to the mechanism

$$\begin{array}{c} \mathrm{EH} \\ \Vert K_\mathrm{A} \\ \mathrm{E^-} \underset{k_{-1}}{\overset{k_1 x}{\rightleftharpoons}} \mathrm{EX^-} \xrightarrow{\tilde{k}_0} \mathrm{E^-} - \mathrm{X} \end{array} \tag{61}$$

the rate equation becomes

$$k_0 = \frac{\tilde{k}_0}{1 + (\tilde{K}_\mathrm{m}/x)(1 + [\mathrm{H^+}]/K_\mathrm{A})} \tag{62}$$

where $\tilde{K}_\mathrm{m} = (k_{-1} + k_2)/k_1$.

Systems obeying this mechanism can be distinguished from those obeying that shown in Eq. (56) because Eq. (57) predicts that k_0 will be a linear function of x whereas Eq. (62) predicts that the dependence will be hyperbolic.[49] Thus K_A can be evaluated graphically by using the reciprocal form of Eq. (62).

$$\frac{1}{k_0} = \frac{\tilde{K}_\mathrm{m}}{\tilde{k}_0}\left(1 + \frac{[\mathrm{H^+}]}{K_\mathrm{A}}\right)\frac{1}{x} + \frac{1}{\tilde{k}_0} \tag{63}$$

Thus a graph of $1/k_0$ against $1/x$ at a series of concentrations of $\mathrm{H^+}$ will yield a family of straight lines that intersect on the $1/k_0$ axis at a value corresponding to $1/\tilde{k}_0$, and replots of the slopes of these lines against the concentration of $\mathrm{H^+}$ will extend to cut the baseline at $-K_\mathrm{A}$. An alternative procedure, suggested by Schmidt and Westheimer,[34] is to use such a low inhibitor concentration that Eq. (62) simplifies to

$$k_0 = \frac{\tilde{k}_0}{(\tilde{K}_\mathrm{m}/x)(1 + [\mathrm{H^+}]/K_\mathrm{A})} \tag{64}$$

when K_A can be determined directly from the effect of pH on k_0. If the

---

[49] R. Kitz and I. B. Wilson, *J. Biol. Chem.* **237**, 3245 (1962).

intermediate enzyme–inhibitor complex is able to ionize

$$\begin{array}{ccc} \text{EH} & & \text{EXH} \\ \Updownarrow K_A & & \Updownarrow K_B \\ \text{E}^- & \underset{k_{-1}}{\overset{k_1}{\rightleftharpoons}} \text{EX}^- & \overset{\tilde{k}_o}{\longrightarrow} \text{E}^- - \text{X} \end{array} \qquad (65)$$

the rate equation becomes

$$k_0 = \frac{\tilde{k}_0}{(\tilde{K}_m/x)(1 + [\text{H}^+]/K_A) + 1 + [\text{H}^+]/K_B} \qquad (66)$$

Thus plots of $1/k_0$ against $1/x$ at a series of different H^+ concentrations will yield a family of lines that intersect to the left of the vertical axis (mixed inhibition). The values of K_A and K_B can be found from replots of the slopes and vertical axis intercepts against $[\text{H}^+]$.

If the reasonable assumption is made that the protonated form of the enzyme can also bind the inhibitor, then the system becomes

$$\begin{array}{ccc} \text{EH} & \underset{k_{-2}}{\overset{k_2 x}{\rightleftharpoons}} & \text{EHX} \\ \Updownarrow K_A & & \Updownarrow K_A' \\ \text{E}^- & \underset{k_{-1}}{\overset{k_1 x}{\rightleftharpoons}} \text{EX}^- & \overset{k_0}{\longrightarrow} \text{E}^- - \text{X} \end{array} \qquad (67)$$

As in the case in which more than one form of the enzyme could bind substrate [see Eqs. (39) and (40)], steady-state treatment of this system gives a complex equation [Eq. (68)].

$$k_0 = \frac{\tilde{k}_0}{1 + [\text{H}^+]/K_A'} \bigg/ \left[1 + \frac{(K_m + k_{-2}[\text{H}^+]/k_1 K_A')(1 + [\text{H}^+]/K_A)}{(1 + k_2[\text{H}^+]/k_1 K_A)(1 + [\text{H}^+]/K_A') \, x} \right] \qquad (68)$$

This equation does not predict a simple dependence of k_0 upon pH and cannot be used to determine either of the pK values. If, however, it is assumed that all the binding processes remain in equilibrium, i.e., that all proton transfer steps are very fast and $k_0 \ll k_{-1}$, the equation simplifies to Eq. (66) with K_m replaced by K_s (k_{-1}/k_1). If it is assumed that the proton transfer reactions are fast but $k_0 \not\ll k_{-1}$, the equation becomes [cf. Eq. (40)]

$$k_0 = \frac{k_0}{1 + [\text{H}^+]/K_A'} \bigg/ \left[1 + \frac{k_2 + k_{-1}(1 + [\text{H}^+]/K_A)}{k_1 (1 + [\text{H}^+]/K_A')} \frac{1}{x} \right] \qquad (69)$$

which would allow pK_A' to be determined but results in a displacement of

pK_A from the true value. It can be argued[21] that k_1 is likely to exceed 10^6 M^{-1} sec^{-1} when the inhibitor is a substrate analog, so that observed values of much under this for the pH-independent value that k_0 approaches at its pH optimum imply that $k_0 < k_{-1}$ so that K_A may be determined. Clearly it is possible to devise more complicated reaction schemes, which can result in displacement of both pK values. Thus, as with studies on the effects of pH on the kinetic parameters of the overall enzyme-catalyzed reaction, this method will give true pK values only if certain simplifying assumptions are made about the kinetic mechanism operative.

The Competitive Labeling Method

Whereas the previous method required a reagent with a high specificity in reacting with a single residue of the enzyme, the competitive labeling method[50] uses a reagent of low specificity that may react with all amino acid residues of a given type. An electrophilic reagent is radioactively labeled and allowed to react with a mixture of the enzyme and a standard nucleophile. So little reagent is used that the proportions of the nucleophiles of different reactivities that are present do not change in the course of the reaction, since only a small fraction of each is used up. The reagent concentration therefore falls to zero with pseudo-first-order kinetics. The fraction of any group modified is thus proportional to the second-order rate constant for its reaction with the reagent. After this reaction is complete, an excess of unlabeled reagent is added to complete the conversion of each nucleophile present into its modified form. The fraction modified in the first reaction is now represented by the degree of labeling of the modified form.

The enzyme is digested enzymically, its peptides are separated, and the specific radioactivity of each of the modified residues is determined. If the rate constant for modification of the standard nucleophile is known, the rate constant can be calculated for the reaction of each group with the reagent, from the ratio of labeling of the modified group to that of the modified standard nucleophile.

The procedure is repeated at a number of different pH values, and correction is made for the titration of the standard nucleophile. Hence the apparent rate constant for reaction of each group can be plotted against pH, to give both the pK of the group and the rate constant for modification of its unprotonated form.

The method was first used for the acetylation of amino groups with acetic anhydride[50] and the pK values of several of these groups in elastase

[50] H. Kaplan, K. J. Stevenson, and B. S. Hartley, *Biochem. J.* **124,** 289 (1971).

were found. The method assumes that the degree of binding of the reagent to specific sites is small, i.e., that the reaction is first-order in reagent. This can be checked by the constancy of relative labeling as the concentration of the reagent is diminished. It also assumes that no pH-dependent structural transition occurs in the enzyme over the operative pH range. In fact, however, Kaplan et al.[50] found a marked increase in the rate constant with pH just above the range. Fortunately, the rate of increase with pH had fallen enough (because of passing the pK) before the onset of the second increase (attributed to unfolding of the enzyme molecule), to allow pK values to be assigned. When, however, Cruikshank and Kaplan[51] applied the method to the modification of histidine residues in chymotrypsin by 1-fluoro-2,4-dinitrobenzene, a fall of reactivity followed the rise as the pH was raised. They attributed the first slowing in the rate of increase to approach of a pK and the later fall to a change of folding, but it is not clear that the phenomena were well enough separated to allow great reliance on the pK values found.

Although the degree of binding must be small, and the fact that it is small can be checked, binding forces may make important contributions to the rate constants. Thus the conclusion that two histidine residues of chymotrypsin, found to be more reactive than expected from their pK values, are specifically activated by chemical means[51] may be unjustified, since preferential binding of the hydrophobic reagent to a protein surface (or parts of it) may be all that is needed to explain the enhanced reactivity.

These two difficulties in interpretation illustrate some limitations of this powerful method.

Limitations of the Methods

The kinetic study of the effects of hydrogen ions as activators or inhibitors of enzyme-catalyzed reactions can yield useful information on the kinetic mechanism followed. As in all kinetic studies it is not possible to conclude that a given kinetic mechanism operates, since several mechanisms can usually be devised that will fit the results. It is necessary to select the simplest mechanism that does so and to recognize that more complex models may have to be considered when further results are obtained.

The first problem in using kinetic results on pK values for identifying the roles of ionizing groups is that simplifying assumptions must be made about the kinetic mechanisms that apply. There is usually no satisfactory justification for these assumptions. Hence although kinetic results can provide a useful indication of the types of groups involved, they can do

---

[51] W. H. Cruikshank and H. Kaplan, *Biochem. J.* **147**, 411 (1975).

little more than this. The pK values found may not be ionizations at all; if they are, they may be distorted by rate constants, and they will be molecular values to which more than one group may contribute. Even true pK values of groups may be far displaced by the environment from the normal values of the groups concerned.

It is sometimes assumed that pK values that affect enzyme activity are those of groups near the active site of the molecule. The ionization of a distant group, however, like that of the terminal isoleucine of chymotrypsin[35,52] may result in a reversible conformational change that converts the enzyme into an inactive form. The coupling of an ionization to a conformational change may, however, raise its temperature coefficient enough to reveal such coupling.

Practical Aspects

Effects of pH on the Stability of Enzymes

Many enzymes are irreversibly denatured at extreme pH values; unless this is allowed for, it may be mistaken for a reversible ionization. The stability is more easily investigated by preincubating samples of the enzyme at different pH values. The duration of the preincubation should be at least as long as the usual assay time. The enzyme is then readjusted to a pH at which it is known to be stable for assay. Possible results that could be obtained in such studies are shown in Fig. 10. In Fig. 10a the enzyme is stable for the required period over the relevant range of pH, so the effects of pH on activity can be interpreted in terms of fully reversible ionizations. In Fig. 10b there is an irreversible loss of activity on the alkaline side of the optimum, so that results in this region should not be attributed to reversible ionizations. Although the amount of active enzyme left at the end of the preincubation falls rather sharply with increasing pH after 50% inactivation in the example shown in Fig. 10b (where it is assumed that the rate of inactivation is proportional to [OH$^-$]), the total activity shown over the assay period (Fig. 10b, solid line) follows a curve like that of a reversible titration and could easily be mistaken for one unless the stability check is made.

In the experimental test just outlined it is important that the conditions for preincubation should resemble those used in the assay itself. Enzyme concentrations should be comparable, because many enzymes become less stable on dilution. The ionic strength should also be similar. It may be convenient to work at such a high value of ionic strength that changes in buffers and substrate concentrations make a negligible difference to its total value. Alternatively, the ionic strength should be adjusted to the

[52] A. Himoe, P. C. Parks, and G. P. Hess, *J. Biol. Chem.* **242**, 919 (1967).

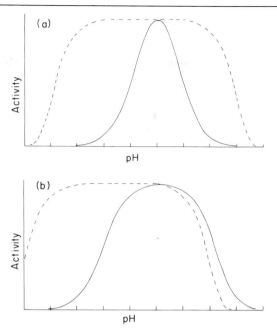

FIG. 10. Checks of enzyme stability. Solid lines represent assay at the pH indicated; dashed lines represent the results of assay at the optimum pH after incubation at the pH indicated for a period equal to the length of the assay. The result in (a) shows that the enzyme is almost fully active after preincubation at any pH at which it showed appreciable activity in the assay, so the assay can be taken to be valid in showing reversible effects. On the alkaline side of (b), however, the falloff in activity proves not to be a measure of reversible effects because the dashed line shows that the enzyme is irreversibly inactivated during assay. The form of irreversible inactivation shown by the dashed line assumes that the activity left at the end of the incubation is e^{-kt} of its value at the start, and that k is proportional to [OH$^-$]. The solid line on the alkaline side of (b) has been plotted on the assumption that there is no reversible loss of activity over the pH range. The amount of product produced over the course of the assay during which the enzyme is being inactivated as a fraction of what would have been formed if the enzyme had been stable is $(1 - e^{-kt})/k$. This can be substantial even when most of the enzyme is inactive by the end of the assay.

same value in each experiment, or the effect of ionic strength on enzyme activity and stability may be separately determined.

Effects of pH on Substrates

Stability of Substrate

Although it is generally recognized that enzymes may be unstable at extremes of pH, it is sometimes forgotten that the same may be true of substrates. Not only may their concentrations be diminished by breakdown, but their breakdown products are likely to be enzyme inhibitors since they share some molecular features with the substrates. The pH stability of the substrate can be checked in much the same way as that of the enzyme. If one of the substrates is unstable at extremes of pH, a study of the time course of its breakdown should indicate whether it would still be possible to determine true initial rates near these extremes when the reaction is initiated by addition of that substrate. Examples of common substrates that are unstable at some pH values include NAD^+, which decomposes at high pH, and NADH, which decomposes at low pH.[53]

Ionizations of Substrates

The simple interpretation of the effects of pH on enzyme activity assumed that the effects were due to ionizations of the free enzyme or of enzyme–substrate complexes and that the substrate did not ionize over the pH range considered. In fact every ionization attributed above to the free enzyme could equally well be due to free substrate. Similarly, ionizations of the enzyme–substrate complex may be due to ionization of a group on the bound substrate. The pK values of the substrates should be determined so that they may be recognized if they should appear in kinetic plots. It is best to titrate the substrate under the conditions used, since a literature value for its pK may apply to a different temperature or ionic strength.

Further, it may not be easy to avoid systematic errors associated with a particular electrode assembly, but these will not greatly matter if all pK values of a given study are self-consistent.

If the following mechanism applies

$$\begin{array}{ccc}
E + SH & \underset{k'_{-1}}{\overset{k'_1}{\rightleftharpoons}} & ESH \\
\updownarrow K_A^S & & \updownarrow K_A^{ES} \\
E + S^- & \underset{k_{-1}}{\overset{k_1}{\rightleftharpoons}} & ES^- \xrightarrow{k_2} E + P
\end{array} \qquad (70)$$

then this is essentially that of Eq. (39) except that the substrate ionization constant K_A^S replaces that of the free enzyme K_A^E. All the same conclusions apply; e.g., Eq. (23) will apply with K^S for K^E if substrate binding is in equilibrium. In effect the protonated form of the substrate is acting as a reversible inhibitor of the enzyme, since sequestration of some enzyme as ESH lowers the concentration of ES^-.

[53] O. H. Lowry, J. V. Passonneau, and M. K. Rock, *J. Biol. Chem.* **236**, 2756 (1961).

Enzymes often bind only one ionic form of the substrate; more strictly, binding greatly favors one form. Thus an enzyme acting on a phosphorylated substrate may bind only the dianion R—O—PO$_3^{2-}$ (i.e., S^{2-}) and not R—O—PO(OH)—O$^-$ (i.e., SH$^-$). The favored binding of the dianion lowers the pK of the bound form of substrate, possibly to below the pH range under study; indeed it may not be possible to obtain an appreciable concentration of bound SH$^-$ at attainable values of the pH and substrate concentration. Thus in the scheme

$$
\begin{array}{ccc}
E + SH^- & \underset{}{\overset{K'}{\rightleftharpoons}} & ESH^- \\
\updownarrow K_A & & \updownarrow K'_A \\
E + S^{2-} & \underset{}{\overset{K}{\rightleftharpoons}} & ES^{2-} \longrightarrow E + P
\end{array}
\qquad (71)
$$

$K'_A \gg K_A$ and $K' \gg K$. We can therefore ignore K'_A and K' as well as [ESH$^-$]. If the enzyme follows the simple scheme of Eq. (22), K_A^S will be seen in V/K_m [Eq. (27)] but not in V [Eq. (25)], so the pK of the substrate will have disappeared from the enzyme–substrate complex, since this complex does not protonate appreciably over the pH range under study.

The complications of several overlapping protonation ranges are increased by substrate ionizations, so it is prudent, if possible, to use substrates that do not change their degree of ionization over the pH range under study. The important point, however, is to allow for substrate ionizations.

A further complication can occur if the true substrate for a reaction is the complex between the substrate and a metal ion. Ionization of the substrate may then alter its affinity for the metal ion, and this effect must be allowed for in calculating substrate concentrations. This problem has been considered by Fromm,[54] and details of a computer program for the calculation of the various species in solution have been published by Storer and Cornish-Bowden.[55]

Substrate Concentrations

An accurate determination of the effects of pH on an enzyme requires the determination of the apparent values of K_m and V at a series of different pH values. A number of workers have sought to diminish the

---

[54] H. J. Fromm, "Initial Rate Enzyme Kinetics," p. 55. Springer-Verlag, Berlin and New York, 1975.
[55] A. C. Storer and A. Cornish-Bowden, *Biochem. J.* **159**, 1 (1976).

amount of work involved by studying the effects of pH at only two substrate concentrations: very high and very low. By inspection of the Michaelis–Menten equation $v = Vs/(s + K_m)$ it can be seen that when $s \gg K_m$ then $v \to V$, so that variation of initial velocity with pH will correspond to the variation of V. Likewise at a very low substrate concentration when $s \ll K_m$ then $v \to Vs/K_m$, so that the variation in initial velocity will correspond to that of V/K_m. The disadvantage of using this method is that K_m would be expected to vary with pH so that the required condition that $s \gg K_m$ or that $s \ll K_m$ may hold for a fixed substrate concentration at some pH values but not at others. Hence the enzyme may not remain saturated over the whole pH range at the higher substrate concentration, and its degree of saturation may not remain negligible at the lower. Further, it may not be possible to use sufficiently high substrate concentrations for the condition $s \gg K_m$ because of insolubility or because of inhibition by high substrate concentrations. Similarly low substrate concentration may render measurements imprecise because of insensitivity of the assay. A full kinetic study would therefore involve less ambiguity and would give more accuracy than the use of only two substrate concentrations.

Enzyme Reaction Involving More Than One Substrate

The theory developed in the preceding section has, for simplicity, dealt with an enzyme mechanism that involves only one substrate. The rate equations to give the effects of pH on the kinetics of reactions of two or more substrates may be readily derived if it is assumed that proton transfers occur so fast that they remain in equilibrium; a simplified procedure for deriving rate equations involving some steps at equilibrium has been developed by Cha.[56] Since a complete kinetic analysis of the effects of pH on such enzymes would take a long time, a high concentration of one of the substrates is often used so that the reaction may be treated by the equations that describe the single-substrate case. If this is done it is important to check that the concentration of the fixed substrate is sufficiently high to maintain saturation at all pH values. In the interpretation of the effects of pH on the kinetics of these reactions, it is also important to check that the kinetic mechanism does not change with pH; this necessitates a full kinetic analysis at the different pH values.

Effects of pH on the Assay Method

In all studies on the effects of pH on enzyme activity, it is important to ensure that the assay method used gives a true measure of the rate of the

---

[56] S. Cha, *J. Biol. Chem.* **243**, 820 (1968).

catalyzed reaction over the whole pH range studied. Complications that could occur include effects of pH on the activity of a coupling enzyme or on the absorbance of a product formed. If a stopped assay is used, i.e., one in which the enzyme-catalyzed reaction must be stopped to allow determination of the amount of product formed, then a check must be made at each pH that the rate of reaction is constant over the time of incubation; a check at one pH is insufficient.

Buffers

There is a wide choice of buffers that may be used in studying enzymes.[57-60] A valuable list of pK values from which buffers may be chosen is given by Jencks and Regenstein.[61] The equation

$$\text{pH} = \text{p}K + \log[\text{buffer base}] - \log[\text{buffer acid}]$$

shows that both the acid and its conjugate base must be in high concentration if the pH is not to be much changed by a small addition or withdrawal of protons, which turns one into the other. Hence the pH must not be far from the pK of the buffer, else one of these concentrations will be small and therefore easily changed. The useful buffering range extends about a unit on either side of the pK. It will therefore be necessary to use several buffers to cover a wide pH range. Their ranges should overlap so that assays can be made at the same pH in different buffers to check for possible inhibitory effects of specific buffer components. Sometimes a mixture of buffers has been used to cover a wide pH range, but this could lead to a failure to recognize that one of the components was inhibitory.

Individual buffers may inhibit a particular enzyme in a number of ways. One component might be a specific inhibitor. If only one ionic form was inhibitory the enzyme activity would decline with the titration curve of the buffer. This could be detected by absence of such a decline over the same pH range with another buffer. The inhibitor might be of any of the known types, competitive, truly noncompetitive, uncompetitive, and mixed. A less specific inhibition could be due to ligation by the buffer of an essential component of the reaction. Many buffer components are che-

---

[57] G. Gomori, this series, Vol. 1, p. 138.
[58] N. E. Good and S. Izawa, this series, Vol. 24, p. 53.
[59] N. E. Good, G. D. Winget, W. Winter, T. N. Connolly, S. Izawa, and R. M. M. Singh, *Biochemistry* **5**, 467 (1966).
[60] R. M. C. Dawson, D. C. Elliott, W. H. Elliott, and K. F. Jones, "Data for Biochemical Research," 2nd ed., p. 474. Oxford Univ. Press, London and New York, 1969.
[61] W. P. Jencks and J. Regenstein, *in* "Handbook of Biochemistry and Molecular Biology," 3rd ed. "Physical and Chemical Data" (G. D. Fasman, ed.), Vol. 1 p. 305. Chem. Rubber Publ. Co., Cleveland, Ohio, 1976.

lating agents and could thereby inhibit enzymes that require metal ions for activity. Since protons usually compete with metal ions for chelating agents, such an inhibitory effect could vary with the pH. Alternatively, a buffer component may appear to activate an enzyme if it removes an inhibitory metal ion. Lists of chelating ability[62] are useful to check this, and Good and co-workers[58,59] have listed several buffers with little chelating ability. A buffer may also affect an enzyme by supplying an essential or nonessential activator. Thus, for example, several enzymes are activated by potassium ions, so a change from potassium phosphate to sodium phosphate buffer would appear inhibitory unless another potassium salt were present. An even less specific effect might be mediated by a change of ionic strength, which should be kept constant or at least within a range where it has been shown not to affect the enzyme. These considerations indicate that a preliminary survey of the effects of several buffers and of ionic strength on the enzyme under study is an important preliminary to the more detailed investigation of the effect of pH.

Just as the ionization constants of groups in proteins can vary with temperature [Eq. (48)], changes in temperature will affect the pK values of buffer components and hence the pH values of buffer solutions; it is important to allow for this if studies are made at different temperatures. The pK values of amines have particularly large temperature coefficients (see the table); thus Tris buffers, for example, fall in pH by 0.3 per 10° rise in temperature. A glass electrode calibrated at one temperature needs care in operating at others because its temperature coefficient will depend on details of the measurement procedure, according to whether the solution at the new temperature brings to its own temperature only the one half-cell consisting of the glass membrane and the solution and electrode inside it, or also the second half-cell consisting of the reference electrode.

If pH measurements are to be made with added organic solvents, one of the approaches indicated above under effects of dielectric constant must be used. It is important to specify exactly what has been done, because various (and probably equally valid) definitions of pH are used for media other than water.

A particular change of solvent that may need to be considered is the substitution of ^2H_2O for water. If the reading of a pH meter calibrated in water is used directly, then 0.4 has to be added to give the reading that the meter would show in water with the same concentration of strong acid, or in mixtures the amount $(0.3139\alpha + 0.0894\alpha^2)$ has to be added, where α is the atom fraction of deuterium, i.e., $[^2H]/([^2H] + [^1H])$.[63] Since pK values

---

[62] L. G. Sillén and A. E. Martell, *Chem. Soc. Spec. Publ.* **17**, 1964; **25**, 1971.
[63] L. Pentz and E. R. Thornton, *J. Am. Chem. Soc.* **89**, 6931 (1967).

are affected by substitution of deuterium, other systems (such as specifying the pH the buffer mixture would show in water, or using uncorrected meter readings) may be just as valid, provided that what is actually done is clearly specified.

Since dissociation of a buffer acid involves a change of charge, the pK is likely to be affected by the ionic strength and hence by the buffer concentration. This effect is most marked with buffers (like phosphate) that contain multivalent species, both because the activity coefficients of such species are more sensitive to ionic strength than those of univalent species, and because multivalent species and their counterions contribute more to the ionic strength. Thus a buffer of 0.1 M NaH_2PO_4 and 0.1 M Na_2HPO_4 exhibits a pH of about 6.7 and this rises by about 0.2 on 10-fold dilution and 0.1 on a further 10-fold dilution.

A simple point too often overlooked is the importance of checking the pH of the reaction mixture rather than that of the added buffer, since other components may donate protons or combine with them. It may be possible to adjust all solutions to the pH of the assay system before they are mixed, but this may be inconvenient, especially if some of the components are not very stable under these conditions. It is also important to check the pH again after the reaction to ensure that the buffering has been effective.

Examples of pH Studies

Carnitine Acetyltransferase

This enzyme (EC 2.3.1.7), which is responsible for maintaining equilibrium in the transfer of acetyl groups between carnitine and CoA,[64] shows fairly straightforward kinetics. Initial rate and product inhibition studies are consistent with a random-order mechanism in which substrate binding is at, or close to, thermodynamic equilibrium,[65] and this conclusion has been supported by estimates of dissociation constants for the complexes of the enzyme with individual substrates and products obtained in studies of the optical rotatory dispersion of the enzyme.[66] Further, the binding site for the acetyl group is so located with respect to those for carnitine and CoA that a bromoacetyl group on bound carnitine alkylates the thiol group of bound CoA.[67]

Chase[68] studied the effect of pH on the activity of this enzyme over the

---

[64] D. J. Pearson and P. K. Tubbs, *Biochem. J.* **105**, 953 (1967).
[65] J. F. A. Chase and P. K. Tubbs, *Biochem. J.* **99**, 32 (1966).
[66] K. F. Tipton and J. F. A. Chase, *Biochem. J.* **115**, 517 (1969).
[67] J. F. A. Chase and P. K. Tubbs, *Biochem. J.* **111**, 225 (1969).
[68] J. F. A. Chase, *Biochem. J.* **104**, 503 (1967).

range 6–9. He took the precautions mentioned above, such as checking several buffers and measuring substrate pK values. The results with V proved to be simple: it was constant over the range studied. Hence there is no group in the enzyme–substrate complex with a pK within or close to this range whose ionization has appreciable effect on the catalyzed reaction. The plots of pK_m against pH for both carnitine and acetylcarnitine showed unit slope at low pH and zero slope at high pH, with the bend at pH 7.1–7.2. Since these substrates have no pK near this value there must be a group in the enzyme whose deprotonated form is required for binding (although we should note the ever-present kinetic ambiguity: we only know the molecular pK, so the group may need to be protonated provided that some other group predominantly protonated over the range studied needs simultaneously to be deprotonated). In view of the likelihood that the trimethylammonium group of these substrates is bound to a negatively charged group in the protein, the finding that protonation prevents binding is not surprising. The group evidently has the pK of 7.1–7.2 in the free enzyme and cannot be protonated above pH 6 in the enzyme–carnitine complex.

The plots of pK_m against pH for CoA and acetyl-CoA were more complex, and again similar to each other. The pK_m again rose with pH (with unit slope) at low pH and showed a bend to zero slope with a pK of 6.4. This corresponded to the pK of the phosphate group of CoA, as determined by titration. Evidently the dianion was the form preferentially bound. Chase checked this elegantly by measuring the pK_m for dephospho-CoA, which gave the same value of V as CoA. With it there was no change in pK_m from pH 6 up to above pH 7.

It may seem strange that removal of one negative charge from the dianionic form of the phosphate group of CoA prevents binding, but removal of both in preparing dephospho-CoA does not. It should, however, be realized that the slope of unity from pH 6 upward with CoA only implies that the dianion is the predominant form bound. If stability had allowed measurements to be continued below pH 6, a further bend in the curve would be expected when the predominance of monoanion in solution would overcome its lesser affinity for the enzyme; in other words, the enzyme–substrate complex would exhibit a pK. We do not expect this pK to be very low, since the dianion of CoA is favored over dephospho-CoA in binding by only about 40-fold.

As the pH is further raised, the pK_m for CoA or acetyl-CoA turns down with a pK of 7.85 and up again to zero slope with a pK of 8.25. Hence these pK values exist in the enzyme and enzyme–substrate complex, respectively. Since that in the enzyme–substrate complex is not seen in V, it does not appreciably affect the reaction catalyzed. Evidently some group in the enzyme of pK 7.85 needs to be deprotonated for optimal binding, but the smallness of the shift of pK shows that binding is only 2.5-fold worse if it is deprotonated. When dephospho-CoA was used, the turndown at 7.85 was similar, diminishing the already low affinity of this substrate. Measurements were not continued to check whether the horizontal slope was regained at pH 8.25 and above.

Bound bromoacetylcarnitine alkylates a histidine residue of the enzyme provided no CoA is bound (because its thiol group would be preferentially alkylated).[69] There is therefore an ionizing group close to the active site, and it shows nucleophilic reactivity at pH 7.2. It is not possible to attribute any of the kinetic pK values to it; possibly its pK is below the range studied.

Fumarate Hydratase

This enzyme (fumarase, EC 4.2.1.2) might appear simpler than carnitine acetyltransferase because it has only a single substrate (malate) in one direction and water is one of the two substrates in the other direction. In fact, however, it shows far more complexities. Cleland[29] has reviewed its action and included the effects of pH.

Its pH dependence has been studied in detail by Alberty and his co-workers.[4,70] The pH dependence of V/K_m gave pK values of 5.8 and 7.1, which on simple theory would be those of the free enzyme. Wigler and Alberty,[71] however, had already found that several dicarboxylic acids were competitive inhibitors of the enzyme and that their binding was diminished by protonation (pK 6.9). The shift of pK from these values for the free enzyme to those of the enzyme–inhibitor complex varied, as is to be expected, with the nature of the inhibitor. Thus inhibitor binding provides correct pK values for the free enzyme, which V/K_m did not.

The discrepancy might be explained by the rapid action of the enzyme; it approaches the diffusion-controlled limit,[70] so substrate binding cannot be in equilibrium. Although this could shift the pK values seen in V/K_m from those of the free enzyme, the shift observed is not that predicted by Eq. (40). As discussed earlier, this mechanism would give two pK values for the free enzyme plus two distorted values; if all forms of enzyme combined with substrate at the same rate ($k_1 = k_1' = k_1''$) only the two distorted pK values would be seen but the shift would be symmetrical, with pK_A lowered as much as pK_B is raised. Since this is not found, there must be some other complexity, which can, in fact, be shown to be that protona-

[69] J. F. A. Chase and P. K. Tubbs, *Biochem. J.* **116**, 713 (1970).
[70] D. A. Brant, L. B. Barnett, and R. A. Alberty, *J. Am. Chem. Soc.* **85**, 2204 (1963).
[71] P. W. Wingler and R. A. Alberty, *J. Am. Chem. Soc.* **82**, 5482 (1960).

tion of the enzyme–substrate complex is relatively slow. Exchange of ^{2}H or ^{3}H between water and malate can, at high substrate concentrations, be markedly slower than ^{14}C exchange between malate and fumarate.[30] It thus appears that the H$^+$ ion removed from malate to form fumarate can remain bound to the enzyme while the fumarate molecule dissociates and is replaced, so the reversible protonation of the enzyme–substrate complex is relatively slow.

Cleland[29] has pointed out that the molecular pK values for the free enzyme of 6.3 and 6.9 are those that would be expected for groups of identical pK of 6.6 if there is no interaction between the groups. This is a useful warning against allotting these pK values to different groups. Usually negative interaction of closely situated groups would be expected because of electrostatic interactions, but the closeness of the molecular pK values shows that it is absent here.

The pK values shown in V in the direction from malate, i.e., those attributable to the enzyme–malate complex, are 6.4 and 9.1.[70] Those in the reverse direction are 4.9 and 7.0. Those of 4.9 and 9.1 have high temperature coefficients, and those of 7.0 and 6.4 low ones. Cleland[29] has the reasonable interpretation that a histidine residue (pK 9.1) has to be protonated and a carboxylate (pK 6.4) deprotonated to attack malate. It is reasonable that in the enzyme–substrate complex the pK of the imidazole is raised from that of the free enzyme because its protonated form is stabilized by interaction with the hydroxyl group of malate; i.e., the proton is ready to assist the hydroxyl to leave. However, in the predominant enzyme–substrate complex

the H$^+$ ion that the carboxylate will accept is presumably still a carbon-bound proton of the substrate and so is incapable of stabilizing the carboxylate form of the group. Hence this pK is not much changed from that of the free enzyme. Likewise in the fumarate complex it is the imidazole group that contributes most to the lower pK (4.9); its deprotonated form is stabilized by interaction with the water molecule that can attack fumarate, and again the carboxyl group, which requires to be protonated, is little changed in pK.

This mechanism looks reasonable, but in view of the serious drawbacks to ascribing pK values on the basis of ΔH values (because shifts in pK are likely to shift ΔH) it must be taken as tentative. The study illustrates both the dangers of inferring much from pK values alone and also the way they can usefully supplement other studies.

Lysozyme

Much of the study of this enzyme (EC 3.2.1.17) has been made against the background of its molecular structure determined by X-ray crystallography and that of its complexes with inhibitors.[72] A mechanism of action was proposed with this structure,[72,73] according to which the carboxyl group of a glutamic residue (Glu-35) protonates the R—O⁻ leaving group of the substrate (thus assisting its departure) and the carboxylate form of an aspartic residue (Asp-52) stabilizes the oxonium ion formed [Eq. (72)].

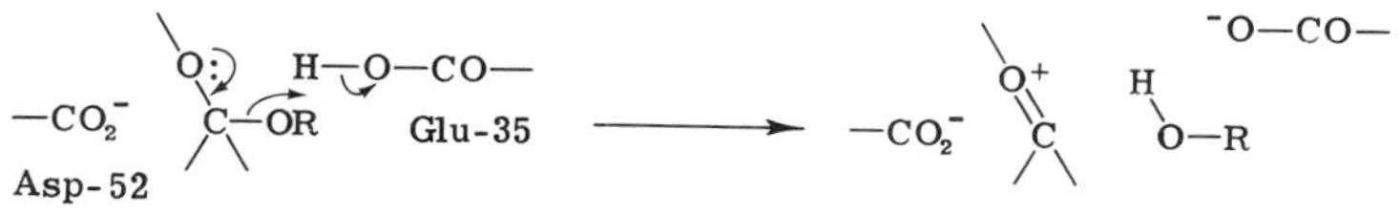

$$\text{(72)}$$

Reverse of this process with water in place of R—OH completes the hydrolysis [Eq. (73)].

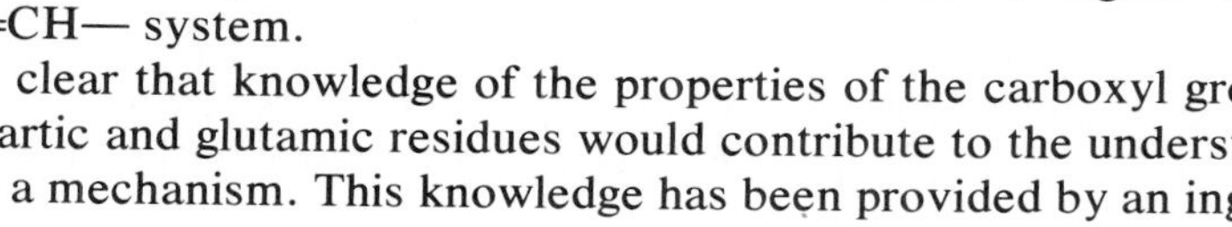

$$\text{(73)}$$

Many variants of this mechanism have been discussed,[74,75] .g., that the carboxylate of Asp-52 forms a covalent bond to some degree with the —O⁺=CH— system.

It is clear that knowledge of the properties of the carboxyl groups of the aspartic and glutamic residues would contribute to the understanding of such a mechanism. This knowledge has been provided by an ingenious

---

[72] C. C. F. Blake, L. N. Johnson, G. A. Mair, A. C. T. North, D. C. Phillips, and V. R. Sarma, *Proc. R. Soc. London, Ser. B* **167**, 378 (1967).
[73] C. A. Vernon, *Proc. R. Soc. London, Ser. B* **167**, 389 (1967).
[74] G. Lowe, *Proc. R. Soc. London, Ser. B* **167**, 431 (1967).
[75] B. Dunn and T. C. Bruice, *Adv. Enzymol.* **37**, 1 (1973).

study by Raftery and colleagues. First they treated lysozyme at pH 4.5 with the Et_3O^+ ion and isolated from the products a derivative in which one carboxyl group was esterified. Two effects contribute to the specificity of the modification: (1) at pH 4.5 more powerful nucleophiles in the protein are mainly protonated, and their higher basicity than that of carboxylate increases their protonation (and hence diminishes their reactivity) to a greater extent than it raises their intrinsic reactivity[19]; (2) the reagent, being a cation, has particularly high reactivity with anionic nucleophiles. The product proved[76] to have Asp-52 esterified; the ester group had low reactivity to nucleophiles (e.g., borohydride and hydrazine), and this may have assisted the isolation of a peptide with the ester bond intact, which allowed characterization of the product.

Parsons and Raftery[36,76] then measured the difference between the titration curvex of lysozyme and that of the esterified derivative. The difference fitted the sum of two titration curves of pK values 4.4 and 6.1 less one of pK 5.2, all three being of one proton per enzyme molecule, so the results could be interpreted as showing that free lysozyme had two molecular pK values, which were replaced in the ester by one of 5.2. A dibasic acid has the same titration curve as an equimolar mixture of two monobasic acids provided that its protonation is not positively cooperative. It follows from the analysis of Simms[77] that if the pK values of the equivalent monobasic acids are called p$K^* \pm \log p$ (this defines pK^* and p), then the molecular pK values of the dibasic acid to which they are equivalent are p$K^* \pm \log (p + 1/p)$.[5a] Hence the pK values of 4.4 and 6.1 need correction in this way to provide true molecular constants of the system, but they are so far apart that this correction is less than 0.01. It thus appears that Asp-52 contributes to the system of molecular pK values 4.4 and 6.1, and so does another group, whose pK is 5.2 when Asp-52 is esterified. This other group is convincingly identified as Glu-35, mainly because it is the only group that dissociates in the required pH range that is close enough to be likely to interact with Asp-52.

It is, as we have indicated, impossible to determine the group pK values that make up molecular values. Nevertheless argument by analogy is possible. Parsons and Raftery assumed that the pK of Glu-35 seen when Asp-52 was esterified would also be its pK when Asp-52 was free but protonated. This identity cannot be exact, but it provides the most reasonable estimate possible. Using it gives the pK of Glu-35 a value of 6.0 when Asp-52 is dissociated, almost the molecular pK of 6.1. This conclusion that the molecular pK of 6.1 is almost entirely due to Glu-35 holds even if

[76] S. M. Parsons and M. A. Raftery, *Biochemistry* **8**, 4199 (1969).
[77] H. S. Simms, *J. Am. Chem. Soc.* **48**, 1239 (1926).

5.2 should not be a very accurate estimate of the pK of Glu-35 when Asp-52 is protonated; what matters is that the value is well below 6.1. Likewise the molecular pK of 4.4 is almost entirely attributable to the group pK of Asp-52 when Glu-35 is protonated, which also has a value of 4.4.

The fact that the titration difference fitted the theoretical curves over the pH range from 3 to 7 showed that no other groups that titrated over this range appreciably affected the pK values under study.

Parsons and Raftery[36] then proceeded to similar difference titrations in the presence of inhibitors and substrate. They found that the binding of methyl 2-acetamido-2-deoxyglucoside raised the pK of 6.1 to one of 6.6. This agreed well with the effect of pH on the binding of this compound[78] as studied by logarithmic plots. Evidently the inhibitor is bound slightly more tightly when Glu-35 is protonated. The pK of 4.4 was not affected, and this is consistent with the absence of any such pK when pK_s for this ligand was plotted against pH. When, however, substrate of high molecular weight was added, the difference curve no longer fitted the sum of two one-site titrations minus a third; evidently other groups that titrated in the pH range used now affected the pK values of the enzyme–substrate complex. Nevertheless, Parsons and Raftery were able to show that the pK of Glu-35 was raised to above 8, and this agreed well with a kinetic pK previously observed in plots of V against pH.

Several other elegant points were made in this study, especially ones arising from the effect of ionic strength and temperature on the pK values reported, but even without them important conclusions emerge. The first is the possibility of assigning the molecular pK values of 4.4 and 6.1 in the free enzyme predominantly to Asp-52 and Glu-35, respectively. The slightly low value for Asp-52 proved to be largely due to the net positive charge of the molecule. Three factors that contribute to the high value for Glu-35 can be recognized. First, there is the rise to 5.2 despite its presence in a positively charged molecule; conceivably this is due to its somewhat hydrophobic environment. Second, there is the rise from 5.2 to 6.1 due to the negative charge on Asp-52. Third, there is the further rise to above 8 when a long-chain substrate is bound.

In the postulated mechanism Asp-52 is required for its charge; a normal pK guarantees this, and its pK is fairly normal. If Glu-35, however, is to act as a general-acid catalyst and supply a proton, an appreciable fraction of it must be protonated, and the raised pK brings this about. The study as a whole gives excellent examples of the combination of techniques, from chemical modification to kinetics, to achieve the

---

[78] F. W. Dahlquist and M. A. Raftery, *Biochemistry* **7**, 3277 (1968).

extremely difficult task of assigning pK values to particular residues, and using the results to understand something of enzymic action.

Acknowledgments

We are grateful to Dr. Athel Cornish-Bowden for his helpful comments on this manuscript and to Miss Marie Gleeson and Mr. Len Jewitt for help in the preparation of the typescript and figures.

[10] Temperature Effects in Enzyme Kinetics

By KEITH J. LAIDLER and BRANKO F. PETERMAN

If an enzyme-catalyzed reaction is studied over a range of temperature, the overall rate passes through a maximum. The temperature at which the rate is a maximum is known as the optimum temperature and was at one time thought to be characteristic of the enzyme system; it is now known, however, to be an ill-defined quantity and to vary with concentrations and with other factors, such as pH.

The explanation of this behavior, first given by Tammann,[1] is that changing the temperature affects two independent processes, the catalyzed reaction itself and the thermal inactivation of the enzyme. In the lower temperature range, up to 30° or so for a typical enzyme, inactivation is very slow and has no appreciable effect on the rate of the catalyzed reaction; the overall rate therefore increases with rise in temperature, as with ordinary chemical reactions. At higher temperatures inactivation becomes more and more important, so that the concentration of active enzyme falls during the course of reaction. An essential feature of the explanation is that the temperature coefficient of the rate of inactivation must be greater than that of the rate of the catalyzed reaction; in the low-temperature range the rate of inactivation is negligible compared with the rate of the catalyzed reaction, whereas in the high-temperature range it is much higher.

The influence of temperature on the inactivation process is considered later; first we consider enzyme-catalyzed reactions, and treat not only the temperature coefficients of these reactions but also certain other matters (e.g., entropies of activation), information about which is provided by the temperature studies.

The rate law for an enzyme-catalyzed reaction involves at least three

[1] G. Tammann, *Z. Phys. Chem. Stoechiom. Verwandschaftslehre* **18**, 426 (1895).

kinetic constants, each of which has its own temperature dependence. The simple Michaelis–Menten mechanism is

$$\text{E} + \text{S} \underset{k_{-1}}{\overset{k_1}{\rightleftharpoons}} \text{ES} \overset{k_2}{\longrightarrow} \text{E} + \text{X}$$

and the steady-state rate equation is then

$$v = \frac{k_2[\text{E}]_0[\text{S}]}{(k_{-1} + k_2)/k_1 + [\text{S}]} \qquad (1)$$

where $[\text{E}]_0$ and $[\text{S}]$ represent the concentrations of enzyme and substrate, respectively. Each of the three constants k_1, k_{-1}, and k_2 will obey the Arrhenius law to a good approximation, and there are three activation energies, E_1, E_{-1} and E_2, for the individual steps. The Arrhenius law will not necessarily apply to the rate v, although it will do so in some special cases. Sometimes the law applies to quite complicated processes that certainly involve several rate constants.[2]

The Arrhenius law can be written as

$$k = Ae^{-E/RT} \qquad (2)$$

where R is the gas constant, T the absolute temperature, E the energy of activation, and A the frequency factor. In terms of activated-complex theory, the equation, in the case of reactions in solution, can be written as[3]

$$k = \frac{\mathbf{k}T}{h} e^{-\Delta G^\ddagger/RT} = e\frac{\mathbf{k}T}{h} e^{\Delta S^\ddagger/R} e^{-E/RT} \qquad (3)$$

Here $\mathbf{k}$ is the Boltzmann constant and h is Planck's constant. The quantity $\Delta G^\ddagger$ is the Gibbs energy of activation (formerly known as the free energy of activation) and is the change in Gibbs energy as the activated complex is formed from the reactants. The quantity $\Delta S^\ddagger$ is the entropy of activation, which is the corresponding change in entropy. Energies of activation are calculated from plots of $\ln k$ against $1/T$, and the frequency factors and entropies of activation are calculated from the value of E and from k at any temperature.

It follows from what has been said that a detailed study of the temperature dependence of an enzyme-catalyzed reaction requires that the individual constants be separated. Care must also be taken with the control of pH, since temperature affects the degree of ionization.

[2] K. J. Laidler, *J. Chem. Educ.* **49**, 343 (1972).
[3] K. J. Laidler, "Theories of Chemical Reaction Rates." McGraw-Hill, New York, 1969.

Separation of Rate Constants

A number of procedures have been used to separate rate constants in enzyme mechanisms. We shall mention these only briefly, with particular reference to the scheme

$$E + S \underset{k_{-1}}{\overset{k_1}{\rightleftarrows}} ES \underset{X}{\overset{k_2}{\searrow}} ES' \overset{k_3}{\longrightarrow} E + Y$$

which applies to many enzyme systems. We shall refer to this as the *double-intermediate* mechanism.

1. Transient-phase studies, using stopped-flow and T-jump techniques, can lead to a separation of constants.
2. Measurement of concentrations of ES and ES' during the steady state can provide information about the relative values of k_2 and k_3. For example, in the chymotrypsin-catalyzed hydrolysis of *p*-nitrophenyl trimethylacetate,[4] ES is large and ES' small, so that $k_2 \gg k_3$. From this it follows that at high substrate concentrations, when $v = V_{\max} = k_c [E]_0$, the catalytic constant k_c can be identified with k_3.
3. Studies with alternative substrates,[5] such that ES' is the same for all of them, can provide information about the relative magnitudes of k_2 and k_3.
4. Work with alternative nucleophiles[6] can also help to separate k_2 and k_3.

Energies and Entropies of Activation for Single-Substrate Reactions

Relatively few investigations have provided values for k_2 and k_3 over a temperature range. Tables I and II list some kinetic parameters obtained in such studies.

In considering the magnitudes of the entropies of activation listed in Tables I and II it is necessary to take into account two types of effect, which it is convenient to refer to as solvent and structural effects.[7] By the former is meant the interaction between the solvent and the reaction system, an interaction that may change during the course of reaction. By structural effects is meant the possibility that the enzyme itself actually undergoes some reversible change in conformation during the process of reaction.

[4] M. L. Bender, G. R. Schonbaum, and B. Zerner, *J. Am. Chem. Soc.* **84**, 2562 (1962).
[5] I. B. Wilson and E. Cabib, *J. Am. Chem. Soc.* **78**, 202 (1956).
[6] I. Hinberg and K. J. Laidler, *Can. J. Biochem.* **50**, 1334 (1972).
[7] K. J. Laidler, *Discuss. Faraday Soc.* **20**, 83 (1955).

TABLE I
KINETIC PARAMETERS RELATING TO THE CONVERSION OF THE ENZYME–SUBSTRATE COMPLEX INTO THE SECOND INTERMEDIATE AND THE FIRST PRODUCT

Enzyme	Substrate	Temperature (°C)	pH	k_2 (sec^{-1})	E_2 (kcal mol^{-1})	$\Delta S_2^{\ddagger}$ (cal K^{-1} mol^{-1})	Reference
Chymotrypsin	Benzoyl-L-tyrosylglycinamide	25.0	7.5	37	11.5	−19.8	a
	Benzoyl-L-tyrosinamide	25.0	7.5	0.625	14.6	−13.0	b
Trypsin	Benzoyl-L-argininamide	25.0	7.8	270	14.9	−8.2	a

[a] J. A. V. Butler, *J. Am. Chem. Soc.* **63**, 2971 (1941).
[b] S. Kaufman, H. Neurath, and G. W. Schwert, *J. Biol. Chem.* **177**, 792 (1949).

TABLE II
Kinetic Parameters Relating to the Conversion of the Second Intermediate into the Enzyme and the Second Product

Enzyme	Substrate	Temperature (°C)	pH	k_3 (sec^{-1})	E_3 (kcal mol^{-1})	$\Delta S_3^{\ddagger}$ (cal K^{-1} mol^{-1})	Reference
Chymotrypsin	Methyl-L-β-phenyllactate	25.0	7.8	1.38	11.1	−23.4	a
	N-Acetyl-L-tyrosine ethyl ester	25.0	8.7	0.0683	10.9	−13.4	b
	Benzoyl-L-tyrosine ethyl ester	25.0	7.8	78.0	9.2	−21.4	c
	Benzoyl-L-phenylalanine ethyl ester	25.0	7.8	37.4	12.5	−11.0	a
	N-trans-Cinnamoyl-imidazole	25.0	8.7	0.0125	11.8	−29.6	b
Alkaline phosphatase (E. coli)	p-Nitrophenyl phosphate	25.0	8.5	28.0	9.4	−22.8	d
Papain	Furylacryloylimidazole	25.0	7.0	0.02	14.9	−22.7	e

[a] J. E. Snoke and H. Neurath, J. Biol. Chem. **182**, 577 (1950).
[b] M. L. Bender, F. J. Kézdy, and C. R. Gunter, J. Am. Chem. Soc. **86**, 3714 (1964).
[c] S. Kaufman, H. Neurath, and G. W. Schwert, J. Biol. Chem. **177**, 792 (1949).
[d] C. Lazdunski and M. Lazdunski, Biochim. Biophys. Acta **113**, 551 (1966).
[e] P. M. Hinkle and J. F. Kirsch, Biochemistry **9**, 4633 (1970).

The possibility of such structural changes was first proposed[8] in an attempt to interpret the effects of pressure on enzyme reactions. It was suggested that in some cases the enzyme molecule assumes a more open conformation when it forms a complex, with an increase in entropy, and that when it undergoes subsequent reactions there is a refolding of the enzyme molecule, with a loss of entropy. Such a picture seems reasonable since it brings the enzyme reactions into line with what takes place during enzyme inactivation, in which it seems clear that unfolding occurs.

This explanation leads to the conclusion that the entropy of activation will be negative for reactions (2) and (3), and Tables I and II show that this prediction is correct in all cases.

Reference must also be made to another type of conformational change for which there is some evidence in certain enzyme systems. In the above discussion it has been assumed that bonds (perhaps hydrogen bonds) are broken, so that the unfolded molecule possesses more freedom of movement, and therefore more entropy, than the folded molecule. A process of a different kind could also occur during complex formation, and give rise to an entropy increase. When the fully extended (β) form of a protein molecule becomes converted into the less extended (α) form, without the formation or breaking of bonds, there is an increase of entropy; such a process may conveniently be referred to as a "rubberlike coiling," since a similar change occurs during the shortening of stretched rubber. A structural change of this type has been considered by Morales and Botts[9] for the myosin–ATP system and is supported by the temperature studies of Ouellet et al.,[10] the solvent studies of Laidler and Ethier,[11] and the pressure studies of Laidler and Beardall.[12] It is supposed that myosin is normally maintained in an elongated form by charges on the protein molecule, and that when these charges are neutralized on complex formation the myosin contracts with an increase of entropy.

The interpretation of entropies of activation in terms of solvent effect is as follows. During the course of reaction there may be polarity changes that will result in either an increase or a decrease in solvent binding. If charges are formed during a reaction, for example, the solvent will be bound more firmly, and there will be a loss of entropy. If, on the other hand, charges are neutralized during the reaction, there will be a release of solvent molecules and a corresponding gain of entropy.

The values in Tables I and II for $\Delta S_2^\ddagger$ and $\Delta S_3^\ddagger$ are all negative, and

[8] K. J. Laidler, *Arch. Biochem.* **30**, 226 (1951).
[9] M. F. Morales and J. Botts, *Arch. Biochem. Biophys.* **37**, 283 (1952).
[10] L. Ouellet, K. J. Laidler, and M. F. Morales, *Arch. Biochem. Biophys.* **39**, 37 (1952).
[11] K. J. Laidler and M. Ethier, *Arch. Biochem. Biophys.* **43**, 338 (1953).
[12] K. J. Laidler and A. J. Beardall, *Arch. Biochem. Biophys.* **55**, 138 (1955).

could therefore be explained on the hypothesis that there are increases in polarity during the processes ES → ES' + X and ES' → E + Y. Such increases are not unreasonable; in reactions involving solvent molecules such as water, it is commonly found that there are increases in polarity when the activated complex is formed.[13,14]

There are only a few cases for which the kinetic constant k_1 for the bimolecular interaction between enzyme and substrate has been obtained at more than one temperature. This was done for catalase and peroxidase, for which Chance has shown that the overall rate constants correspond to k_1 for the initial step. Using a rapid titration method, Bonnichsen, Chance, and Theorell[15] have determined rate constants at two temperatures. Their results for catalase are summarized in Table III. Rates at two temperatures do not, of course, permit a test of the Arrhenius law, but if the validity of the law is assumed, the figures in Table IV can be calculated. For horse blood catalase, the average value of E is 1.7 kcal. This is an unusually low activation energy for any reaction and, as shown in Table IV, is much lower than for the uncatalyzed hydrogen peroxide decomposition and for the reaction with other catalysts. The negative entropies of activation are consistent with an increase in polarity during reaction, or with some tightening of the enzyme structure. The results for the formation of the peroxide–hydrogen peroxide complex are quite similar. The activation energy is zero, and the entropy of activation about -28 cal K^{-1} mol^{-1}.

The steady-state treatment of the double-intermediate mechanism leads to the conclusion that the limiting rate at low substrate concentra-

TABLE III
KINETIC PARAMETERS FOR THE DECOMPOSITION OF
HYDROGEN PEROXIDE CATALYZED BY CATALASE[a]

Type of catalase	Temperature (°C)	$k \times 10^{-7}$ (dm^3 mol^{-1} sec^{-1})	E_1 (kcal mol^{-1})	$\Delta S_1^\ddagger$ (cal K^{-1} mol^{-1})
Horse blood	25.5	3.50	1.9	−19.6
	23.5	3.50	1.5	−20.9
	22.0	3.54	1.8	−19.9
	23.5	3.50	1.8	−19.9
Horse liver	22.0	3.00	1.3	−21.9

[a] Data from R. K. Bonnichsen, B. Chance, and H. Theorell, *Acta Chem. Scand.* **1**, 685 (1947).

[13] K. J. Laidler and P. A. Landskroener, *Trans. Faraday Soc.* **52**, 200 (1956).
[14] K. J. Laidler and D. T. Y. Chen, *Trans. Faraday Soc.* **54**, 1026 (1958).
[15] R. K. Bonnichsen, B. Chance, and H. Theorell, *Acta Chem. Scand.* **1**, 685 (1947).

TABLE IV
Comparison of Activation Energies for Hydrogen Peroxide Decompositions

Catalyst	E (kcal)	Reference
Horse blood catalase	1.7	a
Platinum	11.7	b
Fe^{2+}	10.1	c
I^-	13.5	d
No catalyst	17–18	e

[a] R. K. Bonnichsen, B. Chance, and H. Theorell, *Acta Chem. Scand.* **1**, 685 (1947).
[b] G. Bredig and R. M. von Berneck, *Z. Phys. Chem. Stoechiomn. Verwandschaftslehre* **31**, 258 (1899).
[c] J. H. Baxendale, M. G. Evans, and G. S. Park, *Trans. Faraday Soc.* **42**, 155 (1946).
[d] J. H. Walton, *Z. Phys. Chem. Stoechiom. Verwandschaftslehre* **47**, 185 (1904).
[e] J. Williams, *Trans. Faraday Soc.* **24**, 245 (1928); C. Pana, *Trans. Faraday Soc.* **24**, 486 (1928).

tions is

$$v_0 = k_0[E]_0[S] = \frac{k_1 k_2}{k_{-1} + k_2}[E]_0[S] \tag{4}$$

The second-order rate constant k_0, equal to $k_1 k_2/(k_{-1} + k_2)$ is composite, and the Arrhenius law will not necessarily apply to it directly. There are, however, two special cases under which this composite constant will obey the law, as follows:

  a. When $k_2 \gg k_{-1}$. If this is the case k_{-1} may be neglected in comparison with k_2, and k_0 is equal to k_1. This means that the rate constant for the overall reaction at low substrate concentrations is simply that for the first step, the formation of the complex. If that is so, the Arrhenius law should apply to k_0, and the corresponding activation energy will be E_1, that for the initial complex formation.

  b. When $k_{-1} \gg k_2$. The constant k_0 is now equal to $k_2 k_1/k_{-1}$. The ratio k_1/k_{-1} is simply the equilibrium constant for the complex formation:

$$E + S \underset{k_{-1}}{\overset{k_1}{\rightleftharpoons}} ES$$

Its variation with temperature is given by

$$\frac{k_1}{k_{-1}} = \frac{A_1 e^{-E_1/RT}}{A_{-1} e^{-E_{-1}/RT}} = \frac{A_1}{A_{-1}} e^{-\Delta E/RT} \tag{5}$$

where A_1 and A_{-1} are temperature-independent frequency factors and ΔE

is the increase in energy per mole for the change from E + S to ES. Since k_2 varies exponentially with temperature according to the Arrhenius law

$$k_2 = A_2 e^{-E_2/RT} \quad (6)$$

it follows that

$$\frac{k_1 k_2}{k_{-1}} = \frac{A_2 A_1}{A_{-1}} e^{-(E_2+E_1-E_{-1})RT} \quad (7)$$

$$= \frac{A_1 A_2}{A_{-1}} e^{-(E_2+\Delta E)/RT} \quad (8)$$

In other words, the Arrhenius law should apply to this composite constant $k_1 k_2/k_{-1}$, but the activation energy does not correspond to a single elementary step; it is the sum of the activation energy for the second step (ES → ES′ + X) and the total energy increase for the first step (E + S → ES).

These relationships are illustrated by the energy diagrams shown in Fig. 1. In case (a), which corresponds to $k_2 \gg k_{-1}$ and $k_0 = k_1$, the highest energy barrier over which the system must pass corresponds to ES‡, the activated complex for the reaction E + S → ES. The measured activation energy corresponding to k_0, the overall rate constant at low substrate concentrations, is thus E_1, corresponding to this initial step. In case (b), on the other hand, the highest energy barrier corresponds to ES‡‡, the ac-

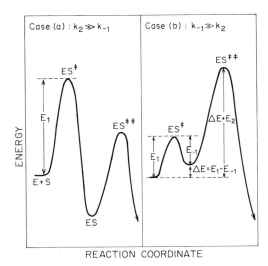

FIG. 1. Energy diagram for a reaction proceeding by the simple Michaelis–Menten mechanism, showing two extreme cases.

tivated complex for the second step, ES → ES' + X. In order to reach this level the system must first reach the level of ES, where the energy is ΔE higher than that of E + S, and then acquire an additional E_2; the activation energy corresponding to k_0 is thus $\Delta E + E_2$. In this case the overall rate is in no way dependent on the rate of the initial step, so that the data can give no information about k_1 or E_1; these quantities can then be obtained only by the use of special techniques, such as those of transient-phase investigations.

When the situation corresponds to neither of these special cases the plots of $\log_{10} k_0$ against $1/T$ will not necessarily be linear, but energies of activation can still be calculated from the slope at any one temperature. The activation energy E_0 at any temperature is defined by

$$E_0 = RT^2 \frac{d \ln k_0}{dT} \tag{9}$$

By Eq. (4)

$$\ln k_0 = \ln k_1 + \ln k_2 - \ln(k_{-1} + k_2) \tag{10}$$

and therefore

$$\frac{E_0}{RT^2} = \frac{d \ln k_0}{dT} = \frac{d \ln k_1}{dT} + \frac{1}{k_2}\frac{dk_2}{dT} - \frac{1}{k_{-1} + k_2}\frac{d(k_{-1} + k_2)}{dT} \tag{11}$$

$$= \frac{d \ln k_1}{dT} + \frac{1}{k_2}\frac{dk_2}{dT} - \frac{1}{k_{-1} + k_2}\left(\frac{dk_{-1}}{dT} + \frac{dk_2}{dT}\right) \tag{12}$$

$$= \frac{d \ln k_1}{dT} + \frac{k_{-1}}{k_{-1} + k_2}\frac{d \ln k_2}{dT} - \frac{k_{-1}}{k_{-1} + k_2}\frac{d \ln k_{-1}}{dT} \tag{13}$$

The individual activation energies are defined by

$$E_1 = RT^2 \frac{d \ln k_1}{dT}, \quad E_{-1} = RT^2 \frac{d \ln k_{-1}}{dT}, \quad E_2 = RT^2 \frac{d \ln k_2}{dT} \tag{14}$$

and it follows that

$$E_0 = \frac{k_{-1}(E_1 + E_2 - E_{-1}) + k_2 E_1}{k_{-1} + k_2} \tag{15}$$

The overall activation energy E_0 is thus the weighted mean of the values E_1 and $E_1 + E_2 - E_{-1}$, the weighting factors being $k_2/(k_{-1} + k_2)$ and $k_{-1}/(k_{-1} + k_2)$.

Some values of E_0 and $\Delta S_0^{\ddagger}$, corresponding to k_0, are shown in Table V. The entropies of activation fall into two main groups. With pepsin, trypsin, and myosin the values are positive, whereas with chymotrypsin, carboxypepidase, and urease they are negative. Light is thrown on these

TABLE V
KINETIC VALUES RELATING TO THE FORMATION OF THE ENZYME–SUBSTRATE COMPLEX

Enzyme	Substrate	Temperature (°C)	pH	k_0 (dm³ mol⁻¹ sec⁻¹)	E_0	$\Delta S_0^{\ddagger}$	Reference
Pepsin	Carbobenzoxy-L-glutamyl-L-tyrosine ethyl ester	31.6	4.0	0.57	23.1	14.1	a
	Carbobenzoxy-L-glutamyl-L-tyrosine	31.6	4.0	0.79	20.2	2.6	a
Trypsin	Chymotrypsinogen	19.6	7.5	2900	16.3	8.5	b
Chymotrypsin	Methyl hydrocinnamate	25.0	7.8	6.66	11.5	−23.2	c
	Methyl-DL-α-chloro-β-phenylpropionate	25.0	7.8	11.2	6.9	−33.0	c
	Methyl-D-β-phenyllactate	25.0	7.8	4.0	3.1	−47.2	c
	Methyl-L-β-phenyllactate	25.0	7.8	138	3.8	−38.5	c
	Benzoyl-L-tyrosine ethyl ester	25.0	7.8	19500	0.8	−38.5	d
	Benzoyl-L-tyrosinamide	25.0	7.8	14.9	3.7	−43.0	d
Carboxypeptidase	Carbobenzoxyglycyl-L-tryptophan	25.0	7.5	17444	9.9	−8.5	c
	Carbobenzoxyglycyl-L-phenylalanine	25.0	7.5	27900	9.6	−8.5	c
	Carbobenzoxyglycyl-L-leucine	25.0	7.5	3920	11.0	−7.5	e
Urease	Urea	20.8	7.1	5.0×10^6	6.8	−6.8	f
Adenosine triphosphatase (myosin)	Adenosine triphosphate	25.0	7.0	8.2×10^6	21.0	44.0	g

[a] E. J. Casey and K. J. Laidler, *J. Am. Chem. Soc.* **72**, 2159 (1950).
[b] J. A. V. Butler, *J. Am. Chem. Soc.* **63**, 2971 (1941).
[c] J. E. Snoke and H. Neurath, *J. Biol. Chem.* **182**, 577 (1950); *Arch. Biochem.* **21**, 351 (1949).
[d] S. Kaufman, H. Neurath, and G. W. Schwert, *J. Biol. Chem.* **177**, 792 (1949).
[e] R. Lumry, E. I. Smith, and R. R. Glantz, *J. Am. Chem. Soc.* **73**, 4330 (1951).
[f] M. C. Wall and K. J. Laidler, *Arch. Biochem. Biophys.* **43**, 299 (1953).
[g] L. Ouellet, K. J. Laidler, and M. F. Morales, *Arch. Biochem. Biophys.* **39**, 37 (1952).

differences by consideration of the types of substrates that are hydrolyzed by these six enzymes. Chymotrypsin and urease invariably act upon electrically neutral molecules. Adenosine triphosphatase, on the other hand, is itself positively charged, and it acts upon a negatively charged substrate. Pepsin will act only upon substrates that contain a free $—CO_2H$ group, and at pH 4 this exists at least in part as a negatively charged group. Carboxypeptidase also acts only on substrates containing the $—CO_2H$ group, which will be ionized. The specificity requirement of trypsin is for a free $—NH_2$ group, and this will exist as the positively charged $—NH_3^+$ group. It is thus likely that with ATPase, pepsin, carboxypeptidase, and trypsin the interaction between enzyme and substrate involves a charge neutralization, whereas with chymotrypsin and urease there is no such neutralization.

Reaction between an enzyme and an uncharged substrate involves certain electron shifts as a result of which the activated complex will be more polar than the reactants; there is therefore an increase in electrostriction and a corresponding negative entropy of activation. With ATPase, pepsin, and trypsin, on the other hand, this effect is presumably counteracted by the charge neutralization that occurs when the enzyme and substrate come together. This neutralization will lead to a release of water molecules, and there will be a corresponding increase of entropy. This explanation of the signs of the $\Delta S^{\ddagger}$ values derives some support from the general correlation between the sign of the entropy change and whether the substrate is charged or not. Carboxypeptidase, however, appears to be an exception, in that the substrates are charged but the entropies of activation are negative; a possible explanation is that the $—CO_2^-$ group on the substrate does not come into contact with a positive group on the enzyme, and is therefore not neutralized. The ionic strength effects for this enzyme tend to suggest that there is an approach of like (negative) charges.

Some of the entropy changes shown in Table V seem to be too large to be explained in terms of electrostatic effects alone, and structural effects must play a role. For chymotrypsin, where there is no charge neutralization, the entropies of activation are strongly negative, and the solvent studies of Barnard and Laidler[16] suggest that the chymotrypsin unfolds during complex formation. Similarly, the solvent studies of Laidler and Ethier[11] suggest that in the myosin system the large positive entropies of activation is partly due to electrostatic effects and partly to structural effects, the enzyme undergoing a contraction of the coiled helix when the substrate becomes attached.

[16] M. L. Barnard and K. J. Laidler, *J. Am. Chem. Soc.*, 74, 6099 (1952).

For a further discussion of the solvent studies, in which mixed solvents are used, the reader is referred to Chapter 7 of the book by Laidler and Bunting.[17]

Reactions Involving More Than One Substrate

When there is more than one substrate the general scheme describing the one-substrate system has to be modified to include the binding and release steps of the additional substrate or substrates. In addition, some other interconversion steps must be considered.

Similar general principles apply to multisubstrate enzyme systems as to single-substrate systems. Quantities like V_{max} and K_m will not necessarily follow the Arrhenius law, since they are combinations of individual rate constants, but if certain rate constants predominate the law may be obeyed.

Little temperature-dependence work has been done on reactions involving more than one substrate. We will limit our discussion to some work on lactate dehydrogenase carried out by Borgmann et al.[18-20] This enzyme catalyzes the overall reaction

$$E + NAD^+ + \text{lactate} \rightleftharpoons E + NADH + \text{pyruvate}$$

and was studied in both directions. The pre-steady-state and steady-state kinetics in the lactate → pyruvate direction[18] are consistent with the four-step mechanism

$$E + A \underset{k_{-1}}{\overset{k_1[A]}{\rightleftharpoons}} EA \underset{k_{-2}}{\overset{k_2[B]}{\rightleftharpoons}} EAB \underset{k_{-3}[X]}{\overset{k_3}{\rightleftharpoons}} EY \underset{k_{-4}[Y]}{\overset{k_4}{\rightleftharpoons}} E + Y$$

where A and B are NAD^+ and lactate, respectively, and X and Y are pyruvate and NADH, respectively. It turned out that the above scheme was an oversimplification. Borgmann et al.[20] showed that when pre-steady state data in the forward direction are compared with similar data in the reverse reaction, it is necessary to include an additional step

$$EAB \rightleftharpoons EXY$$

We shall first discuss the results on the basis of the four-step model, and shall later consider the implications of adding the additional step.

The studies of the temperature dependence of the lactate dehy-

---

[17] K. J. Laidler and P. S. Bunting, "The Chemical Kinetics of Enzyme Action." Oxford Univ. Press (Clarendon), London and New York, 1973.
[18] U. Borgmann, T. W. Moon, and K. J. Laidler, *Biochemistry* **13**, 5152 (1974).
[19] U. Borgmann, K. J. Laidler, and T. W. Moon, *Can. J. Biochem.* **53**, 1196 (1975).
[20] U. Borgmann, K. J. Laidler, and T. W. Moon, *Can. J. Biochem.* **54**, 915 (1976).

drogenase action[18] reveal that the individual rate constants obey the Arrhenius law. The maximum steady-state velocities in the lactate → pyruvate direction are given by

$$\vec{V}_{max} = \frac{k_3 k_4 [E]_0}{k_3 + k_4} \quad (16)$$

and in the pyruvate → lactate direction the equation is

$$\tilde{V}_{max} = \frac{k_{-1} k_{-2} [E]_0}{k_{-1} + k_{-2}} \quad (17)$$

Straight lines were obtained when log $\vec{V}_{max}$ and log $\tilde{V}_{max}$ were plotted against $1/T$. This suggests that the maximum steady-state velocity each direction is controlled by one rate constant. Further study showed that $\vec{V}_{max}$ is equal to $k_4[E]_0$. The Michaelis constants K_m with respect to lactate and pyruvate do not give straight lines when log K_m is plotted against $1/T$, as shown in Fig. 2, suggesting that the K_m's are controlled by more than one rate constant. For the methods of analysis of these curves, the reader is referred to the original paper.[18]

From the temperature-dependence measurements in the pre-steady-

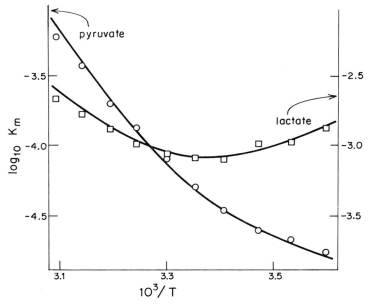

FIG. 2. Plots of log K_m against $1/T$ for the lactate dehydrogenase system, for the forward and reverse reactions. From U. Borgmann, T. W. Moon, and K. J. Laidler, *Biochemistry* **13**, 5152 (1974).

state and steady-state studies on this system, it was possible to construct the enthalpy, entropy, and Gibbs energy profiles shown in Fig. 3. The values for $\Delta H^\ddagger$, $\Delta S^\ddagger$, and $\Delta G^\ddagger$ are given in Table VI. The ternary complex EAB has a lower Gibbs energy than any other state, which means that the ternary complex is the most stable and predominant under standard conditions. The entropy profiles are of special interest, since they can provide insight into the state of the protein–substrate complex. The low entropy for EAB suggests a more folded structure. The entropy profile also suggests that the binding of NAD^+ produces a more folded structure, and that the binding of NADH results in partial unfolding of the protein.

The studies of the pre-steady-state and the steady-state kinetics in the lactate → pyruvate direction did not provide any evidence for more than four steps. However, when the work was done on the reverse reaction[20] it was found necessary to include the additional step EAB ⇌ EXY, so that

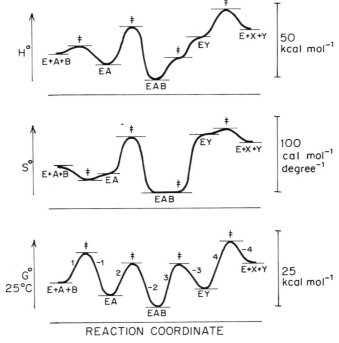

FIG. 3. Profiles showing the changes in standard enthalpy, entropy and Gibbs energy for the lactate dehydrogenase system at 25°. This diagram is based on the results obtained in the lactate → pyruvate direction, which leads to the four-step mechanism. From U. Borgmann, T. W. Moon, and K. J. Laidler, *Biochemistry* **13**, 5152 (1974).

TABLE VI
Enthalpies, Entropies, and Gibbs Energies

	ΔH^0 or $\Delta H^{\ddagger}$ (kcal mol^{-1})	ΔS^0 or $\Delta S^{\ddagger}$ (cal K^{-1} mol^{-1})	ΔG^0 or $\Delta G^{\ddagger}$ at 25° (kcal mol^{-1})
k_1	4.1	−18.2	9.5
k_{-1}	12.5	−5.8	14.2
k_2	25.6	50.0	10.7
k_{-2}	36.2	74.2	14.1
k_3	13.9	−0.2	14.0
k_{-3}	−14.5	−74.8	7.9
k_4	17.5	7.6	15.2
k_{-4}	12.8	17.8	7.4
k_1/k_{-1}	−8.4	−12.4	−4.7
k_2/k_{-2}	−10.6	−24.2	−3.4
k_3/k_{-3}	28.4	74.6	6.1
k_4/k_{-4}	4.7	−10.2	7.8
K_{eq}	14.1	27.8	5.8

the mechanism is now

$$E + A \rightleftharpoons EA \rightleftharpoons EAB \rightleftharpoons EXY \rightleftharpoons EY \rightleftharpoons E + Y$$

The addition of this step affects almost all the thermodynamic parameters and the energy profiles. How the insertion of a step affects the enthalpy profile can be seen in Fig. 4. Comparison of Fig. 4A with 4B shows that interpretation of lactate–pyruvate data in terms of four-step model misplaces the barrier that is really between EXY and EY, placing it between EA and EAB. The use of the incomplete model does not affect the heights of the barriers and intermediates in the enthalpy profile, but it misplaces the barriers.

It is of interest to see what relationship exists between the adaptation of an animal to its thermal environment and the thermodynamic and kinetic parameters for the enzyme reactions. Ectothermic animals, such as fish, conform to environmental temperatures; endothermic animals, such as most mammals, maintain a relatively constant body temperature, which is more or less independent of the ambient temperature. Since ectothermic animals normally operate at lower temperatures than endothermic animals it might be expected that the kinetic parameters would favor more rapid reaction at lower temperatures, but this seems not to be the case. Low et al.[21] compared enzymes from endothermic and ectothermic animals, but their work was confined to high substrate concentrations and

[21] P. S. Low, J. L. Bada, and G. N. Somaro, *Proc. Natl. Acad. Sci. U.S.A.* **70**, 430 (1973).

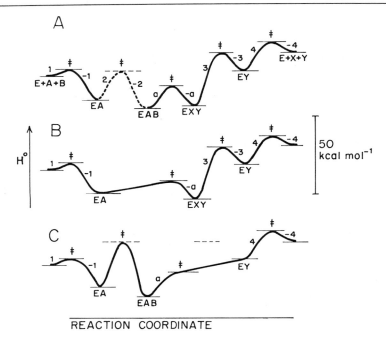

FIG. 4. Enthalpy profiles for the five-step mechanism (A) and the four-step mechanism (B and C). Profile B is based on the data for the pyruvate → lactate reaction, profile C for the reverse reaction. From U. Borgmann, K. J. Laidler, and T. W. Moon, *Can. J. Biochem.* **54,** 915 (1976).

therefore concerned only with V_{max} values; under physiological conditions the substrate concentrations are always too low to saturate the enzymes.[22]

Borgmann et al.[19] compared the individual thermodynamic and kinetic parameters for lactate dehydrogenase from beef heart and beef muscle (endothermic animals) and from flounder muscle (ectothermic). The comparison was based on the four-step model which, although incomplete, still provides a valuable characterization of the trends brought about by thermal adaptation. Figure 5 shows the Gibbs energy profiles for the three enzymes, and Fig. 6 compares the enthalpy and entropy profiles. It is particularly significant that there are greater differences between the thermodynamic parameters for the stable complexes AE, EAB, and EY than for the activated complexes. The changes affecting the stable and activated

---

[22] P. W. Hochachka and G. N. Somero, "Strategies of Biochemical Adaptation," p. 11. Saunders, Philadelphia, 1973.

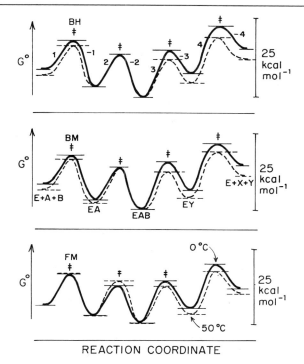

FIG. 5. Gibbs energy profiles for three forms of lactate dehydrogenase at 0° (boldface line) and 50° (dashed line). The three types of enzyme are from beef heart (BH), beef muscle (BM), and flounder muscle (FM). From U. Borgmann, K. J. Laidler, and T. W. Moon, *Can. J. Biochem.* **53**, 1196 (1975).

complexes can be interpreted in terms of the numbers of weak bonds (e.g., hydrogen bonds) formed or broken during the formation of the complexes.[19] The main conclusion is that the number of weak bonds is greater for the beef heart muscle enzyme and smaller for the flounder muscle enzyme. This is consistent with the suggestion of Low *et al.*[21] that endothermic enzymes have, through adaptability and selectivity, developed higher structural stability and can therefore more effectively resist higher temperatures.

Another interesting comparison is related to overall rates under physiological conditions, when pyruvate concentrations are about 0.05 mM. At this concentration the rate of the pyruvate → lactate conversion with beef heart enzyme goes through a maximum at about 35°, which is close to body temperature.[18] The reason for this maximum is related to the temperature variation of K_m; at lower temperatures the enzyme is almost completely saturated with substrate, but at higher temperatures the sub-

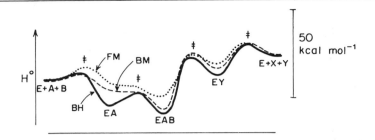

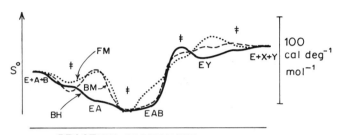

Fig. 6. Enthalpy and entropy profiles, superimposed for the three forms of lactate dehydrogenase (compare Fig. 5). From U. Borgmann, K. J. Laidler, and T. W. Moon, *Can. J. Biochem.* **53**, 1196 (1975).

strate concentration is well below the K_m, and the rate falls off. With the flounder muscle enzyme, on the other hand, the maximum temperature under physiological conditions is lower; this is significant in view of the lower temperature of the fish.

Diffusional Effects in Enzyme Systems

Certain enzyme processes, such as the binding of enzymes to substrates, sometimes occur very rapidly. When this is the case, the question arises whether the rates are influenced by the rates of diffusion of the reacting substances. This has an important bearing on temperature coefficients and will now be briefly considered.

If the rate of the chemical interaction between two substances A and B is very much greater than the rate of diffusion, the rate of the process will be equal to the rate with which A and B diffuse together and the rate constant is then

$$k_D = 4\pi(D_A + D_B) d_{AB} \tag{18}$$

Here D_A and D_B are the diffusion coefficients of A and B and d_{AB} is the distance between the centers of A and B when reaction occurs. If the reactant molecules A and B are large compared with the solvent molecules (which is the case with enzyme reactions), Stokes's law will apply to a good approximation, and if d_{AB} is taken to be the sum of the molecular radii r_A and r_B, Eq. (18) becomes

$$k_D = \frac{2\,\mathbf{k}\,T}{3\eta}\frac{(r_A + r_B)^2}{r_A r_B} \tag{19}$$

where **k** is the Boltzmann constant and η is the viscosity of the medium. For aqueous solutions at 25°, and with typical values for the radii, this expression leads to

$$k_D = 7.0 \times 10^9 \text{ dm}^3 \text{ mol}^{-1} \text{ sec}^{-1}$$

for the rate constant of a fully diffusion-controlled reaction. This value is modified if electrostatic effects are involved[23]; equal and opposite unit charges lead to an increase by a factor of about 4.

If the rate of the chemical interaction is not very much larger than the rate of diffusion there will be only partial diffusion control. If the rate constant for the chemical interaction is k_{chem}, the overall rate constant is given by

$$k = (k_D k_{chem})/(k_D + k_{chem}) \tag{20}$$

If $k_{chem} \gg k_D$, the rate constant is k_D, as previously discussed; if $k_D \gg k_{chem}$, the rate constant is k_{chem}. The latter situation applies to most chemical reactions. Figure 7 shows a plot of k against k_{chem}, with k_D taken as 7.0×10^9 dm^3 mol^{-1} sec^{-1}, the value given by Eq. (19) for water at 25°. When $k_{chem} = 7.0 \times 10^9$ dm^3 mol^{-1} sec^{-1} the value of k is 3.5×10^9, one half of the value for full diffusion control, and one can say that there is 50% diffusion control. We reach 90% diffusion control only when k_{chem} is 6.3×10^{10} dm^3 mol^{-1} sec^{-1}.

Inspection of the data given in Tables I–VI shows that in no case is the rate constant large enough for there to be any significant amount of diffusion control. The highest rate constant recorded in the tables is for horse blood catalase (Table III), for which $k_1 = 3.54 \times 10^7$ dm^3 mol^{-1} sec^{-1} at 22°C. This corresponds to a diffusional influence of less than 1%. This conclusion is consistent with the very low activation energies shown in Table III; they are much lower than the values (4–5 kcal) characteristic of diffusion in water.

Although there is no evidence of diffusion control in the examples

---

[23] R. A. Alberty and G. G. Hammes, *J. Phys. Chem.* **62**, 154 (1958).

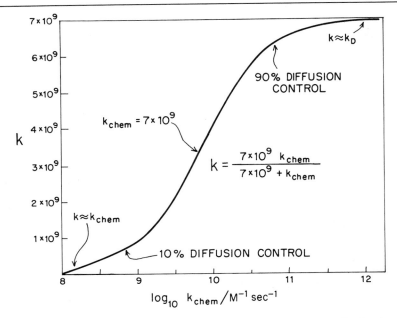

FIG. 7. Plots of the overall rate constant k against the rate constant k_{chem} for the chemical interaction, on the basis of Eq. (20). The rate constant for the diffusion process, k_D, is taken to be 7.0×10^9 dm^3 mol^{-1} sec^{-1}, a typical value for water at 25°.

given in Tables I–VI, there have been some cases reported where the enzyme reaction occurs at diffusion-controlled rate.[23,24] With immobilized enzymes, there may be substantial diffusional control, as discussed in Vol. 64 [9].

Enzyme Inactivation

We have noted that the temperature optima found in studies of enzyme-catalyzed reactions are due to the inactivation of the enzyme at higher temperatures. The kinetics of enzyme inactivations, which are protein denaturations arising from conformational changes, have been investigated extensively, and a number of temperature studies have been made. Space does not permit a detailed treatment of the topic; the interested reader is referred to Chapter 13 of the book by Laidler and Bunting,[17] where references to other articles and reviews are given.

It is necessary to consider both temperature and pH effects on enzyme inactivations, since the two effects are closely related. It is usually found

---

[24] B. Jonsson and H. Wennerstrom, *Biophys. Chem.* **7**, 285 (1978).

that rates of inactivation pass through a minimum as the pH is varied, and the behavior is frequently quite complex. A satisfactory treatment of pH and temperature effects has been given by Levy and Benaglia[25] on the basis of the various ionized states, P_1, P_2, P_3, etc., in which a protein can occur. Suppose that P_1 is the form in which the protein exists in the most acid solutions, and that P_2 is P_1 which has lost a proton, etc:

$$P_1 \xrightleftharpoons{K_1} P_2 \xrightleftharpoons{K_2} P_3 \xrightleftharpoons{K_3} P_4 \xrightleftharpoons{K_4} P_5, \text{ etc.}$$
$$\downarrow k_1 \quad \downarrow k_2 \quad \downarrow k_3 \quad \downarrow k_4 \quad \downarrow k_5$$
$$D \quad\quad D \quad\quad D \quad\quad D \quad\quad D$$

Each form can go into a denatured form D with a rate constant k_1, k_2, etc. If one of the intermediate rate constants, such as k_3, is smaller than the others, the overall rate of denaturation will pass through a minimum at the pH at which the corresponding form (e.g., P_3) is predominant.

A satisfactory treatment of temperature effects obviously requires that the rate constants k_1, k_2, etc. and the dissociation constants K_1, K_2, etc. are separated, on the basis of a pH study, and that their values are determined over a range of temperatures. This has been done in only a few cases, the most complete study being for the denaturation of ricin.[25] On the basis of the above scheme, the form P_4 is the one that undergoes deactivation most slowly, the kinetic parameters for the process being, at 65°,

$\Delta G^{\ddagger} = 38.9$ kcal mol^{-1}; $\Delta H^{\ddagger} = 53$ kcal mol^{-1}; $\Delta S^{\ddagger} = 42$ cal K^{-1} mol^{-1}

The form P_3 undergoes more rapid denaturation, with

$\Delta G^{\ddagger} = 32.3$ kcal mol^{-1}; $\quad \Delta H^{\ddagger} = 89$ kcal mol^{-1};
$$\Delta S^{\ddagger} = 169 \text{ cal K}^{-1} \text{ mol}^{-1}$$

It is of interest that the more rapid rate for P_3 is associated with a very much larger entropy of activation, and that the enthalpy of activation is higher. The remaining forms, P_1, P_2, P_3, etc. undergo denaturation even more rapidly, but the kinetic parameters cannot be determined from the data available. For the process $P_3 \rightarrow P_4$ the thermodynamic parameters, at 65°, are

$\Delta G = 11.7$ kcal mol^{-1}; $\quad \Delta H = 30$ kcal mol^{-1}; $\quad \Delta S = 54$ cal K^{-1} mol^{-1}

Of special interest are the very large enthalpy and entropy values associated with these processes. Further examples are given in Table VII, which shows the energies and entropies of activation obtained in studies at a fixed pH. If a process has a very high energy of activation and occurs

---

[25] M. Levy and A E. Benaglia, *J. Biol. Chem.* **186**, 829 (1950).

TABLE VII
ENERGIES AND ENTROPIES OF ACTIVATION FOR ENZYME INACTIVATIONS

Enzyme	pH	Energy of activation (kcal mol^{-1})	Entropy of activation (cal K^{-1} mol^{-1})	Reference
Pancreatic lipase	6.0	46.0	68.2	a
Trypsin	6.5	40.8	44.7	b
ATPase	7.0	70.0	150.0	c

[a] I. H. McGillivray, *Biochem. J.* **24**, 891 (1930).
[b] J. Pace, *Biochem. J.* **24**, 606 (1930).
[c] L. Ouellet, K. J. Laidler, and M. F. Morales, *Arch. Biochem. Biophys.* **39**, 37 (1952).

at an appreciable speed at ordinary temperature, it must also have a large positive entropy of activation. It was noted at the beginning of this chapter that the occurrence of a temperature optimum requires that the activation energy for the inactivation process must be greater than that for the enzyme–substrate reaction.

The interpretation of these large enthalpy and entropy changes requires a detailed consideration of individual cases. We will here comment on only a few examples; further discussion is to be found in Chapter 13 of Laidler and Bunting's book.[17] The process $P_3 \rightarrow P_4 + H^+$ in the ricin system involves an entropy increase of 54 cal K^{-1} mol^{-1} at 65°. The splitting-off of a proton can follow two patterns, of which the following are examples:

$$RNH_3^+ \rightleftharpoons RNH_2 + H^+$$
$$ROH \rightleftharpoons RO^- + H^+$$

In the first example the proton leaves a positively charged group, and there is no change in the number of charges when the process occurs. There will be no large change in entropy arising from electrostatic effects, and this conclusion is supported by experimental values obtained for simple ionizations of this kind (e.g., for $NH_4^+ \rightarrow NH_3 + H^+$, $\Delta S° = -0.5$ cal K^{-1} mol^{-1} at 25°). In the second example, two charged species are created when ionization occurs, and negative entropy changes will therefore arise from electrostatic causes and are found experimentally for simple systems (e.g., for $CH_3COOH \rightarrow CH_3COO^- + H^+$, $\Delta S° = -22.0$ cal K^{-1} mol^{-1} at 25°. It follows that the value of 54 cal K^{-1} mol^{-1} cannot be interpreted in terms of electrostatic effects, and must be due to a very profound structural difference between P_3 and P_4.

The entropy of activation of 42 cal K^{-1} mol^{-1} observed for the denaturation of the P_4 form of ricin must be explained by a large structural change when P_4 becomes the activated complex $P_4^{\ddagger}$. Inactivations usually involve a large overall entropy increase, because the denatured protein has a more open and disordered structure. Since the activated complex $P_4^\ddagger$ is on its way toward becoming the denatured protein, it is not surprising that it also has a higher entropy than P_4. The same explanation applies to the values shown in Table VII, and it appears that the positive entropies of activation, and corresponding high energies of activation, are generally associated with enzyme inactivations, as a result of the loosening of the enzyme structure.

[11] Approaches to Kinetic Studies on Metal-Activated Enzymes

By JOHN F. MORRISON

There are two classes of enzyme that require bivalent metal ions for their catalytic activity. They are the metalloenzymes, which contain tightly bound metal ions that do not dissociate during their isolation, and the metal-activated enzymes, which are inactive in the absence of added metal ion. In this chapter discussion will be restricted to enzymes belonging to the latter class.

The largest group of metal-activated enzymes comprises undoubtedly the phosphotransferases, of which over 100 are listed in the 1973 edition of "Enzyme Nomenclature."[1] It is because of their widespread occurrence and metabolic importance that considerable effort has been expended on the determination of the kinetic characteristics of phosphotransferases. Such investigations have been of a very variable quality. Some have been pursued with considerable care whereas others have been carried out with a complete lack of awareness of the difficulties that can be encountered with studies on metal-activated enzymes. Questions about the ionic form of the nucleotide at the pH of the reaction mixture, the complexing of metal ion by the ionic form(s) of the nucleotide, and whether the free or metal-complexed nucleotide participates in the reaction often are not considered. It seemed, therefore, that there would be merit in using reactions catalyzed by phosphotransferases as a means of illustrating the approaches that should be used to gain the maximum amount of information from kinetic investigations on both these and other metal-activated enzymes.

Phosphotransferases catalyze the reversible transfer of the terminal phosphoryl group of a nucleoside triphosphate to an acceptor molecule, which may be an alcohol, carboxylic acid, nitrogenous compound, or

---

[1] "Enzyme Nomenclature," Elsevier, Amsterdam, 1973.

phosphorylated compound. Virtually all reactions occur with ATP as the phosphoryl group donor and have an essential requirement for a bivalent metal ion (M^{2+}). Thus the general reaction can be expressed as

$$\text{ATP} + \text{acceptor} \xrightleftharpoons{M^{2+}} \text{ADP} + \text{phosphorylated acceptor}$$

While Mg^{2+} and Mn^{2+} function as activators of most phosphotransferases, other bivalent metal ions, including Ca^{2+}, Sr^{2+}, Ba^{2+}, Fe^{2+}, Co^{2+}, and Ni^{2+}, have been shown to activate different phosphotransferases.[2] Some enzymes, such as pyruvate kinase, also require the presence of a monovalent cation for activity.[3] This requirement can be satisfied by K^+, NH_4^+, Tl^+ or Rb^+. There is considerable evidence to support the idea that bivalent metal–nucleotide complexes function as substrates for those phosphotransferases, such as arginine kinase,[4] creatine kinase,[5] galactokinase,[6] hexokinase,[7] nucleoside diphosphokinase,[8] and 3-phosphoglycerate kinase,[9] which have been subjected to careful detailed investigations. Thus it seems likely that all phosphotransferases utilize metal–nucleotide complexes as substrates. It has also become apparent that phosphotransferases may be inhibited by concentrations of M^{2+} and/or free nucleotide, which, as outlined below, can be present together with the metal–nucleotide complex. Consequently it is of importance to determine how inhibitory either M^{2+} or free nucleotide is for a particular enzyme and to choose experimental conditions that eliminate or minimize their inhibitory effects.

By contrast with the situation for the nucleotide substrate, it seems that the second substrates, like phosphocreatine for the reverse creatine kinase reaction[10] and phosphoenol pyruvate for the reverse pyruvate kinase reaction react in their uncomplexed forms. Both these substrates complex weakly with M^{2+} (cf. this volume [12]). The more descriptive

---

[2] J. F. Morrison and E. Heyde, *Annu. Rev. Biochem.* **41**, 29 (1972).
[3] F. J. Kayne, in "The Enzymes" (P. D. Boyer, ed.), 3rd ed., Vol. 8, p. 353. Academic Press, New York, 1973.
[4] J. F. Morrison, in "The Enzymes" (P. D. Boyer, ed.), 3rd ed., Vol. 8, p. 457. Academic Press, New York, 1973.
[5] D. C. Watts, in "The Enzymes" (P. D. Boyer, ed.), 3rd ed., Vol. 8, p. 383. Academic Press, New York, 1973.
[6] J. S. Gulbinsky and W. W. Cleland, *Biochemistry* **7**, 566 (1968).
[7] D. L. Purich, H. J. Fromm, and F. B. Rudolph, *Adv. Enzymol.* **39**, 249 (1973).
[8] E. Garces and W. W. Cleland, *Biochemistry* **8**, 633 (1969).
[9] R. K. Scopes, in "The Enzymes" (P. D. Boyer, ed.), 3rd ed., Vol. 8, p. 335. Academic Press, New York, 1973.
[10] S. A. Kuby and E. A. Noltmann, in "The Enzymes" (P. D. Boyer, H. Lardy, and K. Myrbäck, eds.), 2nd ed., Vol. 6, p. 515. Academic Press, New York, 1962.

general equation for phosphotransferase reactions is

$$\text{MATP} + \text{acceptor} \rightleftharpoons \text{MADP} + \text{phosphorylated acceptor}$$

A particularly interesting example of this formulation is the reaction catalyzed by muscle adenylate kinase where MgADP$^-$ and ADP^{3-} are the substrates for the reaction in the reverse direction.[11,12] By contrast, both the nucleoside triphosphate and nucleoside diphosphate substrates for the nucleoside diphosphokinase reaction must be present as their metal complexes.[8] In the discussions that follow, it will be assumed that the substrates for phosphotransferase reactions in each direction are the metal–nucleotide complex and the uncomplexed form of the second substrate.

Problems Associated with Studies on
  Metal-Activated Enzymes

The difficulties encountered with studies on metal-activated enzymes can be illustrated by first considering the nonenzymic interactions between a bivalent metal ion and nucleoside phosphates. For the most part, ATP will be used as the nucleotide under consideration, but, where appropriate, specific references will be made to ADP. M^{2+} will refer to the free form of the bivalent metal ion. The ionization of ATP and the reaction of M^{2+} with the HATP^{3-} and ATP^{4-} species is shown in Scheme 1, where

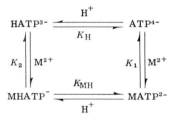

Scheme 1

K_1, K_2, K_H, and K_{MH} represent dissociation constants for the appropriate reactions. It is the usual practice to express the degree to which a metal ion is complexed by a ligand in terms of a stability constant that is the reciprocal of the dissociation constant. Thus a stability constant of 10,000 M^{-1} is equivalent to a dissociation constant of 0.1 mM. Since dissociation constants are used almost invariably by enzymologists as a mea-

---

[11] L. Noda, in "The Enzymes" (P. D. Boyer, ed.), 3rd ed., Vol. 8, p. 279. Academic Press, New York, 1973.
[12] B. D. N. Rao and M. Cohn, J. Biol. Chem. 253, 1149 (1978).

sure of the strength of binding of a substrate to an enzyme, the same term will be used to express the binding of a metal ion by a substrate. In Scheme 1 it is assumed that neither M_2ATP^0 nor $M(ATP)_2^{6-}$ is formed. There is no evidence in support of the formation of $M(ATP)_2^{6-}$ and it is unlikely that such a complex would be formed because its formation would involve the approach of two negatively charged species, $MATP^{2-}$ and ATP^{4-}. On the other hand, there are reports indicating that M_2ATP^0 can be formed,[13-16] but the dissociation constants for such complexes are relatively high. Thus under the ionic strength conditions used for kinetic investigations (0.05–0.15 M) the amount of M_2ATP^0 present in reaction mixtures would be small unless the concentration of M^{2+} were increased to high levels.[17] With zero overall charge it is perhaps unlikely that M_2ATP^0 would compete with $MATP^{2-}$, and any inhibition by high levels of metal ion is likely to be due to reduction in the concentration of $MATP^{2-}$ and/or to the increase in the M^{2+} concentration.

The last two pK_a values for ATP are 3.93 and 6.97,[18] and it is only the latter that will determine the nature of the ionic species present in solution over the usual pH range for studies on phosphotransferases. $HATP^{3-}$ will predominate at pH 6.0, and the ATP will be present largely as ATP^{4-} at pH 8.0. Solutions of M^{2+} and ATP at pH 7.0 will contain four nucleotide species as well as M^{2+}. The relative amounts of each will be determined by the dissociation constants for the $MHATP^-$ and $MATP^{2-}$ complexes, which will depend on the identity of the metal ion as well as temperature and ionic strength (cf. this volume [12]). The kinetic results obtained with a phosphotransferase will then depend on the concentration of the substrate ($MATP^{2-}$) and whether the concentrations of $MHATP^-$, $HATP^{3-}$, ATP^{4-}, and M^{2+} that are present cause activation or inhibition of the enzyme.

Because of the aforementioned complexities, it is of importance that careful consideration be given to the choice of experimental conditions for studying the kinetics of the reaction catalyzed by a metal-activated enzyme. Unless this is done, the resulting data may lead to erroneous conclusions about the kinetic behavior of an enzyme and/or to incorrect values for kinetic parameters. The experimental conditions will, of course, vary with the aims of the investigation. Experiments can be designed to either demonstrate or avoid any inhibition by, for example, ATP^{4-}.

[13] C. M. Frey, J. L. Banyasz, and J. E. Stuehr, *J. Am. Chem. Soc.* **94**, 9198 (1974).
[14] M. S. Mohan and G. A. Rechnitz, *Arch. Biochem. Biophys.* **162**, 194 (1974).
[15] R. N. F. Thorneley and K. R. Willison, *Biochem. J.* **139**, 211 (1974).
[16] G. Noat, J. Richard, M. Borel, and C. Got, *Eur. J. Biochem.* **13**, 347 (1970).
[17] D. D. Perrin and V. S. Sharma, *Biochim. Biophys. Acta* **127**, 35 (1966).
[18] W. J. O'Sullivan and D. D. Perrin, *Biochemistry* **3**, 18 (1964).

Procedures to Be Avoided

There are two unsatisfactory practices which have been used extensively for studies on phosphotransferases. One involves the use of a single concentration of total metal ion, which is added in excess of the highest concentration of nucleotide that is used. Such a procedure does tend to increase the proportion of total nucleotide that is present as its metal complex and to reduce the concentration of free nucleotide in solution (Table I). However, if M^{2+} were inhibitory, the increase in steady-state velocity of a reaction on raising the total nucleotide concentration could be due not only to an increase in the concentration of the metal–nucleotide complex, but also to a reduction in the concentration of M^{2+}. Increasing concentrations of a second substrate which complexes M^{2+} would also reduce the inhibition provided that the complex formed was inert. It is apparent from the data of Table I that the changes in concentration of M^{2+} are relatively small, but this would not be so for higher concentrations of the nucleotide. It will be noted that the concentrations of $MgADP^-$ are considerably lower than those of the added ADP and that the difference is pH dependent. The differences in the concentrations of metal–nucleotide complex and total nucleotide would be less with ATP as $MgATP^{2-}$ has a much lower dissociation constant than does $MgADP^-$.

The second practice, which has been more commonly employed, involves the variation of the concentrations of total metal ion and total nu-

TABLE I
VARIATION IN THE CONCENTRATIONS OF Mg^{2+} AND NUCLEOTIDE SPECIES AS A RESULT OF VARYING THE TOTAL ADP CONCENTRATIONS WITH TOTAL MAGNESIUM HELD CONSTANT AT 2 mM [a]

pH	Total ADP concentration (mM)	Mg^{2+} (mM)	$MgADP^-$ (mM)	ADP^{3-} (mM)	$MgHADP^0$ (mM)	$HADP^{2-}$ (mM)
6.0	0.4	1.765	0.211	0.030	0.024	0.134
	0.2	1.880	0.108	0.014	0.012	0.064
	0.1	1.939	0.055	0.007	0.006	0.032
7.0	0.4	1.671	0.326	0.049	0.004	0.022
	0.2	1.833	0.165	0.023	0.002	0.010
	0.1	1.916	0.083	0.011	0.001	0.005
8.0	0.4	1.654	0.345	0.052	0	0.002
	0.2	1.825	0.175	0.024	0	0.001
	0.1	1.912	0.088	0.012	0	0.001

[a] Values used in the calculations: pK values for ADP, 6.65 and 3.93; dissociation constants for $MgADP^-$ and MgHADP, 0.25 mM and 10 mM, respectively.

cleotide in constant ratio. There are several problems associated with the use of this procedure, especially if the ratio is low and the metal–nucleotide complex has a relatively high dissociation constant. The data of Table II show that when the total magnesium and total nucleotide concentrations are varied in a constant ratio of 1:1, each velocity measurement would be made not only at a different concentration of MgADP$^-$ or MgATP^{2-}, but also at different concentrations of free nucleotides and Mg^{2+}. It will be noted also that the concentration of the substrate, MgADP$^-$ or MgATP^{2-}, may be very different from the concentration of added ADP or ATP and that the difference is pH dependent. Therefore, a plot of v^{-1} against the reciprocal of the total ADP or total ATP concentration would have no physical significance. Indeed, such plots could be non-

TABLE II
Effect of pH and the Variation, in a Constant Ratio of 1:1, of the Total Concentrations of Magnesium and ADP (or ATP) on the Concentrations of Mg^{2+} and Various Nucleotide Species[a]

pH	Total magnesium (mM)	Total ADP (mM)	Mg^{2+} (mM)	MgADP$^-$ (mM)	ADP^{3-} (mM)	MgHADP$^\circ$ (mM)	HADP^{2-} (mM)
6.0	0.4	0.4	0.318	0.073	0.058	0.008	0.258
	0.2	0.2	0.175	0.022	0.032	0.002	0.142
	0.1	0.1	0.093	0.006	0.017	0.001	0.076
7.0	0.4	0.4	0.240	0.158	0.166	0.002	0.074
	0.2	0.2	0.143	0.056	0.099	0.001	0.044
	0.1	0.1	0.081	0.018	0.056	0	0.025
8.0	0.4	0.4	0.218	0.182	0.209	0	0.009
	0.2	0.2	0.133	0.067	0.127	0	0.006
	0.1	0.1	0.077	0.023	0.074	0	0.003

pH	Total magnesium (mM)	Total ATP (mM)	Mg^{2+} (mM)	MgATP^{2-} (mM)	ATP^{4-} (mM)	MgHATP$^-$ (mM)	HATP^{3-} (mM)
6.0	0.4	0.4	0.176	0.205	0.017	0.019	0.157
	0.2	0.2	0.111	0.082	0.011	0.008	0.099
	0.1	0.1	0.067	0.030	0.006	0.003	0.060
7.0	0.4	0.4	0.092	0.305	0.048	0.003	0.045
	0.2	0.2	0.062	0.137	0.032	0.001	0.030
	0.1	0.1	0.041	0.059	0.021	0.001	0.020
8.0	0.4	0.4	0.072	0.328	0.066	0	0.006
	0.2	0.2	0.049	0.151	0.045	0	0.004
	0.1	0.1	0.033	0.067	0.030	0	0.003

[a] Values used in the calculations: pK values for ATP, 6.97 and 3.93; ADP, 6.65 and 3.93; dissociation constants for MgATP^{2-}, 0.0143 mM; MgHATP$^-$, 1.44 mM; MgADP$^-$, 0.25 mM; MgHADP$^\circ$, 10 mM.

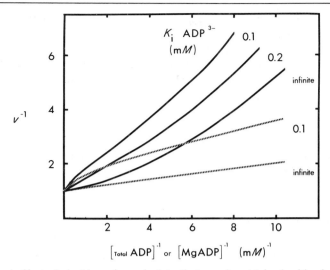

FIG. 1. Artifactual double-reciprocal plots that can be obtained with phosphotransferases. It was assumed that MgADP$^-$ is the substrate for the phosphotransferase reaction ($V_m = 1.0$, $K_m = 0.1$ mM) and that ADP^{3-} may or may not be a competitive inhibitor with respect to MgADP$^-$ ($K_i = 0.2$ mM, 0.1 mM, or ∞). Steady-state velocities were calculated for the case where the total concentrations of magnesium (Mg$_t$) and ADP (ADP$_t$) were maintained in a constant ratio of 1:1 and the concentrations of MgADP$^-$ and ADP^{3-} were determined using a dissociation constant of 0.25 mM for the MgADP$^-$ complex. The curves were constructed by plotting v^{-1} against [ADP$_t$]$^{-1}$ (solid lines) or against [MgADP]$^{-1}$ (dashed lines).

linear, the nonlinearity being also a function of the concentrations of the free nucleotide species as well as the values for their inhibition constants (Fig. 1). From experiments involving the use of total metal and total nucleotide concentration, the concentrations of metal–nucleotide complex could be determined by the method used for the calculations of Table II.[19] A plot could then be made of v^{-1} against the reciprocals of the metal–nucleotide concentrations, but no allowance could be made for any kinetic effects of the free metal ion and free nucleotide species. The complexities associated with this approach are enhanced when a second substrate is also capable of complexing the metal ion.

Recommended Procedure for Kinetic Studies

In this section an outline will be given of the approach adopted in this laboratory some 18 years ago for kinetic studies on the creatine kinase

[19] D. D. Perrin and I. G. Sayce, *Talanta* **14**, 833 (1967).

reaction,[20,21] which has been applied subsequently to studies on other phosphotransferases. Emphasis will be placed on the practical aspects of kinetic studies on metal-activated enzymes and the opportunity taken to treat in detail the theory associated with various cases that could apply. The basic principles of the approach are listed below.

1. Choose conditions that minimize the number of nucleotide species present in solution unless there are specific reasons for not doing so.

2. Use the appropriate dissociation constant for the metal–nucleotide complex to calculate the total concentrations of metal and nucleotide required to give the desired concentrations of metal-nucleotide complex while holding the concentration of free metal ion constant. The free metal ion, rather than the free nucleotide, concentration is usually maintained at a fixed concentration because its concentration does not vary with pH whereas the free nucleotide concentration does. Further, free nucleotide is frequently an inhibitory species.[16,21,22] When this procedure is followed the ATP:MATP ratio is kept constant, and thus Michaelis–Menten kinetics are followed. But it should be noted that if the ratio is high, the values for K_m and V will be low. For the reaction $M^{2+} + ATP^{4-} \rightleftharpoons MATP^{2-}$, it is possible to maintain constant only two of the three reacting species. If $MATP^{2-}$ and M^{2+} are held constant, the concentration of ATP^{4-} would be determined by the relationship $ATP^{4-} = K_1 (MATP)/(M)$.

3. Hold free metal ion at a fixed concentration that is not inhibitory, but is sufficiently high to reduce free nucleotide species to noninhibitory levels.

4. Allow for any complexing by the nucleotide(s) of a univalent cation that is required for catalytic activity or is present as a counterion for acidic substrates.

5. Allow for complexing of the bivalent metal ion by a second substrate, a dead-end inhibitor, or a product inhibitor.

The calculations that must be made in connection with the implementation of the above procedure for various studies on different types of phosphoryl group transfer reactions are given below. The kinetic theory associated with the inhibition and activation of an enzyme by free nucleotide or metal ion is also included.

Reactions Involving MATP Only or Both MATP and a Noncomplexing Second Substrate

The reactions catalyzed by ATPase and hexokinase are examples of reactions where the essential bivalent metal ion is complexed only by the

---

[20] J. F. Morrison, W. J. O'Sullivan, and A. G. Ogston, *Biochim. Biophys. Acta* **52**, 82 (1961).
[21] J. F. Morrison and W. J. O'Sullivan, *Biochem. J.* **94**, 221 (1965).
[22] D. L. Purich and H. J. Fromm, *Biochem. J.* **130**, 63 (1972).

nucleotide substrate. The nonenzymic interactions that can occur are illustrated in Scheme 1. Provided that the interactions are at thermodynamic equilibrium, the relationship $K_1 \cdot K_{MH} = K_2 \cdot K_H$ must hold, because the concentration of each species will be independent of its pathway of formation. It also follows that only three of the four dissociation constants are required to express the concentration of each species; the value of the fourth constant is determined by the relationship between the other three, i.e., $K_{MH} = K_2 \cdot K_H/K_1$.

Calculations for Holding M^{2+} Constant While Varying $MATP^{2-}$

From the equilibrium reactions shown in Scheme 1, it follows that

$$K_1 = \frac{(M)(ATP)}{(MATP)}; \quad K_2 = \frac{(M)(HATP)}{(MHATP)}; \quad K_H = \frac{(H)(ATP)}{(HATP)} \quad (1)$$

In these and subsequent equations, the charge on each species has been omitted for the sake of clarity. The concentration of each nucleotide species may be written in terms of the concentration of $MATP^{2-}$ as

$$ATP = \frac{K_1}{(M)} \cdot MATP; \quad HATP = \frac{K_1}{(M)} \cdot \frac{(H)}{K_H} \cdot MATP;$$

$$MHATP = \frac{K_1}{K_2} \cdot \frac{(H)}{K_H} \cdot MATP \quad (2)$$

The total concentration of metal ion (M_t) and the total concentration of ATP (ATP_t) are given by

$$M_t = M + MATP + MHATP = M + MATP \left[1 + \frac{K_1}{K_2} \cdot \frac{(H)}{K_H}\right] \quad (3)$$

$$ATP_t = MATP + ATP + HATP + MHATP$$
$$= MATP \left[1 + \frac{K_1}{(M)} + \frac{K_1}{(M)} \cdot \frac{(H)}{K_H} + \frac{K_1}{K_2} \cdot \frac{(H)}{K_H}\right] \quad (4)$$

With a knowledge of the values for K_1, K_2, and K_H that apply under the chosen experimental conditions, Eqs. (3) and (4) can be used to calculate the total concentrations of metal ion and ATP required to give particular concentrations of $MATP^{2-}$ at any fixed concentration of M^{2+}.

The relationships given under Eq. (2) permit calculations to be made of the concentrations of other nucleotide species that will be present in reaction mixtures containing different concentrations of $MATP^{2-}$. It will be noted that the concentrations of both $HATP^{3-}$ and $MHATP^-$ are dependent on the hydrogen ion concentration, whereas the concentrations of ATP^{4-} and $HATP^{3-}$ vary with the fixed concentration of M^{2+}. At any particular pH the ratio of $HATP^{3-}$ to ATP^{4-} remains constant and equal to H/K_H. Since $K_1 \ll K_2$, the amounts of MHATP⁻ in solution will usually be negligibly small at pH 7 and above (cf. this volume [12]) (Table III). If the H⁺ concentration is less than the value of K_H by a factor of 10 or more, i.e., the pH of the solution is one or more pH units above the last pK_a value of the terminal phosphoryl group of ATP, then the concentrations of HATP³⁻ and MHATP⁻ tend to zero. The last pK_a values for ATP and ADP are, respectively, in the region of pH 7.0 and 6.7, depending on temperature and ionic strength. Therefore, when kinetic investigations are performed at pH 8.0 and above it is not necessary to take into account any kinetic effects of HATP³⁻ and MHATP⁻. Equations (3) and (4) now simplify to those listed as Eq. (5)

$$M_t = M + MATP; \qquad ATP_t = MATP \left(1 + \frac{K_1}{(M)}\right) \qquad (5)$$

It is for these reasons that a number of investigations on phosphotransferases have been made at pH 8.0.[20-24] Quantitative illustrations of the aforementioned points are given by the data of Table III for both ATP and ADP.

Kinetic Effects of ATP⁴⁻

For experiments conducted at pH 8.0 with M²⁺ held constant at a noninhibitory concentration, the only complicating factor would be inhibition by ATP⁴⁻. It is apparent from the first relationship given under Eq. (2) that the concentration of ATP⁴⁻ would be reduced either by raising the concentration of M²⁺ or by choosing a bivalent metal ion that is chelated more strongly by the oxyanions of the nucleotide. Provided that it is not more inhibitory, there may be merit in choosing Mn²⁺, rather than Mg²⁺, as the activating metal ion. Mn–nucleotide complexes have lower dissociation constants than the corresponding Mg–nucleotide complexes, but frequently Mn²⁺ is more inhibitory and shows cooperative effects. This metal ion also undergoes hydrolysis with a pK_a value of 10.6.[25] When Mn–nucleotide complexes are used for NMR and EPR studies[26] (cf. this volume [12]), the same metal–nucleotide complexes should be used for kinetic investigations.

When the second substrate of a phosphotransferase does not react with a bivalent metal ion, the ideal conditions for initial studies on the

---

[23] J. F. Morrison and E. James, *Biochem. J.* **97**, 37 (1965).
[24] J. P. Infante and J. E. Kinsella, *Int. J. Biochem.* **7**, 483 (1976).
[25] D. D. Perrin, *Pure Appl. Chem.* **20**, 133 (1969).
[26] A. S. Mildvan, in "The Enzymes" (P. D. Boyer, ed.), 3rd ed., Vol. 2, p. 445. Academic Press, New York, 1970.

TABLE III
CONCENTRATION OF NUCLEOTIDE SPECIES IN SOLUTION WHEN MgADP$^-$ (OR MgATP^{2-}) IS VARIED WHILE Mg^{2+} IS HELD CONSTANT AT 2.0 mM^a

pH	MgADP$^-$ (mM)	ADP^{3-} (mM)	MgHADP$^\circ$ (mM)	HADP^{2-} (mM)
6.0	0.4	0.050	0.045	0.223
	0.2	0.025	0.023	0.112
	0.1	0.013	0.011	0.056
7.0	0.4	0.050	0.005	0.022
	0.2	0.025	0.002	0.011
	0.1	0.013	0.001	0.006
8.0	0.4	0.050	0	0.002
	0.2	0.025	0	0.001
	0.1	0.013	0	0.001

pH	MgATP^{2-} (mM)	ATP^{4-} (mM)	MgHATP$^-$ (mM)	HATP^{3-} (mM)
6.0	0.4	0.003	0.005	0.027
	0.2	0.001	0.003	0.013
	0.1	0.001	0.001	0.007
7.0	0.4	0.003	0.001	0.003
	0.2	0.001	0	0.001
	0.1	0.001	0	0.001
8.0	0.4	0.003	0	0
	0.2	0.001	0	0
	0.1	0.001	0	0

^{a} The values used for the calculations were the same as those used in Table II.

kinetics of the reaction involve the use of a metal ion, which is chelated strongly by the nucleotide substrate, at a high noninhibitory concentration. Further, the pH of reaction mixtures should be such as to preclude the formation of other complexed and noncomplexed nucleotide species. But the ideal cannot always be achieved because of the relatively high dissociation constant of the metal–nucleotide substrate being used. For instance, ADP is a far weaker chelator than ATP irrespective of the identity of the metal ion (Table III) (this volume [12]). This being so, it is important to enquire into the kinetic effects of having free nucleotide(s) present in reaction mixtures at significant concentrations.

Effect on Kinetic Parameters of Having ATP^{4-} and MATP^{2-} Varying in Constant Ratio

At pH 8.0, the only species of nucleotide present in significant concentrations will be MATP^{2-} and ATP^{4-}. With M^{2+} held at a fixed noninhibitory concentration, variation of the concentration of $MATP^{2-}$ will cause the concentration of ATP^{4-} to vary in constant ratio with the $MATP^{2-}$ concentration. This follows from the fact that

$$K_1 = \frac{(M)(ATP)}{(MATP)} \quad \text{so} \quad \frac{K_1}{(M)} = \text{constant} = \frac{(ATP)}{(MATP)} \tag{6}$$

If the value of $K_1/(M)$ is such that ATP^{4-} is present as a significant proportion of $MATP^{2-}$ and functions as a linear competitive inhibitor with respect to the nucleotide substrate, then the determined values for the maximum velocity and Michaelis constant for MATP will be in error. The general equation that describes linear competitive inhibition is

$$v = \frac{VA}{K_a(1 + I/K_i) + A} \tag{7}$$

If A represents $MATP^{2-}$ and I denotes ATP^{4-}, it follows from Eq. (6) that the term $K_1(A)/(M)$ can be substituted for I in Eq. (7). Rearrangement of the resulting equation in reciprocal form gives Eq. (8).

$$\frac{1}{v} = \frac{K_a}{V}\frac{1}{A} + \frac{1}{V}\left[1 + \frac{K_aK_1}{K_i(M)}\right] \tag{8}$$

which shows that a plot of v^{-1} against $[A]^{-1}$ would yield the correct value for K_a/V, but a reduced apparent maximum velocity and a lower apparent Michaelis constant. The determination of the steady-state velocity as a function of the concentration of $MATP^{2-}$ at different fixed concentrations of M^{2+} would yield a family of parallel straight lines when the data are plotted in double-reciprocal form. The apparent activation of the reaction by M^{2+} is due to the effect of the metal ion in lowering the inhibitory concentration of ATP^{4-}. A secondary plot of the vertical intercepts as a function of $[M]^{-1}$ would give a straight line whose vertical intercept and slope would correspond, respectively, to the maximum velocity and value of K_aK_1/K_iV for the particular fixed concentration of a second substrate. From a knowledge of the values for K_a/V and K_1, the value of K_i could be calculated. If there were no inhibition by ATP^{4-}, either because M^{2+} was varied over a range of concentrations that were high compared with K_1 or because the value for K_i was very high, all the lines of the plots of v^{-1} against $[MATP]^{-1}$ at different M^{2+} concentrations would be superimposable. In this instance the correct values for V and K_a would be obtained for one particular concentration of the second substrate.

If the experiments described above were performed at lower pH values, reaction mixtures could contain appreciable concentrations of $HATP^{3-}$, and it is possible that this nucleotide species would also act as a linear competitive inhibitor with respect to $MATP^{2-}$. In this case the intercept term of Eq. (8) would contain the additional term $K_1 K_a(H)/K_j K_H(M)$, where K_j is the inhibition constant for $HATP^{3-}$ [Eq. (2)]. The presence of the extra term in the equation would not influence the qualitative results, but would preclude calculation of values for either inhibition constant.

Determination of the Inhibition by ATP^{4-}

There is a more direct method than that described above for determining the inhibition constant for ATP^{4-} under conditions where the concentration of $HATP^{3-}$ can be neglected. It involves determining the steady-state velocity of the reaction as a function of the $MATP^{2-}$ concentration at different fixed concentrations of ATP^{4-}. The approach to holding ATP^{4-} constant is similar to that used to maintain M^{2+} at a fixed concentration, and the relationships that would apply are

$$M_t = MATP \left(1 + \frac{K_1}{(ATP)}\right); \quad ATP_t = ATP + MATP \quad (9)$$

Under these circumstances the concentrations of M^{2+} vary as the concentrations of $MATP^{2-}$ and ATP^{4-} are varied. But provided that M^{2+} is not an activator or an inhibitor, or is present only at concentrations in which neither the activation nor inhibition is observed, a linear competitive inhibition would be expected. If the reaction under study involved a second substrate and conformed to an ordered kinetic mechanism, only an apparent inhibition constant for ATP^{4-} would be obtained. The true value would have to be determined by taking into account the fixed concentration of the second substrate and its appropriate kinetic parameter. On the other hand, if the two substrates added to the enzyme in a random manner under rapid equilibrium conditions, the directly determined value for the competitive inhibition constant would not represent the dissociation constant for the combination of ATP^{4-} with either the free enzyme or the enzyme–second substrate complex.[27] The values of the inhibition constants for the combination of ATP^{4-} with free enzyme and the enzyme–second substrate complex would have to be determined from the noncompetitive inhibition by ATP^{4-} with respect to the second substrate, taking into account the fixed concentration of $MATP^{2-}$ and its appropriate kinetic parameters.

[27] E. James and J. F. Morrison, *J. Biol. Chem.* **241**, 4758, (1966).

Kinetic Effects of M^{2+}

M^{2+} as an Inhibitor of the Reaction

It is possible for M^{2+} to behave as a linear competitive inhibitor with respect to $MATP^{2-}$ by combining, like ATP^{4-}, at the same site on the enzyme as $MATP^{2-}$. In this case the inhibition constant(s) associated with M^{2+} would be determined as described for ATP^{4-}.

M^{2+} as an Essential Activator of the Reaction

It has been demonstrated that for the reactions catalyzed by choline kinase[24] and pyruvate carboxylase[28] Mg^{2+} is an activator while $MgATP^{2-}$ functions as a substrate. A similar situation holds for the pyruvate kinase reaction, but in this case the dissociation constant for the enzyme–Mg complex is much lower than that for $MgADP^-$ and so there is masking of the activation by Mg^{2+}.[26] Consideration is now given to the kinetic results that would be expected with various models involving M^{2+} as an essential activator and $MATP^{2-}$ as a substrate. As shown in Scheme 2, M^{2+} could

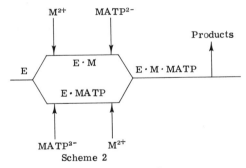

Scheme 2

add in an ordered sequence either before (upper pathway) or after (lower pathway) the addition of $MATP^{2-}$. Alternatively, M^{2+} and $MATP^{2-}$ could add to the enzyme in a random fashion via the two pathways of Scheme 2. If the random combination occurred under rapid equilibrium conditions, then the steady-state rate equation in the absence of products would have the same general form as those for the two ordered sequences. The specific equation that describes the ordered addition of M^{2+} followed by $MATP^{2-}$ or the random addition of M^{2+} and $MATP^{2-}$ is

$$v = \frac{V(M)(MATP)}{K_{im}K_a + K_a(M) + K_m(MATP) + (M)(MATP)} \quad (10)$$

[28] W. R. McClure, H. A. Lardy, and H. P. Kneifel, *J. Biol. Chem.* **246**, 3569 (1971).

where K_{im} is the dissociation constant for the EM complex; K_a and K_m are Michaelis constants for MATP^{2-} and M^{2+}, respectively; V is the maximum velocity. When the concentration of ATP^{4-} is maintained constant, the ratio of M^{2+}:MATP^{2-} will remain constant and equal to $K_1/$(ATP). Substitution into Eq. (10) of M = K_1(MATP)/(ATP) gives

$$v = \frac{V(\text{MATP})^2}{(\text{MATP})^2 + [K_a + K_m(\text{ATP})/K_1](\text{MATP}) + K_{im}K_a(\text{ATP})/K_1} \quad (11)$$

which is an equation of a sigmoid curve that gives rise to a parabola for a plot of v^{-1} against [MATP]$^{-1}$. From the reciprocal form of the equation, it is apparent that the curvature of the parabola increases as a function of the fixed concentration of ATP^{4-}.

M^{2+} as an Essential Activator, ATP^{4-} as a Competitive Inhibitor

The equation that describes the activation of an enzyme by M^{2+} and its inhibition by ATP^{4-} will differ according to the kinetic mechanism. In the following section three models will be considered, and for the sake of simplicity it will be assumed that the concentration of the acceptor is present either at a fixed nonsaturating concentration or at a saturating concentration. In the former case only apparent values would be obtained for those kinetic constants that can be determined.

Model a. Ordered Mechanism with M^{2+} Combining before MATP^{2-}

In this model inhibition by ATP^{4-} would occur because it combines with the E-M form of enzyme with which MATP^{2-} reacts (cf. Scheme 2, upper pathway). The inhibition by ATP^{4-} would be accounted for by modification of the K_a(M) term of Eq. (10) by the factor [1 + (ATP)/K_i]. The resulting equation in reciprocal form would be

$$\frac{1}{v} = \frac{1}{V}\left[\frac{K_{im}K_a}{(\text{M})(\text{MATP})} + \frac{K_a}{(\text{MATP})} + \frac{K_a(\text{ATP})}{K_i(\text{MATP})} + \frac{K_m}{(\text{M})} + 1\right] \quad (12)$$

When the concentration of M^{2+} is fixed so that ATP^{4-} and MATP^{2-} vary in constant ratio equal to $K_1/$M, then Eq. (12) may be presented as Eq. (13) with MATP^{2-} as the variable substrate. Thus

$$\frac{1}{v} = \frac{K_a}{V}\left[\frac{K_{im}}{(\text{M})} + 1\right]\frac{1}{(\text{MATP})} + \frac{1}{V}\left[\frac{K_m}{(\text{M})} + \frac{K_aK_1}{K_i(\text{M})} + 1\right] \quad (13)$$

If a series of experiments are performed by determining the steady-state velocity with MATP^{2-} as the variable substrate (in constant ratio with ATP^{4-}) at different fixed concentrations of M^{2+}, an intersecting pattern will be obtained when the data are plotted in double-reciprocal form. This pattern is obtained because both the slopes and the vertical intercepts of the plot will vary as a function of M^{2+} concentration. It should be noted that as the changing fixed concentration of M^{2+} is increased, the value of K_1/M decreases as does the ATP^{4-}:MATP^{2-} ratio. Thus at any fixed concentration of MATP^{2-}, the concentration of ATP^{4-} decreases, as does the inhibition, and the steady-state velocity increases. Secondary plots of the slopes against [M]$^{-1}$ [Eq. (13)] would give a value for K_{im} while a replot of the vertical intercepts against [M]$^{-1}$ would yield a value for the maximum velocity only. If M^{2+} activates at low concentrations so that $K_{im} \simeq K_m \simeq 0$, Eq. (13) would reduce to Eq. (8) and plots of v^{-1} against (MATP)$^{-1}$ at different concentrations of M^{2+} would consist of a family of parallel straight lines. An intersecting pattern would also be obtained for a double-reciprocal plot of velocity as a function of the concentration of M^{2+} at different fixed concentrations of MATP^{2-} [Eq. (14)].

$$\frac{1}{v} = \frac{K_m}{V}\left[\frac{K_{im}K_a}{K_m(\text{MATP})} + \frac{K_aK_1}{K_mK_i} + 1\right]\frac{1}{(\text{M})} + \frac{1}{V}\left[\frac{K_a}{(\text{MATP})} + 1\right] \quad (14)$$

A secondary plot of vertical intercepts against [MATP]$^{-1}$ would give values for K_a and the maximum velocity. In practice the values for the kinetic parameters should be determined by making a least squares fit of the velocity data to an equation that describes the kinetic mechanism (cf. this volume [6]).

If ATP^{4-}, rather than M^{2+}, were held at different fixed concentrations with MATP^{2-} as the variable substrate, then M^{2+} and MATP^{2-} would vary in constant ratio equal to K_1/(ATP). The reciprocal form of Eq. (12) would then be

$$\frac{1}{v} = \frac{K_{im}K_a(\text{ATP})}{VK_1}\left(\frac{1}{\text{MATP}}\right)^2 + \frac{K_a}{V}\left[1 + \frac{(\text{ATP})}{K_i} + \frac{K_m(\text{ATP})}{K_aK_1}\right]\left(\frac{1}{\text{MATP}}\right) + \frac{1}{V} \quad (15)$$

which indicates that plots of v^{-1} against [MATP]$^{-1}$ are parabolic, the curvature of the plots increasing as the concentration of ATP^{4-} increases. The parabolicity of the plot is not dependent on ATP^{4-} being a competitive inhibitor.

If the combination of M^{2+} with the free form of enzyme were at thermodynamic equilibrium,[29,30] then the equation describing the kinetic

[29] J. F. Morrison and K. E. Ebner, *J. Biol. Chem.* **246**, 3977 (1971).
[30] W. W. Cleland, in "The Enzymes" (P. D. Boyer, ed.), 3rd ed., Vol. 2, p. 1. Academic Press, New York, 1970.

mechanism would be similar to Eq. (12) except that the $K_m/(M)$ term would be missing. Under these circumstances and with M^{2+} present at a fixed concentration, Eqs. (16) and (17) would be obtained.

$$\frac{1}{v} = \frac{K_a}{V}\left[\frac{K_{im}}{(M)} + 1\right]\frac{1}{(MATP)} + \frac{1}{V}\left[\frac{K_aK_1}{K_i(M)} + 1\right] \quad (16)$$

$$\frac{1}{v} = \frac{K_{im}}{V}\left[\frac{K_a}{(MATP)} + \frac{K_aK_1}{K_iK_{im}}\right]\frac{1}{(M)} + \frac{1}{V}\left[\frac{K_a}{(MATP)} + 1\right] \quad (17)$$

Both these equations predict that the steady-state velocity patterns in the absence of products would be of the symmetrical intersecting type. However, if ATP^{4-} were not inhibitory so that K_i was equal to infinity, then the steady-state velocity patterns would be asymmetric. A plot of v^{-1} against $[MATP]^{-1}$ at different concentrations of M^{2+} would yield a pattern that intersects on the vertical ordinate, while a pattern that intersects to the left of the vertical ordinate would be obtained with a plot of v^{-1} against $[M]^{-1}$ at different fixed concentrations of $MATP^{2-}$. For the latter pattern, a replot of the slopes of the lines against $[MATP]^{-1}$ would be linear and pass through the origin.

Model b. Ordered Mechanism with $MATP^{2-}$ Combining before M^{2+}

With a reaction following the lower pathway of Scheme 2, inhibition by ATP^{4-} would be due to the formation of an E–ATP complex. This mechanism is described by Eq. (18)

$$v = \frac{V(M)(MATP)}{[K_{ia}K_m + K_a(M)][1 + (ATP)/K_i] + K_m(MATP) + (M)(MATP)} \quad (18)$$

where K_{ia} and K_i are dissociation constants for the E-MATP and E-ATP complexes, respectively; K_a and K_m are Michaelis constants for $MATP^{2-}$ and M^{2+}, respectively. Rearrangement of Eq. (18) with $MATP^{2-}$ and M^{2+} as variable reactants for the condition that $(ATP)/(MATP) = K_1/(M)$ gives Eqs. (19) and (20), respectively.

$$\frac{1}{v} = \frac{K_a}{V}\left[\frac{K_{ia}K_m}{K_a(M)} + 1\right]\frac{1}{(MATP)} + \frac{1}{V}\left[\frac{K_{ia}K_mK_1}{K_i(M)^2} + \frac{K_aK_1}{K_i(M)} + \frac{K_m}{(M)} + 1\right] \quad (19)$$

$$\frac{1}{v} = \frac{K_{ia}K_mK_1}{VK_i}\left(\frac{1}{M}\right)^2 + \frac{K_m}{V}\left[\frac{K_{ia}}{(MATP)} + \frac{K_aK_1}{K_mK_i} + 1\right]\left(\frac{1}{M}\right)$$

$$+ \frac{1}{V}\left[\frac{K_a}{(MATP)} + 1\right] \quad (20)$$

Equation (19) predicts that plots of v^{-1} against $[MATP]^{-1}$ at different fixed concentrations of M^{2+} would give a family of straight lines that do not have a common intersection point to the left of the vertical ordinate. Secondary plots of slopes against $[M]^{-1}$ would be linear, while replots of vertical intercepts against $[M]^{-1}$ would be parabolic. In the absence of inhibition by ATP^{4-}, i.e., $K_i = \infty$, the primary plots corresponding to Eq. (19) would have a common point of intersection and secondary plots for the vertical intercepts would be linear. This mechanism is characterized further by the fact that plots of v^{-1} against $[M]^{-1}$ are parabolic, the parabolicity and vertical intercepts decreasing as a function of the $MATP^{2-}$ concentration [Eq. (20)]. Again, if $K_i = \infty$, double-reciprocal and secondary plots would be linear.

If for a reaction conforming to the above mechanism the concentration of ATP^{4-} were maintained at different fixed concentrations, a plot of v^{-1} against $[MATP]^{-1}$ would be parabolic. Substitution of $M = K_1(MATP)/(ATP)$ into Eq. (18) gives Eq. (21), which shows that the parabolicity of the curves can be very sensitive to the ATP^{4-} concentration.

$$\frac{1}{v} = \frac{K_{ia}K_m(ATP)}{K_1V}\left[1 + \frac{(ATP)}{K_i}\right]\left(\frac{1}{MATP}\right)^2$$
$$+ \frac{K_a}{V}\left[1 + \frac{K_m(ATP)}{K_aK_1} + \frac{(ATP)}{K_i}\right]\left(\frac{1}{MATP}\right) + \frac{1}{V} \quad (21)$$

Parabolic double-reciprocal plots would still be obtained under conditions where there was no inhibition by ATP^{4-}.

If, for this same mechanism, the interaction of M^{2+} with the $E-MATP$ complex were at thermodynamic equilibrium, the steady-state velocity of the reaction would be described by Eq. (18) although K_m would now be a dissociation constant for the release of M^{2+} from the $E-M-MATP$ complex. Since there is no change in the form of the rate equation, the use of M^{2+} and $MATP^{2-}$ as variable reactants will not yield results indicating that the combination of M^{2+} with $E-MATP$ takes place under conditions of thermodynamic equilibrium.

If the addition of $MATP^{2-}$ to the free enzyme were at thermodynamic equilibrium, Eq. (18) would not contain a term in K_a, and double-reciprocal plots with $MATP^{2-}$ and M^{2+} as variable reactants would be intersecting and described by Eqs. (22) and (23).

$$\frac{1}{v} = \frac{K_{ia}}{V}\left[\frac{K_m}{M}\right]\frac{1}{(MATP)} + \frac{1}{V}\left[\frac{K_{ia}K_mK_1}{K_i(M)^2} + \frac{K_m}{(M)} + 1\right] \quad (22)$$

$$\frac{1}{v} = \frac{K_{ia}K_mK_1}{VK_i}\left(\frac{1}{M}\right)^2 + \frac{K_m}{V}\left[\frac{K_{ia}}{(MATP)} + 1\right]\left(\frac{1}{M}\right) + \frac{1}{V} \quad (23)$$

Equation (22) indicates that a secondary plot of slopes against $[M]^{-1}$ would pass through the origin while a secondary replot of vertical intercepts against $[M]^{-1}$ would be parabolic. Equation (23) predicts that a plot of v^{-1} against $[M]^{-1}$ would also be parabolic. If ATP^{4-} were not inhibitory, all primary and secondary plots would be linear and the steady-state velocity pattern would be of the typical asymmetric type for an equilibrium-ordered kinetic mechanism.

Model c. Random Addition of M^{2+} and $MATP^{2-}$

The random addition of M^{2+} and $MATP^{2-}$ occurs when an E–M–MATP complex can be formed by both pathways of Scheme 2. With ATP^{4-} acting as an inhibitory analog of $MATP^{2-}$, it would combine with both free enzyme and the E–M complex. These interactions are described by Eq. (24)

$$v = \frac{V(M)(MATP)}{K_{ia}K_m[1 + (ATP)/K_i] + K_a[1 + (ATP)/K_I](M) + K_m(MATP) + (M)(MATP)} \quad (24)$$

where K_{ia} represents the dissociation constant of the E–MATP complex; K_a and K_m are Michaelis constants for $MATP^{2-}$ and M^{2+}, respectively; K_i and K_I denote dissociation constants for the E–ATP and E–M–ATP complexes, respectively. Under conditions where either M^{2+} or ATP^{4-} is maintained at different fixed concentrations, the double-reciprocal form of the equations derived from Eq. (24) will correspond to those of Eqs. (19)–(21).

The steady-state velocity patterns for the models discussed above are summarized in Table IV. Inspection of these patterns shows that they can differ according to the order of reactant addition to the enzyme, the conditions under which free metal ion combines with the enzyme and the inhibition of the enzyme by free substrate. However, steady-state velocity patterns alone do not allow a distinction to be made between the ordered addition of $MATP^{2-}$, followed by M^{2+}, and the rapid equilibrium, random addition of these two reactants.

Many more models can be formulated for reactions involving a second substrate, but it is beyond the scope of this article to discuss extensively the kinetic consequences of the metal ion activation of reactions involving $MATP^{2-}$ and a second substrate.

Two-Substrate Reactions Involving $MATP^{2-}$ and the Free Form of a Substrate That Complexes M^{2+}

The reverse reactions catalyzed by creatine kinase and acetate kinase are examples of phosphotransferase reactions whose nonnucleotide substrates complex bivalent metal ions. Thus kinetic studies on these en-

TABLE IV
STEADY-STATE VELOCITY PATTERNS TO BE EXPECTED WITH M^{2+} AS AN ACTIVATOR WITH AND WITHOUT COMPETITIVE INHIBITION BY ATP^{4-} IN RELATION TO $MATP^{2-}$

Mechanism	Inhibition[a] by ATP^{4-}	Steady-state velocity patterns[b] with variable reactant pairs				Equation No.
		$MATP/M^c$	M/MATP	MATP/ATP		
Ordered (M before MATP)	+ or −	Intersecting[d]	Intersecting	—		(13), (14)
	+ or −	—	—	Para-comp[e] inhibition		(15)
Equilibrium ordered (M before MATP)[g]	+	Intersecting	Intersecting	—		(16), (17)
	−	Intersecting on vertical ordinate	Intersecting[f]	—		(16), (17)
	+ or −	—	—	Para-comp inhibition		(18)
Ordered (MATP before M)	+	Intersecting, no common intersection point	Parabolic curves variable intercepts	—		(19), (20)
	−	Intersecting	Intersecting	—		(19), (20)
	+ or −	—	—	Para-comp inhibition		(21)

Equilibrium Ordered (MATP before M)[g]	+	Intersecting,[f] no common intersection point	Parabolic curves intersecting on vertical ordinate	—	(22), (23)
	—	Intersecting[f]	Intersecting on vertical ordinate	—	(22), (23)
	+ or —	—	—	Para-comp inhibition	(21)
Rapid equilibrium random	+	Intersecting, no common intersection point	Parabolic curves, variable intercepts	—	(24)
	—	Intersecting	—	—	(24)
	+ or —	—	—	Para-comp inhibition	(24)

[a] When ATP^{4-} does not inhibit, $K_1 = \infty$.
[b] Steady-state velocity patterns are the family of curves that would be obtained for a particular mechanism when v^{-1} is plotted against the reciprocals of the concentrations of the variable reactant at different fixed concentrations of the other reactant. When M^{2+} (or ATP^{4-}) is held constant, $MATP^{2-}$ and ATP^{4-} (or M^{2+}) vary in constant ratio.
[c] The variable reactant is given before, and the changing fixed reactant after, the slant line.
[d] Unless stated otherwise, intersecting patterns have a common point of intersection to the left of the vertical ordinate.
[e] Para-comp inhibition indicates that the lines are curved and have a common point of intersection of the vertical ordinate.
[f] Slope replot passes through the origin.
[g] The addition of the first reactant occurs at thermodynamic equilibrium.

zymes require variation of MADP⁻ and the free concentration of either phosphocreatine or acetyl phosphate. The latter two substrates have relatively weak abilities to complex metal ions (cf. this volume [12]) so that the fraction of metal bound substrate is only a small proportion of the total substrate added. Nevertheless, if this substrate must be varied over a range of concentrations that is high compared with the total metal ion present in solution, a considerable reduction in the concentration of MADP⁻ could occur. It is generally assumed that the metal complexes of the nonnucleotide substrate are not inhibitory or are present only at concentrations well below the values of their inhibition constants. The evidence in support of these conclusions for the creatine kinase reaction has been discussed in detail.[10] In the absence of evidence to the contrary, the assumption that metal–nonnucleotide complexes are inert appears to be reasonable, especially when the complexes carry zero charge rather than the two negative charges of the uncomplexed substrate.

For studies at pH 8.0 on reactions falling into the category under discussion, the following relationships will hold:

$$K_3 = \frac{(M)(ADP)}{(MADP)}; \quad K_4 = \frac{(M)(X)}{(MX)} \tag{25}$$

where X and MX denote, respectively, the free and metal-complexed forms of the phosphorylated substrate. Since

$$ADP = \frac{K_3}{(M)} \cdot (MADP) \quad \text{and} \quad MX = \frac{(M)}{K_4} \cdot X \tag{26}$$

$$ADP_t = MADP + ADP = MADP \left[1 + \frac{K_3}{(M)}\right] \tag{27}$$

$$M_t = M + MADP + MX = MADP + M\left(1 + \frac{X}{K_4}\right) \tag{28}$$

$$X_t = X + MX = X\left[1 + \frac{(M)}{K_4}\right] \tag{29}$$

For fixed concentrations of M^{2+}, Eqs. (27)–(29) permit calculations of the total concentrations of ADP, metal ion and phosphorylated substrate to give the desired concentrations of MADP⁻ and free X.

Two-Substrate Reactions for Which a Univalent Cation
  Is Also Required

The reverse reaction catalyzed by mammalian pyruvate kinase is similar to those catalyzed by creatine and acetate kinases except that, in addition to a second tightly bound metal ion, the enzyme requires a univalent cation such as K^+ for catalytic activity.[3] Although the interaction of univalent cations with ADP^{3-} (or ATP^{4-}) is quite weak (cf. this volume [12]), maximum activation of pyruvate kinase requires that potassium ions be present at concentrations up to 100 mM. Consequently the addition of potassium ions can cause considerable reduction in the concentration of ADP^{3-} available for reaction with M^{2+}, and allowance must be made for this effect. In most circumstances the proportional reduction in the potassium ion concentration is sufficiently small so that the K^+ concentration can be taken as being equal to the total potassium concentration. The additional equilibrium relationship to be taken into account is

$$K_5 = \frac{(P)(ADP)}{(PADP)} \tag{30}$$

where P represents potassium ions. It follows that

$$PADP = \frac{(P)}{K_5}(ADP) = \frac{(P)}{K_5}\frac{K_3}{(M)}(MADP) \tag{31}$$

so that for the pyruvate kinase reaction

$$ADP_t = MADP + ADP + PADP = MADP\left[1 + \frac{K_3}{(M)} + \frac{K_3}{(M)}\frac{(P)}{K_5}\right] \tag{32}$$

Product and Dead-End Inhibition by Metal–Nucleotide Complexes

It is essential that product inhibition studies on phosphotransferases be undertaken with the metal–nucleotide complex that functions as the substrate for the reaction in the reverse direction. For studies of the product inhibition of the pyruvate kinase reaction by $MATP^{2-}$ it would be necessary to consider the additional equilibrium relationships

$$K_1 = \frac{(M)(ATP)}{(MATP)}; \qquad K_6 = \frac{(P)(ATP)}{(PATP)} \tag{33}$$

where P again denotes K^+. From these relationships it follows that

$$ATP = \frac{K_1}{(M)}(MATP); \quad PATP = \frac{(P)}{K_6}(ATP) = \frac{K_1}{(M)}\frac{(P)}{K_6}(MATP) \tag{34}$$

and hence that

$$M_t = M + MADP + MX + MATP$$

$$= MADP + MATP + (M)\left(1 + \frac{X}{K_4}\right) \tag{35}$$

$$\text{ADP}_t = \text{MADP}\left[1 + \frac{K_3}{(M)} + \frac{K_3}{(M)}\frac{(P)}{K_5}\right] \quad (36)$$

$$\text{ATP}_t = \text{MATP} + \text{ATP} + \text{PATP} = \text{MATP}\left[1 + \frac{K_1}{(M)} + \frac{K_1}{(M)}\frac{(P)}{K_6}\right] \quad (37)$$

It is important that the quantitative aspects of the product inhibition not be complicated by the presence of significant concentrations of the free form of nucleotide whose metal complex is used as a product inhibitor. This is usually not a problem with MATP^{2-} as the product inhibitor because of the relatively strong complexing of M^{2+} by ATP^{4-}. There can be more of a problem using MADP^- as a product inhibitor. Because of the higher dissociation constant for this metal–nucleotide complex, inhibitory concentrations of ADP^{3-} could be present, depending on the fixed M^{2+} concentration. Compensations can occur as MADP^- is frequently a stronger product inhibitor than MATP^{2-}, which means that it is used at lower concentrations at which the absolute concentration of ADP^{3-} may be less than its inhibition constant.

An approach similar to that described above is used to study dead-end inhibition by metal–nucleotide complexes. For these studies, however, the complications associated with the dissociation of a dead-end metal–nucleotide inhibitor can be circumvented through use of chromium–ADP and chromium–ATP analogs. With these complexes the ligand exchange rate is such that the complexes are stable under the appropriate experimental conditions.

Effect of pH on Values for the Kinetic Parameters of Phosphotransferases

The determination of the variation with pH of the values for the kinetic parameters of a phosphotransferase involves a formidable amount of work, and this is undoubtedly the reason why detailed investigations of pH effects have not been undertaken. It seemed that it would be appropriate in a volume of this type to outline briefly an approach that might be made in connection with the elucidation of amino acid residues involved in catalysis and the binding of nucleotide and nonnucleotide substrates to a phosphotransferase.

The approach can be illustrated by reference to the reaction catalyzed by creatine kinase. This reaction conforms to a rapid equilibrium, random mechanism in which the interconversion of the central ternary complexes is slower than all other steps in the reaction sequence.[23] The reverse reaction in the absence of products at pH 8.0 or higher pH values can be illus-

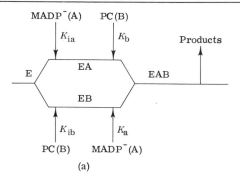
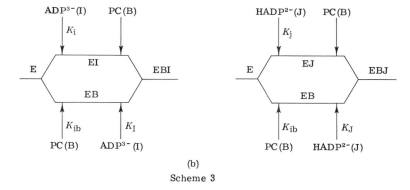

Scheme 3

trated as in Scheme 3. In Scheme 3a it is assumed that there is no inhibition by ADP^{3-}. However, when there is inhibition by ADP^{3-}, and $HADP^{2-}$ at lower pH values, allowance must be made for the formation of four dead-end complexes: E–ADP, E–HADP, E–ADP–PC, and E–HADP–PC (Scheme 3b). ADP^{3-} functions as a competitive inhibitor with respect to $MgADP^-$,[21] and it is likely that $HADP^{2-}$ would act in a similar manner. $MgHADP^0$ is considered to be neither an inhibitor nor a substrate.

In the absence of any inhibition, the rate equation for the creatine kinase reaction can be expressed as

$$v = \frac{VAB}{K_{ia}K_b + K_bA + K_aB + AB} \tag{38}$$

where the symbols are as defined in Scheme 3 and where the terms $K_{ia}K_b$, K_bA, K_aB, and AB represent free enzyme, EA, EB, and EAB, respectively. The mutually exclusive combination of $ADP^{3-}(I)$ and $HADP^{2-}(J)$

with the E and EB complexes (cf. Scheme 3b) leads to the equation

$$v = \frac{VAB}{K_{ia}K_b(1 + I/K_i + J/K_j) + K_aB(1 + I/K_I + J/K_J) + K_aB + AB} \quad (39)$$

From relationships similar to those given in Eq. (2), I and J can be replaced by the terms $K_3(A)/(M)$ and $K_3(H)(A)/K_H(M)$, respectively, to give Eq. (40).

$$v = VAB \bigg/ \bigg(K_{ia}K_b \left\{ 1 + \frac{K_3(A)}{(M)} \left[\frac{1}{K_i} + \frac{(H)}{K_j K_H} \right] \right\}$$
$$+ K_aB \left\{ 1 + \frac{K_3(A)}{(M)} \left[\frac{1}{K_I} + \frac{(H)}{K_J K_H} \right] \right\} + K_bA + AB \bigg) \quad (40)$$

It is apparent that, as M^{2+} tends to infinity, Eq. (40) reduces to Eq. (38). Thus, provided that M^{2+} is not inhibitory and is present at a concentration which is high relative to K_3, true values will be obtained for each of the kinetic parameters by application of the usual analyses (cf. this volume [6]). In the event of higher concentrations of M^{2+} causing inhibition, it would be necessary to determine apparent values for K_{ia} and K_a at different fixed concentrations of M^{2+} and to determine their true values from plots of the reciprocals of the apparent values against $[M]^{-1}$. The kinetic theory associated with determinations of K_{ia} and K_a as a function of pH can be illustrated as follows.

With B as the variable substrate Eq. (40) can be rearranged as

$$\frac{1}{v} = \frac{K_b}{V} \left\{ \frac{K_{ia}}{A} \left[1 + \frac{K_3(A)}{K_i(M)} + \frac{K_3(H)(A)}{K_j K_H(M)} \right] + 1 \right\} \frac{1}{B}$$
$$+ \frac{1}{V} \left\{ \frac{K_a}{A} \left[1 + \frac{K_3(A)}{K_I(M)} + \frac{K_3(H)(A)}{K_J K_H(M)} \right] + 1 \right\} \quad (41)$$

A replot of slope as a function of $[A]^{-1}$ yields an apparent value for K_{ia} from which the true value at each pH can be determined from a tertiary plot of (app K_{ia})$^{-1}$ against $[M]^{-1}$ as indicated in Eq. (42).

$$\frac{1}{\text{app } K_{ia}} = K_3 \left\{ \frac{1}{K_i} + \frac{(H)}{K_j K_H} \right\} \frac{1}{(M)} + \frac{1}{K_{ia}} \quad (42)$$

Similar replots of vertical intercepts would yield values for K_a and V. It should be noted that values cannot be determined for K_i, K_j, K_I or K_J because of the proportional relationship that always exists between ADP^{3-}(I) and $HADP^{2-}$(J). Further, it should be noted that for this model the directly determined values for the parameters associated with B are true values, which would not vary with the concentration of M^{2+}.

If interest centered only on the amino acid residues involved with the binding to the enzyme of the substrate with the lower values for its kinetic parameters, then it may be possible to use a more simplified approach for the kinetic studies. This approach involves use of the substrate with the higher kinetic parameters at a concentration that is low relative to the values for its kinetic parameters. Equation (38) can be rearranged as

$$v = \frac{VAB}{K_a[(K_{ia}K_b/K_a) + B] + A(K_b + B)} \tag{43}$$

When B is equal to or less than one-tenth of the values for K_b and $K_{ia}K_b/K_a$ (which is equal to K_{ib} for a rapid equilibrium, random mechanism; see Scheme 3), Eq. (43) reduces to an equation, which, in reciprocal form, can be expressed as

$$\frac{1}{v} = \frac{K_{ia}}{V}\left(\frac{K_b}{B}\right)\frac{1}{A} + \frac{1}{V}\left(\frac{K_b}{B}\right) \tag{44}$$

In the absence of inhibition by free nucleotides, the value for K_{ia} would be given by the horizontal intercept of a plot of v^{-1} against $[A]^{-1}$. If there were inhibition by free nucleotide species, it would be necessary to perform experiments at a series of M^{2+} concentrations, as described above, to determine the true K_{ia} value at any pH. The validity of the assumption that $[B] \ll K_b$ and K_{ib} must, of course, be checked at each pH value. The analysis assumes that substrate A undergoes reaction with the free enzyme at all pH values.

Practical Aspects

Purification and Standardization of Metal Salt Solutions

To obtain data that reflect the kinetic effects of a particular bivalent metal ion, it is essential that the metal ion salt used in the study be free of inhibitory ions. It is especially important that solutions of the activating metal ion be free of heavy-metal ions such as Hg^{2+} and Pb^{2+}, which are strong inhibitors of enzymic activity. The procedure used in this laboratory for the purification of magnesium, calcium, and manganese salts is as follows. A 2 M solution of the metal salt is extracted repeatedly in a separating funnel with an equal volume of a 0.001% solution of dithizone in CCl_4 until there is no color change in the CCl_4 layer. The metal ion solution is then washed twice with CCl_4 to remove any dithizone, and the CCl_4 is removed by aeration. This solution may be used directly, but the practice has been to concentrate the solution to obtain crystalline material for storage in the dry state. It should be noted that up to ten extractions of

MnCl$_2$ solutions have been required before dithizone solutions cease to turn cherry red. Therefore, it is conceivable that some of the reported inhibitions of enzymes that have been obtained with unpurified solutions of MnCl$_2$ may be due to heavy-metal ions rather than to Mn^{2+} (or Cl$^-$) ions.

Solutions of bivalent metal ion salts may be readily standardized by passage of a sample of the solution through a column of a cation exchanger (H$^+$ form), which is then washed with water before the effluent is titrated with standard alkali. Stock solutions prepared by dilution should be checked to determine that the concentrations are correct and then stored in the frozen state. Solutions of manganese salts should be kept under slightly acid conditions to avoid any hydrolysis of Mn^{2+}. Solutions of metal ion salts may also be standardized by use of atomic absorption spectroscopy in conjunction with certified standards for the appropriate metal ion.

Removal of Metal Ions from Components of Reaction Mixtures

The determination of whether or not a bivalent metal ion is essential for activity, rather than being an activator, requires that the other components of the reaction mixture be free of metal ions that may cause activation. If a metal ion is essential for activity the residual activity in the absence of added metal ion should be zero. To perform this check and to remove any inhibitory metal ions, anionic buffers and solutions of the reactants may be passed through a column of Chelex 100 resin (Bio-Rad Laboratories). Contaminating cations may be exchanged for Na$^+$, K$^+$, (CH$_3$)$_4$N$^+$, or (C$_2$H$_5$)$_4$N$^+$ ions. Chelex resin cannot be used for removing metal ions from cationic buffers, and alternative methods such as recrystallization from EDTA solutions or distillation must be used for this purpose.

There is negligible complexing of (CH$_3$)$_4$N$^+$ and (C$_2$H$_5$)$_4$N$^+$ ions by nucleotides. Thus it may be advantageous to use one or the other as a replacement for Na$^+$ or K$^+$ ions when the latter ions are inhibitory or their concentration is sufficiently high to cause significant depletion of the free nucleotide concentration through the formation of Na– or K–nucleotide complexes. Such replacements can be made by use of a cation exchanger.

It is a useful precaution to include in reaction mixtures low (10 μM) concentrations of EDTA, which would chelate and inactivate any heavy metal ions that may have escaped removal. The suggested concentration of EDTA will almost certainly be negligibly small compared to the total bivalent metal ion concentration present in reaction mixtures for studies on metal-activated enzymes. The storage of metal ion solutions, buffers, reactants, and other reagents in Pyrex glass bottles is to be preferred.

Choice of Buffers

It is essential for all studies on metal-activated enzymes that there be negligible complexing of added metal ions by the buffer. The concentration of buffer present in reaction mixtures (50–150 mM) is invariably very much higher than the total metal ion concentration. Therefore, even though the buffer may have only a weak ability to complex metal ions, its concentration is such that a high proportion of an added metal ion can be converted to a metal–buffer complex. The dissociation constant for the Mn–Tris complex is high at 250 mM, but in a solution of 100 mM Tris and 2 mM Mn^{2+} almost 29% of the metal ion would be chelated. Other buffers to be avoided include phosphate, dicarboxylic acids, and glycylglycine.

N-Ethylmorpholine · HCl has been used extensively in this laboratory as a buffer for studies on reactions that require metal ions. N-Ethylmorpholine is nonchelating, inhibits few enzymes, and can be purified readily by distillation under reduced pressure (15 mm) at 35°. Triethanolamine · HCl has also proved to be a satisfactory buffer. HEPES, TES, BES, and PIPES have been reported to have negligible abilities to complex Mg^{2+}, Ca^{2+}, and Mn^{2+}, but some caution about their use is necessary, as the same publication[31] indicates that the binding of Mn^{2+} by Tris is negligible. Consideration should be given to the acid used to bring cationic buffer ions to the desired pH. Two points for consideration are the abilities of the anionic counterion of the buffer to complex the metal ion and to inhibit the enzyme. Nitrate and perchlorate ions do not coordinate the activating bivalent metal ions but may inhibit the enzyme. Acetate is usually not inhibitory but does coordinate Mg^{2+}.[32] No accurate value is available for the dissociation constant of the MgCl$^+$ complex although it has been reported to be in the vicinity of 300 mM.[33] SO$_4^{2-}$ ions can also be inhibitory. Thus it is apparent that preliminary experiments must be undertaken to determine which buffer and which metal salt are the best to use for a particular investigation. In this connection it should be mentioned that the kinetic properties of an enzyme can be altered by the ions present in reaction mixtures.[5]

Establishment of the Identity of the Substrate for Metal-Activated Enzymes

When there is negligible interaction between an essential bivalent metal ion and the substrate for an enzyme-catalyzed reaction, the free

---

[31] N. E. Good, G. D. Winget, W. Winter, T. N. Connolly, S. Izawa, and R. M. M. Singh, *Biochemistry* **5**, 467 (1966).

[32] L. G. Sillen and A. E. Martell, "Stability Constants of Metal Ion Complexes," Suppl. 1, *Chem. Soc. Spec. Publ.* **n25** (1971).

[33] J. Blair, *Eur. J. Biochem.* **13**, 384 (1970).

concentrations of both the metal ion and substrate can be taken as being identical with their total concentrations. In this event, kinetic investigations would simply involve variation of the total concentrations of each reactant and the application of standard kinetic theory for activation. On the other hand, when interaction does occur between an essential bivalent metal ion and a substrate, it is necessary to determine whether or not the metal–substrate complex functions as the substrate if use is to be made of the approaches and theory elaborated above.

If the metal–substrate complex does not function as a substrate, then reaction could proceed by the addition of free metal ion and free substrate to the enzyme in an ordered or rapid equilibrium, random manner. In these cases, symmetrical intersecting steady-state velocity patterns would be obtained when free metal ion and free substrate are used as the variable reactants. The metal–substrate complex could be inert because the metal ion and free substrate combine at different, separated sites on the enzyme at which the complex cannot react. If the metal ion were to add to the enzyme before the free substrate and if the formation of the enzyme–metal complex were at thermodynamic equilibrium the steady-state velocity patterns would then be asymmetric. In any kinetic mechanism involving free metal ions and free substrate for which the metal–substrate complex was inert, increasing concentrations of one reactant, at a fixed concentration of the other, would lead first to an increase and then to a reduction in reaction velocity. Such a reduction would occur because the concentration of either free metal or free substrate would decrease with the rise in the concentration of the metal–substrate complex.

If a reaction proceeds at a fixed concentration of metal–substrate complex in the virtual absence of free metal ions or free substrate, then the metal–substrate complex must be acting as a substrate. If the presence of excess metal ion or excess free substrate does not result in any decrease in the reaction velocity, it may be concluded that neither of these species is inhibitory. In this case, a plot of v^{-1} against [total substrate]$^{-1}$ at a certain total metal ion concentration would yield a curve that could be superimposed on one obtained by plotting v^{-1} against [total metal ion]$^{-1}$ at the same total substrate concentration. The curves will not be linear unless the concentrations of total substrate and total metal are high relative to the dissociation constant of the metal–substrate complex (cf. data of columns 2 and 4 of Table I). If either the free metal ion or the free substrate is inhibitory, the curves will not be superimposable.

It is possible that the same enzyme–metal–substrate complex could be formed by the random addition to the enzyme of free metal ion and free substrate as well as by the direct combination of the metal–substrate with the enzyme. However, if all the reactions occur under rapid equilibrium conditions, evidence for the formation of an enzyme–metal–substrate complex by all three pathways cannot be obtained from steady-state kinetic studies.[20] Such information can be gained from the use of rapid reaction techniques.[34] It might be noted that nonlinear double-reciprocal plots may be observed if the random addition of two reactants does not occur under rapid equilibrium conditions. The occurrence of multiple pathways of reaction does not preclude studies on the enzyme reaction using the metal–substrate complex as the variable reactant at one or more fixed concentrations of M^{2+}.

Metal Ion Buffers

An alternative method for maintaining M^{2+} ions at a constant concentration in reaction mixtures is to use a metal ion buffer.[35,36] The method involves use of a ligand that does not influence the activity of the enzyme, but chelates the metal ion to a reasonable degree; the principle is, in effect, the same as for a buffer. For the reaction $M + L \overset{K^*}{\rightleftharpoons} ML$

$$pM = pK^* + \log \frac{(L)}{(ML)} \qquad (45)$$

where $pM = -\log [M]$. Equation (45) has a formal similarity to the Henderson–Hasselbach equation. The concentration of M^{2+} can be set at a low value and be effectively buffered against changes in concentration on the addition of nucleotide substrates by having ML and L present at concentrations that are relatively high compared with those of the nucleotides. Difficulties may arise with this approach because the reaction is affected by L and/or ML. Further, it is necessary to consider the degree to which the concentration of the uncomplexed nonnucleotide substrate is reduced as a result of metal-complex formation. The procedure does not have any marked advantages over that elaborated above for kinetic studies on phosphotransferases unless M^{2+} must be kept at low concentrations.

Choice of Experimental Conditions

In the aforegoing discussion attention has been drawn to the fact that in addition to forming a metal–nucleotide complex that acts as the substrate for an enzyme, a bivalent metal ion may also function as an acti-

---

[34] G. G. Hammes and J. K. Hurst, *Biochemistry* **8**, 1083 (1969).
[35] P. D. Boyer, *Biochem. Biophys. Res. Commun.* **34**, 702 (1969).
[36] D. D. Perrin and B. Dempsey, "Buffers for pH and Metal Ion Control." Chapman & Hall, London, 1974.

vator or inhibitor of the reaction. By contrast, free substrate is likely to be inhibitory because of the formation of a dead-end complex through reaction with the enzyme at the binding site for the metal–substrate complex. However, the concentrations of free substrate in reaction mixtures may be such that inhibition is not observed and free metal ions may be inert. In preliminary experiments it is necessary to determine the kinetic effects of free metal ions and free substrate by determining the steady-state velocity of the reaction as a function of the metal–substrate concentration at different fixed concentrations of free metal or free substrate concentrations. If free substrate is an inhibitor while M^{2+} is inert or an activator, the system may be simplified by holding M^{2+} at a sufficiently high concentration so as to reduce the concentration of free substrate and eliminate the inhibition. Of course, under these conditions, no information is obtained about the kinetics of the activation by M^{2+}, and such information must come from studies in which M^{2+} is used as a variable reactant. If higher concentrations of both M^{2+} and free substrate are inhibitory, a compromise has to be reached and note taken of the kinetic consequences. But from experience with phosphotransferases, it appears that there are few problems associated with the maintenance of Mg^{2+} at concentrations from 1 to 5 mM. This, together with the fact that free nucleotides are frequently strong inhibitors of phosphotransferases, are the reasons why it has become general practice to keep M^{2+} constant. In this connection, it is of importance to note that the higher the dissociation constant of the metal–nucleotide complex, the higher must be the concentration of M^{2+} to reduce the free nucleotide to negligible concentrations.

The emphasis in the Theory section has been on the results to be expected when kinetic investigations on phosphotransferases are made at pH 8.0 or above. At these pH values all nucleoside di- and triphosphates can be considered to exist in their fully ionized form and hence only two ionic species of nucleotide must be taken into account. There has developed an attitude among some who study enzymes that reactions should be investigated at their pH optima, as though this pH had some physical significance. This is not the case when, irrespective of pH, the same arbitrary concentrations of reactants are used and no cognizance is taken of the effect that pH changes will undoubtedly have on the values for the maximum velocity and other kinetic parameters. Enzyme-catalyzed reactions should be studied at a pH that is suitable for a particular purpose. If a phosphotransferase reaction is to be studied at a single pH only, it is recommended that a pH of 8.0 be used with free metal ion held constant at a high, noninhibitory concentration. Under these conditions the most meaningful results are likely to be obtained. When reactions are studied at different pH values attention must be paid to any ionization of the substrate and to the maintenance of a constant ionic strength if a single dissociation constant is to be used in all calculations associated with each metal–substrate complex.

Preparation of Reaction Mixtures

When kinetic experiments on phosphotransferases are conducted under conditions where the concentration of M^{2+} is held constant, the concentrations of $MADP^-$ and free phosphorylated acceptor are varied, and allowance is made for the interaction of ADP with Na^+ ions present in the sodium salt of the phosphorylated acceptor, a large number of calculations must be made in connection with the preparation of reaction mixtures. The calculations are not difficult, but they become tedious because of the number involved. To overcome the tedium, computer programs, written in FOCAL, have been used to make the calculations on a PDP 8/I computer. The program used for studies on the reverse reaction catalyzed by creatine kinase has been rewritten in FORTRAN and is reproduced in Table V; the output is illustrated in Table VI. The user supplies information about the fixed concentration of M^{2+}, the highest concentrations of each of $MgADP^-$ and free phosphocreatine to be used, and the pH values at which steady-state velocities are to be determined. The program itself (Table V) incorporates values of the dissociation constants for $MADP^-$, $NaADP^{2-}$, and M–phosphocreatine as well as the last pK_a value for ADP. The output (Table VI) lists, for each chosen pH, the total concentrations of metal ion, ADP, and phosphocreatine required to give five different concentrations of $MADP^-$ at five different concentrations of phosphocreatine. In addition, the output contains the concentrations of $HADP^{2-}$, ADP^{3-}, $MHADP^0$, M–phosphocreatine, and $NaADP^{2-}$ that would be present in each reaction mixture. It is apparent from the data of Table VI how the relative concentrations of $MADP^-$, $HADP^{2-}$, ADP^{3-}, and $MHADP^0$ remain constant. At pH 7.0 both $HADP^{2-}$ and ADP^{3-} are present at concentrations that might be inhibitory. A similar program can be written for experiments that require ADP^{3-} to be held at constant concentration.

Analysis of Data

The equations describing the activation by M^{2+} and/or the inhibition by free nucleotide of phosphotransferases for which the metal–nucleotide complex is the variable substrate have standard forms. Hence there are no problems associated with the least squares fitting of data to these equations (cf. this volume [6]). Standard procedures are also used for analysis

TABLE V
FORTRAN Program for Studies on the Reverse Reaction Catalyzed by Creatine Kinase

```
         WRITE(6,5)
5        FORMAT(//' TOTAL CONCENTRATIONS OF ADP, PHOSPOCREATINE'
        2' AND METAL REQUIRED'/' TO GIVE DESIRED CONCENTRATIONS'
        3' OF MADP(-) AND FREE PHOSPHOCREATINE'/' AT A FIXED'
        4' CONCENTRATION OF M(2+).')
         WRITE(6,45)
45       FORMAT(/' THE FOLLOWING CONSTANTS ARE ASSUMED'/' DISSOCIATION'
        2' CONSTANTS'/
        3'        MADP     0.25 MM'/'     MHADP    10.0 MM'/
        4'        NAADP    67.0 MM'/'     MPC       0.25 MM'/
        5' PKA FOR ADP  6.8'//)
         WRITE(6,55)
55       FORMAT(/' INPUT METAL MADP AND PC, THEN PH1 PHIN AND PH2'
        2' (3F5.0)'/)
         READ(5,10)ZMETAL,ZMADP,PHCR
10       FORMAT(3F5.0)
         READ(5,10)PH1,PHIN,PH2
         WRITE(6,20)ZMETAL
20       FORMAT(//' FREE METAL =',F6.4/)
         PH=PH1
30       CONTINUE
         WRITE(6,35)PH
35       FORMAT(//'>>>>> PH=',F4.1)
         CONST=EXP(-PH*2.303)*EXP(6.8*2.303)
         DO 200 I=1,5
         XMADP=ZMADP/(FLOAT(I))
         ADP=.25*XMADP/ZMETAL
         HADP=CONST*ADP
         ZMHADP=ZMETAL*HADP/10.
         WRITE(6,40)XMADP,HADP,ADP,ZMHADP
40       FORMAT(//' MADP =',F6.4,'    HADP=',F6.4,'    ADP=',F6.4,
        2'    MHADP=',F6.4)
         WRITE(6,50)
50       FORMAT(/'   PC       T M      T ADP    T PC     MPC      NADP'/)
         DO 100 J=1,5
         XPHCR=PHCR/FLOAT(J)
         ZMPHCR=ZMETAL*XPHCR/25.
         ZNAADP=(2.*XPHCR+ADP+XMADP+ZMHADP+HADP)*ADP/67.
         TMETAL=ZMETAL+XMADP+ZMHADP+ZMPHCR
         TADP=ADP+XMADP+HADP+ZNAADP+ZMHADP
         TPHCR=XPHCR+ZMPHCR
         WRITE(6,60)XPHCR,TMETAL,TADP,TPHCR,ZMPHCR,ZNAADP
60       FORMAT(6F8.4)
100      CONTINUE
200      CONTINUE
         IF(ABS(PH-PH2)-1.E-5)300,300,290
290      PH=PH+PHIN
         GO TO 30
300      CONTINUE
         CALL EXIT
         END
```

TABLE VI
Output of FORTRAN Program of Table V

TOTAL CONCENTRATIONS OF ADP, PHOSPOCREATINE AND METAL REQUIRED
TO GIVE DESIRED CONCENTRATIONS OF MADP(-) AND FREE PHOSPHOCREATINE
AT A FIXED CONCENTRATION OF M(2+).

THE FOLLOWING CONSTANTS ARE ASSUMED
DISSOCIATION CONSTANTS
```
      MADP    0.25 MM
     MHADP   10.0  MM
     NAADP   67.0  MM
       MPC    0.25 MM
PKA FOR ADP  6.8
```

INPUT METAL MADP AND PC. THEN PH1 PHIN AND PH2 (3F5.0)
```
>    2.   .4  20.
>    7.   1.   8.
```

FREE METAL = 2.0000

>>>> PH= 7.0

MADP = .4000   HADP= .0315   ADP= .0500   MHADP= .0063

PC	T M	T ADP	T PC	MPC	NADP
20.0000	4.0063	.5181	21.6000	1.6000	.0302
10.0000	3.2063	.5031	10.8000	.8000	.0153
6.6667	2.9396	.4982	7.2000	.5333	.0103
5.0000	2.8063	.4957	5.4000	.4000	.0078
4.0000	2.7263	.4942	4.3200	.3200	.0063

MADP = .2000   HADP= .0158   ADP= .0250   MHADP= .0032

PC	T M	T ADP	T PC	MPC	NADP
20.0000	3.8032	.2589	21.6000	1.6000	.0150
10.0000	3.0032	.2515	10.8000	.8000	.0076
6.6667	2.7365	.2490	7.2000	.5333	.0051
5.0000	2.6032	.2477	5.4000	.4000	.0038
4.0000	2.5232	.2470	4.3200	.3200	.0031

MADP = .1333   HADP= .0105   ADP= .0167   MHADP= .0021

PC	T M	T ADP	T PC	MPC	NADP
20.0000	3.7354	.1726	21.6000	1.6000	.0100
10.0000	2.9354	.1676	10.8000	.8000	.0050
6.6667	2.6688	.1660	7.2000	.5333	.0034
5.0000	2.5354	.1651	5.4000	.4000	.0025
4.0000	2.4554	.1646	4.3200	.3200	.0020

of data obtained from kinetic investigations for which metal–nucleotide complexes and uncomplexed second substrate are the variable reactants.[23] When the metal–nucleotide complex is the variable substrate, this should be indicated clearly on any plot.

Examples of the Utilization of the Recommended Approaches

It is not intended that this section constitute a catalog of references to studies on phosphotransferases for which some or all of the above procedures have been utilized. Rather, the purpose is to make known some key references that may be consulted for the conditions chosen for kinetic investigations on particular phosphotransferases and that would provide guidelines for those without experience in the area of metal-activated enzymes.

The kinetic mechanisms of the reactions catalyzed by several phosphotransferases have been elucidated by studies at pH 8.0 that involve holding Mg^{2+} constant at 1 mM while varying the concentrations of $MgADP^-$ and $MgATP^{2-}$ both as substrates and products. Included among the enzymes subjected to this approach are arginine kinase,[37,38] choline kinase,[24] creatine kinase,[23,27] galactokinase,[6] and hexokinase.[39] For studies on the reverse reactions, velocities were determined as a function of the concentration of the uncomplexed phosphorylated acceptor. By contrast, the kinetic studies on nucleoside diphosphokinase were performed by varying the concentrations of the magnesium complexes of both the nucleoside di- and triphosphates.[8] The competitive inhibition by free nucleotide with respect to the metal–nucleotide substrate has been demonstrated with creatine kinase[21] and hexokinase,[13–16] and studies on creatine kinase have shown that higher concentrations of $MgCl_2$ cause noncompetitive inhibitions with respect to both $MgADP^-$ and phosphocreatine.[21] The possibility that this type of inhibition is due to the inhibitory effects of chloride ions has not been excluded.

The kinetic investigations on the choline kinase reaction have indicated that (a) $MgATP^{2-}$ acts as a substrate; (b) Mg^{2+}, in excess of that required to form the substrate, functions as an activator; (c) ATP^{4-} does not cause inhibition; and (d) there is an ordered addition of choline, $MgATP^{2-}$ and Mg^{2+} to the enzyme with the addition of $MgATP^{2-}$ being at thermodynamic equilibrium.[24] The conclusions relating to the kinetic ef-

[37] E. Smith and J. F. Morrison, *J. Biol. Chem.* **246** 7764 (1971).
[38] E. Smith and J. F. Morrison, *J. Biol. Chem.* **244**, 4224 (1969).
[39] D. L. Purich and H. J. Fromm, *J. Biol. Chem.* **247**, 249 (1972).

fects of Mg^{2+}, ATP^{4-}, and $MgATP^{2-}$ are in accord with the findings that a linear, asymmetric intersecting steady-state velocity pattern is obtained with $MgATP^{2-}$ (varied in constant ratio with ATP^{4-}) and Mg^{2+} as the variable reactants (cf. Table IV). Such a linear pattern would not have been observed if ATP^{4-} were acting as a competitive inhibitor. The conclusions are also consistent with the result that a double-reciprocal plot of velocity as a function of $MgATP^{2-}$ (varied in constant ratio with Mg^{2+}) gives a parabola whose curvature increases with increasing concentrations of ATP^{4-}. For this reaction, the increased curvature is not due to inhibition by ATP^{4-}, but rather to the effect of increasing concentrations of ATP^{4-} in decreasing the concentration of, and activation by, Mg^{2+}.

It appears that there are no examples of a free nucleotide acting as an allosteric activator of a reaction catalyzed by a phosphotransferase. However, it has been shown that free isocitrate functions as an activator while Mg–isocitrate acts as a substrate for a plant isocitrate dehydrogenase.[40]

Conclusion

The primary aim of this chapter has been to draw attention to the approaches that have been developed for studies on metal-activated enzymes and have been applied to kinetic investigations on phosphotransferases. The number of phosphotransferases subjected to such investigations is limited. Nevertheless it has become apparent that, by the utilization of the basic principles and the judicious choice of experimental conditions, artifactual results can be avoided and meaningful information obtained about the kinetic properties of the enzymes. No attempt has been made to cover every possible situation or combination of kinetic effects that could arise. However, the number of examples given should be sufficiently large to provide a foundation for any further development. By contrast, the theory associated with the activation of a phosphotransferase by free metal ions and its inhibition by free nucleotide while the metal–nucleotide complex functions as a substrate has not been subjected to extensive experimental investigation. The comments made here may well provide the necessary stimulus for additional studies.

The proposed procedures are based on the premise that metal–nucleotide complexes act as substrates for all phosphotransferases and there is certainly considerable evidence to support this contention. But the investigator must be mindful that this assumption may not be true for every phosphotransferase and be watchful for results that cast doubt on the assumption.

[40] R. G. Duggleby and D. T. Dennis, *J. Biol. Chem.* **245**, 3745 (1970).

It is unfortunate that studies on many metal-activated enzymes, especially those subject to allosteric activation or inhibition, have been undertaken using conditions that preclude interpretation of the data. It might be argued that the conditions used approximate more closely to those that occur within the living cell. However, it is only when the kinetic mechanism of an enzyme has been determined and values obtained for its kinetic parameters that predictions can be made of the likely metabolic effects of variation in the intracellular concentrations of reactants and modifiers.

[12] Stability Constants for Biologically Important Metal–Ligand Complexes

By WILLIAM J. O'SULLIVAN and GEOFFREY W. SMITHERS

The catalytic activity of a large number of enzymes is dependent upon the presence of a metal ion. The metal ion may form complexes with the enzyme, the substrate, or both for activity to occur and may also interact with other components of the reaction mixture. Thus, an essential requirement to the accurate kinetic analysis of enzyme reactions in which metal–ligand substrates or inhibitors are involved is the exact determination of the concentration of the various metal complexes in solution.[1] In order to obtain such information, a measure of the affinity of the various ligands for the metal ion, expressed as the stability constant of the metal–ligand complex, must be evaluated.

This chapter is concerned with the stability constants of complexes between metal ions and substrates containing phospho-oxyanions, because of their overwhelming importance in enzymology, though the discussion is relevant to other metal–substrate interactions. The emphasis is on the nucleoside di- and triphosphates, taking ADP and ATP, or rather $MADP^-$ and $MATP^{2-}$, as the prototypes, though some attention is given to other compounds, in particular phosphoribosyl pyrophosphate (PRPP). Further, the principal concern is with those metal ions that are normally associated with these substrates, i.e., the divalent metal ions Mg^{2+}, Ca^{2+}, Mn^{2+} and the monovalent cations K^+ and Na^+. Of these, manganese is biologically the least important but is worth consideration because of its increasing use as a paramagnetic probe in kinase-type reactions.[2-4]

The chapter is intended to provide a guide to the calculation of the

[1] J. F. Morrison, this volume [11].
[2] A. S. Mildvan and M. Cohn, *Adv. Enzymol.* **33**, 1 (1970).
[3] A. S. Mildvan, *in* "The Enzymes" (P. D. Boyer, ed.), 3rd ed. Vol. 2, p. 445. Academic Press, New York, 1970.
[4] W. J. O'Sullivan, *in* "Inorganic Biochemistry" (G. L. Eichhorn, ed.), p. 582. Elsevier, Amsterdam, 1973.

concentration of various metal–ligand species in solution. Thus the stability constant determination is considered not as an end in itself, but as an essential adjunct to obtaining accurate kinetic or equilibrium data related to the interaction of the enzyme with its substrates and other components of the reaction. For this reason the primary emphasis is on stability constants immediately relevant to the conditions of the working enzyme, rather than values extrapolated to zero ionic strength in order to extract orthodox thermodynamic functions.[5-8]

The organization of the chapter falls into three categories: (1) general theoretical considerations, (2) discussion of the determination of stability constants, and (3) some guidelines toward the selection of constants from the literature and to their application.

Theoretical Considerations

In general terms, the stability constant for the interaction between a metal ion, M, and a ligand, L,

$$M + L \rightleftarrows ML \qquad (1)$$

is given by

$$K_{ML} = \frac{[ML]}{[M][L]} \qquad (2)$$

and is normally expressed as M^{-1}, the inverse of the more familiar dissociation constant. As values for such stability constants may vary by 30 orders of magnitude, they are expressed as log values in many compilations.[9,10]

For preliminary discussion, ATP is taken as the prototype of the compounds to be considered in this chapter. Neglecting the two low pK values (<2),[11] species that could be relevant for biochemical systems are H_2ATP^{2-}, $HATP^{3-}$, and ATP^{4-}. The pK values for the interconversion of these species are approximately 4 and 7, respectively.[9,12,13] In the pres-

[5] P. George, R. C. Phillips, and R. J. Rutman, *Biochemistry* **2**, 508 (1963).
[6] R. C. Phillips, P. George, and R. J. Rutman, *J. Am. Chem. Soc.* **88**, 2631 (1966).
[7] M. S. Mohan and G. A. Rechnitz, *Arch. Biochem. Biophys.* **162**, 194 (1974).
[8] M. S. Mohan and G. A. Rechnitz, *J. Am. Chem. Soc.* **92**, 5839 (1970).
[9] L. G. Sillen and A. R. Martell, "Stability Constants" [*Chem. Soc. Spec. Publ.* **n17** (1964)]; and "Stability Constants," Suppl. 1 [*Chem. Soc. Spec. Publ.* **n25** (1971)].
[10] W. J. O'Sullivan, in "Data for Biochemical Research" (R. M. C. Dawson, D. C. Elliott, W. H. Elliott, and K. M. Jones, eds.), p. 523. Oxford Univ. Press (Clarendon), London and New York, 1969.
[11] R. H. Symons, in "Data for Biochemical Research" (R. M. C. Dawson, D. C. Elliott, W. H. Elliott, and K. M. Jones, eds.), p. 145. Oxford Univ. Press (Clarendon), London and New York, 1969.
[12] W. J. O'Sullivan and D. D. Perrin, *Biochemistry* **3**, 18 (1964).
[13] R. C. Phillips, P. George, and R. J. Rutman, *Biochemistry* **2**, 501 (1963).

ence of a divalent metal ion (M^{2+}), the equilibria set up are shown in Eq. (3).

$$
\begin{array}{ccc}
H_2ATP^{2-} \xrightleftharpoons[]{pK_{a_1}} & HATP^{3-} \xrightleftharpoons[]{pK_{a_2}} & ATP^{4-} \\
K_{H_2ML} \updownarrow & K_{HML} \updownarrow & K_{ML} \updownarrow \\
MH_2ATP \xrightleftharpoons[]{} & MHATP^{-} \xrightleftharpoons[]{} & MATP^{2-}
\end{array}
\qquad (3)
$$

pK_{a_1} and pK_{a_2} are the dissociation constants for the species H_2ATP^{2-} and $HATP^{3-}$, respectively. The stability constants for the respective metal complexes are given by

$$K_{H_2ML} = \frac{[MH_2ATP]}{[M^{2+}][H_2ATP^{2-}]} \qquad (4)$$

$$K_{HML} = \frac{[MHATP^{-}]}{[M^{2+}][HATP^{3-}]} \qquad (5)$$

$$K_{ML} = \frac{[MATP^{2-}]}{[M^{2+}][ATP^{4-}]} \qquad (6)$$

The reactions in Eq. (3) represent the logical sequence in a pH titration experiment, both in the absence and in the presence of M^{2+}. The analysis of data from such experiments is commonly used for the determination of pK_a values and of stability constants of metal complexes.[14,15] Because the equations used to describe the above equilibria for ATP and related compounds have proved difficult to solve without suitable computer programs, a more pragmatic approach has been to determine an "apparent" or "effective" stability constant under conditions relevant to the anticipated enzyme investigation. While not as thermodynamically sound in a classic physicochemical sense, such apparent constants are of more immediate applicability to the determination of kinetic parameters.

In this case, the interaction may be represented as

$$K'_{ML} = \frac{[M]}{[M][L]_T} \qquad (7)$$

where $[L]_T$ is the sum of forms of L, other than ML, under the specified conditions.

Equivalent equations can be set up for the nucleoside diphosphates, except that the respective species will have one less charge, viz.

---

[14] A. Albert and E. P. Seargent, "The Determination of Ionization Constants," 2nd ed. Chapman & Hall, London, 1971.
[15] F. J. C. Rossotti and H. Rossotti, "The Determination of Stability Constants and Other Equilibrium Constants in Solution." McGraw-Hill, New York, 1961.

MH_2ADP^+, MHADP, and $MADP^-$. The most protonated species, viz. MH_2ATP and MH_2ADP^+, can be neglected except under extreme conditions, such as low pH and high metal ion concentration (see Fig. 3).[15a]

PRPP merits special consideration, insofar as metal complexation could occur independently at the pyrophosphate and the phosphate moiety with the further possibility of one metal ion acting as a bridge between the two moieties. The most detailed analysis of this system has come from the laboratory of Spivey[16] and is discussed separately below.

Practical Aspects

General

Of the various techniques employed for the determination of stability constants, two in particular appear to rest on a solid theoretical foundation and to be of wide application. These are pH titration and spectrophotometry using mixed ligands, as exemplified by the use of 8-hydroxyquinoline.[17]

The latter approach has proved to be particularly applicable to metal–ligand complexes relevant to enzyme kinetic experiments and is therefore discussed in some detail. A résumé of the application of pH titration and a discussion of the use of ion-specific electrodes is also presented. The latter has a substantial appeal in principle, but appears to suffer from particular problems and has not been widely applied. Finally, brief résumés of a number of other published techniques are also presented. These include ion exchange, fluorescence, gel filtration, Raman spectroscopy, and electron paramagnetic resonance.

Nature of Supporting Medium

Ideally, the determination of a stability constant should be carried out under conditions that, if not identical to the anticipated enzyme experiment, can be regarded as closely approximating to it. This is particularly pertinent to the choice and assessment of conditions used to determine the constant. As an illustration, pH titration experiments are normally carried out in a supporting medium of moderate ionic strength $(0.1 - 0.15 \, M)$ in order to minimize changes in activity coefficients.[14] It has

---

[15a] As the rate of complex formation is some orders of magnitude greater than the relevant enzymic reaction, it is never the rate-limiting step in the enzymic reaction.

[16] R. E. Thompson, E. L.-F. Li, H. O. Spivey, J. P. Chandler, A. J. Katz, and J. R. Appleman, *Bioinorg. Chem.* **9**, 35 (1978).

[17] K. Burton, *Biochem. J.* **71**, 388 (1959).

been assumed, though not always justifiably, that the nature of the supporting medium does not affect the measurement or application of the results obtained. Several areas of particular concern are as follows:

1. Measurements have often been carried out in the presence of tetraalkylammonium ions and any interaction with the ligand has been considered trivial. Although this conclusion has been queried in the literature, it would appear from the results of Smith and Alberty with monovalent cations and adenosine nucleotides, that the above assumption is justified.[18] A recent rigorous analysis of pH titration data for the interaction of tetrapropylammonium bromide and ATP has indicated that quaternary ammonium ions do not significantly complex with ATP.[19]

2. A great deal of work has been carried out in the presence of potassium or sodium ions, usually $0.1-0.2$ M K^+ or Na^+, in an attempt to simulate physiological conditions and/or provide standard experimental conditions for comparative estimations. It should be noted that there is ample evidence for the formation of $KATP^{3-}$ and $NaATP^{3-}$,[8,12,18,20] and also for $KADP^{2-}$ and $NaADP^{2-}$.[18,20] Thus, such interactions may have to be taken into account if the results are to be applied to conditions without these cations or if exact concentrations of the nucleotide species are required. In practice they have often been ignored or considered irrelevant to the study. (The addition of K^+ as KOH in pH titration experiments, and of Na^+ as the salt of ATP, is probably small enough so as not to seriously perturb the final results, though some workers have taken the precaution of exchanging either or both with quaternary ammonium salts.[13])

3. By their nature, all buffers may be considered as potentially able to complex metal ions. Interactions between the alkaline-earth metal ions and the nitrogen base buffers, such as N-ethylmorpholine, triethanolamine, and many of the sulfonic acid morpholine and piperazine derivatives,[21] are very small to negligible. In contrast, for other commonly used buffers, such as Tris, phosphate, and glycylglycine, there may be a small but measurable degree of interaction, which can become significant under the conditions of the normally high concentration of the buffer species compared to the metal ion and ligand. Further, there is also the potential of interaction between the cationic form of a buffer and a species such as ATP^{4-}.

If possible, phosphate buffers and similar compounds, which clearly

---

[18] R. M. Smith and R. A. Alberty, *J. Phys. Chem.* **60,** 180 (1956).
[19] L.-F. Li, M. Sc. Thesis, Oklahoma State Univ., 1977.
[20] N. C. Melchior, *J. Biol. Chem.* **208,** 615 (1954).
[21] N. E. Good, G. D. Winget, W. Winter, T. N. Connolly, S. Izawa, and R. M. M. Singh, *Biochemistry* **5,** 467 (1966).

do complex with divalent metal ions, should be avoided. If the use of phosphate is unavoidable it can be allowed for using a suitable iterative computer program (see the Appendix).

The effect of pH, ionic strength, and temperature was discussed in the preceding chapter[1] and is also treated below. With the obvious exception of a pH titration experiment, the choice of pH will be determined by the nature of the conditions under which the constant is to be applied and also by the pK_a values of the ligand species. In the determination of an apparent constant, it is highly desirable to minimize the concentration of all but one species. For example, a suitable range for ATP, with a pK_a of approximately 7, would be pH 8.0–8.5 where ATP^{4-} will comprise 89–98% of the total nucleotide. Realistic values of ionic strength are in the range 0.05–0.15 M, and of temperature, 20–40°.

Materials

For the ligand under investigation, pure and well characterized salts are preferable. This is a particularly stringent condition for pH titration, and most commercial preparations require some degree of purification. Recrystallization is usually sufficient for ATP,[22] though separation on suitable columns may be necessary for other nucleotides[23] and for PRPP.[24] The latter compound merits special concern, as it has a greater lability than the nucleotides. The purity of commercial preparations received in this laboratory has varied from 60% to 90%. Moreover, one of the breakdown products, pyrophosphate, is a relatively stable compound and a stronger metal ion complexing agent than PRPP.[9]

The standardization of metal ion solutions may also present a problem. Many of the more common water-soluble salts (e.g., $MgCl_2$) are hygroscopic with variable water contents, so that accurate solutions cannot be prepared by gravimetric procedures. Solutions made up to approximate concentrations may be subsequently analyzed by atomic absorption spectrometry[25] or by titration with standard EDTA using suitable colorimetric indicators (e.g., Eriochrome Black T for Mg^{2+}).[26] A convenient technique is to pass measured volumes of the metal ion solution through a Dowex 50-W column in the H^+ form and titrate the effluent acid

[22] L. Berger, *Biochim. Biophys. Acta* **20**, 23 (1956).
[23] W. E. Cohn, this series, Vol. 3, p. 724.
[24] G. W. Smithers and W. J. O'Sullivan, unpublished results.
[25] "Analytical Methods for Flame Spectroscopy," a practical manual printed by Varian, 1972.
[26] A. I. Vogel, "A Text-book of Quantitative Inorganic Analysis," 3rd ed. Longmans, Green, New York, 1961.

with standard hydroxide. An alternative procedure is to use a characterized salt (e.g., MgO, $CaCO_3$) dissolved in acid and adjusted to neutrality. It may be necessary, in some instances, to remove trace amounts of contaminating heavy-metal ions, for example, with dithizone,[27] though any interference with the determination of the stability constant would be less than the potential for enzyme inhibition.

Base for accurate pH titration must be prepared in carbonate-free water, maintained free of CO_2, and carefully standardized.[14,26,28]

Specific Techniques

Competition with 8-Hydroxyquinoline

Rationale. Spectrophotometric or fluorescent metal ion indicators, which act in competition with the ligand of interest for metal ions in solution, have been extensively used for the determination of stability constants.[15] Of the competing ligands, the most successful application to nucleotides and other phosphorylated compounds has been that first elaborated by Burton with 8-hydroxyquinoline (oxine).[17] The 1:1 complex formed by this compound with Mg^{2+} (or Ca^{2+}) is accompanied by a change in the absorption maximum from 310 nm to 350 nm, the maximum change occurring at approximately 360 nm (Fig. 1). Thus, the change in extinction at 360 nm is a measure of the metal complex formed. In the presence of another compound that complexes Mg^{2+}, the amount of metal ion needed to obtain a particular extinction will be increased. The *extra* Mg^{2+} that has to be added is equivalent to that complexed to the second ligand. By algebraic manipulation, it is possible to determine the concentration of the major species and thus to calculate the stability constant of the metal–ligand species of interest.

Experimental. Small amounts of a concentrated solution of $MgCl_2$ are added to a solution containing 8-hydroxyquinoline in a suitable buffer system, both in the presence and in the absence of the ligand. The solution is maintained at constant temperature, and the spectral change at 360 nm is recorded. It is preferable to carry out the whole reaction in cuvettes in a thermostatted holder in a spectrophotometer, so that the solution is not disturbed from the light path during additions of $MgCl_2$.

The concentration of 8-hydroxyquinoline to be used should be established by prior experimentation; the authors have found the range of 0.2 to 0.8 mM to be most useful. The compound is not very soluble, and in preparing stock solutions (5–10 times the desired final concentration) it is

---

[27] J. F. Morrison and M. L. Uhr, *Biochim. Biophys. Acta* **122**, 57 (1966).
[28] G. Schwarzenbach and W. Biedermann, *Helv. Chim. Acta* **31**, 331 (1948).

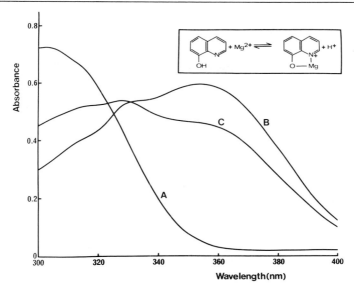

FIG. 1. Absorption spectra of 8-hydroxyquinoline (0.26 mM) in 25 mM HEPES-KOH, pH 8.0, at 20°. Curve A, Alone; B, with 500 mM MgCl$_2$ (ϵ_{360} = 2.2 × 10^3 M^{-1} cm^{-1}); C, with 500 mM CaCl$_2$ (ϵ_{360} = 1.7 × 10^3 M^{-1} cm^{-1}). Cell, 1.0 cm light path. *Inset:* Structure of 8-hydroxyquinoline and the formation of its magnesium complex.

usually necessary to stir the solution for some hours. Thence, it should be stored in the absence of light.

Generally, careful serial additions of a stock solution of the ligand is satisfactory. If a characterized salt is available (e.g., Na$_2$ATP · 4H$_2$O), greater accuracy can be obtained by weighing the salt directly into the cuvette. In this situation, all other components can be transferred to the cuvette in a single addition, say, 3.0 ml from a bulb pipette, minimizing errors from multiple additions. All solutions should be brought to the desired temperature just prior to measurement.

Results from a typical experiment for the determination of the stability constant for MgADP$^-$ are illustrated in Fig. 2. The conditions of the experiment were: 8-hydroxyquinoline, 0.2 mM; N-ethylmorpholine-HCl, pH 8.0, 50 mM; in 3.0 ml in a 1-cm light path cuvette. MgCl$_2$ (0.1 M) was added in 5-μl portions from an Agla syringe. The solution was carefully mixed (without removing the cuvette), and the change in extinction at 360 nm was measured against a reagent blank (this need not be disturbed during the course of the experiment). Generally, 1–2 min is sufficient for the system to come to equilibrium. The procedure is repeated in the presence of various concentrations of the ligand, usually 4 to 6. The standard is repeated approximately after every third test. Finally, the extinction of the Mg-8-hydroxyquinoline complex, under the experimental conditions, is determined by the addition of a large excess (approximately 100-fold) of MgCl$_2$ (component AB in Fig. 2). For ATP, which is a strong complexing agent, the spectral changes seen in a cell of 1-cm light path are insufficient for accurate measurements. It is therefore necessary to use a larger cell, e.g., a 4-cm light path cuvette, with a total volume of ~20.0 ml.

Analysis. The shaded area in Fig. 2 indicates the amount of Mg-8-hydroxyquinoline complex present in solution. For the same extinction, the total magnesium added is distributed as follows: AB = magnesium bound to 8-hydroxyquinoline, BC = free Mg^{2+}, CD = magnesium bound to ADP.

The stability constant, K, for MgADP$^-$ is given by

$$K = \frac{1}{[\text{Mg}^{2+}]_{1/2}} \tag{8}$$

where $[\text{Mg}^{2+}]_{1/2}$ is the free-magnesium concentration when $[\text{MgADP}^-] = \frac{1}{2}[\text{ADP}]_\text{T}$; i.e., $K = 1/\text{BC}$ at $\text{CD} = [\text{ADP}]_\text{T}$. The value of $[\text{Mg}^{2+}]_{1/2}$ depends on the concentration of $[\text{ADP}]_\text{T}$, so that K can be represented by,

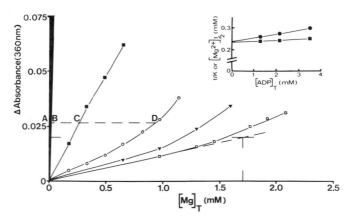

FIG. 2. Titration of 8-hydroxyquinoline with magnesium chloride in the absence (■—■) and in the presence of ADP (1.32 mM, ○—○; 2.16 mM, ▼—▼; 3.5 mM, □—□). The shaded area represents the concentration of Mg-8-hydroxyquinoline present to give the observed extinction. For the first concentration of ADP, the dashed line, CD, represents half the concentration of nucleotide (0.66 mM). At this concentration, the $[\text{Mg}^{2+}]_{1/2}$ value is given by BC, the contribution of AB, from the Mg-8-hydroxyquinoline being taken into account. The initial slope, for the highest concentration of ADP, is represented by the dashed line drawn asymptotically to the lower curve. *Inset:* Determination of the apparent stability constant for MgADP$^-$. $[\text{Mg}^{2+}]_{1/2}$ values (●—●) and initial slope values (■—■) have been plotted as a function of $[\text{ADP}]_\text{T}$ and extrapolated to zero nucleotide concentration.

$$K = \frac{1}{[\text{Mg}^{2+}]_{1/2}} + k_1[\text{ADP}]_T \tag{9}$$

where k_1 is a factor introduced by Burton[17] to allow for this dependence.

Values for $[\text{Mg}^{2+}]_{1/2}$ are easily obtained from careful plotting of the experimental points (Fig. 2). For example, at the lowest concentration of $[\text{ADP}]_T$ used, 1.32 mM, the $[\text{Mg}]_{1/2}$ value (BC), at CD equivalent to 0.66 mM, is obtained directly from Fig. 2 as 0.27 mM. Similarly, the $[\text{Mg}^{2+}]_{1/2}$ values at $[\text{ADP}]_T$ concentrations of 2.16 mM and 3.50 mM are 0.28 mM and 0.30 mM, respectively. A plot of the $[\text{Mg}^{2+}]_{1/2}$ values against $[\text{ADP}]_T$ (inset, Fig. 2), when extrapolated to zero nucleotide concentration, gives a value of 0.24 mM, which corresponds to a stability constant of approximately 4000 M^{-1}. Four or five data points are usually required to obtain an accurate estimate of the stability constant, though only three points have been used in Fig. 2 (inset) to illustrate the procedure.

For some complexes, the $[\text{Mg}^{2+}]_{1/2}$ values are difficult to obtain and an alternative approach, which used the initial slope of the titration curve, can be used. The procedure is illustrated for the highest concentration of ADP used in Fig. 2. The initial slope is indicated by the dashed line tangent to the curve. At $\Delta\epsilon = 0.020$, the $[\text{Mg}]_T$ value is 1.7 mM. Of this, $[\text{Mg}^{2+}]$ represents 0.2 mM; so that $[\text{MgADP}^-]$ is 1.5 mM. As $[\text{ADP}]_T$ is 3.5 mM, free ADP^{3-} is 2.0 mM. Thus $1/K = (0.2 \times 2.0)/1.5 = 0.27$ mM. As for the $[\text{Mg}^{2+}]_{1/2}$ values, there is a slight dependence of K on the total concentration of ADP so that the stability constant is obtained by extrapolation to zero nucleotide concentration, of a plot of $1/K$ as a function of $[\text{ADP}]_T$ (inset, Fig. 2).

The use of the initial slope portion of the curve, though not as reliable as $[\text{Mg}^{2+}]_{1/2}$ values, is useful for weaker complexing agents and where low concentrations of nucleotide must be used. An example of the former situation is the determination of the stability constant for Mg-glucose-1-phosphate[29] and an example of the latter, the formation of the europium-ADP complex.[30]

pH Titration

Rationale. The ligand is titrated with base in the presence and in the absence of metal ion. The difference in the pH change with respect to added base in the presence of added metal ion reflects the amount of ligand bound by metal ion, and thus the stability constant of the metal complex.

[29] W. J. Ray, Jr. and G. A. Roscelli, *J. Biol. Chem.* **241**, 2596 (1966).
[30] K. J. Ellis and J. F. Morrison, *Biochim. Biophys. Acta* **362**, 201 (1974).

This technique has been widely used for the determination of stability constants of metal-ligand complexes.[15] It is, however, a relatively difficult technique to apply to complex systems, such as ATP, ADP. Thus only an outline of its application to ATP is given below.

Experimental. Small aliquots of standardized base are added in stepwise fashion to a solution of the ligand, and the pH is recorded after each addition. In a typical experiment by one of the authors,[12] 0.5 ml of 0.1 M KOH was added to 50 ml of a 0.525 mM solution of the disodium salt of ATP, in 0.1 M tetraethylammonium bromide, maintained at 30°. The initial pH was approximately 3.8, though this will vary slightly according to the concentration of ATP, temperature, etc. Additions were made in 10-μl portions from an Agla syringe, the system was allowed to come to equilibrium after each addition, and the pH was recorded. An inert atmosphere should be maintained above the solution, with, for example, scrubbed nitrogen. The pH meter should be carefully calibrated at the temperature required; freshly prepared 50 mM potassium hydrogen phthalate (pH 4.011 at 30°) and 50 mM sodium borate (pH 9.142 at 30°) are suitable standards.

The titration is carried out in the absence of added divalent ion in order to determine pK_a values. It is repeated in the presence of one or more concentrations of metal ion, e.g., equimolar and a moderate excess, five to ten times the ligand concentration.

Analysis. The pK_a values can be determined directly from the titration curve.[14] A limited refinement for the determination of the pK_a values has recently appeared.[31] Also, an expanded version of the approach described by Sayce[32] has been applied to the titrations of PRPP.[16]

Linear equations describing the equilibria established during the titration may be set up in terms of total ATP, total Mg, and Σ (positive charges) = Σ (negative charges).

$$[MgH_2ATP] + [MgHATP^-] + [MgATP^{2-}] + [H_2ATP^{2-}] + [HATP^{3-}] + [ATP^{4-}] = [ATP]_T \quad (10)$$

$$[Mg^{2+}] + [MgH_2ATP] + [MgHATP^-] + [MgATP^{2-}] = [Mg]_T \quad (11)$$

$$[K^+] + [H^+] + 2[ATP]_T + 2[Mg^{2+}] = [MgHATP^-] + 2[MgATP^{2-}] + 2[H_2ATP^{2-}] + 3[HATP^{3-}] + 4[ATP^{4-}] + [OH^-] + 2[Mg]_T \quad (12)$$

(N.B. The 2[ATP]$_T$ term arises because ATP is added as the disodium salt and the 2[Mg]$_T$ term because magnesium is added as the chloride.)

These equations contain two more unknowns than there are measurable quantities and thus are incapable of exact solution at a given point.

[31] T. N. Briggs and J. E. Stuehr, *Anal. Chem.* **46**, 1517 (1974).
[32] I. G. Sayce, *Talanta* **15**, 1397 (1968).

However, various iterative computer programs have been successfully applied, combining data at two or three independent points.[12,19,33] More recently, generally applicable methods using all data points have been developed (e.g., Sayce[32]). Approximation methods, which have neglected minor species, have not given reliable results.[34]

Ion-Specific Electrodes

Rationale. The introduction and development of cation-selective electrodes[35] responsive to the activities of uncomplexed metal ions in solution (at low ionic strength) has made possible, in principle, the direct measurement of thermodynamic stability constants for metal–ligand complexes. Both monovalent and divalent cation-selective electrodes have been applied to the measurement of free metal ions in equilibrium with several ligands (including ATP), allowing the subsequent determination of the metal–ligand stability constant using suitable Nernstian and mass conservation relationships.[7,8,36–38]

Experimental. The equilibrium potential developed by the ion-selective electrode is directly related to the logarithm of the metal ion activity under investigation.[35] The electrode is initially calibrated with suitable standard solutions of known metal ion activity, and the data are fitted by a linear least squares regression procedure. The resultant working curve of electrode potential versus logarithm of metal ion activity is used for interpolating the metal ion activity of unknown solutions.

Potentiometric measurements should be carried out in a temperature-controlled vessel under an atmosphere of scrubbed inert gas, such as nitrogen or argon. Metal ion activity and pH are monitored using suitable ion-selective and glass electrodes, respectively. A résumé of the experimental procedure involved in the determination of the stability constants for $KATP^{3-}$ and $MgATP^{2-}$, as type examples of monovalent and divalent metal ion complexes, respectively, is given below.[7,8]

ASSOCIATION OF K^+ AND ATP^{4-}. Both the valinomycin potassium-selective electrode (Orion)[39] and the monovalent cation selective glass

---

[33] D. D. Perrin and V. S. Sharma, *Biochim. Biophys. Acta* **127,** 35 (1966).
[34] R. M. Smith and R. A. Alberty, *J. Am. Chem. Soc.* **78,** 2376 (1956).
[35] R. A. Durst (ed.), "Ion-Selective Electrodes." National Bureau of Standards, U.S.A., 1969.
[36] J. De Moura, D. Le Tourneau, and A. C. Wiese, *Arch. Biochem. Biophys.* **134,** 258 (1969).
[37] G. A. Rechnitz and M. S. Mohan, *Science* **168,** 1460 (1970).
[38] M. S. Mohan and G. A. Rechnitz, *J. Am. Chem. Soc.* **94,** 1714 (1972).
[39] Orion Research Inc., Cambridge, Massachusetts.

electrode (Corning)[40] are suitable for monitoring K^+ activities. The electrode is calibrated by adding small aliquots of a standard 25 mM KCl solution from an Agla syringe to a known volume of water (~50 ml). Sufficient 50 mM KOH is also added to keep the pH above 9. After calibration, potentiometric measurements are made following the addition of small aliquots of an accurately known fresh solution of K_2H_2ATP (~10 mM) and sufficient 50 mM KOH to 50 ml of water so that the pH remained above 9. The K^+ activity of the solution is interpolated from the electrode calibration curve after each addition of ligand and base.

ASSOCIATION OF Mg^{2+} AND ATP^{4-}. The commercially available divalent cation-selective electrode (Orion)[39] is suitable for monitoring the activity of a variety of divalent cations, including Mg^{2+}. The electrode is initially calibrated with various concentrations of $MgCl_2$ as described above for KCl. Potentiometric changes are then measured following the addition of small aliquots of an accurately known fresh solution of K_2H_2ATP (~10 mM) and 10 mM $MgCl_2$, sufficient KOH being added to maintain the pH above 9. After each addition of ligand, metal ion, and base, the Mg^{2+} activity of the solution is interpolated from the calibration curve. A suitable correction must be made for the presence of the $KATP^{3-}$ complex due to the introduction of K^+ ions (as KOH).

Analysis. The thermodynamic stability constant for the association of a metal ion, M, with a charged ligand, L [Eq. (1)], is related to the apparent stability constant [Eq. (7)] in the following manner:

$$K^\circ_{ML} = K'_{ML} \frac{\gamma_{ML}}{\gamma_M \gamma_L} \tag{13}$$

where γ_i represents the activity coefficient of the ion, i, which relates the activity (a_i) and concentration (c_i) of an ion, i.e.,

$$a_i = c_i \gamma_i \tag{14}$$

These equations are applicable only at low ionic strength, where the activity of ions in solution can be accurately predicted using the Debye–Hückel theory.[41] Under these conditions the activity coefficient of an ion can be calculated using the Davies equation.[42]

$$-\log \gamma_i = Az_i^2 \left(\frac{I^{1/2}}{1 + I^{1/2}} - 0.3I \right) \tag{15}$$

[40] Glass Works, Corning, New York.
[41] R. A. Robinson and R. H. Stokes, "Electrolyte Solutions: the Measurement and Interpretation of Conductance, Chemical Potential and Diffusion in Solutions of Simple Electrolytes," 3rd ed. Butterworth, London, 1965.
[42] C. W. Davies, "Ion Association." Butterworth, London, 1962.

z_i is the valency of the ion i, A is the Debye-Hückel parameter (0.512 in water at 25°), and I is the ionic strength of the solution, defined as

$$I = \tfrac{1}{2}\Sigma c_i z_i^2 \tag{16}$$

The equilibrium potential (E) developed by the ion-selective electrode, when used in combination with a suitable reference (e.g., saturated calomel electrode), is related to a_i by the Nernst equation,

$$E = E° + S \log a_i \tag{17}$$

where $E°$ represents the electrode potential developed in a standard solution of the metal ion ($a_i = 1.0$) and S represents the change in potential with a 10-fold change in the activity of the metal ion. After suitable calibration of the electrode with standard metal ion solutions, a plot of E versus $\log a_i$ gives a linear correlation and serves as the basis for interpolating the activity of unknown solutions.

Application of Eq. (14), in conjunction with suitable mass conservation relationships, leads to the following formulas, which are used to determine the concentrations of the relevant ionic species in solution, for the formation of $KATP^{3-}$.

$$[K^+] = a_{K^+}/\gamma_{K^+} \tag{18}$$

$$[KATP^{3-}] = [K]_T - [K^+] \tag{19}$$

$$[ATP^{4-}] = [ATP]_T - [KATP^{3-}] \tag{20}$$

Before these equations can be applied, an evaluation of γ_{K^+} must be made using Eq. (15). This requires an accurate estimate of the ionic strength of the solution using Eq. (16), i.e.,

$$I = \tfrac{1}{2}\{16[ATP^{4-}] + 9[KATP^{3-}] + [K^+] + [OH^-]\} \tag{21}$$

By initially setting $\gamma_{K^+} = 1.0$ and then approximating the values for $[K^+]$, $[KATP^{3-}]$, and $[ATP^{4-}]$ using Eqs. (18), (19), and (20), respectively, an estimate for the ionic strength of the solution can be made [Eq. (21)] and thence a better estimate for γ_{K^+} using Eq. (15). Such an iterative procedure is repeated until the values obtained for $[K^+]$, $[KATP^{3-}]$, and $[ATP^{4-}]$ become self-converging. Finally, an application of Eqs. (15) and (13) leads to an evaluation of the thermodynamic stability constant at infinite dilution for the metal–nucleotide complex.

The iterative procedure described here has been incorporated into a FORTRAN IV computer program, which has been useful in the analysis of monovalent metal ion interactions with ATP. A listing, together with a sample output, is given in the Appendix.

Comment. Estimations of the stability constants for the complexes of several monovalent and divalent metal ions with ATP represent the major application of ion-selective electrodes in this area.[7,8,37,38] These investigations have invariably been carried out at low ionic strength (10–20 mM) where the activity coefficients of the ions can be accurately predicted, and the reported values appear to be much higher than those obtained by other techniques. Extrapolations from the conditions of measurement to higher ionic strengths, however, cannot be made with confidence because of the limitations of the Debye–Hückel theory. Mohan and Rechnitz claim that an approximate calculation [Eq. (15)] demonstrates that an increase in ionic strength from 0.01 to 0.2 M would decrease the stability constant for an MATP^{3-} complex by a factor of 2.[8] For example, the reported thermodynamic constant of 220 M^{-1} for KATP^{3-} would extrapolate to a value of 110 M^{-1} at an ionic strength of 0.2 M. Independent calculations in this laboratory have indicated that the factor involved is closer to 5. Extrapolation would therefore give a value for KATP^{3-} of approximately 40 M^{-1} at an ionic strength of 0.2 M, which is within a factor of 2–3 of previously published estimates (Table IV). Further, an extrapolation of the reported thermodynamic stability constant for MgATP^{2-} (1.15 × 10^6 M^{-1})[7] to higher ionic strengths (0.1 M), using the empirical equation of Phillips et al.,[6] yields a value (61,000 M^{-1}), consistent with previous determinations (Table I).

In summary, ion-specific electrodes appear to give reliable values for stability constants at low ionic strength. Extrapolation to higher (and more realistic) ionic strengths, though hazardous on theoretical grounds, may be achieved if satisfactory empirical equations are available.

Ion Exchange Resins

Rationale. The method is based on the competition between a metal ion and a suitable anion exchange resin for a free ligand in solution.[43] The technique relies upon the measurement of ligand distribution between the resin and solution, both in the absence and in the presence of several concentrations of metal ion. The metal–ligand stability constant is subsequently evaluated using a formula relating the equilibrium of the ligand between the metal ion and the resin.

The method has been most extensively applied to the investigation of complex formation between a variety of divalent metal ions and several biologically important molecules, including the nucleoside mono-, di-, and triphosphates.[6,44–46]

[43] J. Schubert, *Methods Biochem. Anal.* **3**, 247 (1963).
[44] E. Walaas, *Acta Chem. Scand.* **12**, 528 (1958).
[45] L. B. Nanninga, *Biochim. Biophys. Acta* **54**, 330 (1961).
[46] J.-M. Jallon and M. Cohn, *Biochim. Biophys. Acta* **222**, 542 (1970).

Experimental. Experimental design is based upon the equilibration of a ligand solution with a known quantity of suitable exchange resin in a suitably buffered medium, both in the absence and in the presence of several concentrations of metal ion. After equilibration, the distribution of the ligand between the resin and the solution is measured.

In a typical experiment, ten 25-ml glass-stoppered flasks are used: two flasks containing ligand in the absence of the exchange resin (blanks), two containing ligand in the presence of the exchange resin but without added metal ion, and six flasks containing different concentrations of the metal ion together with the ligand and exchange resin. The flasks are shaken for 3–5 hr at constant temperature and the distribution of the ligand is subsequently measured. The reader is referred to Walaas[44] for a detailed account of the materials and experimental procedure used in this technique.

Analysis. The basic relationship used in the analysis of data takes the form[43]

$$\log \left(\frac{D°}{D} - 1 \right) = n \log [M^{2+}] + \log K \tag{22}$$

where n is the number of molecules of the metal ion (M^{2+}) in the metal–ligand complex, K is the apparent stability constant of the complex, and D and $D°$ represent the distribution coefficients of the ligand (% resin bound ligand)/(% ligand in solution) × (volume of resin)/(mass of resin) in the presence and in the absence, respectively, of M^{2+}. A plot of $\log [(D°/D) - 1]$ versus $\log [M^{2+}]$ using the measured values for D at different concentrations of metal ion, will yield a linear correlation with slop n and an ordinate intercept, $\log [M^{2+}] = 0$, of $\log K$.

With one apparent exception,[6] it appears that results from this technique have been based on the use of total metal ion, $[M]_T$, rather than free metal ion, $[M^{2+}]$, concentrations. An assumption therefore exists that upon extrapolation to $\log [M^{2+}] = 0$, the concentration of free metal ion is equivalent to the concentration of the total metal ion added, giving indistinguishable $\log K$ values, whether the plot is based on $\log [M^{2+}]$ or $\log [M]_T$ data points. This assumption is justified by the fact that in the presence of a large excess of metal, $\log [M^{2+}] = 0$, only a very small proportion of the total metal ion is complexed by the ligand; i.e., upon extrapolation $[M^{2+}] \approx [M]_T$. Although these assumptions have not been clearly stated in previous publications, the technique has given useful comparative results.

Fluorescence

Changes in the fluorescence intensity of the magnesium–8-hydroxyquinoline complex at 530 nm (after activation at 360 nm) has been used to monitor the formation of Mg–nucleotides.[47] The principle is essentially the same as that described above for following spectral changes with magnesium–8-hydroxyquinoline at 360 nm.[12,17,30] However, it is possible to work at much lower concentrations of all the reactants, as the technique is somewhat more sensitive.

Competition with Arsenazo III

The spectrophotometric metal ion indicator arsenazo III [o-(1,8-dihydroxy-3,6-disulfonaphthylene-2,7-bisazo)bisbenzenearsonic acid] forms a specific 1:1 complex with Ca^{2+} accompanied by a large absorption shift, the maximum change occurring at 650–660 nm. The compound has recently been applied to the measurement of Ca^{2+} in equilibrium with ATP^{4-}, the principle being essentially the same as that described above for 8-hydroxyquinoline.[48] The reported stability constant for $CaATP^{2-}$ in the presence of 0.1 M NaCl at pH 7.4 and 25° (9000 M^{-1}) is in substantial agreement with previous estimations under similar conditions.[12,49] Although commercial preparations of arsenazo III require some degree of purification,[50] the use of this dye in the determination of Ca–ligand stability constants appears promising.

Gel Filtration

Gel filtration on cross-linked dextran (Sephadex) has been used in the quantitative analysis of the reversible interaction between macromolecules and low molecular weight ligands.[51] The application of this technique to the measurement of metal–ligand stability constants depends upon the separation of the metal–ligand complex from the free metal ion, based upon the difference in molecular weight between the two species. A strongly cross-linked gel material, such as Sephadex G-10, facilitates the successful determination of the affinity of metal ions for ligands with molecular weights >450 (e.g., ADP, ATP).

With this technique, the metal–ligand interaction may be studied under physicochemical conditions identical to the proposed enzyme kinetic experiments. The method has been applied to the interaction of Mn^{2+} with several nucleoside diphosphates, NAD, NADH, NADP, and NADPH.[52]

---

[47] S. Watanabe, T. Trosper, M. Lynn, and L. Evenson, *J. Biochem. (Tokyo)* **54**, 17 (1963).
[48] N. C. Kendrick, R. W. Ratzlaff, and M. P. Blaustein, *Anal. Biochem.* **83**, 433 (1977).
[49] M. M. Taqui Khan and A. E. Martell, *J. Phys. Chem.* **66**, 10 (1962).
[50] N. C. Kendrick, *Anal. Biochem.* **76**, 487 (1976).
[51] J. P. Hummel and W. J. Dreyer, *Biochim. Biophys. Acta* **63**, 530 (1962).
[52] R. F. Colman, *Anal. Biochem.* **46**, 358 (1972).

Raman Spectroscopy

The frequency of the phosphate moiety Raman line has been used to monitor the formation of various metal complexes of ATP.[53] It was necessary to work at relatively high concentrations of ATP (0.02–0.3 M), which increased the possibility of the formation of complexes other than those of 1:1 stoichiometry. Nevertheless, a value for MgATP^{2-} (32,000 M^{-1}), in reasonable agreement with other reports, was obtained. A value of 26 M^{-1} for NaATP^{3-} was also reported.

Electron Spin Resonance (ESR)

The solvated manganous ion, Mn(H$_2$O)$_6^{2+}$, produces a characteristic hyperfine six-line ESR signal. The amplitude of the observed peaks is a linear function of the manganous ion concentration over the range of at least 10 μM to 10 mM. While many manganous complexes retain some of the six-line character, the amplitude of the signal is usually considerably reduced. Thus, this decrease in amplitude on the addition of a ligand to a manganous ion solution, is a direct measure of the bound metal ion, provided that the contribution of the complex to the spectrum is negligible. The approach was first exploited by Cohn and Townsend, who reported a stability constant for Mn–glucose 1-phosphate of 200 M^{-1}.[54]

The use of ESR has proved particularly useful in determining weak stability constants, where the contribution of the complexed species to the overall spectrum is minimal. Examples are given by O'Sullivan and Cohn[55] and Chapman et al.[56] (A small consistent error has recently been observed for complexes that retain a net negative charge and could lead to a 10–20% error in the final stability constant.[57])

Selection of Constants

General Considerations

The primary objective of this chapter is to provide the reader with a guide to the determination of the concentration of various species in solution for application to enzyme kinetic experiments. An examination of the literature demonstrates that considerable variation exists in the magnitude of reported values for the stability constants of biologically impor-

---

[53] M. E. Heyde and L. Rimai, *Biochemistry* **10**, 1121 (1971).
[54] M. Cohn and J. Townsend, *Nature (London)* **173**, 1090 (1954).
[55] W. J. O'Sullivan and M. Cohn, *J. Biol. Chem.* **241**, 3104 (1966).
[56] B. E. Chapman, W. J. O'Sullivan, R. K. Scopes, and G. H. Reed, *Biochemistry* **16**, 1005 (1977).
[57] A. Hu, M. Cohn, and G. H. Reed, unpublished observations.

tant ligands with a variety of metal ions. Much of this variation can be attributed to differences in the physicochemical conditions of measurement, such as ionic strength, temperature, pH, and nature of the supporting medium, and there is considerable convergence of these values when they are "normalized" to similar conditions.[58]

To illustrate these remarks, a selection of constants from the literature for $MgATP^{2-}$ is presented in Table I,[6,7,12,17,47,49] and for $MgADP^-$ in Table II.[6,12,17,47,59] The conditions of measurement are indicated, and the tables include values "normalized" to pH 8.0, 30°, $I = 50$ mM in a noninteracting supporting medium, using the following relationships: pK_a of $HATP^{3-}$ as 6.95,[9] pK_a of $HADP^{2-}$ as 6.65,[9] temperature variation and effect of ionic strength as indicated by the data and empirical equations, respectively, of Phillips et al,[6]; $KATP^{3-}$ as 15 M^{-1},[12] $NaATP^{3-}$ as 14 M^{-1},[12] $KADP^{2-}$ and $NaADP^{2-}$ as 5 M^{-1}.[18,20] The amount of $MgATP^{2-}$ or $MgADP^-$ present at 1 mM of both added constituents is also included; the variation is approximately 2% for $MgATP^{2-}$ and 14% for $MgADP^-$.

The procedures used to "normalize" the constants are elaborated below.

Normalization Procedures

a. pH. Differences in the pH under which the $MgATP^{2-}$ and $MgADP^-$ stability constants were measured will lead to differences in the proportion of ATP, present as ATP^{4-}, and of ADP, present as ADP^{3-}. The reported apparent stability constants were corrected to pH 8 using the relationship

$$K'_{pH8} = K'(\delta) \qquad (23)$$

where K' and K'_{pH8} are the apparent and corrected stability constants, respectively, and δ is a factor used in correction to pH 8, defined as (percent total nucleotide present in fully ionized form at pH 8)/(percent total nucleotide present in fully ionized form at pH of stability constant measurement), i.e.,

$$\delta = (1 + 10^{(pK_a-pH)})/(1 + 10^{(pK_a-8)}) \qquad (24)$$

---

[58] It might be noted that the mode of using log K values is probably a better initial guide than absolute values. The former does provide a more reliable indication of the differences between the constants. For example, a stability constant of 100,000 M^{-1} for $MgATP^{2-}$ does not indicate twice as much metal–nucleotide complex as does a value of 50,000 M^{-1}. At 1 mM of added $[ATP]_T$ and $[Mg]_T$ the difference is only 4%, i.e., 0.91 mM to 0.87 mM $[MgATP^{2-}]$.

[59] M. M. Taqui Khan and A. E. Martell, *J. Am. Chem. Soc.* **84**, 3037 (1962).

TABLE I
SELECTION OF STABILITY CONSTANTS FOR MgATP^{2-}

Method	Conditions of measurement				Reported apparent stability constant (M^{-1})	"Normalized" stability constant[a] (M^{-1})	[MgATP^{2-}] at 1 mM of both added constituents[b] (mM)
	Buffer and pH	Supporting electrolyte	Ionic strength (M)	Temp. (°C)			
Competition in solution with 8-hydroxyquinoline (spectral changes at 360 nm)[12]	N-Ethylmorpholine · HCl, pH 8.0	—	~0.05	30	73,000	73,000	0.89
Competition in solution with 8-hydroxyquinoline (spectral changes at 360 nm)[17]	Triethanolamine · HCl, pH 8.4	Tributylethylammonium bromide	0.1	25	38,000	57,000	0.88
Competition in solution with 8-hydroxyquinoline (fluorescence changes at 530 nm)[47]	Triethanolamine · HCl, pH 8.2	—	~0.05	25	80,000	85,000	0.90
Competition with anion exchange resin[6]	Tris · HCl, pH 8.7	Tetra-n-propyl ammonium bromide	0.1	25	43,000	63,000	0.88
pH titration[49]	—	KNO$_3$	0.1	25	16,600	65,000	0.88
Divalent cation-selective electrode[7]	~9.2 (maintained with KOH)	—	0.002–0.008 (corrected to infinite dilution)	25	1,150,000	88,000	0.90

[a] $I = 50$ mM, pH 8.0, in a noninteracting supporting medium at 30°.
[b] The normalized stability constant was used to calculate the concentration of MgATP^{2-} in a solution containing 1 mM added [ATP]$_T$ and [Mg]$_T$.

TABLE II
SELECTION OF STABILITY CONSTANTS FOR MgADP$^-$

Method	Conditions of measurement				Reported apparent stability constant (M^{-1})	"Normalized" stability constant[a] (M^{-1})	[MgADP$^-$] at 1 mM of both added constituents[b] (mM)
	Buffer and pH	Supporting electrolyte	Ionic strength (M)	Temp. (°C)			
Competition in solution with 8-hydroxyquinoline (spectral changes at 360 nm)[12]	N-Ethylmorpholine · HCl, pH 8.0	—	~0.05	30	4000	4000	0.61
Competition in solution with 8-hydroxyquinoline (spectral changes at 360 nm)[17]	Triethanolamine · HCl, pH 7.9	Tributylethylammonium bromide	0.1	25	2200	3000	0.57
Competition in solution with 8-hydroxyquinoline (fluorescence changes at 530 nm)[47]	Triethanolamine · HCl, pH 8.1	—	~0.05	25	6000	6600	0.68
Competition with anion exchange resin[6]	Tris · HCl, pH 8.7	Tetra-n-propyl ammonium bromide	0.1	25	2800	3650	0.59
pH titration[59]	—	KNO$_3$	0.1	25	1500	3000	0.57

[a] I = 50 mM, pH 8.0, in a noninteracting supporting medium at 30°.
[b] The normalized stability constant was used to calculate the concentration of MgADP$^-$ in a solution containing 1 mM added [ADP]$_T$ and [Mg]$_T$.

TABLE III
Factors (δ) Involved in the Correction of Apparent Stability Constants for MgATP^{2-} and MgADP$^-$ to pH 8[a]

pH	δ	
	MgATP^{2-}	MgADP$^-$
7.5	1.177	1.092
7.6	1.124	1.065
7.7	1.082	1.0425
7.8	1.048	1.025
7.9	1.021	1.011
8.0	1.000	1.000
8.1	0.983	0.991
8.2	0.970	0.984
8.3	0.959	0.979
8.4	0.951	0.974
8.5	0.944	0.971

[a] Values for δ are based upon a pK_a of 6.95 for HATP^{3-} and of 6.65 for HADP^{2-}.

A selection of values for δ, based upon the above pK_a values for the ionization of the terminal phosphate group of ATP and ADP, are collected in Table III.

Example

The value of 43,000 M^{-1}, determined by Phillips et al.[6] for the MgATP^{2-} complex at pH 8.7 is corrected to pH 8 using the appropriate value for δ [Eq. (24)] and substitution in Eq. (23),

$$K'_{pH8} = 43,000 \, (0.935) = 40,000 \, M^{-1}.$$

b. Temperature. The effect of temperature variation on both the MgATP^{2-} and MgADP$^-$ stability constants can be calculated from a plot of log (apparent stability constant) versus 1/(absolute temperature).

The data of Phillips et al.[6] were treated in this way, and the lines were fitted by linear least squares regression.[60]

The following empirical factors derived from the plots were used to correct the reported apparent stability constants to 30°.

---

[60] From the limited data available it appears that the stability constants for complexes with other divalent metal ions (Ba, Sr, Ca, Co, Mn, Zn, Ni) show an opposite trend in magnitude as a function of temperature.[46,61] The reader is referred to the article by Khan and Martell[61] for guidance on the temperature variation of the ATP complexes.

[61] M. M. Taqui Khan and A. E. Martell, *J. Am. Chem. Soc.* **88**, 668 (1966).

MgATP^{2-}: Correlation coefficient for linear fit of data = -0.998.

$$\Delta \log K = 750(\Delta T) \tag{25}$$

where $\Delta \log K$ is the change in the apparent stability constant with a change in temperature from the measured (T_m) to the corrected (T_c) value of T, expressed as °K, i.e.,

$$\Delta T = 1/T_m - 1/T_c \tag{26}$$

MgADP$^-$: Correlation coefficient for linear fit of data = -0.9997.

$$\Delta \log K = 810(\Delta T) \tag{27}$$

where $\Delta \log K$ and ΔT are as above [Eq. (25)].

EXAMPLE

Burton[17] found that the apparent stability constant for the MgADP$^-$ complex was 6900 M^{-1} at a temperature of 64°. This value is corrected to 30° in the following manner,

Temperature of measurement $(T_m) = 64°$ (337°K)
Corrected temperature $(T_c) = 30°$ (303°K)
Therefore $\Delta T = 1/337 - 1/303 = -3.33 \times 10^{-4}$.

By substitution in Eq. (27),

$\Delta \log K = -0.2697$ for this change in temperature,
i.e., $K = 3700$ M^{-1} at 30°.

c. *Ionic Strength.*[62] Apparent stability constants could be extrapolated to an ionic strength of 50 mM using the following empirical equations, originally derived by Phillips *et al*[6]:

MgATP^{2-}:

$$\log K° = \log K' + 6.10 I^{1/2} - 8.74 I + \frac{2.04 I^{1/2}}{1 + 6.02 I^{1/2}} \tag{28}$$

where $K°$ and K' represent values for the stability constant at an ionic strength of zero and I, respectively.

MgADP$^-$:

$$\log K° = \log K' + 4.06 I^{1/2} - 6.36 I + \frac{2.04 I^{1/2}}{1 + 6.02 I^{1/2}} \tag{29}$$

where $K°$ and K' are as above (Eq. 28).

[62] There is a paucity of data available concerning the variation of the magnitude of the stability constants for metal–nucleotide complexes, other than those involving magnesium, with changes in ionic strength. The reader is referred to Phillips *et al*.[6] for details concerning the derivation of the type of equations shown [see Eqs. (28) and (29)].

EXAMPLE

The stability constant for the MgATP^{2-} complex, evaluated by Mohan and Rechnitz[7] at infinite dilution ($I = 0$), viz. $1.15 \times 10^6\ M^{-1}$, can be corrected to a higher ionic strength (e.g., $0.1\ M$) by substituting the following values in Eq. (28);

$$K° = 1.15 \times 10^6\ M^{-1}$$
$$\log K° = 6.061$$
$$I = 0.10\ M$$

therefore $\log K' = 4.784$, $K' = 61{,}000\ M^{-1}$ ($I = 0.1\ M$).

d. *Supporting Electrolyte* (K^+ *or* Na^+). Where applicable, the apparent stability constants were corrected for the presence of sodium or potassium ions using Eq. (30).

$$K'_{corr} = K'(1 + k[M^+]) \tag{30}$$

where K'_{corr} and K' are the corrected (in the absence of Na^+ or K^+) and apparent (measured in the presence of Na^+ or K^+) stability constants, respectively, for the metal–nucleotide complex. M^+ represents Na^+ or K^+, and k the apparent stability constant for the MATP^{3-} or MADP^{2-} complex.

EXAMPLE

A value of $16{,}600\ M^{-1}$ was reported by Taqui Khan and Martell[49] for the apparent stability constant of the MgATP^{2-} complex in the presence of $0.1\ M$ KNO$_3$. Correction of the reported value for the presence of K^+ was made by substituting the following values in Eq. (30),

$$K' = 16{,}600\ M^{-1}$$
$$k_{K^+} = 15\ M^{-1}$$
$$[K^+] = 0.1\ M$$

therefore $K'_{corr} = 40{,}000\ M^{-1}$.

Recommended Values

On the basis of these considerations, useful values for the stability constants of Mg^{2+}, Ca^{2+}, and Mn^{2+} with ATP and ADP, under the above-specified conditions, have been collected in Tables IV and V. The selection is influenced by the experience of the authors, and are in substantial agreement with much of the literature. The list includes a few values, indicated in parentheses, for which an empirical estimate has been made as no suitable data in the literature were available. The reader is re-

TABLE IV
RECOMMENDED STABILITY CONSTANTS FOR COMPLEXES OF ATP

Complex	Apparent stability constant (M^{-1})
H_2ATP^{2-}	1.0×10^{4a}
$HATP^{3-}$	8.9×10^{6b}
MgH_2ATP	20
$MgHATP^-$	500
$MgATP^{2-}$	73,000
CaH_2ATP	20
$CaHATP^-$	400
$CaATP^{2-}$	35,000
MnH_2ATP	$(100)^c$
$MnHATP^-$	$(1000)^c$
$MnATP^{2-}$	100,000
$KATP^{3-}$	15
$NaATP^{3-}$	14

[a] $pK_a = 4.0$.
[b] $pK_a = 6.95$.
[c] Estimated value.

ferred to various compilations for variations.[9,10,63,64] The values reported in Tables IV and V have been used in the calculations presented below, and the effect of moderate error in their selection is also considered.

The magnitude of the stability constants for the other nucleotides will not vary significantly from those for adenosine analogs.[44] This will in general be true also for other esters of phosphate, pyrophosphate, and triphosphate provided that the major site of interaction for the metal ion is the phosphate moiety. For example, the stability constants of metal–thiamine triphosphate (TPP) complexes will be substantially the same as for similar metal–ATP complexes. Such considerations would also be valid for compounds such as the intermediates of cholesterol biosynthesis. Thus, the stability constants for Mg–phosphomevalonic acid and Mg–pyrophosphomevalonic acid will be substantially the same as for the MgAMP and MgADP$^-$ complexes, respectively. The validity of such a conclusion is supported by the fact that Mg^{2+} interacts with the pyrophosphate moiety of PRPP to an extent equivalent to that with ADP.[16]

Stability constants for a number of other phosphorylated compounds of biological interest are collected in Table VI.

[63] R. C. Phillips, *Chem. Rev.* **66**, 501 (1966).
[64] R. M. Izatt, J. J. Christensen, and J. H. Rytting, *Chem. Rev.* **71**, 439 (1971).

TABLE V
RECOMMENDED STABILITY CONSTANTS FOR COMPLEXES OF ADP

Complex	Apparent stability constant (M^{-1})
H_2ADP^-	$8.0 \times 10^{3\,a}$
$HADP^{2-}$	$4.5 \times 10^{6\,b}$
MgHADP	100
MgADP$^-$	4000
CaHADP	(80)[c]
CaADP$^-$	2000
MnHADP	(500)[c]
MnADP$^-$	30,000
KADP^{2-}	5
NaADP^{2-}	5

[a] $pK_a = 3.9$.
[b] $pK_a = 6.65$.
[c] Estimated value.

TABLE VI
METAL–LIGAND STABILITY CONSTANTS FOR VARIOUS
BIOLOGICALLY IMPORTANT PHOSPHORYLATED COMPOUNDS[a]

Ligand	Apparent stability constant (M^{-1})		
	Mg^{2+}	Ca^{2+}	Mn^{2+}
AMP	70	40	200
Glycerol 1-phosphate	60	50	
Hexose 1-phosphate	70		160
Hexose 6-phosphate	50	30	(200)
Phosphorylarginine	100	(50)	
Phosphorylcreatine	40	20	150
3-Phosphoglycerate	(40)		800[b]
$HP_2O_7^{3-}$	1200	(500)	
HPO_4^{2-}	500	200	

[a] Reported values in parentheses represent estimates based on the experience of the authors, as no suitable data were available in the literature.
[b] The greater affinity of Mn^{2+} for 3-phosphoglycerate, in comparison with the other monophosphate esters, may reflect metal ion interaction with the phosphoglycerate carboxy group.

Higher-Order Complexes

The evidence for complex species such as $M(ATP)_2^{6-}$ and M_2ATP, involving the alkaline earth metal ions, is generally indirect and unconvincing. If these complexes do exist, it is doubtful that they ever reach concentrations such that they are of relevance in kinetic experiments. Such complexes, including those where interaction between the metal ion and purine ring nitrogens occurs, either in intermolecular or intramolecular complexes, do become increasingly important for the transition metal ions, such as Mn, Co, and Cu. With the exception of Mn, these metal ions are of minor interest in enzyme kinetic experiments. While there is evidence that complexes involving Mn^{2+}, other than those of simple 1:1 interaction with the phospho-oxyanion, can occur, they are of little relevance to the determination of kinetic constants.

5-Phosphoribosyl-α-1-Pyrophosphate (PRPP)

The importance of PRPP in many biosynthetic reactions has been well documented, as has the requirement for Mg^{2+} in the activation of these reactions. However, the evaluation of accurate kinetic parameters on the phosphoribosyltransferase (PRPP utilizing) enzymes has been hindered by the lack of data concerning proton and metal ion–PRPP interactions, particularly "Mg—PRPP," the form proposed as the active ribose donor in the transferase reaction. The analysis of such complex formation poses special problems due to the individual phosphate (C-5) and pyrophosphate (C-1) moieties separated on the ribose sugar of the molecule.

At the time of writing, the work of Thompson et al.,[16] using the pH titration method, represents the only rigorous investigation of the stability constants of H^+ and Mg^{2+} complexes with PRPP.[65] The relevant equilibria used in their investigation are shown in Eqs. (31).

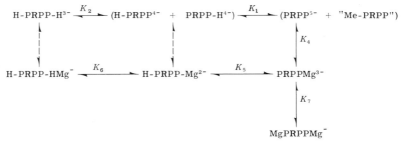

---

[65] Morton and Parsons[66] used competition with 8-hydroxyquinoline to obtain an apparent stability constant of approximately 4000 M^{-1} for Mg-PRPP in 0.1 M Tris, 0.15 M KCl, pH 8.5 at 25°. It is almost certain that what they determined was the combination constant, $K_4 K_7$.

[66] D. P. Morton and S. M. Parsons, Arch. Biochem. Biophys. **175**, 677 (1976).

Only the species deduced to achieve a significant concentration over the pH and [Mg^{2+}] range covered in their experiments are shown. Symbols to the right and left of PRPP represent proton or Mg^{2+} complex formation to the C-1 pyrophosphate and C-5 phosphate groups, respectively. "Me-PRPP" represents the sum of supporting electrolyte (K^+ or Na^+) complexes with PRPP. Values obtained for the relevant stability constants at 25° and an ionic strength of ~0.2 M (0.17 M Na^+ or K^+) are collected in Table VII.[16]

Effects of Ionic Strength, Supporting Electrolyte and Temperature on the Analysis of PRPP Complex Species in Solution. In order to extrapolate the constants determined by Thompson *et al.*[16] to other physicochemical conditions, it is assumed that the pyrophosphate and phosphate binding sites for Mg^{2+} act independently and, to a first approximation, in a similar manner to the phospho-oxyanion moieties of ADP and AMP, respectively. This appears to be justified, as extrapolation of the values for $MgADP^-$ (Table V) and MgAMP (Table VI), to the conditions of Thompson *et al.* yield estimates in good agreement with their values for K_4 and K_7, respectively.[67] Further support comes from the similarity between the pK_a values for $PRPPH^{4-}$ (6.56) and $HADP^{2-}$ (6.65, Table V). On the basis of these arguments, the effect of temperature, using Eq. (27), and of ionic strength, using Eq. (29), on the magnitude of K_4 may be calculated.

To allow for the effect of monovalent cations, the stability constants for K^+ or Na^+ binding to the pyrophosphate and phosphate groups of PRPP were assumed to mimic the values for $MADP^{2-}$ (5 M^{-1}, Table V) and $MAMP^-$ (2 M^{-1}),[18] respectively, using Eq. (30).

On the basis of these considerations, extrapolated values for the step stability constants, K_4 and K_7, under the physicochemical conditions specified in Tables I and II (I = 50 mM in a noninteracting supporting medium at 30°) have been included in Table VII.

Application of Constants

The emphasis of the chapter has been on the determination and selection from the literature of the most applicable stability constants for relevant metal-ligand species. In this section, the recommended constants

---

[67] The stability constant for the interaction of Mg^{2+} with the diphosphate group of ADP^{3-}, when extrapolated to the conditions of 0.2 M ionic strength (0.17 M Na^+ or K^+) using Eqs. (29) and (30), gives a value (log K = 3.14) consistent with the step stability constant for the association of Mg^{2+} with $PRPP^{5-}$ (log K_4 = 3.19). A similar extrapolation of data for the complex formation between Mg^{2+} and the monophosphate group of AMP^{2-} gives an average log K value of 1.87, in reasonable agreement with the step stability constant for the formation of $Mg-PRPP-Mg^-$ (log K_7 = 1.67).

TABLE VII
STABILITY CONSTANTS FOR PROTON AND MAGNESIUM COMPLEXES WITH PRPP

Complex	Apparent step stability constant[a] (M^{-1})	"Normalized" step stability constant[b] (M^{-1})	Apparent overall stability constant[a]
(HPRPP^{4-} + PRPPH^{4-})	$K_1 = 3.63 \times 10^6$		$K_1 = 3.63 \times 10^6\ M^{-1}$
HPRPPH^{3-}	$K_2 = 6.92 \times 10^5$		$K_1 \times K_2 = 2.51 \times 10^{12}\ M^{-2}$
PRPPMg^{3-}	$K_4 = 1.54 \times 10^3$	3300	$K_4 = 1.54 \times 10^3\ M^{-1}$
HPRPPMg^{2-}	$K_5 = 1.86 \times 10^6$		$K_4 \times K_5 = 2.86 \times 10^9\ M^{-2}$
HPRPPHMg$^-$	$K_6 = 1.10 \times 10^4$		$K_4 \times K_5 \times K_6 = 3.15 \times 10^{13}\ M^{-3}$
MgPRPPMg$^-$	$K_7 = 47.0$	63	$K_4 \times K_7 = 7.24 \times 10^4\ M^{-2}$

[a] Compiled from the data of Thompson et al.[16]
[b] Reported apparent step stability constant corrected to 50 mM ionic strength in a noninteracting supporting medium at 30°. (Refer to text for detailed procedure.)

reported in Tables IV–VII are used to illustrate several manipulations involving these stability constants. The variation in the concentration of metal–ligand species as a function of pH is illustrated for ATP, ADP, and PRPP, and recommendations as to the optimization of the concentration of the metal–nucleotide "active" species are made. In addition, there is a brief consideration of the use of the constants in the evaluation of enzyme kinetic parameters.

Calculation of the concentration of the various free metal ion, free ligand, and metal–ligand complex species, using the relevant stability constants and the total concentration of the various components added, was made possible using the FORTRAN IV iterative computer program presented in the Appendix.

Variation of Species with pH

The variation in the concentration of complex species present in mixtures of ATP and $MgCl_2$, as a function of pH, is illustrated in Fig. 3A. The results of a similar analysis for ADP and PRPP are presented in Fig. 3B

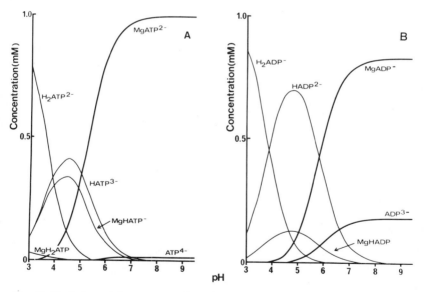

FIG. 3. Variation of the concentration of complex species present in mixtures of the adenosine tri- and diphosphate nucleotides and magnesium chloride as a function of pH. (A) Concentrations calculated for a mixture containing 1 mM [ATP]$_T$, 2 mM [Mg]$_T$ and having pH values as indicated, using the stability constants reported in Table IV. (B) Concentrations calculated for a mixture containing 1 mM [ADP]$_T$, 2 mM [Mg]$_T$ and pH values as indicated, using the constants reported in Table V.

and Fig. 4, respectively. Calculations were based on a total ligand concentration of 1 mM and, with the exception of Fig. 4B, a total magnesium concentration of 2 mM. The results in Fig. 4B were based on $[Mg]_T = 10$ mM. (The importance of maintaining an excess of metal ion over that of the ligand species is elaborated in the following section.) The salient features of Fig. 3 are as follows:

1. Above pH 7, the only ATP species of any significance is $MgATP^{2-}$.

2. In contrast, while $MgADP^-$ reaches a maximum value above pH 8, ADP^{3-} is still a very significant component. This is of particular relevance as ADP^{3-} has often been demonstrated to be a good competitive inhibitor with respect to $MgADP^-$. In contrast, ATP^{4-} is more often a weak inhibitor with respect to $MgATP^{2-}$.[1] An increase in $[Mg]_T$ would decrease the concentration of ADP^{3-}; e.g., at 5 mM $[Mg]_T$, ADP^{3-} represents only 5% of the total nucleotide at pH 8.

3. The monoprotonated species, particularly $HADP^{2-}$, make a significant contribution only in the range pH 4.5–6.0. Such species, however, may be relevant in isolated circumstances. For example, adenosine

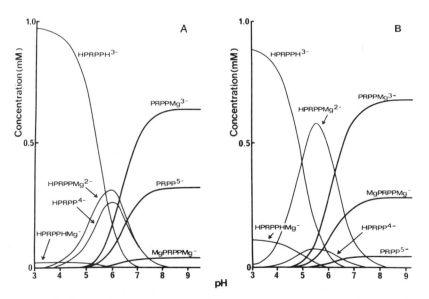

FIG. 4. Variation of the concentration of complex species present in mixtures of PRPP and magnesium chloride as a function of pH. The stability constants reported in the last column of Table VII were used for all calculations. (A) Concentrations determined for a mixture containing 1 mM $[PRPP]_T$ and 2 mM $[Mg]_T$. (B) Concentrations calculated for a mixture containing 1 mM $[PRPP]_T$ and 10 mM $[Mg]_T$.

kinase from erythrocytes appears to have a pH optimum between 5.1 and 6.0 dependent upon the conditions of assay.[68]

The analysis of the species present in mixtures of PRPP and $MgCl_2$ (Fig. 4) shows that below pH 7.5 the composition of the solution varies in a complicated way with at least five, and up to seven, metal ion and protonated PRPP species needing consideration. At pH values above 8, however, only three species become significant, i.e., $PRPP^{5-}$, $PRPPMg^{3-}$, and $MgPRPPMg^-$, making enzyme kinetic analysis much simpler in this region. A similar analysis of a mixture of 1 mM PRPP and 10 mM $MgCl_2$ demonstrates that the concentration of the dimagnesium complex ($MgPRPPMg^-$) is highly dependent upon the concentration of Mg^{2+} (Fig. 4B).

The use of the program presented in the Appendix can be extended to take into account a variety of other species that may be present in solution. An example of an extended version of Fig. 3A has been given by Storer and Cornish-Bowden in a consideration of the components present in the glucokinase reaction mixture.[69]

Optimization of Active Species

For many kinetic investigations of metal–ligand-utilizing enzymes, the major requirement is to maximize the concentration of the metal–substrate active species, while maintaining a small and preferably constant concentration of the free ligand and metal ion, both potential modifiers of the reaction. To obtain this situation, for the metal–nucleotide complexes, it is preferable to maintain a *constant excess* of metal ion over the total nucleotide concentration rather than a constant stoichiometric ratio between the two. A comparison of both these approaches is illustrated for $MgATP^{2-}$ and $MgADP^-$ in Figs. 5 and 6, respectively.

The maintenance of a constant excess of $[Mg]_T$ over that of $[ATP]_T$, ensures that the proportion of ATP present as $MgATP^{2-}$ is maximized and remains constant over a large range of ATP concentrations (Fig. 5A). For example, at a constant metal ion excess of 1 mM, the percentage of total ATP present as $MgATP^{2-}$ remains constant at 99%, over the range 0.1–1 mM $[ATP]_T$. Further, at this constant $[Mg]_T$ excess, $MgATP^{2-}$ still represents 98.6% of the total nucleotide at 1 μM $[ATP]_T$. Such a high and relatively constant concentration of $MgATP^{2-}$ over a wide range of $[ATP]_T$ concentrations ensures a very low and negligible level of ATP^{4-} and a constant concentration of Mg^{2+} (inset, Fig. 5A).

[68] J. Kyd and A. S. Bagnara, personal communication.
[69] A. C. Storer and A. Cornish-Bowden, *Biochem. J.* **159**, 1 (1976).

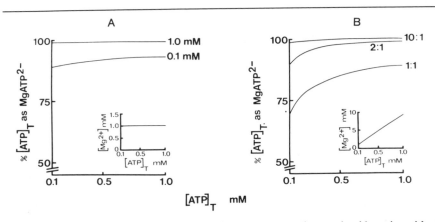

FIG. 5. Comparison of the maintenance of a constant metal to nucleotide ratio, with a constant excess of metal ion, on the proportion of total ATP present as MgATP^{2-} at pH 8. (A) Concentrations of MgATP^{2-} were calculated for mixtures containing a constant excess (0.1 and 1.0 mM) of [Mg]$_T$ over and above that of [ATP]$_T$. *Inset:* Variation of the concentration of free magnesium in mixtures of ATP and MgCl$_2$ over the range 0.1–1.0 mM [ATP]$_T$ at a constant excess of [Mg]$_T$ (1.0 mM). (B) Concentrations of MgATP^{2-} were calculated for mixtures containing [Mg]$_T$:[ATP]$_T$ ratios of 1:1, 2:1 and 10:1. *Inset:* Variation of the concentration of free magnesium in mixtures of ATP and MgCl$_2$ over the range 0.1–1.0 mM [ATP]$_T$ at an [Mg]$_T$:[ATP]$_T$ ratio of 10:1.

In contrast, variation of the total concentrations of ATP and MgCl$_2$, in constant ratio, does not give a constant proportion of any species (Fig. 5B). For example, at an [Mg]$_T$:[ATP]$_T$ ratio of 1:1, the percentage of total ATP present as MgATP^{2-} varies from 69% at 0.1 mM [ATP]$_T$ to 89% at 1 mM [ATP]$_T$. Although at an [Mg]$_T$:[ATP]$_T$ ratio of 10:1 the concentration of MgATP^{2-} remains high and relatively constant over this range of nucleotide concentrations (Fig. 5B), the situation becomes progressively worse at lower ATP concentrations. For example, at 1 μM [ATP]$_T$, the percentage of [ATP]$_T$ present as MgATP^{2-} is only 41%. Further implications of this approach are a large variation in [ATP^{4-}] and a large and variable concentration of Mg^{2+} (inset, Fig. 5B).

The proportion of [ADP]$_T$, present as MgADP$^-$, for various mixtures of MgCl$_2$ and ADP at pH 8, is shown in Fig. 6. Again, maintenance of a constant excess of [Mg]$_T$ over that of [ADP]$_T$, ensures that the proportion of ADP present as MgADP$^-$ is maximized and remains constant, although the constant metal ion excess must be increased from that required for MgATP^{2-} because of the lower affinity of ADP^{3-} for Mg^{2+}. For example, the percentage of total ADP present as MgADP$^-$ remains constant at 95%, over the range 0.1–1 mM [ADP]$_T$, at a constant [Mg]$_T$ excess of 5 mM (Fig. 6A). This ensures a relatively small proportion of [ADP^{3-}] in solution and a constant concentration of Mg^{2+} (inset, Fig. 6A).

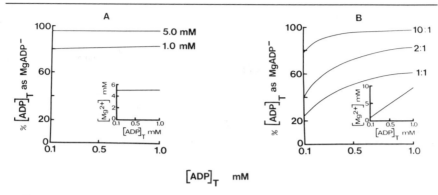

FIG. 6. Comparison of the maintenance of a constant metal to nucleotide ratio with a constant excess of metal ion on the proportion of total ADP present as MgADP$^-$ at pH 8. (A) Concentrations of MgADP$^-$ were calculated for mixtures containing a constant excess (1.0 mM and 5.0 mM) of $[Mg]_T$ over that of $[ADP]_T$. *Inset:* Variation of the concentration of free magnesium in mixtures of ADP and MgCl$_2$ over the range 0.1–1.0 mM $[ADP]_T$ at a constant excess of $[Mg]_T$ (5.0 mM). (B) Concentrations of MgADP$^-$ were calculated for mixtures containing $[Mg]_T:[ADP]_T$ ratios of 1:1, 2:1, and 10:1. *Inset:* Variation of the concentration of free magnesium in mixtures of ADP and MgCl$_2$ over the range 0.1–1.0 mM $[ADP]_T$ at an $[Mg]_T:[ADP]_T$ ratio of 10:1.

Variation of $[ADP]_T$ and $[Mg]_T$ in constant ratio gives poorer results than the comparable approach with ATP. At a $[Mg]_T:[ADP]_T$ ratio of 1:1, the percentage of the total nucleotide present as MgADP$^-$ varies from 23% at 0.1 mM $[ADP]_T$ to 61% at 1 mM $[ADP]_T$ (Fig. 6B). The situation becomes worse at lower nucleotide concentrations; e.g., at 1 μM $[ADP]_T$, MgADP$^-$ represents only 4% of the total nucleotide, even at an $[Mg]_T:[ADP]_T$ ratio as high as 10:1. Further, the concentration of ADP^{3-} and also of Mg^{2+} (inset, Fig. 6B) show a large variation.

Although the above considerations have been applied in particular to nucleoside di- and triphosphates, they also serve as a guide to the maximization of other "active" metal tri- and diphosphate ester species, using approaches similar to those used for MgATP^{2-} and MgADP$^-$, respectively.

Application to Enzyme Kinetic Data: The Creatine Kinase Reaction

As an illustration of the ultimate effect on the magnitude of the various kinetic constants, introduced by the presence of other metal ions and ligands, or by possible "errors" in the evaluation or selection of the relevant metal–substrate stability constant, recent kinetic data obtained in this laboratory for kangaroo muscle creatine kinase have been analyzed.[70]

[70] G. Grossman and W. J. O'Sullivan, unpublished results.

The enzyme catalyzes the reversible transfer of a phosphoryl group from free phosphorylcreatine to MgADP$^-$ to form creatine and MgATP^{2-}. Two sets of data have been treated: first, initial velocity measurements in the direction of MgATP^{2-} formation; and, second, product inhibition studies with MgATP^{2-} as a competitive inhibitor with respect to MgADP$^-$.

A. Initial velocity measurements of the reaction were analyzed with consideration given to (1) MgADP$^-$ as the only relevant species; (2) the effect of the formation of the complexes, Mg-phosphorylcreatine, and KADP^{2-} (37 mM [K$^+$] was introduced in the assay mixture as the base component of 50 mM HEPES–KOH buffer, pH 8.0), and (3) a value for the stability constant of MgADP$^-$ in "error," e.g., half the value reported in Table V. Such an "error" could arise for one or more of the following reasons: (a) choice of a stability constant from the literature that is not applicable to the particular enzyme reaction conditions; (b) disregard for any variation in the magnitude of the constant with changes in temperature or ionic strength; (c) disregard for other ligand species, introduced by changes in pH, e.g., HATP^{3-} or HADP^{2-}, and their subsequent variable affinity for the activating metal ion (Tables IV and V). The relevant kinetic constants for the reaction were evaluated using a modification of the SEQUEN computer program originally developed by Cleland[71] and are summarized in Table VIII.

Comparison of the tabulated values shows that corrections made for the presence of minor species (Mg-phosphorylcreatine and KADP^{2-}), in calculating the concentration of MgADP$^-$, produces only a slight refinement of the kinetic constants. An "error" of 50% in the stability constant for MgADP$^-$, however, does alter the value for K_a by up to 30%, but would not change the overall interpretation of the kinetic mechanism of the reaction.

B. Competitive product inhibition (MgATP^{2-} with respect to MgADP$^-$) measurements of the reaction catalyzed by creatine kinase were similarly analyzed with consideration given to (1) MgADP$^-$ and MgATP^{2-} as the only relevant species; (2) the effect of the complexes Mg–phosphorylcreatine, KADP^{2-}, and KATP^{3-}; and (3) a value for the stability constant of MgATP^{2-} in "error," e.g., half the value reported in Table IV. The kinetic constant for the inhibition (K_{iq}, the dissociation constant for the MgATP–enzyme complex) was evaluated using a modification of the COMP computer program.[71] Assuming MgADP$^-$ and MgATP^{2-} to be the only complex species present, K_{iq} was determined as 2.14 ± 0.17 mM.

Corrections made for the presence of minor complex species produced

---

[71] W. W. Cleland, *Nature (London)* **198**, 463 (1963).

TABLE VIII
EFFECT OF OTHER SPECIES AND VARIATION IN THE MgADP⁻
STABILITY CONSTANT ON KINETIC DATA FOR CREATINE KINASE

Kinetic constant[a]	Metal complexes considered in analysis (apparent stability constant)		
	MgADP⁻ (4000 M^{-1})	MgADP⁻ (4000 M^{-1}) MgPC (40 M^{-1})[b] KADP²⁻ (5 M^{-1})	MgADP⁻ (2000 M^{-1}) MgPC (40 M^{-1})[b] KADP²⁻ (5 M^{-1})
	(mM ± SE)		
K_a	0.119 (± 0.017)	0.10 (± 0.015)	0.075 (± 0.012)
K_{ia}	0.26 (± 0.06)	0.265 (± 0.06)	0.23 (± 0.05)
K_b	2.30 (± 0.30)	2.17 (± 0.275)	2.15 (± 0.27)
K_{ib}	5.0 (± 1.3)	5.8 (± 1.5)	6.6 (± 1.7)

[a] K_a and K_{ia} are the constants for the dissociation of MgADP⁻ from the MgADP-enzyme-phosphorylcreatine and the MgADP-enzyme complexes, respectively; K_b and K_{ib} are the constants for the dissociation of phosphorylcreatine from the MgADP-enzyme–phosphorylcreatine and the enzyme–phosphorylcreatine complexes, respectively.
[b] MgPC, Mg-phosphorylcreatine.

only a slight refinement of the inhibition constant (2.10 ± 0.16 mM). Further, a 50% "error" in the value for the MgATP²⁻ stability constant altered K_{iq} by only approximately 3% (2.04 ± 0.16 mM).

In summary, the greatest accuracy is always desirable in any experimental approach. However, for a simple kinase reaction, errors of up to 50% in the stability constants of metal nucleotide complexes produced significant but not gross errors. The larger the stability constant for the metal–substrate species, the smaller the effect of possible "errors" in the stability constant on the relevant kinetic constants.

Concluding Remarks

A variety of physicochemical techniques are available for the determination of metal–ligand stability constants. Not all are easily accessible to the nonspecialist. It is probable that the most generally useful are spectrophotometric techniques, using mixed ligands, of which the use of 8-hydroxyquinoline as advocated by Burton[17] may be recommended. Of the other techniques described, ion exchange methods are relatively straightforward though pH titration remains a relatively specialized technique. Further, the use of ion selective electrodes has provided a useful adjunct

```
PROGRAM CECCS                  73/74    OPT=1  MANTRAP                    FTN 4.6.1+439                       78/11/15. 12.50.28

         PROGRAM CECCS(INPUT,OUTPUT,PFILE,WFILE,OFILE,SKFILE,TAPE 5=
        CINPUT,TAPE 6=OUTPUT,TAPE 7=PFILE,TAPE 8=WFILE,TAPE 9=OFILE,
        CTAPE 10=SKFILE)
C    *************************************
C    *                                   *
C    *  THIS PROGRAM CARRIES OUT AN ITERATIVE PROCESS TO CALCULATE THE
C    *  FREE CONCS OF COMPONENTS,AND THEN THE CONCNS OF
C    *  COMPLEXES,AT EQUILIBRIUM,IN AN INTERACTING SYSTEM.
C    *                                   *
C    *                                   *
C    *  N=NO. COMPONENTS IN SYSTEM(<=10)
C    *  M=NO. COMPLEXES IN SYSTEM(<=10)
C    *  PH=PHI=TOT.CONCN. OF COMPONENT(S) WERE MEASURED
C    *  AA(I)=TOT.CONCN. OF I TH COMPONENT(MILLIMOLAR)
C    *  SK(J)=STAB. CONST.(1/MILLIMOLAR) OF J TH COMPLEX
C    *  ALPHA(I,J)=WEIGHTING FACTOR,I.E. NO. MOLECULES/IONS OF I TH
C    *  COMPONENT IN J TH COMPLEX(MILLIMOLAR)
C    *  X(J)=CONCN. IN OF J TH COMPLEX(MILLIMOLAR)
C    *  XRECIP(J)=1/X(J) (1/MILLIMOLAR)
C    *  OAA(I)=PREVIOUS VALUE OF AA(I) DURING ITERATIVE PROCESS
C    *  FLAG(I)=TEST PARAMETER FOR CONVERGENCE
C    *************************************
C
C  HERE STARTS THE PROGRAM PROPER
      LOGICAL PROTON
      INTEGER ALPHA(10,10),FLAG(10)
      DIMENSION A(10),AA(10),SK(10),X(10),OAA(10),AHFD(10),AAHED(10),
     CXHED(10),VV(10),N*M,PH,XRECIP(10),XHFD(10)
      READ(5,1)N,M,PH
      IF(N.GT.10.OR.M.GT.10)GO TO 100
      READ(5,2)(SK(J),J=1,M)
      READ(5,3)(ALPHA(I,J),I=1,N),J=1,M)
C
C  IF THE FREE CONCN. OF ONE COMPONENT IN THE SYSTEM IS CONST.
C  (E.G. FREE PROTON CONCN. IN BUFFERED SYSTEM) TYPE IN COLUMNS 1-4
C  THE WORD "TRUE", IF NOT, TYPE IN COLUMNS 1-5 THE WORD "FALSE".
      READ(5,65)PROTON
C
C   READ IN HEADINGS FOR A(I)
      READ(5,4)(AHFD(I),I=1,N)
C
C   READ IN HEADINGS FOR AA(I)
      READ(5,4)(AAHED(I),I=1,N)
C
C   READ IN HEADINGS FOR X(J)
      READ(5,4)(XHFD(I),I=1,M)
C
C   READ IN HEADINGS FOR XRECIP(J)
      READ(5,4)(XPHED(I),I=1,M)
      IF(N.GE.6.OR.M.GE.6)GO TO 30
```

```
PROGRAM CECCS              73/74    OPT=1  MANTRAP                       FTN 4.6.1+439        78/11/15. 12.50.28

      C
      C  WRITE OUT MAIN HEADINGS
 60         WRITE(6,5)PH
            WRITE(6,6)(AHED(I),I=1,N)
            IT=N*12
            WRITE(6,7)IT,(AAHED(T),I=1,N)
 65         WRITE(7,8)PH
            WRITE(7,6)(XHED(J),J=1,M)
            IT=M*12
            WRITE(7,7)ITT,(XPHED(J),J=1,M)
            GO TO 9
      C
      C  WRITE OUT HEADINGS FOR SITUATION WHEN N >= 6 OR M >= 6
 70   30    WRITE(6,31)PH
            WRITE(6,6)(AHED(I),T=1,N)
            WRITE(7,33)PH
 75         WRITE(7,6)(AAHED(I),T=1,N)
            WRITE(8,6)(XHED(J),J=1,M)
            WRITE(9,34)PH
            WRITE(9,6)(XRHED(J),J=1,M)
      C
      C  READ IN VALUES FOR TOT. CONCN. WITH THE FINAL VALUE READ BEING
 80   C  THE COMPONENT WITH CONST. FREE CONCN. .IF PRESENT
      9     READ(5,2)(A(T),I=1,N)
            IF(EOF(5))99,91
      C
      C  SET FIRST VALUE AA(I)=A(T)
 85   91    DO 10 I=1,N
            AA(I)=A(I)
            CAA(I)=AA(T)
 10         CONTINUE
            II=0
 90   C
      C  CALCULATE NEW AA(I) USING EQN(7) IN BIOCHEM. J.,159,1-5(1976)
 95   11    DO 14 I=1,N
            FLAG(I)=0
            VV(I)=0.0
            DO 13 J=1,M
            V(J)=1.0
 100        DO 12 K=1,N
            V(J)=V(J)*AA(K)**ALPHA(K,J)
 12         CONTINUE
            VV(I)=VV(I)+ALPHA(I,J)*SK(J)*V(J)
 13         CONTINUE
 105        IF(I.EQ.N)GO TO 70
            AA(I)=A(I)/(AA(T)+VV(I))
 14         CONTINUE
      C
      C  CALCULATE NEW A(T) FOR COMPONENT WITH CONST. AA(I) USING EQN(6)
            IN BIOCHEM. J.,159,1-5(1976)
 110  70    IF(.PROTON)AA(N)=A(N)+VV(N)
            IF(.NOT.PROTON)AA(N)=A(N)/(AA(N)+VV(N))
```

```
PROGRAM CFCCS          73/74    OPT=1 MANTRAP                  FTN 4.6.1+439                 79/11/15. 12.50.28

      C TEST FOR CONVERGENCE
            ISUM=0
115         DO 15 I=1,N
            IF(ABS(OAA(I)-AA(I)).LE.0.00005)FLAG(I)=1
            ISUM=ISUM+FLAG(I)
120      15 CONTINUE
            IF(ISUM.EQ.N)GO TO 17
            II=II+1
      C
      C WRITE MESSAGE IF NO CONVERGENCE OF AA(I) TO 0.005% OF OAA(I)
125   C IN 500 ITERATIONS
            IF(II.EQ.500)GO TO 22
            DO 16 I=1,N
            OAA(I)=AA(I)
130      16 CONTINUE
            GO TO 11
      C
      C CALCULATE CONCN OF COMPLEXES AND THEIR RECIPROCALS
         17 DO 19 J=1,M
135         SIGMA=1.0
            DO 18 K=1,N
            SIGMA=SIGMA*AA(K)**ALPHA(K,J)
         18 CONTINUE
            X(J)=SK(J)*SIGMA
140         XRECIP(J)=1.0/X(J)
         19 CONTINUE
            IF(N.GE.6.OR.M.GE.6)GO TO 60
      C
      C WRITE OUT VALUES FOR A(T) AND AA(I)
145         WRITE(6,20)(A(I),I=1,N)
            WRITE(6,21)TT,(AA(I),I=1,N)
      C
      C WRITE OUT VALUES FOR X(J) AND XRECIP(J)
            WRITE(7,20)(X(J),J=1,M)
150         WRITE(7,35)TT,(XRECIP(J),J=1,M)
            GO TO 9
      C
      C WRITE OUT VALUES FOR A(T),AA(I),X(J) AND XRECIP(J) WHEN
      C N >= 6 OR M >= 6
155      60 WRITE(6,20)(A(I),T=1,N)
            WRITE(6,20)(AA(I),T=1,N)
            WRITE(8,20)(X(J),J=1,M)
            WRITE(9,62)(XRECIP(J),J=1,M)
            GO TO 9
160      22 WRITE(6,23)
            WRITE(7,23)
            WRITE(8,23)
            WRITE(9,23)
            GO TO 9
165     100 WRITE(6,90)
            GO TO 85
      C
      C WRITE OUT VALUES FOR STAR. CONST(S) USED IN THE ABOVE CALCULATIONS
         99 WRITE(10,40)(XHED(J),J=1,M)
170         WRITE(10,41)(XHED(J),J=1,M)
            WRITE(10,42)(SK(L),L=1,M)
```

STABILITY CONSTANTS FOR METAL–LIGAND COMPLEXES

```
        85 WRITE(10,43)
  175      CONTINUE
           STOP
        C
        C
  180    1 FORMAT(2I3,F5.2)
         2 FORMAT(10F10.4)
         3 FORMAT(40I2)
         4 FORMAT(10A10)
         5 FORMAT(1H1,30X,45HCALCULATED CONCENTRATIONS OF COMPONENTS AT PH,
          CF5.2)
         6 FORMAT(//,10(2X,A10))
  185    7 FORMAT(1H ,T=5X,8(2X,A10))
         8 FORMAT(1H1,30X,44HCALCULATED CONCENTRATIONS OF COMPLEXES AT PH,
          CF5.2)
  20     FORMAT(1X,10(1PE12.5))
         21 FORMAT(1H ,T=5X,8(1PE12.5))
         23 FORMAT(25X,42HPROCESS DID NOT CONVERGE IN 500 ITERATIONS)
  190    31 FORMAT(1H1,30X,40HTOTAL CONCENTRATIONS OF COMPONENTS AT PH,
          CF5.2)
         32 FORMAT(1H1,30X,33HCONCENTRATIONS OF COMPLEXES AT PH,F5.2)
         33 FORMAT(1H1,30X,39HFREE CONCENTRATIONS OF COMPONENTS AT PH,F5.2)
         34 FORMAT(1H1,30X,44HRECIPROCAL CONCENTRATIONS OF COMPLEXES AT PH,
          CF5.2)
  195    35 FORMAT(1H ,T=5X,8(1PE12.3))
         40 FORMAT(1H1,30X,52HSTABILITY CONSTANT(S) USED IN THE ABOVE CALCULAT
          CIONS,//)
         41 FORMAT(6X,8HCOMPLEX ,10(2X,A10))
  200    42 FORMAT(1X,13HSTAB. CONST.:,10(1PE12.3))
         43 FORMAT(1X,14H(1/MILLIMOLAR))
         62 FORMAT(1X,10(1PE12.3))
         65 FORMAT(L5)
         90 FORMAT(1H1,35X,39HTOO MANY COMPONENTS/COMPLEXES IN SYSTEM)
  205    END

              CALCULATED CONCENTRATIONS OF COMPONENTS AT PH 8.00

   ADP-TOT      MG-TOT       PC-TOT       K-TOT        ADP-FREE     MG-FREE      PC-FREE      K-FREE
  2.4000E-01   4.5000E+00   5.0000E-01   3.6900E+01   3.8950E-02   3.2423E-02   2.3814E-01   3.6902E+01
  2.4000E-01   4.5000E+00   1.0000E+00   3.6900E+01   3.9238E-02   1.2329E-01   4.7649E-01   3.6928E+01
  2.4000E-01   4.5000E+00   2.0000E+00   3.6900E+01   3.9811E-02   2.1209E-01   9.5380E-01   3.6927E+01
  2.4000E-01   4.5000E+00   2.0000E+00   3.6900E+01   4.0947E-02   1.1691E+00   1.9105E+00   3.6924E+01

              CALCULATED CONCENTRATIONS OF COMPLEXES AT PH 8.00

   MG-ADP       MG-PC        K-ADP        1/MG-ADP     1/MG-PC      1/K-ADP
  1.9386E-01   1.1852E-02   7.1849E-03   5.1587E+00   8.4371E+01   1.3927E+02
  1.9352E-01   2.3500E-02   7.2380E-03   5.1678E+00   4.2525E+01   1.3827E+02
  1.9284E-01   4.6200E-02   7.3437E-03   5.1856E+00   2.1645E+01   1.3624E+02
  1.9149E-01   8.9352E-02   7.5532E-03   5.2226E+00   1.1186E+01   1.3243E+02

              STABILITY CONSTANT(S) USED IN THE ABOVE CALCULATIONS


      COMPLEX:      MG-ADP       MG-PC        K-ADP

   STAB. CONST.:   4.000E+00    4.000E-02    5.000E-03
   (1/MILLIMOLAR)
```

```
      PROGRAM IONSEL                            73/74    OPT=1  MANTRAP                    FTN 4.6.1+439                    78/11/15.

  1         PROGRAM IONSFL(INPUT,OUTPUT,TAPE 5=INPUT,TAPE 6=OUTPUT)
            DIMENSION COMP(10),COMPT(10),GAMMA(10),TITLE(10),OCOMP(10)
            INTEGER Z(10),FLAG(10)
            READ(5,2)(TITLE(I),I=1,10)
  5         WRITE(6,51)(TITLE(I),I=1,10)
      C     COMP(1) = FREE METAL
      C     COMP(2) = FREE LIGAND
      C     COMP(3) = HYDROXIDE
      C     READ IN VALENCY OF THESE SPECIES
 10         READ(5,3)(Z(I),I=1,4)
      C     READ IN VALUES FOR TOT. METAL AND TOT. LIGAND (MOLAR)
        86  READ(5,1)(COMPT(I),I=1,2)
            IF(EOF(5))99,90
 15      90 READ(5,1)AM
      C     READ IN VALUE FOR ACTIVITY OF METAL ION
      C     READ IN PH
            READ(5,4)PH
            POH=14.-PH
 20         COMP(4)=10.**(-POH)
            WRITE(6,52)COMPT(1)
            WRITE(6,53)COMPT(2)
            WRITE(6,54)PH
            DO 11 I=1,3
 25      11 COMP(I)=0.
            FLAG(I)=0
            GAMMA(1)=1.
         10 COMP(1)=AM/GAMMA(1)
            COMP(2)=COMPT(1)-COMP(1)
 30         COMP(3)=COMPT(2)-COMP(2)
            SIGMA=0.
         20 MU=SIGMA/2.
            SIGMA=SIGMA+COMP(I)*Z(I)**2
            DO 30 I=1,3
 35      30 GAMMA(I)=10.**(-0.512*Z(I)**2*(SQRT(MU)/(1.+SQRT(MU))-0.3*MU))
            WRITE(6,56)MU,COMP(1),COMP(2),COMP(3),COMP(2)
            ISUM=0
            DO 40 I=1,3
         40 ISUM=ISUM+FLAG(I)
 40         IF(ABS(OCOMP(I)-COMP(I)).LE.1.E-12)FLAG(I)=1
            IF(ISUM.EQ.3)GO TO 100
            III=II+1
            IF(III.E0.100)GO TO 99
            DO 45 I=1,3
 45      45 OCOMP(I)=COMP(I)
            GO TO 10
        100 SK=(COMP(2)/(COMP(1)*COMP(3)))
            SK0=SK*(GAMMA(2)/(GAMMA(1)*GAMMA(3)))
            WRITE(6,80)SK*MU
 50         WRITE(6,81)SK0
            GO TO 86
         99 CONTINUE
            STOP
```

```
PROGRAM IONSEL       73/74      OPT=1 MANTRAP              FTN 4.6.1439         78/11/15.

   1 FORMAT(2E15.6)
   2 FORMAT(10A6)
   3 FORMAT(4I2)
60  4 FORMAT(1H1,25X,10A6)
   51 FORMAT(1H1,25X,12HTOTAL METAL:,1PE15.6,6H MOLAR/)
   52 FORMAT(9X,13HTOTAL LIGAND:,1PE15.6,6H MOLAR/)
   53 FORMAT(10X,3HPH:,F11.6//)
   54 FORMAT(19X,17HIONIC STRENGTH(M),5X,13HFREE METAL(M),5X,
65 C14HFREE LIGAND(M),5X,1PE15.6,3X1PE15.6,4X,15HMETAL-LIGAND(M)/)
   56 FORMAT(/,1PE15.6,3X1PE15.6,4X,1PE15.6,5X,1PE15.6)
   80 FORMAT(/,30X,15HSTAB. CONST. =,F10.2,8H 1/MOLAR)
   81 FORMAT(/,30X,15HSTAR. CONST. =,F10.2,8H 1/MOLAR,
70 C18H AT IONIC STRENGTH)
   C21H AT INFINITE DILUTION)
   END

     STAB. CONST. FOR NA-ATP USING CORNING GLASS ELECTRODE

TOTAL METAL:   2.024000E-03 MOLAR
TOTAL LIGAND:  4.500000E-04 MOLAR
       PH:     9.292000

IONIC STRENGTH(M)    FREE METAL(M)      FREE LIGAND(M)      METAL-LIGAND(M)

3.773794E-03         1.796000E-03       2.300000E-04        2.280000E-04
4.270818E-03         1.920256E-03       3.542594E-04        1.037441E-04
4.309200E-03         1.927781E-03       3.617814E-04        6.218640E-05
4.308679E-03         1.927721E-03       3.622121E-04        5.778850E-05
4.302671E-03         1.922447E-03       3.622461E-04        5.775320E-05
4.302781E-03         1.922448E-03       3.622483E-04        5.775170E-05
4.302788E-03         1.922448E-03       3.622484E-04        5.751620E-05
4.302788E-03         1.922484E-03       3.622484E-04        5.751610E-05
4.302788E-03         1.922484E-03       3.622484E-04        5.751610E-05

          STAB. CONST. =   137.08 1/MOLAR AT IONIC STRENGTH  4.3028E-03 MOLAR
          STAR. CONST. =   242.01 1/MOLAR AT INFINITE DILUTION
```

to the available techniques used for the evaluation of metal–ligand stability constants but should be used with caution because of their restriction to very low ionic strength and the difficulties encountered in extrapolation to conditions of higher ionic strength.

There is now available in the literature a great deal of information on various metal–nucleotide complexes. Most enzymologists should be able to select values relevant to their desired experimental conditions provided attention is paid to variation in such parameters as pH, temperature, ionic strength, and the nature of the supporting medium. The recommended stability constants for the adenosine nucleotides and several other phosphorylated compounds reported in Tables IV–VI provide a basis from which extrapolations for other similar ligands can be made.

Finally, an attempt has been made to assess the errors in final kinetic parameters that might arise from the use of inappropriate stability constants. A moderate degree of "error" in a single stability constant is unlikely to produce gross errors in the magnitude of the kinetic constants. However, substantial misinterpretation could arise if the concentrations of potential inhibitory species were significantly altered.

Appendix (pp. 300–335)

A. Listing and sample output of the FORTRAN IV computer program used in the calculation of the equilibrium concentration of free and complex species present in an interacting system.

B. Listing and sample output of the FORTRAN IV computer program used in the analysis of cation specific electrode measurements for the interaction of monovalent ions with charged ligands (e.g., ATP^{4-}).

Further details of the use of the programs can be obtained from the authors.

Acknowledgments

We are much indebted to Dr. H. O. Spivey for sending us details of his work on PRPP prior to publication. We are also grateful for support from the National Health and Medical Research Council of Australia and the Australian Research Grants Committee.

[13] Cryoenzymology: The Study of Enzyme Catalysis at Subzero Temperatures[1]

By ANTHONY L. FINK and MICHAEL A. GEEVES

For the purposes of this chapter we will define cryoenzymology as the study of enzymes and enzyme catalysis at subzero temperatures in fluid

---

[1] Supported in part by grants from the National Science Foundation and the National Institutes of Health. M. A. G. was the recipient of a travel grant from the Wellcome Foundation.

solvents. Since recent reviews[2-7] have dealt with the underlying theory, applications, and results, we will emphasize the practical methodology of the technique, especially as it relates to kinetic studies of enzyme catalysis. We will begin with a brief overview of the advantages and limitations of cryoenzymology, then discuss the general approach used in its application to enzyme mechanistic studies. The remainder of the chapter will be taken up with descriptions of the procedures and equipment used.

In order to obtain a detailed quantitative and qualitative understanding of the factors responsible for the efficiency of enzyme catalysis, it is necessary to have a detailed knowledge of all the intermediates and transition states along the productive catalytic pathway. The short lifetimes, and often low concentrations, of intermediate species present during catalysis mean that conventional techniques and approaches for obtaining structural and kinetic information are often unsuitable when applied to the dynamic processes of enzyme catalysis. One potential means of circumventing these inherent problems is to carry out the reaction at a sufficiently low temperature so that the lifetimes of individual enzyme–substrate intermediates will be long enough for structural as well as kinetic data to be obtained. In order to prevent rate-limiting enzyme–substrate diffusion, which would occur in a frozen medium, it is necessary to use cryosolvents that will remain fluid to the requisite low temperatures. Thus the basis of cryoenzymology is the initiation of the enzyme-catalyzed reaction by mixing enzyme and substrate at very low temperatures in fluid cryosolvents.

If the low-temperature findings are to be of relevance to those under normal conditions, one must first demonstrate that neither the cosolvent nor the subzero temperatures have any adverse effects on the structural or catalytic properties of the enzyme, and, second, the correspondence (particularly kinetic) between the reaction at low temperatures with that under normal conditions. Thus cryoenzymological studies of a particular enzyme can be classified into two major categories, those concerned with determining whether the combination of cosolvent and low temperature has any adverse effect on the catalytic and structural properties of the enzyme, and those connected with studies of the catalytic mechanism itself. The latter will usually take the form of the detection, accumulation, and stabilization of discrete intermediates, their characterization with respect to kinetics, thermodynamics, and structure, and demonstration that the

[2] A. L. Fink, *Acc. Chem. Res.* **10**, 233 (1977).
[3] P. Douzou, "Cryobiochemistry." Academic Press, New York, 1977.
[4] M. W. Makinen and A. L. Fink, *Annu. Rev. Biophys. Bioeng.* **6**, 301 (1977).
[5] P. Douzou, *Adv. Enzymol.* **45**, 157 (1977).
[6] A. L. Fink, *J. Theor. Biol.* **61**, 419 (1976).
[7] P. Douzou, *Methods Biochem. Anal.* **22**, 401 (1974).

TABLE I
SOME ENZYMES AND RELATED PROTEINS STUDIED BY CRYOENZYMOLOGY

Enzyme	Cosolvent[a]	Enzyme	Cosolvent[a]
Chymotrypsin[b-d]	MeOH, DMSO	Glyceraldehye-3-phosphate	MeOH/EtGly
Trypsin[e,f]	DMSO	dehydrogenase[u]	
Papain[g]	DMSO, EtOH	Dihydrofolate reductase[v]	MeOH, EtOH
Elastase[h]	MeOH, DMSO	Bacterial luciferase[w,x]	EtGly
Subtilisin[i]	DMSO, MeOH	D-Amino acid oxidase[y]	EtGly
Carboxypeptidase A[j]	MeOH, EtGly	Glucose oxidase[z]	EtGly-MeOH
β-Glucosidase[k,l]	DMSO	Cytochrome oxidase[aa]	EtGly
β-Galactosidase[m,n]	DMSO, MeOH	Xanthine oxidase[bb]	EtGly
Lysozyme[o,p]	MeOH, DMSO	Peroxidase[cc,dd]	MeOH, Gly, EtGly, DMF
Ribonuclease A[q]	MeOH, EtOH		
Alkaline phosphatase[r]	MeOH, Gly	Catalase[ee]	Gly
Myosin Sl[s]	EtGly	Hemoglobin[ff]	MeOH, EtGly
Alcohol dehydrogenase[t]	DMSO, DMF	Cytochrome P-450[gg,hh]	Gly, EtGly
		Lactoglobin[ii]	

[a] Abbreviations: MeOH, methanol; EtGly, ethylene glycol; EtOH, ethanol; Gly, glycerol.
[b] B. Bielski and S. Freed, *Biochim. Biophys. Acta* **89**, 316 (1964).
[c] A. L. Fink, *Biochemistry* **12**, 1736 (1973).
[d] A. L. Fink, *Biochemistry* **15**, 1580 (1976).
[e] A. L. Fink, *J. Biol. Chem.* **249**, 5027 (1974).
[f] P. Maurel, G. Hui Bon Hoa, and P. Douzou, *J. Biol. Chem.* **250**, 1376 (1975).
[g] A. L. Fink and K. J. Angelides, *Biochemistry* **15**, 5287 (1976).
[h] A. L. Fink and A. I. Ahmed, *Nature (London)* **236**, 294 (1976).
[i] A. L. Fink and A. Tsai, unpublished results.
[j] M. W. Makinen, K. Yamamura, and E. T. Kaiser, *Proc. Natl. Acad. Sci. U.S.A.* **73**, 3882 (1976).
[k] A. L. Fink and N. Good, *Biochem. Biophys. Res. Commun.* **58**, 126 (1974).
[l] A. L. Fink and J. W. Weber, personal communication.
[m] A. L. Fink and K. J. Angelides, *Biochem. Biophys. Res. Commun.* **64**, 701 (1975).
[n] A. L. Fink and K. Magnusdottir, personal communication.
[o] P. Douzou, G. Hui Bon Hoa, and G. Petsko, *J. Mol. Biol.* **96**, 367 (1975).
[p] A. L. Fink, R. Homer, and J. Weber, unpublished results.
[q] A. L. Fink and B. Grey, in "Biomolecular Structure and Function" (P. F. Agris, R. N. Loeppky, and B. Sykes, eds.), p. 471. Academic Press, New York, 1978.
[r] V. P. Maier, A. L. Tappel, and D. H. Volman, *J. Am. Chem. Soc.* **77**, 1278 (1955).
[s] D. Trenthan, *Biochem. Soc. Trans.* **5**, 5 (1977).
[t] M. A. Geeves and A. L. Fink, unpublished results.
[u] A. L. Fink, M. A. Geeves, and A. MacGibbon, unpublished results.
[v] A. L. Fink, M. E. Russell, and A. Howard, unpublished results.
[w] J. W. Hastings, C. Balney, C. Le Peuch, and P. Douzou, *Proc. Natl. Acad. Sci. U.S.A.* **70**, 3468 (1973).
[x] C. Balny and J. W. Hastings, *Biochemistry* **14**, 4719 (1975).
[y] T. Shiga, M. Layani, and P. Douzou, in "Flavins and Flavoproteins" (K. Yagi, ed.), p. 140. Univ. of Tokyo Press, Tokyo, 1968).
[z] A. L. Fink and A. I. Ahmed, unpublished results.
[aa] B. Chance, N. Graham and V. Legallais, *Anal. Biochem.* **67**, 552 (1975).

detected intermediate is on, or at least consistent with being on, the productive catalytic pathway.

The major advantages of cryoenzymology over more traditional approaches include the following: the most specific substrate(s) may be used, intermediates may be trapped in high concentrations (much higher than in freeze quenching-type experiments), detailed kinetic and thermodynamic information about the elementary steps in the reaction can be readily obtained, and high-resolution structural information about individual intermediates can be acquired. Essentially all the standard chemical–physical–biophysical techniques normally used in the study of proteins and enzymes may also be used at subzero temperatures. For example, the long lifetimes of trapped intermediates at appropriately low temperatures permit the use of techniques such as column chromatography, X-ray diffraction, and nuclear magnetic resonance. Furthermore the great reduction in reaction rate means that processes occurring on a time-scale of microseconds under normal conditions will happen on a time-scale of seconds at appropriately low temperature. Thus reactions normally detectable only by relaxation techniques may become detectable by rapid mixing or even by conventional techniques.

Limitations to the cryoenzymology approach include technical problems due to the low temperatures involved, the necessity of using aqueous–organic solvents (or alternative freezing-point depressants), the possibility that the potential energy surface may be different at the low temperature, and hence the reaction pathway might be different, the fact that intermediates can be accumulated only if they break down more slowly than they are formed, and possible difficulties in ascertaining that intermediates detected at low temperatures are indeed on the productive catalytic pathway under normal conditions.

In spite of these limitations, the method has great potential as demonstrated in a variety of results. In recent years the procedure has been applied to at least thirty enzymes (Table I) and in each case a suitable cryosolvent has been found. The variety of classes of enzymes and macromolecular complexes which have been successfully studied in this manner suggests the general applicability of the technique to both monomeric and

---

[bb] R. Sireix, J. Canva, and P. Douzou, *C. R. Acad. Sci. Ser.* **270**, 557 (1970).
[cc] V. P. Maier and A. L. Tappel, *Anal. Chem.* **26**, 564 (1954).
[dd] P. Douzou, R. Sireix, and F. Travers, *Proc. Natl. Acad. Sci. U.S.A.* **66**, 787 (1970).
[ee] G. K. Strother and E. Ackerman, *Biochim. Biophys. Acta* **47**, 317 (1961).
[ff] R., Banerjee, P. Douzou, and A. Lombard, *Nature (London)* **217**, 23 (1968).
[gg] P. Debey and P. Douzou, *FEBS Lett.* **39**, 271 (1974).
[hh] P. Debey, G. Hui Bon Hoa, and P. Douzou, *FEBS Lett.* **32**, 227 (1973).
[ii] S. Guinard, G. Hui Bon Hoa, and C. Pantaloni, *Biochimie* **56**, 863 (1974).

oligomeric enzymes; the list includes proteases, glycosidases, phosphatases, dehydrogenases, oxidases, reductases, luciferases, cytochromes, microsomal enzymes, and photosynthetic proteins.

Among the more notable results of cryoenzymological studies are the following: high-resolution X-ray diffraction studies on trapped crystalline intermediates leading to detailed structures of enzyme–substrate intermediates,[8,9] the detection and accumulation of tetrahedral intermediates in protease catalysis,[10] the detection of previously undetected intermediates,[11,12] the isolation of productive enzyme–substrate intermediates,[13,14] and the demonstration that the kinetic and spectral properties of an intermediate detected for the first time at subzero temperatures correlate very well with those obtained in rapid-mixing studies under normal conditions.[10]

Theory

The underlying basis of cryoenzymology is, of course, the Arrhenius expression, $k = Ae^{-E_a/RT}$, and the fact that elementary steps in an enzyme-catalyzed reaction have energies of activation usually in 6–20 kcal mol^{-1} range (excepting the initial binding step). Furthermore different elementary steps are likely to, and in fact do,[10] have different energies of activation. Thus, with exception of the initial bimolecular binding step, rate reductions of the order of 10^4 to 10^{10} may be obtained at temperatures of the order of $-100°$. In addition to slowing down intermediate transformations, the low temperatures would be expected to have some effect on the structure of the protein. The fact that the dielectric constant of a solution increases with decreasing temperature means that as the temperature is reduced polar interactions (including hydrogen bonds) will be strengthened, and correspondingly the nonpolar, hydrophobic, interactions will be weakened.[6] It appears in fact that the net effect of the *combination* of cosolvent and low temperature is in *many* cases minor, judging by the *demonstrated* stability of many proteins under these conditions.

The general approach used is shown in Table II. The first five require-

[8] A. L. Fink and A. I. Ahmed, *Nature (London)* **263**, 294 (1976).
[9] T. Alber, G. A. Petsko, and D. Tsernoglou, *Nature (London)* **263**, 297 (1976).
[10] K. J. Angelides and A. L. Fink, *Biochemistry* **18**, 2363 (1979).
[11] A. L. Fink and N. Good, *Biochem. Biophys. Res. Commun.* **58**, 126 (1974).
[12] M. W. Makinen, K. Yamamura, and E. T. Kaiser, *Proc. Natl. Acad. Sci. U. S. A.* **73**, 3882 (1976).
[13] J. W. Hastings, C. Balny, C. Le Peuch, and P. Douzou, *Proc. Natl. Acad. Sci. U. S. A.* **70**, 3468 (1973).
[14] A. L. Fink, *Arch. Biochem. Biophys.* **155**, 473 (1973).

TABLE II
PROTOCOL FOR CRYOENZYMOLOGICAL STUDIES

1. Preliminary tests to identify possible cryosolvent(s)
2. Determination of the effect of cosolvent on the catalytic properties
3. Determination of the effect of cosolvent on the structural properties
4. Determination of the effect of subzero temperature on the catalytic properties
5. Determination of the effect of subzero temperature on the structural properties
6. Detection of intermediates by initiating catalytic reaction at subzero temperature
7. Kinetic, thermodynamic, and spectral characterization of detected intermediates
8. Correlation of low-temperature findings with those under "normal" conditions
9. Structural studies on trapped intermediates

ments are necessary to demonstrate whether or not the cryosolvent is devoid of deleterious effects on the enzyme. The remainder are connected with mechanistic investigations. Until it has been adequately demonstrated that the cryosolvent does not have any adverse effects on the enzyme, any findings of a mechanistic nature must be considered questionable in terms of their relevance to the mechanism under normal conditions. Each of the steps taken in determining the choice and suitability of a cryosolvent will now be considered in detail.

Choice of Cryosolvent

Although several means of depressing the freezing point of aqueous solutions exist, e.g., high salt concentration, and supercooling by suspending microdroplets in a nonmiscible organic solvent,[15] the most suitable method for cryoenzymology studies is the use of mixed aqueous–organic solvent systems. These cryosolvents will usually have 50–80% by volume of the organic cosolvent for the desired temperature range. Because of their low viscosities, methanol-based cryosolvents are most desirable. Unfortunately, many proteins seem to be very sensitive to methanol, and alternative solvents are necessary. The ones that seem to be of most general applicability are those based on dimethyl sulfoxide, dimethyl formamide, ethylene glycol, ethylene glycol–methanol, and ethanol. Details concerning cryosolvent properties are given in a subsequent section. The initial test of a possible cryosolvent is carried out in the fol-

[15] P. Douzou, P. Debey, and F. Franks, *Nature (London)* **268**, 466 (1977).

lowing way. A suitable assay method is chosen to monitor catalytic activity. If possible this should be a direct assay in which the conversion of substrate to product is monitored continuously. Spectrophotometric assays using specific substrates are best. The advantages of a continuous monitoring assay is that the reaction can be followed to completion, thus yielding additional data, such as the occurrence of gradual loss of activity, or change in K_m. The assay is preferably carried out under pseudo-first-order conditions, i.e., below substrate saturation, since the cosolvents usually cause an increase in K_m (see below). The assay should be carried out at 25° in aqueous solution as a control, at an appropriate pH, then under similar conditions except at 1° or 2°.

The assay is then repeated at 1° or 2° using a 25% (v/v) cryosolvent solution (preparation of cryosolvents is considered subsequently). It is convenient to screen two or three cosolvents at the same time. The shape of the reaction curve, and rate of the reaction, in the cryosolvent should be compared to that of the reaction in aqueous solution under similar conditions. A first-order semilog plot of the data should be made and compared to that for the corresponding aqueous reaction. Contraindications of the cosolvent include deviations from first-order kinetics, lack of reaction, and incomplete reaction, i.e., a first-order reaction is observed, but the amount of product formed (or substrate used up) does not correspond to the total. Either identical kinetics, or a slower first-order reaction involving total conversion of substrate to product are good signs. In most cases a decreased rate relative to the aqueous control will be observed, since we have found that in almost all cases the cosolvent exerts a negative effect on substrate binding (probably due to a hydrophobic partitioning effect on the substrate).[16,17,17a] This effect on substrate or inhibitor binding results in an exponential increase in K_m (or exponential decrease in k_{cat}/K_m), thus plots of log K_m (or log k_{cat}/K_m) vs cosolvent concentration will be linear. If the cryosolvent system looks good at 25% cosolvent, the assay at 1° can be repeated using 50% cosolvent. If, based on comparison with the aqueous assay, the system looks good at this concentration also, then 70% cosolvent (65% for dimethyl sulfoxide)[18] can be tried.

If, at 1°, the particular cosolvent appears suitable at 50%, let us say, but causes inactivation at 70%, there are two options available. One can try intervening concentrations of the cosolvent at 1° and find the max-

---

[16] A. L. Fink, *Biochemistry* **12**, 1736 (1973).
[17] A. L. Fink and K. J. Angelides, *Biochemistry* **15**, 5287 (1976).
[17a] P. Maurel, *J. Biol. Chem.* **253**, 1677 (1978).
[18] The lowest temperature for liquid solution on the phase diagram for aqueous DMSO mixtures corresponds to 65% by volume.

imum concentration that appears satisfactory for these conditions; and/or one can repeat the assays at lower temperatures. For example, the assay can be repeated in 50% cosolvent at $-15°$, and then tried at this temperature using 60% cosolvent. The basis of this lowering of the temperature is that the native ⇌ denatured transition shifts to lower temperatures as the cosolvent concentration increases.[19] Except for DMSO[20] (65%) the maximum cosolvent concentration used is usually 80%.

On the basis of this simple approach, it is usually easy to choose a potentially suitable cryosolvent rather quickly. At this stage of our understanding of protein–solvent interactions, the process is still an empirical one, involving trial-and-error experiments of the sort just considered. There are a number of factors to keep in mind in experiments to find a suitable cryosolvent. If the enzyme requires a coenzyme it is best to add the coenzyme to the aqueous stock enzyme solution prior to addition to the cryosolvent, as the presence of the coenzyme often stabilizes the enzyme structure toward the cosolvent. Not only does the T_m for the thermal denaturation of the protein decrease with increasing cosolvent concentration, but so too does the solubility of the enzyme. In addition, the solubility is very temperature dependent. A more detailed discussion of the effects on protein solubility has been given by Douzou.[7] The combination of high cosolvent concentration and low temperature thus often results in quite limited protein solubility. This may be cause for investigating several possible cryosolvents, since marked variations in solubility occur with different cryosolvents. In cases of limited solubility of the enzyme it is most important to distinguish between the possible spectral changes brought about by aggregation, as contrasted to those due to the catalytic reaction. Since aggregation involves a light-scattering effect, it is possible to monitor it in a spectral region where neither enzyme nor substrate absorb. Thus, if a control experiment indicates that spectral changes occur in a region where none of the components absorb it is an indication of an artifact, most commonly aggregation (which often appears as a first-order kinetic process). If the enzyme under consideration is oligomeric, it is necessary also to be concerned with its state of association. From the above experiments it should be possible to decide on a likely cryosolvent.

---

[19] A. L. Fink and B. L. Grey, *in* M. W. Makinen and A. L. Fink, *Annu. Rev. Biophys. Bioeng.* **6**, 301 (1977).

[20] Abbreviations used: DMSO, dimethyl sulfoxide; DMF, dimethyl formamide; pH*, the apparent pH in the aqueous–organic solvent; NMR, nuclear magnetic resonance; ESR, electron spin resonance; UV, ultraviolet.

Enzyme Stability in the Cryosolvent

The stability of the enzyme in the chosen cryosolvent can then be checked at various temperatures and pH* values as follows. The enzyme (with coenzyme if necessary) is incubated in the desired cryosolvent at 0° (or lower if necessary). Aliquots are assayed periodically both at 0° in the cryosolvent and at 25° in aqueous solution, suggested time intervals for this being 0.25, 0.5, 1, 2, 5, 12, and 24 hr. It is useful to map out the pH* limits of stability in this way by carrying out the incubations at a variety of pH* values. In this manner both irreversible and reversible inactivation may be determined. The effect of various ligands and the enzyme concentration on the stability of the protein can be ascertained in this fashion.

It is also very useful to determine the position of the native ⇌ denatured transition in the cryosolvent. Since most proteins show a change in their UV spectral properties on denaturation, this can usually be done as follows[17]: the enzyme is introduced to the cryosolvent at a temperature at least 20° below that at which it is known from the above experiments to be stable (e.g., −20° or below) and the UV absorbance, fluorescence, or circular dichroism signal at a suitable wavelength is monitored as the temperature is slowly raised (in practice one usually raises the temperature 1–5° at a time, depending on whether one is below or in the transition zone, and waits until no further changes in the signal occurs before raising the temperature again). The rate of denaturation at subzero temperatures is often slow owing to the high energies of activation, so these experiments take considerable time. Plots of the incremental signal changes vs temperature will reveal the region over which the denaturation occurs.

Cosolvent Effects on Catalysis

The following experiments should be carried out at 0° (or lower if the enzyme is not stable in the cryosolvent at 0°). The effects of the cosolvents on the catalytic properties of the enzyme can be ascertained by determining the effects on the kinetic parameters as a function of cosolvent concentration. If the values of K_m, and the substrate solubility, permit, the values of the turnover parameters k_{cat} and K_m should be measured at 10% or 15% increments of cosolvent concentration, to the maximum.[16,17] If it is not possible to determine both k_{cat} and K_m separately, their ratio should be determined from pseudo-first-order conditions. If the reaction under study is a two-substrate one, these experiments can be carried out with excess concentrations of each substrate successively. Plots of k_{cat}, K_m, and log K_m against cosolvent concentration can be used to judge whether the cosolvent has any untoward effects on the catalysis. For example, if the reaction is a hydrolysis a linear decrease in k_{cat} with decreasing water concentration is to be expected.[16] The previously mentioned hydrophobic partitioning effect on substrate binding causes plots of K_m to be nonlinear with cosolvent concentration, whereas plots of log K_m or log (k_{cat}/K_m) will be linear.[16,17,17a] Deviations from linearity suggest that an undesirable process is occurring. Confirmation that the exponential effect of cosolvent on K_m results from the binding step may be obtained by determining K_i for a competitive inhibitor under the same conditions.[21] In some cases the cosolvent itself may be a competitive inhibitor. This will result in curvature superimposed on the log K_m plots. Depending on the magnitude of the inhibition constant, the curvature may or may not be readily visible. If such curvature is obtained with a particular cosolvent, then comparison with a cosolvent that is unlikely to be an inhibitor is in order.

An additional complication arises in the case of hydrolytic enzymes in alcohol-based cryosolvents, namely, that of transferase activity, in which the alcohol rather than water acts as the acceptor. In such cases the interpretation of the effect of the cosolvent on the catalytic parameters is somewhat more complicated if the rate-determining step involves the attack of water. If there is no specific binding site for the alcohol, then a linear relationship between cosolvent concentration and the increased rate (k_{cat}) should result, since the alcohols are usually more effective attacking agents because of their better nucleophilicity. The increased rate brought about by the alcohol, however, can also result in a change in the rate-determining step. Problems of this type may be circumvented by using a different substrate, in which the rate-limiting step precedes the involvement of water.

If the data from the above experiments look satisfactory, one can then examine the pH*-rate profile for the catalytic reaction in the cryosolvent, again at 0° (or below if necessary). Unless the substrate is a polyelectrolyte (in which case a large change in the profile may occur) similar pH dependence to that observed in aqueous solution should be observed if no adverse effects arise from the cosolvent. This statement is based on empirical observation, rather than theoretical prediction. One would, in fact, expect that the cosolvent presence would result in significant perturbation of the pK's of ionizable groups in the enzyme, owing to the lower polarity of the cosolvent compared to water, as is observed with the weak acids and bases used as buffers (see next section). A satisfactory explanation as to why the enzyme pK's are not substantially perturbed is not yet available. However, it is possible, in fact probable, that the hydration

[21] A. L. Fink, *Biochemistry* **13**, 277 (1974).

shell around the protein is not significantly disturbed by the presence of the cosolvent, and that the environment about the key ionizable catalytic groups may be quite similar in the cryosolvent to that in aqueous solution, resulting in similar proton activities in the vicinity of the active-site groups. A contributing factor may also be that many of the cryosolvents used have dielectric constants not too far from 80 even at temperatures as high as 0° (see section on cryosolvent properties). The most common ionizable groups found in active sites have heats of ionization in the 0 to 10 kcal mol^{-1} range, so that appreciable shifts in the observed pK's may arise owing to temperature effects also.

If the reaction involves a coenzyme, or two substrates, the effects of the cosolvent can be determined separately for the coenzyme. For example, binding of coenzyme usually results in spectral perturbations, from which one can measure both the kinetics and (binding) association constant as a function of the cosolvent concentration. Furthermore, if some of the elementary reactions in the overall catalysis can be readily monitored (e.g., acylation/deacylation; glycosylation/deglycosylation), it may be desirable to determine the effects of the cosolvent individually on these steps.

At this point one should have amassed enough data to have a pretty clear picture as to whether the cryosolvent has any adverse effects on the catalysis. Further confirmation can and should be obtained by examination of the effects of the cosolvent on the structural properties of the enzyme.

Effects of Cosolvent on Structure

In general the effect of the cosolvent on the structure of the protein can be determined using the intrinsic spectral properties of the enzyme. The underlying basis is that linear or smooth monotonic changes in the spectral property, as a function of cosolvent concentration, reflect solvent effects on exposed side-chain residues, whereas sharp breaks in such plots indicate structural perturbations.[16,17] The particular spectral properties found most useful in this regard are the UV absorbance, fluorescence, and circular dichroism spectra. Some cosolvents, e.g., DMSO, themselves absorb in the UV region and may preclude investigations of the peptide-backbone contributions. Fluorescence energy-transfer experiments, e.g., exciting at a wavelength where Tyr absorbs, but monitoring Trp emission, are particularly useful in detecting small structural changes (Fig. 1). For enzymes containing visible chromophores, e.g., flavin enzymes, the visible absorbance spectrum may also be used.

For relatively small, soluble enzymes proton magnetic resonance pro-

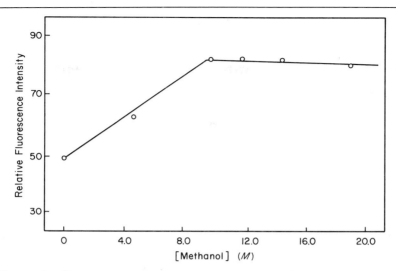

FIG. 1. The effect of methanol on the intrinsic fluorescence emission of papain, as monitored using Tyr → Trp energy transfer. Conditions were 0°, pH* 7.0. Excitation was a 260 nm, emission at 330 nm (λ_{max} for Trp emission). The sharp break at 8.4 M methanol signifies a structural rearrangement. Unpublished data of A. L. Fink and K. J. Angelides.

vides an excellent means of detecting cosolvent-induced structural changes. The NMR spectrum of the enzyme in D_2O at 1° can be compared with that of the enzyme in the deuterated cryosolvent under otherwise similar conditions. Laser Raman could be used in a similar fashion for particularly soluble enzymes.

For oligomeric enzymes the state of association in the cryosolvent may be ascertained using exclusion chromatography.[22] The choice of packing is optional to some extent; however, a noncompressible type of material is essential. Porous glass, e.g., CPG-glycophase, is an excellent packing, with the exception that some proteins are rather strongly adsorbed to it. The minimal additional equipment necessitated by the low temperatures is a means of thermostatting the column (e.g., a jacket or cold chamber). The column is basically set up in the normal fashion for molecular weight determinations. Using UV absorbance (or other appropriate means) to monitor the eluent, the enzyme is run through the column initially in aqueous solution at 25°, then at 1°, then in the cryosolvent at 1°, and whichever other subzero temperatures are deemed necessary. If subunit dissociation does occur, it may be necessary to run the column at several different temperatures to cover the extremes of complete associa-

[22] A. L. Fink and J. Weber, unpublished results.

tion to complete dissociation. In some cases it may be feasible to use ultrafiltration or ultracentrifugation for such purposes.

Effects of Low Temperature on the Structure

As mentioned previously the transition involving the thermal denaturation of proteins occurs at lower temperatures in the presence of organic cosolvents. Interestingly, in most of the cases examined so far in which a suitable cryosolvent has been found the midpoint of the transition occurs in the vicinity of 0°. This is the reason for carrying out the investigations of cryosolvent at 0° or below. In order to determine whether subzero temperatures have any effects on the structure of the enzyme, one again can make use of the intrinsic spectral properties of the protein. For example, temperature-induced conformational changes will usually be detectable by deviations from linearity or smooth curvature in plots of the spectral signal against temperature.[23] Well-suited techniques include fluorescence, UV absorbance, and circular dichroism. NMR may also be used in this manner.

Effects of Low Temperature on the Catalysis

Undoubtedly the catalytic properties of the enzyme are the most sensitive probes of structural perturbations. Thus, if an Arrhenius plot is made of the rate of the turnover reaction to as low a temperature as is feasible, it may provide several useful pieces of information. The value of the energy of activation should compare favorably to that in the absence of the cosolvent. The value of the turnover rate extrapolated to 25°, and corrected for the cosolvent effect on K_m if necessary, should be in good agreement with that measured directly under such conditions. Deviations from linearity in the Arrhenius plot are most likely due to changes in the rate-determining step or a structural change in the enzyme.[24] If a structural change in the enzyme is responsible for the nonlinearity of the Arrhenius plot, it should be possible to document it by the above-mentioned procedures.

A very useful measure of the catalytic competency of the enzyme at low temperature in the cryosolvent may be obtained for those enzymes for which active-site titration procedures can be applied.[17] For example catalysis by many hydrolases involves the intermediacy of a covalent intermediate such as an acyl enzyme or glycosyl enzyme. In these cases it

[23] A. L. Fink, *J. Biol. Chem.* **249**, 5027 (1974).
[24] A. L. Fink, R. Homer, and J. Weber, unpublished observations.

is often possible to measure a "burst" of chromophoric product liberated in a 1:1 stoichiometry with the enzyme, corresponding to the buildup of the covalent intermediate followed by its slow breakdown.[11,17] This procedure will work only under conditions where the breakdown of the intermediate is the rate-limiting step.

Although the above experiments are clearly very time consuming, they are necessary and essential if one wants to have confidence that the combination of low temperatures and organic cosolvent do not cause any undesirable effects on the enzyme. In general the conclusions to be drawn from such studies are quite straightforward: either the cryosolvent is suitable for studying the catalytic mechanism or it is not. An overview of the application of these principles to particular enzymes may be obtained by consulting appropriate articles.[6,16,17,21-23,25]

Detection of Intermediates

The underlying premise in experiments aimed at detecting intermediates on the catalytic reaction pathway is that, if suitable monitoring probes (or "reporter groups") are used, differences in the environment about the probe in different enzyme–substrate intermediates will result in different signals from the probe. It is therefore important that probes be chosen carefully so as to maximize the probability that their environment will be different in different intermediates, and also that the probe will be sufficiently sensitive to the change in environment so that a detectable change in its signal will occur.

Ideally one would like to be able to incorporate, or use, environment-sensitive probes in each part of the substrate and active site. Unfortunately, the high degree of specificity in many cases severely restricts the opportunities available for incorporating "probes" into the substrate. A wide variety of environment-sensitive probes may be used, including UV and visible chromophores, fluorescent groups, "spin-labels," such as nitroxides, and groups suitable for NMR studies, such as ^{31}P, ^{19}F, and ^{13}C. Often the intrinsic spectral characteristics of the enzyme itself, e.g., fluorescence emission, are very valuable in this regard.[26] Ultimately the particular enzyme and its specificity will determine which groups can be used as monitoring probes.

Having chosen the substrate and cryosolvent to be used, the most straightforward procedure to detect enzyme–substrate intermediates involves the following steps. Using the methodology outlined subsequently,

---

[25] A. L. Fink and K. Magnusdottir, personal communication.
[26] A. L. Fink and E. Wildi, *J. Biol. Chem.* **249**, 6087 (1974).

the enzyme and substrate are mixed together at some appropriately low temperature and the signal from the environment-sensitive probe is monitored as a function of time. There are two general approaches one can use. One is to start at the lowest possible temperature; the other is to start at a high temperature, around 0°. The advantage of the latter is that one can observe the turnover reaction, that is, a reaction of *known* identity. In either approach the idea is to carry out a series of experiments in which only the temperature and the pH* are varied and in which only time-dependent changes are monitored. Eventually one should have a series of experiments at 15–20° intervals covering the accessible range of temperature (typically 0° to −90°), at a series of pH* values, usually at 0.5–1 pH unit apart. Because of the underlying expressions governing the effects of temperature and pH on reaction rates, it is possible to keep track of a particular reaction as these conditions vary. Thus if increases in intensity of the monitoring signal occur, it is possible to distinguish whether they reflect only one elementary step or more.[27-29]

The following series of steps are recommended for the detection of enzyme–substrate intermediates if one starts with a very low temperature and works toward higher ones: (1) dissolving the enzyme or substrate in the cryosolvent at 0° and cooling to the lowest temperature at which the solvent is fluid; (2) scanning the resulting spectrum using UV, visible, fluorescence, circular dichroism, NMR, ESR, or Raman spectroscopy, whichever is (are) appropriate for the substrate and enzyme being used; (3) mixing the substrate and enzyme at this temperature, e.g., −100°; (4) rescanning the spectrum over a period of approximately 2 hr (or if changes are observed, until these cease); (5) raising the temperature an appropriate increment (e.g., 20°) and scanning at frequent intervals until no further changes are observed (or 2–3 hr if no changes appear); (6) lowering the temperature back to the original value (e.g., −100°) and scanning the spectrum twice; (7) raising the temperature by an additional increment above that in step 5 and scanning as in step 5; (8) lowering the temperature to −100° and repeating step 6, repeating steps 7 and 8 until a temperature at which turnover occurs is reached; (9) comparing the series of spectra obtained at the lowest temperature (i.e., steps 2, 4, 6, and 8); any differences between them are indicative of the presence of different enzyme–substrate species, if appropriate controls are run (e.g., omitting enzyme).

[27] A. L. Fink, *Biochemistry* **15**, 1580 (1976).
[28] K. J. Angelides and A. L. Fink, *Biochemistry* **18**, 2355 (1979).
[29] P. Douzou, R. Sireix, and F. Travers, *Proc. Natl. Acad. Sci. U. S. A.* **66**, 787 (1970).

Characterization of Intermediates and Correlation of Subzero and Normal Data

Characterization of detected intermediates involves determining the effect of pH*, temperature, and substrate and enzyme concentration on their rates of formation, and their spectral characteristics.[10,27-29] Usually the kinetics of intermediate transformation will be first-order and the acquisition of the rate constant itself is straightforward. From a knowledge of the dependence of the rate on the substrate concentration (for reactions run under excess substrate conditions), it is possible to calculate the individual microscopic rate constants for elementary steps in the reaction, under nonturnover conditions, as well as estimate the fraction of the total enzyme concentration in the form of a particular intermediate.[6,10,28] The details of such kinetic analyses have been given elsewhere.[6] In this manner it is also possible to construct free-energy diagrams for the reaction pathway, as well as ascertaining the activation parameters for the reaction.[10,27,28]

From knowledge of the effect of the cosolvent on the rate (the decrease in binding affinity for the substrate results in a corresponding decrease in the rates of intermediate transformation if the reaction is carried out under non-substrate-saturating conditions[6]) and extrapolation of the Arrhenius plot, it is possible to estimate the rate for a particular intermediate transformation at 25° (or 1°) and aqueous solution, i.e., normal conditions.[27,28] If this value is such that the rate allows the kinetics of the reaction to be measured by conventional rapid-mixing (e.g., stopped-flow) equipment, then it should be possible to look for the intermediate under such conditions. It is also necessary to check the Arrhenius plots of the preceding and succeeding reactions to make sure that a change in rate-determining step (with respect to that intermediate) does not occur.[22] If the intermediate can be detected under "normal" conditions and the kinetics of its formation and its spectral properties determined, it is possible to provide a quantitative measure of the agreement between the low-temperature findings and those obtained under normal conditions.[10]

It is probably unnecessary to stress the importance of demonstrating that the combination of low temperature and cosolvent have no adverse effect on the conformation of the protein from that in aqueous solution at the normal temperature for the enzyme, and that it is also critical to show that intermediates detected are on the productive catalytic reaction pathway. Only when all these requirements are fulfilled can one have assurance that the information acquired at the low temperatures bears directly on the question of the mechanism under normal conditions. Demonstration that an intermediate detected at low temperature is indeed on the productive catalytic pathway can be done mostly on the basis of kinetic arguments, in conjunction with spectral and other types of information that may be available for that specific system.[10,27]

Structural Determination of Trapped Intermediates

At present the best-suited procedure for yielding structural data on enzyme–substrate intermediates at atomic resolution are X-ray and neutron diffraction analysis for the molecule as a whole and NMR for certain key parts of the molecule. In keeping with the theme of this volume, investigations concerning the crystallographic aspects of cryoenzymology will not be considered here; details may be found elsewhere.[4,8,9,30] In brief, however, the procedures involve growing enzyme crystals in the usual fashion, transferring them to the cryosolvent under appropriate conditions, determining the rate reduction caused by the diffusional restraints of the crystal interstices, mounting the crystal in a flow-cell on the diffractometer, and allowing the substrate to diffuse in at low temperature to form the desired intermediate. An alternative procedure, which is feasible in some cases, is to trap the desired intermediate in the dissolved state, then crystallize it at appropriately low temperature to produce the crystalline trapped intermediate directly.[31,32]

In many cases the identity of a key intermediate revolves around the question of the hybridization state of a particular carbon atom in the substrate. The large differences in chemical shift of trigonal and tetrahedral carbon in ^{13}C NMR makes this an attractive technique to resolve such questions. It is necessary to use highly enriched ^{13}C in the desired position and to determine the necessary conditions to trap the intermediate at sufficiently high concentrations for the NMR experiment using spectral techniques. Similarly ^{31}P NMR may be very valuable in providing structural information in systems involving phosphorous groups.

In some instances particular spectral properties of an intermediate may be very revealing with regard to identifying the state of the substrate in the intermediate.[10] In general the specific system will determine which types of techniques can yield structural information about the intermediate.

---

[30] G. A. Petsko, *J. Mol. Biol.* **96**, 381 (1975).
[31] J. Saint Blanchard, A. Clochard, P. Cozzona, J. Berthou, and P. Jollès, *Biochim. Biophys. Acta* **491**, 354 (1977).
[32] A. L. Fink and R. Homer, unpublished observations, 1977.

Cryosolvents

Owing largely to Douzou and his colleagues, a large body of data is now available on the physical–chemical properties of the most commonly used cryosolvents.[3,5,7,33] There is little point in documenting the data here; the reader is referred to "The Handbook of Biochemistry and Molecular Biology"[33] for a convenient compilation. Rather we will confine ourselves to the effect of these properties on the design of experiments at subzero temperatures.

There are three types of cosolvents in common use; alcohols (methanol, ethanol), polyols (ethylene glycol, glycerol, 2-methyl-2,4-pentanediol), and DMSO and DMF.[20] Methanol is the solvent of choice for many systems, as cryosolvents based on it may remain fluid to $-100°$ and below, and its cryosolvents have low viscosities. Unfortunately, the specific denaturation of many proteins in this cosolvent prevents its widespread use. Most enzymes remain active in solutions of the polyols, but the high viscosities of these cryosolvents restrict their use to solutions of 50% composition and temperatures above $-30°$. These limitations can be overcome by combining a polyol with an alcohol to form a ternary solvent system (Table III). Several enzymes that are denatured by alcohols remain active in DMSO and DMF, and these solvent systems can be used

TABLE III
VISCOSITY AND FREEZING-POINT DATA FOR SOME CRYOSOLVENTS[a]

Solvent system[b]	Temperature (°C) at which $\eta = 50$ cp	Freezing point (°C)
Ethylene glycol/water 30:70	—	-17
50:50	-25	-44
Methanol/water 70:30	-60	-85
Ethylene glycol/methanol/water 40:20:40	-30	-71
25:25:50	-35	-50
10:50:40	-45	-69
10:60:30	-55	SC

[a] Data from P. Douzou, G. Hui Bon Hoa, P. Maurel, and F. Travers in "Handbook of Biochemistry and Molecular Biology," (G. Fasman, ed.), 3rd ed., Vol. I, p. 520. Chem. Rubber Publ. Co., Cleveland, Ohio, 1976.
[b] Volume percent of organic component.

[33] P. Douzou, G. Hui Bon Hoa, P. Maurel, and F. Travers, in "Handbook of Biochemistry and Molecular Biology" (G. Fasman, ed.), 3rd ed., Vol. I, p. 520. Chem. Rubber Publ. Co., Cleveland, Ohio, 1976.

TABLE IV
FREEZING POINTS OF SOME COMMON CRYOSOLVENTS[a]

Solvent system[b]	Freezing point[c] (°C)	Solvent system[b]	Freezing point[c] (°C)
Ethylene glycol/water		DMSO/water	
100:0	−12.5	100:0	18.5
60:40	−69	65:35	SC[d]
40:60	−26	50:50	SC
20:80	−10	40:60	−41
		20:80	−12
Methanol/water			
100:0	−96	DMF/water	
80:20	−105	100:0	−62
70:30	−85	80:20	−100
50:50	−49	70:30	−83
		50:50	−40

[a] From P. Douzou, G. Hui Bon Hoa, P. Maurel, and F. Travers, in "Handbook of Biochemistry and Molecular Biology" (G. Fasman, ed.), 3rd ed., Vol. I, p. 520. Chem. Rubber Publ. Co., Cleveland, Ohio, 1976.
[b] Volume percent.
[c] The exact freezing point will vary with the solute concentration.
[d] Supercooled.

down to −90° or below. Freezing points of some common cryosolvents are shown in Table IV.

Dielectric Constant

The variation in dielectric constant with temperature for all the solvent systems so far investigated fits the equation[34,35] $\log D = a - bT$ where D is the dielectric constant, T is the temperature, and a and b are constants. Thus D increases with decreasing temperature.

The dielectric constant is critical for many biochemical processes since it determines the interaction between charged species. The addition of most organic solvents to an aqueous solution causes a drop in the dielectric constant, which can be reversed by lowering the temperature (Table V). This may *in part* explain the observation that many enzymes are inactive in aqueous–organic solvents at ambient temperatures, but active at lower temperatures. Dielectric constant is not the only factor involved, however, since many proteins are active in solvents of dielectric

[34] G. Akerlof, *J. Am. Chem. Soc.* **54**, 4125 (1932).
[35] F. Travers and P. Douzou, *J. Phys. Chem.* **74**, 2243 (1970).

TABLE V
TEMPERATURE AT WHICH THE DIELECTRIC CONSTANT IS 80 FOR SOME CRYOSOLVENTS[a]

Solvent system	$T (D = 80)$ (°C)
Water 100:0	+20
Methanol/water 50:50	−32
DMSO/water 50:50	+9
DMF/water 50:50	−21
Ethylene glycol/water 50:50	−18

[a] Data from F. Travers and P. Douzou, *Biochimie* **56**, 509 (1974); F. Travers and P. Douzou, *J. Phys. Chem.* **74**, 2243 (1970).

constant much below 80. Those effects due to the dielectric constant probably emanate mostly from ionic interactions involving exposed side-chain residues.

Solubility and Dissociation of Strong and Weak Electrolytes

Since enzyme activity is dependent on the ionic environment of the solvent, it is important to be aware of the limitations imposed by the solvent on the solubility and dissociation of the solutes.

Neutral salts, such as NaCl and KCl, have been shown to be soluble to a concentration of 0.1 M in 50% organic solvent, down to the freezing point of the solvent.[36] Higher ionic strengths can be achieved with the use of KI (0.2 M) but ionic strengths higher than this are difficult to obtain at temperatures approaching the freezing point of the cryosolvent. For example, although ammonium acetate is very soluble in most cryosolvents, it has a similar salting-out effect as ammonium sulfate on proteins.

The solubility of buffer salts can also prove to be a problem in some solvent systems. Generally concentrations greater than 10 mM may precipitate as the temperature is lowered. Conditions leading to precipitation vary depending on the cosolvent, its concentration, the temperature and the concentration of enzyme, substrate, and other ligands present. It is therefore necessary to test a particular solvent system for the solubility of its components as a function of temperature before embarking on any kinetic experiments. Buffer salts found to be most satisfactory with respect to solubility include chloroacetate, acetate, and cacodylate. Other commonly used buffer salts, which are *least* soluble, are phosphate, Tris, and glycine.

Strong electrolytes are probably completely dissociated in the

[36] P. Douzou, *Mol. Cell. Biol.* **1**, 15 (1973).

aqueous–organic mixtures, an observation supported by the indicator titration experiments of Hui Bon Hoa and Douzou.[37] For weak electrolytes the extent of dissociation will depend upon the dielectric constant of the microenvironment around the electrolyte, which may vary from that of the bulk aqueous–organic solvent. Thus the dissociation behavior of an electrolyte cannot be predicted from the observed bulk dielectric constant, and the dissociation of a weak acid must be measured both as a function of cosolvent concentration and temperature. Fortunately, Douzou and his collaborators have examined a wide range of buffer salts for the effect of cosolvent and temperature on their dissociation constants.[33,37] The results show that in all cases investigated the addition of the cosolvent causes an increase in the apparent pH, with the exception of Tris, which in DMSO and DMF solvents shows a decrease in the apparent pH. In cryosolvents with up to 50% cosolvent the pH*-change corresponded to 0.1–1.5 pH* units (Table VI). With higher concentrations of cosolvent the change in pH* can be as much as 3 units (65% DMSO, cacodylate).

The temperature effect on the dissociation of weak electrolytes in aqueous–organic solvent systems is similar to that in aqueous solution and is linear with respect to the reciprocal of the absolute temperature (ln $K = \Delta G°/RT$), and may increase or decrease with decreasing temperature. Some representative data are shown in Table VII.

TABLE VI
Change[a] in pH* on Adding an Aqueous Buffer Solution at Its pK to the Following Cosolvents to Produce a Cryosolvent 10 mM in the Buffer Component[b]

Cosolvent	Buffer		
	Acetate $pK = 4.75$ (H$_2$O, 25°)	Cacodylate $pK = 6.19$ (H$_2$O, 25°)	Phosphate $pK = 7.21$ (H$_2$O, 25°)
50% Methanol	0.70	0.75	1.05
70% Methanol	1.25	1.15	1.60
50% DMSO	0.60	0.45	1.15
50% DMF	2.15	—	2.00
50% Ethylene glycol	0.50	0.45	0.60

[a] All pH* changes are positive.
[b] Data from P. Douzou, G. Hui Bon Hoa, P. Maurel, and F. Travers, in "Handbook of Biochemistry and Molecular Biology," 3rd ed., Vol. I, p. 520 (1976), and G. Hui Bon Hoa and P. Douzou, J. Biol. Chem. **248**, 4689 (1973).

[37] G. Hui Bon Hoa and P. Douzou, J. Biol. Chem. **284**, 4649 (1973).

TABLE VII
CHANGE IN pH* OF SOME CRYOSOLVENTS AS THE TEMPERATURE
IS REDUCED FROM +20° TO −40°

Buffer[a,b]	Acetate	Cacodylate	Phosphate
50% Methanol	0.15[c]	0.10	0.15
70% Methanol	0.25	0.15	0.33
50% DMSO	0.70	0.90	1.30
50% DMF	0.60	—	0.10
50% Ethylene glycol	0.30	0.35	0.50

[a] Data from G. Hui Bon Hoa and P. Douzou, *J. Biol. Chem.* **248,** 4649 (1973).
[b] Concentration of buffer component in the cryosolvent is 10 mM.
[c] All pH* changes are positive.

Protonic Activity

Strictly speaking, the interpretation of measured pH is restricted to aqueous solutions if the pH is to relate to the hydrogen ion concentration in a known way. The meaning and measurement of "pH" is considerably more complex in an aqueous–organic solution. For most purposes the apparent pH, pH*, i.e., $-\log a_H^*+$ where a_H^*+ is the proton activity in the mixed solvent, is very useful both in practical and theoretical terms. As the activity coefficient of the proton will vary with different solvent systems, each cryosolvent will have its own pH* scale. The pH* will vary both with the temperature and the cosolvent concentration. Measurements of pH* can be made in the following ways.

The most convenient, albeit the least accurate, method is to use a conventional pH meter with a glass electrode that has been thoroughly equilibrated with the aqueous–organic solution to be measured. The meter then gives a direct estimate of pH*. Readings can be made over a range of temperatures (above 0°) to provide the temperature dependence of that particular solvent system. Unfortunately the electrode cannot be used below 0° as the solvent inside the electrode must remain fluid. However, since the pH* is linearly related to $1/T$, the pH* at a given subzero temperature may be readily estimated graphically.

A more accurate method to determine pH* is to use the modified electrode system of Larroque et al.[38] In this case, the electrolyte solution of the glass electrode is replaced by 10^{-1} M HCl in a solution of the cosolvent of appropriate concentration. Similarly, the saturated KCl solution in the calomel electrode is replaced by a 0.1 M KCl solution in the cryosolvent. With this electrode system the pH* may be accurately measured at any temperature at which the cryosolvent is fluid.

[38] C. Larroque, P. Maurel, C. Balny, and P. Douzou, *Anal. Biochem.* **73,** 9 (1976).

A third, more laborious method is the spectrophotometric determination of pH* using indicator solutions. The pK's of a series of indicators covering the pH* range 1–12 have been measured in a variety of aqueous and organic solutions,[39] and these have been used to measure the pH* of a number of cryosolvents at subzero temperatures.[37]

Viscosity

The viscosity of cryosolvents is of interest for two reasons; first because of its effect on the rate of diffusion-controlled reactions, and second because of the technical problems in mixing viscous solutions. Very few studies have been made on the effect of viscosity on a diffusion-controlled reaction, but complex effects would be expected. Two factors are involved; increased viscosity will result in a reduced number of collisions between reacting species in a given time interval, but at the same time the average length of time of each collision will be increased, possibly resulting in a higher ratio of productive collision complexes.[40]

It has been suggested that viscosities in excess of 50 cp make efficient mixing of solutions by conventional techniques difficult.[41] Well-designed tangential jet or ball mixers, however, can effectively mix solutions of greater than 100 cp (see below). Unfortunately, relatively few precise data are currently available on the viscosity of many of the commonly used cryosolvents. Some representative data are given in Table III. Generally alcohol and DMF-based cryosolvents have relatively low viscosities and do not present any problem down to their freezing points. Dimethyl sulfoxide solutions at low concentrations do not present problems either, but at concentrations above 50% they become increasingly viscous below −50°. Polyol-based cryosolvents have high viscosities, even at high subzero temperatures.

Preparation of Cryosolvent Solutions

There are several ways in which the aqueous–organic solvent systems can be prepared. Probably the easiest are based on volume/volume or weight/weight bases. Potential problems arise from the volume contraction on mixing such solutions, exothermic heats of mixing, and the temperature dependence of density. To avoid these it is best to use a procedure of the following type. For example, to make a 70% (v/v) aqueous methanol solvent 70 parts by volume of methanol are carefully added to

---

[39] R. Gaboriaud, *C. R. Acad. Sci. Ser., C* **263**, 911 (1966).
[40] G. K. Strother and E. Ackerman, *Biochim. Biophys. Acta* **47**, 317 (1961).
[41] F. Travers, P. Douzou, T. Pederson, and I. Gunsalus, *Biochimie* **57**, 43 (1975).

30 parts by volume of the appropriate aqueous solution in a thermostatted vessel. In the author's laboratory the volumes are usually measured at 25° but mixed slowly at 0°. Naturally the purity of the solvents must be of the highest possible, especially with relatively labile components such as dimethyl sulfoxide and dimethyl formamide. Solvents may be purified by standard procedures. Once prepared the cryosolvents should be stored at $-20°$. Fresh solutions should be made every few days, since changes in the cryosolvent composition occur owing to decomposition of components and evaporation.

In order to obtain a cryosolvent of desired cosolvent content and pH* the aqueous component must be carefully chosen. In particular the effect of the cosolvent on the pK of the buffer component, and of temperature on the pH* of the cryosolvent buffer system, must be known (see above). The buffer systems normally used are HCl, formate, chloroacetate, acetate, phosphate, cacodylate, morpholine, borate, Tris, carbonate, and glycine. The effective pK of these varies depending on the cosolvent and its concentration. We have found it very useful to construct plots of the pH of the aqueous component vs the pH* of the cryosolvent (Fig. 2). Using such plots one can readily prepare the appropriate aqueous buffer. The pH* of the cryosolvent can then be checked using the pH meter (see below). Alternatively, the cryosolvents can be prepared on a more empirical basis by the following method. Let us suppose that a 20% (v/v) solution of DMSO of pH* 7.0 is required. Since organic solvents increase the pK of most buffer components an aqueous solution of pH approximately 6.5 would be prepared, using phosphate, for example (phosphate buffers are not very satisfactory for high cosolvent concentrations owing to limited solubility). Using a thermostatted vessel, 8.0 ml of the aqueous buffer and 2.0 ml of DMSO would be mixed at 0°. After a few minutes the pH* is read using a pH meter (at 0°). If the reading is lower than desired a small aliquot of base (e.g., 0.5 M KOH) of known volume is then added, and the new pH* measured. By varying the concentration of the acid or base used, it is possible to achieve the desired pH* with a minimum of perturbation to the cosolvent concentration, and by adding known amounts of the acid or base one can calculate the exact volume percent of the final solution. The ionic strength can be controlled by the addition of a salt such as KCl or KI, either in solid form to the final solution, or by addition to the aqueous buffer component.

Preparation of Enzyme and Substrate Solutions in Cryosolvents

The exact details of solution preparation will be determined by the particular system under investigation, so the following should be consid-

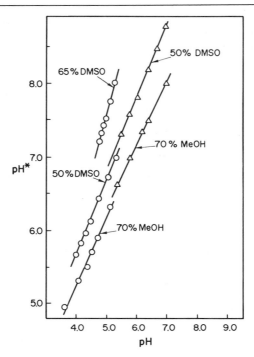

FIG. 2. Relationship between the pH of the aqueous buffer component and the pH* of the resulting cryosolvent for some dimethyl sulfoxide and methanol cryosolvents. In each case the final buffer concentration is 0.1 M. Buffers used were acetate ⊙ and cacodylate △. Temperature was 25°.

ered more as a general guideline. A concentrated stock solution of substrate can often be made up in the cryosolvent itself. Limitations may occur owing to lack of solubility at the desired concentration or to instability, e.g., due to pH. The stock substrate solution may also be prepared in neat cosolvent, or even an alternate solvent if necessary. Such solutions should be stored at appropriate low temperature and made up fresh frequently. Aliquots of the substrate stock solution can then be added to the cryosolvent to give the desired concentration (see below).

For preparation of the enzyme solution in the cryosolvent one would normally start with a concentrated aqueous solution of homogeneous enzyme. In the case of coenzyme-requiring systems the coenzyme may also be present. For most enzymes examined thus far, the following procedure may be used. If the enzyme has been shown to be stable in the cryosolvent at 0°C, then an aliquot of the stock enzyme solution is added to the cryosolvent at 0°C and carefully mixed; the solution is cooled to the desired temperature. If the enzyme is not stable in the cryosolvent at 0° but is stable at, for example, $-25°$, then the aqueous aliquot of the stock enzyme solution is diluted (usually 1:2 to 1:4) with the cryosolvent at 0°; the resulting mixture is then added to the cryosolvent at $-25°$. This procedure is necessary to avoid the problems resulting from trying to add an aqueous solution to one below 0° (which leads to tiny ice crystals).

In some cases, more commonly with crystals for X-ray diffraction experiments, it is necessary to gradually increase the cosolvent concentration while concurrently lowering the temperature. An instrument to do this automatically has been designed by Douzou,[7] but it is usually sufficient to perform the operation as follows. An aqueous solution of the enzyme is cooled to 0°C and gradually enough cosolvent is added to make a 10–15% (v/v) solution. This is then cooled to $-10°$, and a further addition of cosolvent is made to bring the concentration up to 25% by volume. The temperature is again lowered, to about $-20°$, and the procedure is repeated until the desired cosolvent concentration is reached. Some enzymes may be very sensitive to the first step of this procedure, and it is necessary to add the initial cosolvent very carefully so as to avoid local buildups of high concentration. For some enzymes, but not for others, it is possible to take an aqueous stock solution of enzyme and gradually add cosolvent with gentle stirring, at 0°, until the desired cosolvent concentration is reached. In such cases a motor-driven syringe and magnetic stirrer are useful. Above all one must keep in mind that most enzymes will become denatured rather rapidly in the presence of high cosolvent concentrations at temperatures much above 0°.

Mixing Techniques

In most chemical systems the limiting step in measuring the rate of a fast reaction is the time of mixing the reactants. With subzero temperature investigations, the situation is a little more complex. The additional problems stem from the usually higher viscosities encountered at low temperatures, and condensation of atmospheric water vapor on cold surfaces of the sample cell if the sample compartment is opened to the atmosphere to add and mix reactants. The former can be overcome with special mixing techniques and the latter by purging with a suitably high flow rate of *dry* nitrogen (or air).

For many solvent systems at temperatures above $-20°$ a simple "plumper"-type of mechanical (hand) stirring is adequate for mixing a solution within 5 sec, but at lower temperatures, and with the more viscous cryosolvents (especially ethylene glycol, 2-methyl-2,4-pentanediol, and glycerol), such methods are no longer satisfactory. The efficiency of mixing can be improved with a vibrating-reed mixer (60 Hz), but care has to be taken in both its design and use in order to prevent the introduction of air bubbles into the solution. Such bubbles can take considerable time to dissipate in viscous solutions.

The major problems connected with mixing cryosolvent solutions at subzero temperatures are intimately connected with the increased viscosities and, in addition to the presence of included bubbles, are temperature, concentration, and density gradients. In some cases it is possible to have continuous stirring during the tenure of a reaction, either using a magnetic stirrer (usually not possible owing to the viscosity) or a built-in vibrating-reed or mechanically driven rotating stirrer (out of the beam of any radiation being transmitted through the sample cell). We have found that, provided the temperature remains constant and the initial mixing was thorough, no additional stirring is necessary. However if the temperature is changed during the experiment, even by a very small amount, it is necessary to remix the contents when the new temperature equilibrium is established.

The most efficient means of carrying out the initial mixing of enzyme and substrate at a low temperature is to use an injection method involving a mixer based on the design of stopped-flow-type mixing devices. Such systems are capable of delivering up to 3 ml of mixed solution within a fraction of a second even with very viscous solutions and are more than adequate in terms of time resolution for use with conventional spectrophotometers. The device used in the author's laboratory consists of two syringes, one for substrate solution, the other for the enzyme, mounted on a mixing block. The mixed solution exits via a short tube that leads into a precooled sample cell. The whole unit is immersed in a refrigerated bath at the desired temperature. The syringes are driven by a plunger, which can either be operated by hand or by pneumatic piston. The most efficient, but most expensive, type of mixing block is a ball type.[42] However, we have found a 10-jet tangential mixer constructed of stainless steel to be quite satisfactory. Design details are given elsewhere.[43] Care has to be taken in the selection of the pressure used to drive the syringes, as too high a pressure will lead to cavitation. The pressure used depends upon the solvent system and temperature. A note of caution: the majority of artifacts observed in subzero temperature experiments stem from mixing-type problems.

[42] R. L. Berger, B. Balko, and H. F. Chapman, *Rev. Sci. Instrum.* **39**, 493 (1968).
[43] R. L. Berger, in "Rapid Mixing and Sampling Techniques in Biochemistry" (B. Chance, R. H. Eisenhardt, Q. H. Gibson, and K. K. Lonberg-Holm, eds.), p. 33. Academic Press, New York, 1964.

Low-Temperature Production and Control

There are basically two sources of low temperatures used in cryoenzymology, liquid nitrogen (occasionally Dry Ice), and refrigeration compressors, usually in the form of refrigerated constant-temperature circulation baths. Both liquid and gas coolants can be used, each having particular advantage in some situations. The most generally useful liquid coolant is an ethanol–methanol mixture (70–80% ethanol); the best gas circulant is nitrogen (dry), although for anaerobic experiments argon may be desirable.

Since NMR and ESR spectrometers have their own temperature control systems, in this section we will consider systems suited for spectrophotometric and chromatographic situations. A substantial number of refrigerated circulation baths covering the range 0° to −30° are commercially available. Ethanol provides a very satisfactory circulant. Foam-insulated tubing is necessary to prevent excess cooling loss. For temperatures in the −30° to −70° range there are a limited number of commercially available constant-temperature circulation baths, among the companies offering such equipment are Neslab, Heto (London Co.), Lauda, and FTS Systems. Both Lauda and FTS Systems have equipment (three-stage compressor systems) that can provide circulation of constant-temperature fluids in the −70° to −120° region. It has been our experience, however, that gas-coolant systems are preferable for temperatures below −70°.

Although gas-coolant constant temperature systems are commercially available, usually in connection with NMR or ESR spectrometers, one can construct an effective and inexpensive one based on the principle that nitrogen gas is cooled with liquid N_2, and then passed over a thermoregulated heater and into the jacketed sample holder. A typical design is shown in Fig. 3; another has been given by Douzou.[3,7] It is much less expensive to use liquid N_2 as the source of the N_2 gas rather than gaseous N_2. The cold N_2 may be obtained by passing the boil-off gas from a liquid N_2 cylinder through a long coil immersed in a Dewar of liquid N_2, or by directly boiling off N_2 gas from a Dewar of the liquid. Either heavy, thick foam, or vacuum, insulation is necessary between the cold N_2 source and the sample cell. Devices of this sort can readily produce temperatures of −150° and below, with a stability of ±0.5° over many hours, and ±0.1° at higher temperatures. Advantages of the gas-coolant system include lower capital outlay, although greater operating costs, than for refrigerated circulation baths, and the ability to rapidly change the coolant temperature over wide ranges.

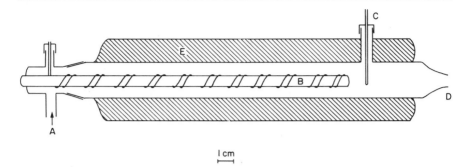

FIG. 3. Inexpensive low-temperature, gas-flow temperature controller. A supply of dry nitrogen gas is cooled by passage through a Dewar flask of liquid nitrogen and enters the temperature controller through port A. The gas then passes over the thermoregulated heating coil, B, and platinum thermistor temperature probe, C, before exiting through port D into the low-temperature sample cell. Temperature regulation is achieved by feedback from the temperature probe to the heater control box. The whole unit is insulated by the vacuum jacket E and a layer of foam insulation (not shown). Temperatures in the range 0° to −120° (±0.1°) can be readily obtained by varying the gas flow rate and the resistance in the heating coil circuit.

Temperature Jumps

Slow temperature jumps may be obtained either by changing the temperature of the gas-flow system, or even more rapidly be using two coolant systems of different temperatures. Temperature changes of 50° or more can be obtained in the following manner in time periods of 1 sec or less. For example, suppose one wishes to follow the renaturation of a denatured protein by changing the temperature from +20° (denatured) to −40° (native) and follow the rate of refolding spectrophotometrically. The apparatus consists of a syringe mounted in a thermostatted vessel at 20°, and connected to a long (2 m), narrow, coiled stainless-steel tube that is immersed in a second thermostatted vessel, this one at −40°. The tip of the tube exits through the bottom of the second vessel and leads into a precooled (−40°) spectrophotometer cell. The syringe is filled with cryosolvent to which the protein is added. After the desired length of time at 20°, the syringe plunger is rapidly pushed in, forcing the liquid through the coiled tube, in which it is rapidly cooled to −40°C, and then into the spectrophotometer cell.[44] Variations on this theme include temperature jumps from lower to higher temperatures, mixing in an additional component just prior to the temperature jump, using the previously described low-temperature syringe, and so on.

[44] A. L. Fink and B. L. Grey, unpublished data.

Temperature Measurement

Measurement of the actual temperature in experiments at subzero temperatures is of course of extreme importance, both in terms of the absolute temperature and also of the maintenance of constant temperature for the duration of the experiment. For the latter reason it is best to continuously monitor the temperature during the course of an experiment. The easiest means of determining the temperature of a reaction mixture is to use either a thermocouple or a temperature-sensitive diode or thermistor. Commercial instruments are available (e.g., Omega Electronics) at reasonable prices which give readouts to $\pm 0.1°$ using a very thin thermocouple. One can also build inexpensive devices. The least expensive system involves two thermocouple wires, one leading to the sample, the other to an ice bath, connected to a millivoltmeter. Reference to standard tables for the particular thermocouple wire composition allows one to determine the temperature to at least $\pm 0.2°$.

Sample Cells

For temperatures down to $-30°$ conventional spectrophotometer cells can be used in conjunction with brass cooling blocks, "sealed" sample compartments, and *dry* gas purging. Access to the sample cell can be achieved via a small opening in the sample compartment lid, directly above the cell and normally closed with a neoprene stopper. In the design of thermostatting blocks, the important features to keep in mind are the need for maximal contact between the cell containing the sample and the cooling block, and the advantage of having the incoming purging gas directed immediately at the optical faces of the cell. It is possible, in fact, to use such cell-holders at temperatures at least as low as $-60°$, with proper precautions. Douzou[7] has also published the details of a gas-cooled spectrophotometer cell system for use at very low temperatures.

In general, however, for temperatures below $-30°$ it is best to use triple-walled cells; the inner compartment serving as the sample chamber, the middle one as the coolant jacket, and the outer compartment functioning as a vacuum insulating jacket. The design of such a cell for absorbance and circular dichroism spectrophotometers is shown in Fig. 4; it is constructed from fused silica and would be mounted on a holder designed to place it in the correct position in the light path. This particular design can be used to make 1-cm path-length cells with volumes in the 2–3-ml range, and to fit all common spectrometers. Such cells may be used to temperatures of $-130°$ and below.

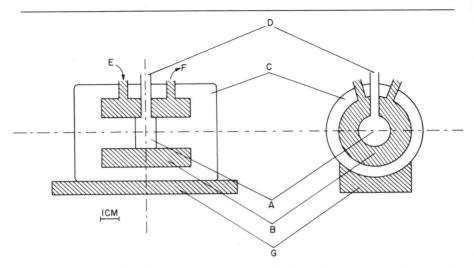

FIG. 4. Triple-walled, low-temperature sample cell. The cell is constructed from fused silica and consists of three concentric cylindrical chambers: A the sample compartment, B the coolant jacket, and C the vacuum insulation chamber. Access to the sample compartment is gained through port D, and coolant is circulated through ports E and F. The whole unit is mounted on a steel block, G, insulated with a rubber strip, and designed for the particular spectrophotometer to be used. A variation of this cell design involving the concentric cylinders in a vertical orientation, i.e., perpendicular to the light path, has also been used.

Column Chromatography at Subzero Temperatures

In several enzyme catalyses relatively long-lived intermediates under normal conditions may be stabilized for almost indefinite periods at relatively high subzero temperatures. Such species are very amenable to isolation and purification by chromatographic techniques at low temperature. The technique has been used to isolate acyl enzymes,[14] and an oxygenated luciferase intermediate,[13] as well as a normally unstable cytochrome P-450.[45] The main differences from column chromatography at ambient temperatures are the need for insulated, jacketed columns with appropriate coolant systems, a cold chamber for collection of eluent at low temperature, and limitations on the packing material due to the (usually) greater viscosity of the elutant.

Packing materials that have been successfully used include Sephadex LH-20, DEAE cellulose (Whatman DE 23), and various forms of porous glass (Corning CPG). We have successfully carried out column separations at temperatures as low as $-45°$. At lower temperatures specially insulated and cooled columns become necessary, and only relatively non-

---

[45] C. Balny, C. Le Peuch, and P. Debey, *Anal. Biochem.* **63**, 321 (1975).

viscous solvents can be used. The simplest systems involve a jacketed column, cooled by circulation from a low-temperature constant-temperature bath using ethanol–methanol as coolant, and insulated with foam rubber or its equivalent. Suitably jacketed columns in a wide variety of sizes are commercially available. Either a pump or gas-pressure solvent reservoir is necessary to force the elutant through the column. Thus an in-line sample injection system is also required. The eluted material may be collected using a fraction collector in a cold chamber. It will usually be necessary to use a fraction collector in which the control unit is separate from the collector itself to prevent malfunctions of the controller due to the cold. The fraction collector can be mounted inside a double-walled, insulated chamber with a cover. Cooling is obtained by circulation of nitrogen gas that has been cooled with liquid nitrogen. Temperature control is achieved with a heater and controller as described in the above section. A small hole in the cover allows the gas to escape and provides entrance for the eluent. The tip of the column should be placed as close to the collection tubes as possible. It is also feasible to build a larger cold chamber and include the unjacketed column itself in the same chamber as the fraction collector.

The column must be packed (except for porous glass and other rigid packings) at the same temperature and pressure as will be used in the experiment, and is usually necessary to maintain a continuous flow through the column. For the formation and separation of an enzyme–substrate intermediate, it is usually convenient to add separately the enzyme and substrate to the top of the column, allow them to react as long as necessary with the column flow stopped, then start the elution.

It is also possible to buy commercial freezers (upright units), which can be readily adapted as cold chambers.

Rapid-Reaction Techniques at Subzero Temperatures

Although the use of temperatures down to $-100°$ results in drastic reductions in the rates of some reactions so that they occur over a time period of seconds or longer, some reactions (for example, those with low energies of activation, such as diffusion-controlled reactions) still remain too fast to be followed by conventional methods. There are a variety of techniques available for monitoring reactions on a millisecond (and to a smaller extent, a microsecond) time scale, and many of these have been adapted to low-temperature studies. The techniques with most potential are those that do not require an initial rapid mixing (e.g., flash photolysis, temperature-jump, pressure-jump), as the viscosity of the solvent often limits the rate of mixing at subzero temperatures. These "nonmixing" techniques require only a minimum of modification of existing equipment to provide adequate thermoregulation over the desired temperature range, and a dry nitrogen purge where necessary. This is not the place to review these methods in detail, but a brief outline of the scope and use of the major procedures will be made.

Flash Photolysis

This approach has been used in recent years to study biologically interesting reactions over a wide range of temperatures using frozen systems down to 2°K.[46-48]

In fluid solutions the reaction between carbon monoxide and several heme proteins has been studied, including hemoglobin[49,50] and cytochrome P-450.[51,52] In each of these cases the experiment consists of photodissociation of the protein–carbon monoxide complex, and then following the reaction as the CO recombines. An extension of this technique has been developed by Chance and his collaborators allowing them to trap cytochrome oxidase–oxygen intermediates.[53,54]

Temperature Jump

This technique, in its usual form, relies on the perturbation of an equilibrated system by a rapid increase in temperature, the relaxation of the system to the new equilibrium position being followed spectrophotometrically. The rapid heating is commonly provided by Joule heating, which can increase the temperature by 5° in 5 μsec. The associated major problems are that the perturbation must be large enough to produce a measurable change in the equilibrium, and that the temperature must revert back to the original at a much slower rate than the reaction rate(s) being observed. This is a particular problem with subzero temperature

---

[46] R. H. Austin, K. Beeson, L. Eisenstein, H. Frauenfelder, and I. C. Gunsalus, *Biochemistry* **14**, 5355 (1975).

[47] N. Alberding, S. S. Chan, L. Eisenstein, H. Frauenfelder, D. Good, I. C. Gunsalus, T. M. Norlund, M. F. Perutz, A. H. Reynolds, and L. B. Sorenson, *Biochemistry* **17**, 43 (1978).

[48] M. Sharrock and T. Yonetani, *Biochim. Biophys. Acta* **434**, 333 (1976).

[49] R. Banerjee, P. Douzou, and A. Lombard, *Nature (London)* **217**, 23 (1968).

[50] M. Bernard, C. Balny, R. Banerjee, and P. Douzou, *Biochim. Biophys. Acta* **393**, 389 (1975).

[51] P. Debey, C. Balny, and P. Douzou, *FEBS Lett.* **35**, 86 (1973).

[52] P. Debey and P. Douzou, *FEBS Lett.* **39**, 271 (1974).

[53] B. Chance, C. Saronio, and J. S. Leigh, *J. Biol. Chem.* **250**, 9226 (1975).

[54] B. Chance, N. Graham, and V. Legallais, *Anal. Biochem.* **67**, 552 (1975).

experiments as the thermal equilibration between the sample and its surroundings must be efficient in order for the temperature to be regulated. Thus the rate of return to the original temperature will limit the rates which can be followed to those of 10 sec^{-1} or faster.

An alternative use of the temperature-jump technique at low temperatures is the "slow temperature" jump, discussed under the section on low temperature and control, as well as by Douzou.[7,55]

Rapid-Mixer Techniques

There are several reports of the use of conventional stopped-flow and quenched-flow apparatus at temperatures down to 3°. The extension of the use of such apparatus to subzero temperatures is difficult without major modification to existing commercial instruments. For successful operation at subzero temperatures several design problems have to be overcome; these include the need for excellent thermal equilibration of the drive syringes, mixing chamber, and observation chamber (or delivery tube in the case of a quenched-flow machine); the need for leak-proof joints between the different components over the entire operating range of temperatures; the avoidance of condensation on the optical faces; and the necessity for the mixing chamber to cope with the viscous solvents usually employed.

Allen et al.[56] built a stopped-flow machine to operate in the 25° to −120° range, but the volume of solution required for each run (2 × 3 ml) and the length of the observation chamber (2.5 mm) make it unsuitable for most biochemical systems. More recently, Douzou has described a stopped-flow instrument with an operating range of 40° to −45°, using 2 × 0.2 ml of solution per run, and with a 1.0-cm observation chamber.[7,57] The machine has a dead-time of 10 msec; thus reactions with rates up to 50 sec^{-1} can readily be observed.

It may be possible to develop a quenched-flow apparatus of similar design, although no reports of such an instrument currently exist. Such a device may be simpler to design than a stopped-flow machine as there is no observation chamber and so small temperature perturbations after mixing will not be such a problem. In stopped-flow apparatus it is critical to maintain the temperature throughout the system within a couple of tenths of a degree, or artifacts will appear.

---

[55] G. Hui Bon Hoa and F. Travers, *J. Chim. Phys.* **69**, 637 (1972).
[56] C. R. Allen, A. J. W. Brook, and E. F. Caldin, *Trans. Faraday Soc.* **56**, 788 (1960).
[57] G. Hui Bon Hoa and P. Douzou, *Anal. Biochem.* **51**, 127 (1973).

Conclusions

Although carrying out enzyme-mechanism studies at subzero temperatures clearly involves difficulties of both a technical and biochemical nature, the potential rewards are great. The cryoenzymological approach offers a realistic means of studying both kinetic and structural aspects of individual elementary steps in enzyme catalysis. By bridging the current gap between kinetic and crystallographic studies, it would appear to be reasonable to expect that this technique will shortly help in resolving many long-standing questions regarding the efficiency of enzyme catalysis.

[14] Anomeric Specificity of Carbohydrate-Utilizing Enzymes

By STEPHEN J. BENKOVIC

In keeping with the theme of this series, the methods of investigation for establishing anomeric specificity will be discussed, rather than a compilation of the specificities of various enzymes that utilize sugars or sugar phosphates as substrates. Such information may be found in recent reviews treating enzymes involved in glycolysis,[1,2] cleavage of di- and polysaccharides,[3,4] and isomerization of carbohydrates.[5]

Background

Hexoses, as pyranose or furanose rings, exist in two major cyclic configurations, referred to as the α- and the β-anomers, which differ only in having the opposite stereochemistry in the cyclic structure at carbons designated C-1 or C-2. These two configurations interconvert readily, as shown in Eq. (1), by way of the acyclic (carbonyl) species of the sugar, which is normally present only in very low concentrations.

---

[1] S. J. Benkovic and K. J. Schray, *Adv. Enzymol. Relat. Areas Mol. Biol.* **44**, 139 (1976).
[2] B. Wurster and B. Hess, *FEBS Lett* **40S**, 112 (1974).
[3] J. A. Thomas, J. E. Spradlin, and S. Dygert, *in* "The Enzymes" (P. D. Boyer, ed.), 3rd ed., Vol. 5, p. 115. Academic Press, New York, 1972.
[4] T. Takagi, H. Toda, and T. Isemura, *in* "The Enzymes" (P. D. Boyer, ed.), 3rd ed., Vol. 5, p. 235. Academic Press, New York, 1972.
[5] I. A. Rose, *Adv. Enzymol. Relat. Areas Mol. Biol.* **43**, 491 (1975).

$$\text{α} \rightleftharpoons \text{(H)R-C=O} \rightleftharpoons \text{β} \quad (1)$$

The table gives the percentages for the anomeric forms of several representative hexoses and hexose phosphates. The method of choice for determining the tautomeric compositions of carbohydrates in solution is ^{13}C nuclear magnetic resonance (NMR), since the sensitivity of the method allows detection of minor forms and the large range of chemical shifts observed for the anomeric carbon atom of the various tautomers makes assignment and accurate quantitative estimation possible.[6] However ^{1}H and ^{31}P NMR, measurements of optical rotation and infrared (IR) spectroscopy are still adjuvants, the latter permitting identification and quantitation of the free carbonyl as opposed to *gem*-diol forms.

It is a relatively simple experiment to elucidate the anomeric specificity of enzymes that utilize sugars whose pure α- and β-anomers are available separately owing to a slow anomerization rate. The finding that D-glucose oxidase is specific for β-D-glucose represents one such study.[7] The slow mutarotation allows an examination of the kinetics for the enzyme acting on a distinct cyclic configuration. Phosphorylated sug-

PROPORTIONS OF PYRANOSE, FURANOSE, AND CARBONYL FORMS OF KETOSES AND ALDOSES AT EQUILIBRIUM IN NEUTRAL AQUEOUS SOLUTIONS

Sugar	%α	%β	% Carbonyl	Temperature (°C)
D-Glucose[a]	37	63	0.0026	20
D-Glucose 6-phosphate[b,c]	38	62	<0.4	0–10
D-Fructose 1,6-bisphosphate[d–g]	15–20	77–80	1.7	30
D-Fructose 6-phosphate[c,d–f]	20	80	4–5	30

[a] W. W. Pilgrim and H. S. Isbell, *Adv. Carbohydr. Chem.* **23**, 11 (1968).
[b] J. M. Bailey, P. H. Fishman, and P. G. Pentchev, *J. Biol. Chem.* **243**, 4827 (1968).
[c] C. A. Swenson and R. Barker, *Biochemistry* **10**, 3151 (1971).
[d] S. J. Benkovic, J. L. Engle, and A. S. Mildvan, *Biochem. Biophys. Res. Commun.* **47**, 852 (1972).
[e] T. A. W. Koerner, Jr., L. W. Cary, M. S. Bhacca, and E. S. Younathan, *Biochem. Biophys. Res. Commun.* **51**, 543 (1973).
[f] C. F. Midelfort, R. K. Gupta, and I. A. Rose, *Biochemistry* **15**, 2178 (1976).
[g] G. R. Gray and R. Barker, *Biochemistry* **9**, 2454 (1970).

[6] G. R. Goray, *Acc. Chem. Res.* **9**, 418 (1976).
[7] D. Keilin and E. F. Hartree, *Biochem. J.* **50**, 341 (1952).

ars, however, generally exhibit more rapid anomerization rates, the phosphoryl moiety apparently acting as an intramolecular general acid-base catalyst—a phenomenon that has its intermolecular counterpart in the phosphate-catalyzed mutarotation of glucose.[8] The rates for anomeric interconversion for phosphorylated sugars may be obtained in favorable cases from the measurement (slow exchange limit) of line broadening of the three anomeric carbon resonances in a ^{13}C NMR spectrum (the α, β, and carbonyl form) owing to a contribution to the resonance frequency arising from the lifetime of the carbon nucleus in each of the three differing chemical and magnetic environments.[9] In this manner the kinetics of mutarotation of D-fructose 1,6-bisphosphate at 25°, pH 7.2 were determined as summarized in Eq. (2)[10] Values for $k_{\alpha \to \beta}$ and $k_{\beta \to \alpha}$ are 8.1 and 1.8 sec^{-1}, respectively.

$$\beta\text{-Fru-1,6-P}_2 \underset{1450 \text{ sec}^{-1}}{\overset{35 \text{ sec}^{-1}}{\rightleftharpoons}} \text{Fru-1,6-P}_2 \text{ (keto)} \underset{8.5 \text{ sec}^{-1}}{\overset{70 \text{ sec}^{-1}}{\rightleftharpoons}} \alpha\text{-Fru-1,6-P}_2 \qquad (2)$$

From the aforementioned data, it should be obvious that experiments designed to probe the anomeric specificity of enzymes that utilize phosphorylated sugars require an experimental time scale in the hundreds of milliseconds. Consequently two kinetic methods, stopped-flow and rapid-quench,[11] have been utilized for the direct kinetic determination of anomeric specificity in many of these cases. However, several other approaches also have been employed in overcoming the experimental difficulties posed by rapid anomerization. The design and merit of these collective procedures are now considered.

Methods

Substrate Analogs

An indirect method avoiding the use of fast kinetic techniques has been the synthesis of substrate analogs that simulate the acyclic configuration or possess the stereochemistry of either the α- or β-anomer of the substrate but are locked chemically into a given configuration. If a given analog functions as a substrate, a favorable comparison of its V and K_m values with those determined for the natural substrate suggests the anomeric preference of the enzyme. Since nonproductive competitive bind-

---

[8] J. M. Bailey, P. H. Fishman, and P. G. Penchev, *Biochemistry* **9**, 1189 (1970).
[9] M. S. Gutowsky and A. Saika, *J. Chem. Phys.* **21**, 1688 (1953).
[10] C. F. Midelfort, R. K. Gupta, and I. A. Rose, *Biochemistry* **15**, 2178 (1976).
[11] B. Chance, R. H. Eisenhardt, O. H. Gibson, and K. K. Lonberg-Holm, eds., "Rapid Mixing and Sampling Techniques in Biochemistry." Academic Press, New York, 1964.

ing of the unreactive anomer often occurs, V for the reactive analog may exceed that for the natural substrate. It is particularly important to test analogs of both anomers and to demonstrate that the opposite configuration is unreactive or inhibitory, especially in cases where the reactive analog is a poor substrate. In this manner fructokinase from beef liver was shown to be specific for the β-furanose anomer of D-fructose (1) since neither 2,6-anhydro-D-mannitol (2) nor 2,6-anhydro-D-glucitol (3) (pyranose analogs) were phosphorylated whereas 2,5-anhydro-D-mannitol (4) and 2,5-anhydro-D-mannose (5) (β-furanose analogs) were converted into their 1-phosphates. The α-furanose analog, 2,5-anhydro-D-glucitol (6), was phosphorylated, but at the 6-carbon, which shows that it is acting as an analog of α-L-sorbofuranose, not of α-D-fructofuranose.[12]

Analog (4) is characterized by V and K_m values 0.5- and 10-fold greater than β-D-fructofuranose.

That phosphofructokinase (rabbit muscle) is totally β-anomer specific and capable of catalyzing phosphorylation of the C-1 hydroxyl without prior ring opening to an enzyme-bound keto form was deduced from experiments with the 6-phosphates of analogs (4) and (6). The rate of reaction with the β-anomer analog, 2,5-anhydro-D-mannitol 6-phosphate, under saturating conditions is about 30% greater than V for D-fructose 6-phosphate despite the approximately 4-fold greater affinity for the latter.[13,14] In contrast, no reaction occurs with the α-anomer analog,

[12] F. M. Raushel and W. W. Cleland, *J. Biol. Chem.* **248**, 8174 (1973).
[13] J. Bar-Tana and W. W. Cleland, *J. Biol. Chem.* **249**, 1263 (1974).
[14] T. A. W. Koerner, Jr., E. S. Younathan, A. L. E. Ashour, and R. J. Voll, *J. Biol. Chem.* **219**, 5749 (1974).

2,5-anhydro-D-glucitol 6-phosphate, although the analog is a weak inhibitor of D-fructose 6-phosphate. Likewise in the reverse reaction, the β-bisphosphate analog replaces D-fructose 1,6-bisphosphate, but the α-bisphosphate analog does not. Methyl β-D-fructofuranoside 6-phosphate, but not the α-fructofuranoside, functions as a substrate, but with a V about 10% that of the native substrate.[15]

The bisphosphate analogs have been employed to probe the anomeric stereospecificity of fructose bisphosphatase (rabbit liver). The enzyme is competitively inhibited by the α- and β-furanose bisphosphates derived from (4) and (6) as well as the corresponding methyl fructosides.[16] A deduction concerning anomeric specificity, based, however, on the relative inhibition constants, is tenuous since both unproductive and productive binding may be contributing. In short, the failure of the analogs to act as substrates means that the interactions between analog and enzyme cannot be optimal.

The use of a substrate analog often is dictated when investigating the reactivity of the free carbonyl form, owing to its low concentration and the corresponding difficulty in its detection in direct kinetic experiments. In cases where one of the anomeric forms is also reactive, the analog approach is the only means of readily demonstrating the reactivity of the acyclic species. The kinetic properties of the acyclic ketose bisphosphate, 5-deoxyfructose 1,6-bisphosphate, are reflected in values of V equal to 2.7-fold and 0.45-fold that for D-fructose 1,6-bisphosphate toward muscle and yeast aldolase, respectively.[10] In fact V/K_m for the analog is 50–60-fold higher with muscle aldolase, a factor[17] expected if the keto form were the only configuration with substrate activity since the two V/K_m values should be in the reciprocal ratio of the fraction keto. However, as described later, this apparently is not the case.

Direct Kinetic Determination

The rationale behind these procedures for the determination of anomeric specificity may be illustrated for the enzyme fructose bisphosphatase (rabbit liver), for which D-fructose 1,6-bisphosphate serves as the substrate. The choice of experimental substrate/enzyme concentration ratios depends on the turnover rate for the enzyme relative to the spontaneous anomerization rates noted above. In the case of an enzyme turn-

---

[15] R. Fishbein, P. A. Benkovic, K. J. Schray, I. J. Siewers, J. J. Steffens, and S. J. Benkovic, *J. Biol. Chem.* **249,** 6047 (1974).

[16] S. J. Benkovic, J. J. Kleinschuster, M. M. deMaine, and I. J. Siewers, *Biochemistry* **10,** 4881 (1971).

[17] I. A. Rose, *Annu. Rev. Biochem.* **35,** 23 (1966).

over number less than the rate of resupply of the reactive anomer by the remaining substrate pool, a one-turnover experiment of a substrate:enzyme concentration ratio approaching unity is demanded in order to obtain data sufficiently biphasic for interpretation. The biphasic nature of plots for the time course of product formation results from the rapid conversion of the reactive anomer to product followed by subsequent slower product formation that is determined by resupply of the reactive anomer through the spontaneous anomerization rate. The experimental time scale, which is determined by the latter rate, must be generally less than 500 msec. With fructose bisphosphatase ($k_{cat} \simeq 10$ sec^{-1}), a substrate:enzyme concentration ratio of approximately 2–3 is optimal. The amount of product formed in the initial phase should be directly related to the fraction of active anomer in the substrate pool. In situations where more than one turnover is required to remove the reactive anomer, the rate of the initial phase is enzyme concentration dependent. The latter phase ideally should be independent of enzyme concentration, if the rate-determining process is simply spontaneous or buffer-catalyzed anomerization, but dependent on its concentration if the enzyme catalyzes the anomerization or lacks anomeric specificity.

A representative plot of D-fructose 6-phosphate formation as a function of time catalyzed by fructose bisphosphatase established by a rapid-quench technique is illustrated in Fig. 1.[18] The initial phase is too rapid to observe; however, the second slower rate extrapolates to an

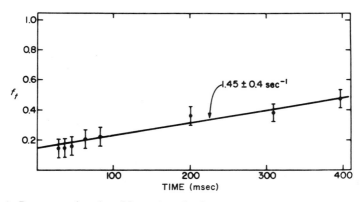

FIG. 1. Representative plot of formation of D-fructose 6-phosphate as a function of time (f_t) from D-fructose 1,6-bisphosphate catalyzed by fructose bisphosphatase. Conditions: [FBPase] = 2.9 μM; [fru-1,6-P$_2$] = 9.65 μM.

[18] W. A. Frey, R. Fishbein, M. M. deMaine, and S. J. Benkovic, *Biochemistry* **16**, 2479 (1977).

average ordinate value of 0.18 ± 0.02. The calculated rate coefficient for the second phase is independent of enzyme concentration with an average value of 1.45 ± 0.4 sec^{-1}, which compares favorably with $k_{\beta-\alpha}$ measured by ^{13}C NMR. The data derived from the rapid-quench experiments can be readily simulated by computer modeling of a reaction system in which the α-anomer is the sole reactive species, despite binding of the β-anomer and free carbonyl form as revealed by studies with substrate analogs. However, the rate constant for the initial rapid phase is greater than 10 sec^{-1} (k_{cat}) and is consistent with our recent finding of a rapid transient phase in approaching the kinetic steady state.[19]

The anomeric specificity of phosphofructokinase (rabbit muscle) was established both by stopped-flow experiments[20] coupling the ADP formed in the kinase-catalyzed process to the oxidation of NADH through reactions catalyzed by pyruvate kinase and lactic dehydrogenase, respectively, and by rapid-quench kinetic procedures.[21] In the former, sufficient phosphofructokinase was present to reduce the active anomer of the substrate to a limiting concentration within 100–300 msec. The progress curve for the reaction is markedly biphasic; the initial rate is accelerated by increasing kinase activity, whereas the second, slower phase is not influenced by alterations in kinase activity. Since in the initial phase approximately 76% of the equilibrium mixture of D-fructose 6-phosphate is consumed, the kinase is β-anomer specific. A similar biphasic progress curve was established by a series of rapid-quench experiments (Fig. 2) in which D-fructose 6-phosphate and [δ-^{32}P]ATP were incubated with sufficent kinase to reduce the β-anomer to a limiting concentration within 100 msec. The reaction was quenched with perchloric acid at preselected time intervals. In these two sets of experiments the substrate:enzyme concentration ratio was much greater than unity since k_{cat} for the kinase is greater than 100 sec^{-1}. The limiting value at $f_t \simeq 0.8$ is consistent with preferred utilization of the β-anomer. Computer simulations utilizing the established kinetic parameters for the kinase[22] are displayed as the solid and dashed lines in Fig. 2. They limit at $f_t = 1.0$ and 0.8, respectively, and assume that either all the initial sugar phosphate is available for reaction or only the β-anomer is reactive. Consequently, the second phase does not result from diminished total substrate concentration or increasing product inhibition, but a limiting anomer concentration. Although the

---

[19] P. A. Benkovic, M. Hegazi, B. A. Cunningham, and S. J. Benkovic, *Biochemistry* **18**, 830 (1979).

[20] B. Wurster and B. Hess, *FEBS Lett.* **38**, 257 (1974).

[21] R. Fishbein, P. A. Benkovic, K. J. Schray, I. J. Siewers, J. J. Steffens, and S. J. Benkovic, *J. Biol. Chem.* **249**, 6047 (1974).

[22] R. L. Hanson, F. B. Rudolph, and H. A. Lardy, *J. Biol. Chem.* **248**, 7852 (1973).

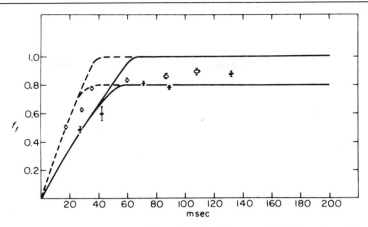

FIG. 2. Plots of the time course of formation of D-fructose 1,6-bisphosphate at two levels of phosphofructokinase: 1102 units/ml (●) and 1920 units/ml (○). Solid and dashed lines were calculated as described in the text.

stopped-flow method offers the advantage of a continuous progress curve, the high concentrations of both the enzyme in question and the coupling enzymes that generally are necessary may give rise to nonspecific binding effects. Thus it is usually not reliable to determine the spontaneous anomerization rate of the remaining sugar species with either the stopped-flow or rapid-quench method in the absence of information about binding of the unreactive anomer. The β-anomeric specificity of phosphofructokinase is in agreement with the substrate analog experiments described above which permitted a more accurate assessment of possible α-anomer reactivity.

The kinase- and phosphatase-catalyzed reactions are irreversible under the experimental conditions. The direct kinetic determination by either rapid-quench or stopped-flow methods of anomeric specificity in cases where an approach to equilibrium is being monitored can be achieved similarly, provided that there is no significant contribution by epimerase activity or unfavored anomer utilization in the fast initial equilibrium formation. Experiments with phosphoglucose isomerase furnish an example of this situation.[23]

In order to establish the anomeric specificity of the enzyme toward D-fructose-6-P, sufficient isomerase is allowed to react with the substrate to rapidly establish an equilibrium between the reactive forms of D-fructose 6-phosphate and D-glucose 6-phosphate in the initial rapid phase.

[23] K. J. Schray, S. J. Benkovic, P. A. Benkovic, and I. A. Rose, *J. Biol. Chem.* **248**, 2219 (1973).

As before, the extent of the initial reaction can be obtained by extrapolation to zero time. This value may then be compared with predictions based on equilibrium values with the various anomeric forms. From the known anomeric distributions and the equilibrium for the interconversion of the two sugar phosphates, the equilibrium proportions of α-D-glucose 6-phosphate : β-D-glucose 6-phosphate : α-D-fructose 6-phosphate : β-D-fructose 6-phosphate is 1:1.63:0.15:0.59, respectively. If α-D-fructose 6-phosphate is the substrate, then $1/1.15 \times 20\%$, or 17% of the total sugar, will react in the initial rapid phase. A similar calculation performed for β-specificity predicts that $1.63/2.22 \times 80\%$, or 77% of the total sugar, will react initially. In this manner the specificity of the isomerase for the α-anomer was demonstrated. The observation that the second phase slope was dependent on the concentration of the isomerase further suggested that the enzyme additionally was able to catalyze a mutarotation process.

Alternatively, the anomeric specificity for a reversible process with a less favorable equilibrium can be determined by employing or trapping reagent in order to attain complete conversion. Such a procedure was employed in a series of rapid-quench experiments with muscle aldolase acting on D-fructose 1,6-[6-^{32}P]bisphosphate using unlabeled glyceraldehyde 3-phosphate as a trapping agent.[24] The equilibrium for the cleavage process is $10^{-4} M$; in the presence of 100-fold excess aldehyde, effectively total cleavage of the sugar bisphosphate was achieved. Since approximately 75% of the D-fructose 1,6-bisphosphate was utilized in the initial rapid phase, the muscle enzyme appears able to utilize the β-anomer as well as the keto carbonyl form, a conclusion also arrived at through the coupled stopped-flow technique.[25] In contrast the 2-keto-3-deoxygluconate aldolase was shown to be specific for the open-chain keto form by the coupled stopped-flow method.[26]

Generation of the Desired Anomer in Situ

The discovery that hexokinase utilized either α- or β-D-glucose made feasible the design of experiments dependent on the availability of the separate anomeric forms of D-glucose 6-phosphate.[27] The β-anomeric specificity of glucose-6-phosphate dehydrogenase was determined by coupling this enzyme (in excess) to the α- or β-D-glucose 6-phosphate

[24] K. J. Schray, R. Fishbein, W. P. Bullard, and S. J. Benkovic, *J. Biol. Chem.* **250**, 4883 (1975).
[25] B. Wurster and B. Hess, *Biochem. Biophys. Res. Commun.* **55**, 985 (1974).
[26] C. F. Midelfort, R. K. Gupta, and H. P. Meloche, *J. Biol. Chem.* **252**, 3486 (1977).
[27] M. Salas, E. Viñuela, and A. Sols, *J. Biol. Chem.* **240**, 561 (1965).

generated by hexokinase from the respective D-glucose anomer. The experiments were conducted at 5° in order to slow the rate of anomerization. Linear kinetics were observed for the coupled system if β-D-glucose served as substrate, whereas an initial lag before the onset of linear kinetics was observed with α-D-glucose. The initial phase was analyzed to yield the anomerization rate of α → β-D-glucose 6-phosphate, a process accelerated by phosphoglucose isomerase owing to its anomerase activity.

The ability to generate β-fructose 1,6-bisphosphate via the phosphorylation process catalyzed by phosphofructokinase was employed to further test the anomeric specificity of muscle aldolase.[28] If the β-form may be used only after spontaneous conversion to the keto species in solution (a possibility indicated by the high reactivity of the acyclic substrate analog) then a lag phase in triose phosphate formation catalyzed by the muscle aldolase may be encountered at low temperatures. Moreover, since the affinity of the enzyme for similar bisphosphate inhibitors is approximately 10 μM, a considerable fraction of the β-species should be bound to the aldolase so that the rate constant characterizing the lag phase would be correspondingly reduced relative to the spontaneous rate for the β→ open step. It was found that under conditions where only approximately 2% of the total D-fructose 1,6-bisphosphate is free, the turnover of the complexed β-D-fructose 1,6-bisphosphate was 20-fold greater than that in which spontaneous ring opening is a required step. Consequently, the ability of muscle aldolase to employ both the β-anomer as well as the free keto form is reaffirmed.

Conclusions

It is apparent that the combination of ^{13}C NMR and rapid kinetic techniques should permit the resolution of most problems involving the anomeric specificity of carbohydrate-utilizing enzymes. The use of substrate analogs, particularly in those situations where specificity for a minor free carbonyl form is suspected, provides an important supplemental method.

---

[28] I. A. Rose and E. L. O'Connell, *J. Biol. Chem.* **252**, 479 (1977).

Section II
Inhibitor and Substrate Effects

[15] Reversible Enzyme Inhibition

By JOHN A. TODHUNTER

The development of enzyme kinetics has placed in the hands of the practicing enzymologist a number of powerful techniques for the study of enzymic catalysis. Of the various enzyme kinetic approaches, the use of reversible inhibitors is perhaps one of the most widely applicable yet can yield a wealth of information to the investigator well versed in their uses and subtleties. Basically, the reversible inhibitor approach is a perturbation method in which the binding of substrate to enzyme is perturbed by the presence of the inhibitor. From studies with inhibitors, information can be obtained as to the relative magnitudes of the various kinetic parameters that describe the system, the order of substrate and product sorption and desorption, and, from the chemical structures of the inhibitors and their relative potencies, the geometry of the active site. In common with other steady-state kinetic methods, no direct information about the bond deformations and isomerizations that result in catalysis can be obtained. With the information culled from reversible inhibitor studies, reasonable models for these latter processes can be formulated for testing by chemical studies.

With this orientation, some limits must be set on the extent of this discussion. The subject of irreversible inhibitors will be covered only tangentially. This is because the techniques to be described assume that the relative proportions of enzyme with bound substrate, S, or inhibitor, I, are determined by the ratios of S to I. It should be apparent that if the inhibitor does not desorb readily from the enzyme, then this assumption is invalid. Where this discussion may touch upon irreversible inhibitors is on the relationship between the binding of these and that of reversible inhibitors.

The subject of multiple inhibition will not be discussed, as it is a specialized area beyond the scope of this chapter.

The topics to be dealt with in this chapter will be the various kinetic types of inhibition and problems associated with distinguishing them, contamination problems in inhibitor studies, pseudo-inhibition, classes of competitive inhibitors, and optimal experimental design and data reduction. It is expected that the reader will already be familiar with the assumptions and limitations of steady-state kinetics, the derivation of rate equations, the general properties of enzyme active and allosteric sites, and the general principles of kinetic experiments. These topics, for the reader unfamiliar with them, are thoroughly covered in Sections I and II of this volume.

Theory and Examples

General Kinetic Theory of Reversible Inhibition

The interaction of a reversible inhibitor, I, with an enzyme can happen in several ways. These can be represented generally by the following scheme, based on the analysis of Webb.[1]

$$\begin{array}{ccc} E & \xrightarrow{K_s} & ES \xrightarrow{k} E + P \\ {\scriptstyle K_i} \downarrow & & \downarrow {\scriptstyle \alpha K_i} \\ EI & \xrightarrow{\alpha K_s} & EIS \xrightarrow{\beta k} EI + P \end{array}$$

The term K_s is the binding constant of substrate to the free enzyme and K_i is the dissociation constant for the EI complex. The alteration in substrate affinity due to bound inhibitor (and vice versa) is expressed by an interaction constant α. The inhibitor effect on the rate of decomposition of the EIS complex to products is denoted by the constant β. This scheme is very general, since appropriate values of α and β will allow the agent I to function as either an inhibitor or an activator of the enzyme reaction. This discussion pertains, also, only to agents that bind reversibly to the enzyme. This is distinguished from an irreversible inhibition in that the binding of I to E is expressed by an equilibrium or by kinetic constants of the order of magnitude of those that express substrate binding. Irreversible inhibition is a time-dependent process, cannot be adequately represented by an equilibrium, and often involves covalent attachment of the inhibitor to the enzyme.

The velocity expression that can be derived for the scheme given above is[1]

$$v_i = V_m \frac{(S)[\alpha K_i + \beta(I)]}{(S)(I) + \alpha[K_i(S) + K_s(I) + K_s K_i]}$$

where v_i is the initial reaction velocity and $V_m = K(E)_{Total}$ is the maximal rate attained when the enzyme active sites are saturated by substrate. A fractional inhibition, i, may be defined as

---

[1] J. L. Webb, "Enzyme and Metabolic Inhibitors," Vols. 1–3. Academic Press, New York, 1963, 1965, 1966.

$$i = \frac{(I)[(S)(1 - \beta) + K_s(\alpha - \beta)]}{(I)[(S) + \alpha K_s] + K_i[\alpha(S) + \alpha K_s]} = 1 - a$$

where a is the fractional activity and is defined as V_i/V_m.

Five types of inhibition may result from the above scheme depending on the values of α and β[1]:

1. Fully competitive inhibition arises when $\alpha = \infty$ and $\beta = 0$. The bound inhibitor completely prevents binding of the substrate.
2. Partially competitive inhibition occurs when $\infty > \alpha > 1$ and $\beta = 1$. Bound inhibitor only partially blocks binding of substrate and does not alter the rate of product formation from substrate.
3. When $\alpha = 1$ and $\beta = 0$, the resulting inhibition is fully noncompetitive. Bound inhibitor does not affect the binding of substrate (and vice versa) but prevents product formation.
4. If $\alpha = 1$ and $0 < \beta < 1$, a partially noncompetitive inhibition results. Substrate binding is not affected, but product formation from EIS is slower than from ES. Kinetically, this is not distinguishable from fully noncompetitive inhibition.
5. If $\alpha < 1$ and $0 < \beta < 1$, then the result is an uncompetitive inhibition. Binding of substrate to free enzyme is not affected, but the rate of product formation is decreased by the presence of inhibitor.

These types of inhibition give rise to characteristic changes in the K_m and V_m values that are determined in kinetic experiments. The nature of these changes, how they arise, and how they can be predicted from reaction schemes, is discussed below.

Characteristics of Different Classes of Inhibition and the Prediction of Inhibition Patterns

The practicing kineticist must always be aware that the kinetic expression of an inhibition is the result of physical processes occurring on the enzyme surface. As stated in the preceding section, inhibitors may compete with the substrate for the active site, may bind to the catalytic complex preventing or modulating further reaction, or both. Depending on what actually happens, different patterns or types of inhibition arise. To understand how these interactions affect the observed kinetic behavior of a system, it must be remembered that in any individual kinetic experiment what is determined is an initial reaction velocity, and, in a set of such experiments, the Michaelis constant, K_m, and the maximal velocity, V_m, are usually determined. It must then be realized that the analysis of kinetic data from these experiments is done, is practice, by the use of double-reciprocal plots (Lineweaver–Burk) or, less commonly, by some other plotting device, such as Hanes plots or Edee–Hofstee plots. The result is that, even with complex inhibitor systems, the investigator is looking for slope and intercept changes in the plots.

The following discussion of the single-substrate reaction will serve to illustrate the kinetic classes of inhibition and elucidate a set of rules for predicting inhibition patterns. These prediction rules were first stated by Cleland in general form[2] but are reformulated here in concise terms. Using Cleland's notation,[2] the single-substrate reaction may be represented by Scheme 1.

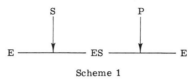

Scheme 1

where E, S, and P represent enzyme, substrate, and product, and ES the catalytic complex. When these symbols are written in parentheses, then the concentration of that species is implied. Since we will be discussing reaction fluxes, the terms downstream and upstream will be used to refer to reaction in the direction of product formation or the reverse, respectively, as per Cleland.[2] The prediction rules will be stated first, and then the theoretical basis for them developed. The rules comprise the following two statements.

Rule 1. If the inhibitor I interacts with the enzyme form to which S binds or to any form upstream of S which is a precursor of the S binding form, then there will be a K_m (slope) effect.

Rule 2. If I interacts with any enzyme form occurring prior to the limiting step of the reaction but to which S cannot bind directly, then there will be a V_m (intercept) effect.

The implementation of these rules will be shown in the following discussion of inhibition patterns (types). For the reaction depicted in Scheme 1, an inhibitor could interact as given in Schemes 1a–1c, where k_1 and k_2 represent the rates of product formation from the ES and ESI complexes, respectively. The value of k_2 may equal 0. These cases will be discussed individually.

For Scheme 1a it is assumed that either S or I may be present on the enzyme surface, but not simultaneously. This may arise because S and I are structurally similar and compete for the active site or from mutually exclusive binding due to allosteric effects or binding of I so as to cover over the active site. All these possibilities are represented in the litera-

---

[2] W. W. Cleland, in "The Enzymes" (P. D. Boyer, ed.), 3rd ed., Vol. 2, p. 1. Academic Press, New York, 1970.

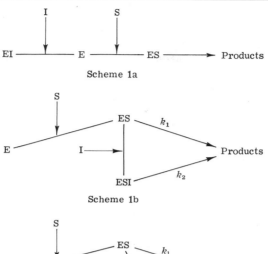

Scheme 1a

Scheme 1b

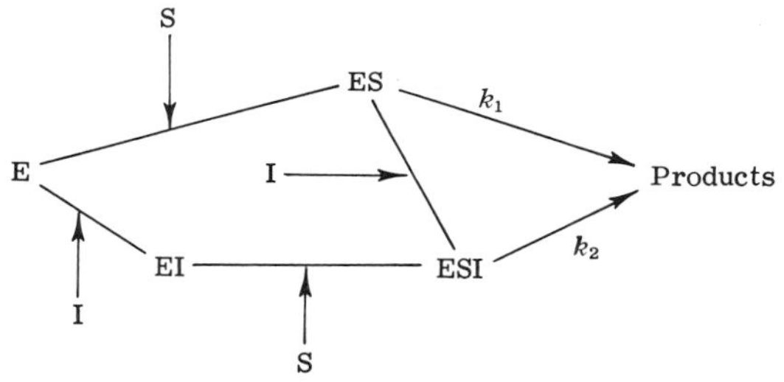

Scheme 1c

ture, yet are impossible to distinguish kinetically. Knowledge of the structures of the inhibitor and substrate and information about the number of inhibitor sites and their affinities can serve as useful guides in such situations. For all situations described by Scheme 1a the slope and intercept effects caused by the presence of I can be determined as follows. The rate equation for Scheme 1a (or any single-substrate reaction) is given in Dalziel[3] form [Eq. (1)]

$$\frac{(E)_0}{v} = \phi_0 + \frac{\phi_s}{(S)} \tag{1}$$

or Cleland[2] form [Eq. (2)].

$$\frac{1}{v} = \frac{1}{V_m} + \frac{K_m}{V_m (S)} \tag{2}$$

Comparing Eqs. (1) and (2), it can be seen that

$$1/V_m = \phi_0/(E)_0; \qquad K_m/V_m = \phi_s/(E)_0$$

[3] K. Dalziel, *Acta. Chem. Scand.* **11**, 1706 (1957).

These expressions represent the intercept and slope terms, respectively, of the double-reciprocal plot for the single-substrate system. In the absence of inhibitor or other processes that may compete for the enzyme, $(E)_0$ is defined as the initial enzyme concentration. This definition is made on the assumption that all the enzyme present is free to combine with S. We will define a more general term, $(E)_{eff}$, which will represent the initial concentration of enzyme available for the binding of S. Clearly, $(E)_{eff} = (E)_0$ only if there are no inhibitors or processes that compete for the enzyme present. The effect of I on the intercept may now be considered. Since

$$v = k_{cat}(ES)$$

it should be apparent that $v = V_m$ only when $(ES) = (E)_0$ if $(E)_0 = (E)_{eff}$. This can be stated in the general form:

$$V_m = \lim_{(S)\to\infty} k_{cat}(ES)$$

For the noninhibited system the value of this expression approaches $k_{cat}(E)_0$. For the inhibition system represented in Scheme 1a the same limiting value is obtained, since $(E)_0 = (E)_{eff} + (EI)$ and $(EI) \to 0$ as $(S) \to \infty$ owing to the relation

$$(EI) = \frac{K_s(ES)(I)}{(S)K_i}$$

where K_s and K_i are the dissociation constants of the ES and EI complexes, respectively, and (ES) is always $\ll$ (S) owing to the assumption of steady-state conditions.[3] This is the mathematical equivalent of the statement that, if S and I bind to the same enzyme form in a mutually exclusive fashion, then by raising (S) sufficiently all of E can be pulled into the ES complex. Since the same limiting value of V_m is obtainable, then there is no intercept effect in Scheme 1a. This is the result that would be predicted by Rule 2, since S can bind directly to the same form of the enzyme to which I binds even though this form occurs before the limiting step of the reaction.

The slope (or K_m) effect can be easily predicted by the expression

$$\text{slope} = \phi_s/(E)_{eff} = \phi_s/[(E)_0 - (EI)]$$

from which it is clear that the slope will always be increased by the presence of an inhibitor which binds to E. The same effect is obtained if I binds to any form of E upstream from S as will be shown later.

For Scheme 1a, then, there is a slope but no intercept effect. This type of inhibition is referred to as fully competitive inhibition and is illustrated by Fig. 1A.

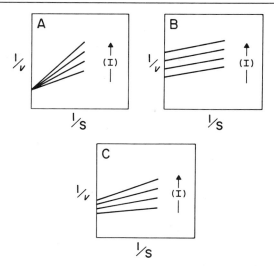

FIG. 1. Representations of slope and intercept effects in double-reciprocal plots as the concentration of inhibitor, (I), is raised for (A) fully competitive inhibition; (B) uncompetitive inhibition; and (C) noncompetitive inhibition.

Scheme 1b presents a different situation. I cannot bind to E, and so $(E)_{eff} = (E)_0$ and no slope effect is possible. The velocity is now represented by the expression

$$v = k_1(ES) = k_2(ESI)$$

and $0 \leq k_2 < k_1$ since I is, by definition, an inhibitor. Under V_m conditions, $(S) = \infty$ and $(E)_0 = (ES) + (ESI)$. Applying the earlier definition of V_m,

$$V_m = \lim_{(S) \to \infty} k_{cat}(ES) = k_{cat}(E)_0$$

for the noninhibited system, it can be seen that this limit becomes $k_{cat}[(E)_0 - (ESI)]$ when inhibitor is present. This predicts that the intercept will increase as (I) increases. This type of inhibition, represented by Scheme 1b, is called uncompetitive inhibition and is shown in Fig. 1B. This type of inhibition is seldom encountered in single-substrate reactions but is common in multisubstrate systems (this volume [19]).

The type of inhibition given by Scheme 1b would have been predicted by the rules given above, since I does not bind to the same enzyme form as S (no K_m effect, therefore, by Rule 1) and binds to an enzyme form at or prior to the rate-limiting step, but not the same form as that to which S binds (V_m effect by Rule 2).

For Scheme 1c, the same arguments apply to the value of V_m in the presence of inhibitor as for Scheme 1b. There is, therefore, an intercept effect, as would have been predicted by Rule 2. For this system, however, $(E)_{eff} = (E)_0 - (EI) - (ESI)$, and, as for Scheme 1a, there is a slope effect. This would also have been predicted from Rule 1. This type of inhibition is called noncompetitive inhibition and is depicted in Fig. 1C.

A complication arises with Scheme 1c if $k_1 = k_2$. In Scheme 1b this was a trivial situation because in that case no effect of I would be detectable. For Scheme 1c the V_m effect would vanish, but a K_m effect would remain (since ESI would be kinetically equivalent to ES). The inhibition would be indistinguishable from that given by Scheme 1a (Fig. 1A). This type of inhibition is called partially competitive and, while it cannot be distinguished from fully competitive inhibition on double-reciprocal plots, it can be differentiated on the basis of plots of $1/v$ vs (I). This sort of plot is linear for fully competitive systems but nonlinear for the partially competitive case (see below). This also points out that, for Rule 2 to be an effective predictor, the binding of I must result in an enzyme form of reduced or no catalytic efficiency.

The above analyses should give the reader a clear definition of the types of inhibition that occur and an example of how the prediction rules for inhibitor effects are applied. When dealing with multisubstrate systems the same types of inhibition patterns are encountered. This is due to the practice, when dealing with such systems, of reducing them to one-substrate approximations by holding one or more of the substrates at constant nonsaturating concentrations (see Sections I and II of this volume). The prediction rules given apply to these systems as well, but the multisubstrate case will not be treated here in detail. As an example only, the inhibition pattern for the Ordered Bi-Bi mechanism will be considered in terms of the prediction rules. For purposes of discussion, I_a and I_b will be competitive inhibitors for substrates A and B, respectively. The system can be represented by the following scheme:

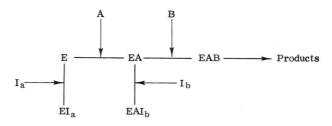

I_a binds to the same enzyme form as A and, by Rule 1, alters K_{ma}. The only enzyme form with which I_a reacts prior to the limiting step is also the form to which A binds, and so, by Rule 2, there is no effect on V_{ma}. I_a is clearly fully competitive with A. The inhibition of I_a relative to B is another matter. Since I_a binds to a precursor of the enzyme form to which B binds, Rule 1 predicts a K_{mb} effect. Note, however, that at saturating B the reaction becomes limited by the formation of EA, since A is nonsaturating. I_a binds prior to this step, but not to the same enzyme form as does B. Rule 2 would, therefore, predict an effect on V_{mb}. I_a is, thus, noncompetitive for B. It is important to realize that to see the proper inhibitor effect the nonvaried substrate must be kept at a nonsaturating concentration.

The inhibition of I_b relative to B is competitive, since it binds to the same form as B (Rule 1, K_{mb} effect), but after the limiting EA formation (Rule 2, no V_{mb} effect). I_b binds prior to the limiting EAB formation when A is varied, but not to the same form as does A. There is, by Rule 2, a V_{ma} effect. I_b does not bind to any enzyme form with which A binds, and so no K_{ma} effect is possible (Rule 1). I_b is, therefore, uncompetitive with A.

The above predicted inhibition patterns for the ordered Bi Bi reaction are, in fact, the observed ones, and it can be seen that the prediction rules are valid for multisubstrate systems.

The Determination of Inhibitor Constants

The preceding section dealt with how typical inhibition patterns arise according to the nature of the inhibitor–enzyme interaction. The strength of these interactions is governed by the dissociation constants of the various inhibitor:enzyme complexes. In general, the inhibitor dissociation constant is written as K_i defined by the equation

$$K_i = \frac{(I)(E)}{(EI)}$$

where the subscript i refers to EI as the first inhibitor complex formed in the reaction stream. Subsequent complexes would be denoted as K_{ii}, K_{iii}, and so on. From the definition of K_i it can be seen that the affinity of I for the enzyme is inversely proportional to the magnitude of K_i. It is extremely important to realize that K_i says nothing about the magnitude of the effect on k_{cat} that may be exerted by an inhibitor binding to the ES, or other catalytic, complex. K_i is therefore an index to the potency of an inhibitor only for strictly competitive inhibitors. Nothing can be said *a priori* about the potency of k_{cat} inhibitors (uncompetitive and noncompetitive) from the K_i value unless the changes in k_{cat} caused by inhibitor binding are comparable.

For any given type of inhibition the values of the various K_i's can be obtained from kinetic experiments. Because of the particular forms of the steady-state rate expressions for the different types of inhibition, the exact protocol for determining K_i values varies. A summary of the protocols to be used with the single-substrate reaction is contained in Table I. Protocols for multisubstrate systems are based on those given in Table I by the expedient of reducing the system to one variable case by holding all substrates but one at constant nonsaturating levels (usually about K_m).

As can be seen from Table I, the first step for all types of inhibition, as defined in the preceding section, is to plot $1/v$ vs (I). For competitive and uncompetitive systems K_i is obtained directly, as the table indicates. For greater accuracy $1/v$ vs (I) should be plotted at various values of (S), and the slopes replotted as a function of $1/(S)$ in the competitive inhibition system. Noncompetitive inhibitors will give linear $1/v$ vs (I) plots, but only an apparent K_i, which contains K_i and K_{ii} terms. These values can be extracted by replots of the slopes and intercepts of the $1/v$ vs $1/(S)$ plots as a function of (I) to determine K_i and K_{ii}, respectively, as shown in Table I. Partial competitive inhibition has the interesting property of giving a nonlinear $1/v$ vs (I) plot, and this is its distinguishing characteristic. The inhibition is self-limiting and approaches the noninhibited rate at very high values of (I), since the bracketed term in the rate expression (Table I) approaches a value of unity as (I) approaches infinity. At low values of (I), the limiting slope of the $1/v$ vs (I) plot can be used to estimate K_i since in this region the form of the expression approaches that of the competitive case and the upper portion of the bracketed term dominates. This is because $K_i < K_{ii}$ for I to act as an inhibitor in the partially competitive case. Using the estimate for K_i so obtained, the value of K_{ii} can be obtained by numerical methods (this volume [6]).

Inhibitor constants for multisubstrate systems are determined by protocols similar to those above once the type of inhibition is established. This area will not be further expanded here.

Design of Inhibitor Compounds

Even though useful inhibitors may at times be found through serendipity, the usual approach is to prepare them specifically for the intended use. We will deal here only with inhibitors useful for kinetic studies and defer the treatment of other applications. Inhibitors used for kinetic studies fall into three broad categories. These are substrate-competitive inhibitors, cofactor competitors, and transition-state analogs. The first two categories of compounds are used primarily in steady-state kinetic studies. Noncompetitive and uncompetitive inhibitors are not designed by

TABLE I
RATE EXPRESSIONS FOR INHIBITION: UNI-REACTANT SYSTEMS AND PROTOCOLS FOR THE DETERMINATION OF INHIBITION CONSTANTS

Inhibition type[a]	Rate expression[b]	Determination of K_i values[c]
FC	$\dfrac{1}{v} = \dfrac{1}{V_m} + \dfrac{K_m}{V_m(S)}[1 + (I)/K_i]$	Plot $1/v$ vs (I); $K_i = \dfrac{K_m}{V_m(S)} \cdot$ slope from $1/v$ vs $1/(S)$ plot slopes vs (I); $K_i = K_m/V_m \cdot$ slope
UC	$\dfrac{1}{v} = \dfrac{1}{V_m} + \dfrac{(I)}{V_m K_{ii}} + \dfrac{K_m}{V_m(S)}$	Plot $1/v$ vs (I); $K_i = \dfrac{1}{V_m} \cdot$ slope from $1/v$ vs $1/(S)$ plot intercepts vs (I); $K_i = 1/V_m \cdot$ slope
NC	$\dfrac{1}{v} = \dfrac{1}{V_m} + \dfrac{(I)}{V_m K_{ii}} + \dfrac{K_m}{V_m(S)} + \dfrac{K_m(I)}{V_m(S)K_i}$	From $1/v$ vs $1/(S)$ plots: plot slopes vs (I); $K_i = \dfrac{K_m}{V_m} \cdot$ slope plot intercepts vs (I); $K_{ii} = 1/V_m \cdot$ slope
PC	$\dfrac{1}{v} = \dfrac{1}{V_m} + \dfrac{K_m}{V_m(S)} \dfrac{1 + (I)/K_i}{1 + [(I)/K_{ii}](K_m/K_m')}$	$1/v$ vs (I) plots nonlinear, replots nonlinear. See text for details of K_i estimation

[a] F, U, N, and P denote fully-, un-, non-, and partially competitive inhibition.
[b] $K_m' = $ (S) (EI)/(EIS).
[c] The term slope refers to the slope of the point indicated.

purpose. For multisubstrate enzymes, these types of inhibition arise by the action of substrate competitive inhibitors. Those of the third group give, in general, more complex inhibition patterns, are often tight binding, and are used both in kinetic studies and in catalytic site mapping.[4-6] We will discuss substrate and cofactor competitors together and then deal with transition-state analogs separately.

Substrate and cofactor competitors function, in general, by reason of structural similarity or isosterism to the true substrate or cofactor. The inhibitor compound is often designed, however, with a nonreactive functional group at the site of chemical reaction on the substrate (the reactive site) or merely lacks any functionality at the reactive site. Typical of the first group are the phosphonate analogs of the nucleoside phosphates, such as β,γ-methylene-adenosine 5′-triphosphate, in which the methylene linkage between the β and γ phosphates is nonhydrolyzable. Compounds of this type find considerable use as inhibitors of ATP-utilizing enzymes.[7-10] Inhibitors lacking a functional group entirely can be represented in the extremes by adenosine disphosphate ribose (ADPR), which is a potent inhibitor of several NADH-dependent dehydrogenases.[11-15] This compound is inhibitory by virtue of its resemblance to the carbohydrate portion of NADH and competes for the cofactor site of the dehydrogenase. Since it lacks the nicotinamide function, it cannot, however, participate in the reaction.

A very large number of enzymes utilize nucleotides, NADH, NADPH, folic acid derivatives, or organic acids as substrates or cofactors. Consequently there are a number of inhibitors that are analogs of these compounds and have been used as inhibitors for several enzymes. For enzymes with more specialized substrates, there are also numerous specific inhibitors. Since this is not a review article, but, rather, a guide for the laboratory investigator, a number of inhibitor compounds are cited

---

[4] J. D. Gass and A. Meister, *Biochemistry* **9**, 1380 (1970).
[5] J. D. Gass and A. Meister, *Biochemistry* **9**, 842 (1970).
[6] A. Meister and S. S. Tate, *Annu. Rev. Biochem.* **45**, 559 (1976).
[7] L. N. Simon and T. C. Meyers, *Biochim. Biophys. Acta* **51**, 178 (1961).
[8] M. R. Atkinson and A. W. Murray, *Biochem. J.* **104**, 10 (1967).
[9] J. W. B. Hershey and R. E. Monroe, *J. Mol. Biol.* **18**, 68 (1966).
[10] T. Ohta, S. Sarkar, and R. E. Thach, *Proc. Natl. Acad. Sci. U. S. A.* **58**, 1638 (1967).
[11] T. K. Li and B. L. Vallee, *J. Biol. Chem.* **239**, 792 (1964).
[12] C. H. Reynolds, D. L. Morris, and J. S. McKinley-McKee, *Eur. J. Biochem.* **14**, 14 (1970).
[13] D. Peacock and D. Boulter, *Biochem. J.* **120**, 763 (1970).
[14] A. McPherson, *J. Mol. Biol.* **51**, 39 (1970).
[15] K. Chandrasekhar, A. McPherson, M. Adams, and M. Rossmann, *J. Mol. Biol.* **76**, 503 (1973).

in Table II along with representative enzymes against which they act and leading references to the primary literature. These will be discussed briefly, by classes, with references to the principles underlying the competition at the active site.

The first group of compounds in Table II comprises the nucleoside phosphonates. These are phosphonic acid analogs of the normal phosphoric acid diester bonds of nucleoside phosphates. The phosphonic acid linkage is not hydrolyzable. As such, these compounds are good inhibitors of reactions that depend on cleavage of the phosphodiester bond. Nucleoside phosphonates with a β,γ-methylene linkage inhibit triphosphatases, kinases, triphosphate-linked synthetases, and other triphosphatase-linked processes.[7-10] If the methylene linkage is from α to β, then pyrophosphate-producing reactions are inhibited. These would be processes such as those catalyzed by adenylyl cyclase, and various nucleotide polymerases.[16,17] Related to the nucleoside phosphonates are the nucleoside phosphoramidates, in which the phosphodiester linkage is replaced by a phosphoramide. This is also a relatively inert linkage, and these compounds have similar inhibitor properties to the nucleoside phosphonates. Interestingly, cyclic phosphonates and phosphoramides have been used as cAMP analogs.[18]

Chromium nucleotides, developed by Cleland and co-workers,[19] constitute yet another useful class of nucleotide inhibitors. The true substrate of reactions involving phosphodiester cleavage is the Mg–nucleotide complex with dissociation of P_i from the Mg–nucleotide product complex. P_i dissociation from Cr^{3+} is very slow,[20] and the Cr–nucleotide complexes are, therefore, relatively stable analogs of the Mg–nucleotide complexes. Cr–nucleotides have found considerable use as kinetic probes[19-22] and can be used for magnetic resonance experiments.

The next class of nucleotide analogs comprises positional isomers of the phosphate groups. In these compounds the phosphate is placed at a different position than in the normal substrate. This serves to alter its position relative to the catalytic groups on the enzyme surface and can also

---

[16] F. Krug, I. Parikh, G. Illiano, and P. Cuatrecasas, *J. Biol. Chem.* **248**, 1203 (1973).
[17] M. J. Chamberlin, in "The Enzymes" (P. D. Boyer, ed.), 3rd ed., Vol. 10, p. 333. Academic Press, New York, 1974.
[18] G. H. Jones, H. P. Albrecht, N. P. Damodaran, and J. G. Moffatt, *J. Am. Chem. Soc.* **92**, 5510 (1970).
[19] C. A. Janson and W. W. Cleland, *J. Biol. Chem.* **249**, 2572 (1974).
[20] M. L. De Pamphilis and W. W. Cleland, *Biochemistry* **12**, 3714 (1973).
[21] J. A. Todhunter, K. B. Reichel, and D. L. Purich, *Arch. Biochem. Biophys.* **174**, 120 (1976).
[22] D. L. Purich, personal communication.

distort the positioning of other groups in the enzyme–substrate complex. The first of the series of staphylococcal nuclease inhibitors listed in Table II illustrates the former effect,[23] and the inhibition of AMP-deaminase by 3'-AMP is exemplary of the latter effect.[24] Note that 5'-dTTP functions against staphylococcal nuclease[23] by virtue of not having a hydrolyzable group on the phosphate and is not a positional isomer. Deoxythymidine-3',5'-disulfate incorporates absence of functionality and isosteric replacement of P by S.[23] The antimetabolite 5'-fluoro-UTP (a metabolite of 5-fluorouracil) similarly inhibits thymidylate synthase by virtue of its isosteric semblance to the product of the reaction, with F replacing —CH_3.[25]

Adenosine diphosphate ribose (ADPR) and N-alkylnicotinamides are inhibitors of NAD-linked dehydrogenases.[11–15,26] They substitute, respectively, for the carbohydrate and nicotinamide functions of NAD. ADPR is inhibitory by virtue of lacking functionality. N-Alkylnicotinamides can, in theory, accept hydride but appear to inhibit here owing to a lack of proper positioning of the function. The well known stereoselectivity of hydride transfer in NAD-linked dehydrogenases[27] speaks to the involvement of very precise binding of the nicotinamide function in these reactions.

Other cofactor analog inhibitors, similar to ADPR, appear to be rather affine and, thus, potent inhibitors. The folate antagonists and the B_{12} analogs exemplify this. Aminopterin, methotrexate, and dichloromethotrexate (Table II) are not readily competed by dihydrofolic acid, and their use as potent inhibitors of dihydrofolate reductase is well established in cancer chemotherapy.[28] B_{12} analogs in which the deoxyadenosyl group of cobalamine is replaced by a variety of coordinating ligands (Table II) bind to the respective apoenzymes in an essentially irreversible fashion.[29]

There are a large number of inhibitory analogs of the three-carbon glycolytic intermediates. These involve simple isosteric analogy, such as between oxamic, oxalic, and pyruvic acid; oxalate and glyoxalate; α-glycerophosphate and glyceraldehyde 3-phosphate; or lactate 2-phosphate, D-tartronate semialdehyde 2-phosphate, and glyceraldehyde 2-phosphate (Table II). Combinations of isosteric effects and positional isomerization are represented by the competition between 3-phosphoglycerate and glyceraldehyde 3-phosphate[30] and by that between β-

[23] P. Cuatrecasas, M. Wilchek, and C. B. Anfinsen, *Biochemistry* **8**, 2277 (1969).
[24] W. Makarewicz, *Comp. Biochem. Physiol.* **29**, 1 (1969).
[25] P. Reyes and C. Heidelberger, *Mol. Pharmacol.* **1**, 14 (1965).
[26] B. M. Anderson and C. M. Anderson, *Biochem. Biophys. Res. Commun.* **16**, 258 (1964).
[27] G. Popjak, in "The Enzymes" (P. D. Boyer, 3rd ed.), Vol. 2, p. 115. Academic Press, New York, 1970.
[28] B. R. Baker, *Cancer Chemother. Rep.* **4**, 1 (1959).
[29] H. P. C. Hogenkamp and T. G. Oikawa, *J. Biol. Chem.* **219**, 1911 (1964).
[30] F. Wold and C. E. Ballou, *J. Biol. Chem.* **227**, 313 (1967).

hydroxypropionate 3-phosphate, 3-aminoenopyruvate 2-phosphate, and glyceraldehyde 2-phosphate.[31]

The amino acids also have their inhibitor analogs. A great many have been developed against L-glutamate (Table II) and are quite illustrative of the principles of simple isosterism (3,5-pyridine dicarboxylate, α-fluoroglutarate), optical isomerism (D-glutamate), absence of functionality (glutaric acid), positional isomerism (β-glutamate), and steric distortion of binding (β-methyl glutamate, γ-methyl glutamate). Inhibitors related to glutamate are β-aminoglutaryl-L-aminobutyrate [glutathione analog[32] and 3-methyl-5-oxoproline (5-oxoproline analog[32]]. Optical isomerism as the basis of inhibition of amino acid utilizing enzymes is also shown by competitive inhibitors to D-α-lysine in the D-α-lysine mutase reaction (Table II). Finally, α-methyl dihydroxyphenylalanine (α-methyl DOPA, Aldomet), which is competitive to dihydroxyphenylalanine (DOPA) in the DOPA-decarboxylase reaction, should be mentioned in view of its importance as an antihypertensive.[33]

Analogs of intermediates of the Krebs cycle are also reasonable inhibitors, and their function embodies all the principles discussed above. A few of these are listed in Table II as leading references.

Urea analogs (which are potent urease inhibitors) fall into two distinct groups. The first are simple isosteres, such as hydroxamic acids,[34] thiourea,[35] and dimethyl sulfoxide.[36] The second class functions by making the amino leaving group a poor electron acceptor (reducing its reactivity). This can be accomplished by linking an aromatic system to nitrogen and is typified by phenylurea.[34]

No new principles of competitive inhibition can be discerned in the various carbohydrate analogs listed in Table II save for the first two. It is apparent that 6-phosphogluconate and 1-phosphoglucuronate cannot form hemiacetal rings. They can, however, assume this conformation and, as such, inhibit a variety of reactions involving glucose 6-phosphate (Table II). The salient point is that inhibitors need not be locked covalently into a certain conformation but often need only have a conformer of the appropriate geometry.[1]

This idea is extended by the competition of spironolactone for the aldosterone receptor,[37] which gives spironolactone its diuretic properties.

---

[31] T. G. Spring, Ph.D. Thesis, Univ. of Minnesota, Minneapolis, 1970.
[32] O. W. Griffith and A. Meister, *Proc. Natl. Acad. Sci. U.S.A.* **74**, 3330 (1977).
[33] M. H. Weil, B. H. Barbour, and R. B. Chesne, *Circulation* **28**, 165 (1963).
[34] K. Kobashi, J. Hare, and T. Komai, *Biochem. Biophys. Res. Commun.* **23**, 34 (1966).
[35] A. J. Lister, *J. Gen. Microbiol.* **14**, 478 (1956).
[36] E. Gerhards and H. Gibian, *Ann. N. Y. Acad. Sci.* **141**, 65 (1967).
[37] G. W. Liddle, *Ann. N. Y. Acad. Sci.* **139**, 466 (1966).

TABLE II
SOME REPRESENTATIVE INHIBITORS USED IN KINETIC STUDIES

Compound type[a]	Analog of	Inhibits	References[b]
Nucleoside phosphonates	*Nucleoside phosphates*	ATPase	7–10
β,γ-Methylene (A)		Kinases	
α,β-Methylene (B)		Synthetases	
		ATP-lyases	
		Cyclases	
		Polymerases	
Nucleoside phosphoramidates	Nucleoside phosphates	ATPase	c
5′-AMP-PNP (C)		Myokinase	
		Myosin ATPase	
Cr^{3+}-Nucleotides	Mg^{2+}-nucleotides	ATPases	19–22
		Acetate kinase	
		Hexokinase	
		Tubulin polymerization	
Nucleotide derivatives	*Nucleotides*	Staphylococcal nuclease	23
Deoxythymidine 3′-fluorophosphate	Deoxythymidyl 3′,5′-diphosphate (D)		
Deoxythymidine 5′-phosphate			
Deoxythymidine 3′-p-nitrophenyl phosphate			
Deoxythymidine 3′,5′-disulfate			
5-Fluoro-UTP	UTP	Thymidylate synthase	25
3′-AMP	5′-AMP	AMP-deaminase	24
Coenzyme analogs	*Coenzymes*	Alcohol dehydrogenase	11–15
Adenosine diphosphate ribose (ADPR)	NAD	Lactate dehydrogenase	
		Formate dehydrogenase	
N-alkylnicotinamides	NAD	Alcohol dehydrogenase	26

Inhibitor	Substrate	Enzyme	Reference
Aminopterin (E)	Dihydrofolate (H)	Dihydrofolate reductase	28
Methotrexate (F)			
Dichloromethotrexate (G)			
5'-Uridylylcobalamine	5'-Deoxyadenylyl cobalamine (cobamide)	Ethanolamine ammonia-lyase	d
5-(1-Methoxy)ribosylcobalamine			
(4-Adenylyl)-1-butylcobalamine			
Methyl cobalamine	Cobamide	Dioldehydrase	e
5'-Deoxyuridylylcobalamine	Cobamide	Glutamate mutase	29
5'-Deoxythymidylcobalamine			
Glycolytic intermediate isosteres			
Oxamic acid	Pyruvate	Lactate dehydrogenase	f, g
Oxalic acid	Pyruvate	Pyruvate kinase	h
Oxalate	Glyoxylate	Isocitrate lyase	30
α-Glycerophosphate	Glyceraldehyde 3-phosphate	Glyceraldehyde-3-phosphate dehydrogenase	
3-Phosphoglycerate	Glyceraldehyde 2-phosphate	Enolase	31
3-Phosphoglycerate			
Lactate 2-phosphate			
β-Hydroxypropionate 3-phosphate			
D-Tartronate semialdehyde phosphate			
3-Aminoenolypyruvate 2-phosphate			
Amino acid analogs			
3,5-Pyridine dicarboxylate	L-Glutamate	L-Glutamate dehydrogenase	i-k
Glutaric acid			
α-Fluoroglutarate			
D-Glutamate			
β-Methyl glutamate	L-Glutamate	γ-Glutamylcysteine synthetase	32
γ-Methyl glutamate		5-Oxoprolionase	
β-Glutamate	L-Glutamate	γ-Glutamylcyclotransferase	32
β-Aminoglutaryl-L-aminobutyrate	Glutathione	γ-Glutamylcyclotransferase	
3-Methyl-5-oxoproline	5-Oxoproline	5-Oxoprolinase	32

(continued)

TABLE II (continued)

Compound type[a]	Analog of	Inhibits	References[b]
L-β-Lysine	D-α-Lysine	D-α-Lysine mutase	l
ε-N-Acetyllysine			
L-Ornithine			
α-Methyl dihydroxy phenylalanine (Aldomet, methyl-DOPA)	Dihydroxyphenylalanine (DOPA)	DOPA-decarboxylase	33
TCA cycle isosteres and derivatives			
Hydroxymalonate	Malonate	Malate dehydrogenases	m
Aminomalonate	Glycine	δ-Aminolevulinate synthetase	n
Malonate	Glyoxylate	Isocitrate lyase	h
Itaconate	Succinate		
Hydroxycitrate	Citrate	ATP:citrate-lyase	o
Fluorocitrate	Citrate	Aconitase	p
Oxalomalate			
γ-Hydroxy-α-ketoglutarate			
2-Oxopropane sulfonate	Acetoacetate	Acetoacetate decarboxylase	q–s
Acetopyruvate			
Acetyl acetone			
Urease inhibitors			
Hydroxamic acids	Urea	Urease	34–36
Thiourea			
DMSO			
Phenylurea			
Carbohydrate analogs			
6-Phosphogluconate	Glucose 6-phosphate	Phosphoglucomutase	t–w
1-Phosphoglucuronate			
Galactose 1,6-diphosphate	Glucose 1,6-diphosphate		
Fructose 1,6-diphosphate			

UDP-3-Deoxyglucose	UDP-glucose	UDP-glucose epimerase	x
Glucose 6-phosphate	Mannose 6-phosphate	Mannose-6-phosphate isomerase	y
Glucosamine 6-phosphate			
Mannitol 1-phosphate			
Galactose 6-phosphate			
2-Deoxy-2,3-dehydro-N-acetylneuraminic acid	Sialic acid carbohydrate	Neuraminadase	z
Tri-N-acetyl chitrotriose	Murein glycan	Lysozyme	aa
β-D-Thiogalactosides	β-D-galactosides	β-Galactosidase	
Steroids			
Spironolactone (J)	Aldosterone (J)	Aldosterone receptor	37
19-Nortestosterone	Δ^5-Androstene-3,17-dione	Δ^5-3-Ketosteroid isomerase	38
17β-Estradiol			
Prostaglandin analogs			
9,11-Azoprosta-5,13-dienoic acid (K)	PGH_2 (K)	Thromboxane synthetase	39
Transition State Analogs			
Methionine sulfoximine phosphate	γ-Glutamyl phosphate	Glutamine synthetase	44, 48
1,5-Gluconolactone		Glycogen phosphorylase	bb
2-Carboxy-D-ribitol 1,5-diphosphate		Ribulose-1,5-diphosphate carboxylase	cc
2-Phosphoglycolate		Triosephosphate isomerase	43
$A(5')P_n(5')A$		Adenylate kinase	dd

[a] Capital letters in parentheses designate structures shown on the accompanying Table of Selected Structures from Table II.
[b] Numbers refer to text footnotes; letters refer to footnotes to this table.
[c] E. R. Stadtman and A. Ginsburg, in "The Enzymes" (P. D. Boyer, ed.), 3rd ed., Vol. 10, p. 755. Academic Press, New York, 1974.
[d] B. M. Babior, *J. Biol. Chem.* **244**, 2917 (1969).
[e] R. H. Abeles and H. A. Lee, *Ann. N. Y. Acad. Sci.* **112**, 695 (1964).

(*continued*)

[f] W. B. Novoa and G. W. Schwert, *J. Biol. Chem.* **236**, 2150 (1961).
[g] G. W. Schwert and Å. D. Winer, *in* "The Enzymes" (P. D. Boyer, H. Lardy, and K. Myrbäck, eds.), 2nd ed., Vol. 7, p. 127. Academic Press, New York, 1963.
[h] H. H. Daron and I. C. Gunsalus, this series, Vol. 6, p. 622.
[i] K. S. Rogers, M. R. Boots, and S. G. Boots, *Biochim. Biophys. Acta* **258**, 343 (1972).
[j] W. S. Caughey, J. D. Smiley, and L. Hellerman, *J. Biol. Chem.* **224**, 591 (1957).
[k] E. Kun and B. Achmatowicz, *J. Biol. Chem.* **240**, 2619 (1965).
[l] C. G. D. Morley and T. C. Stadtman, *Biochemistry* **9**, 4890 (1970).
[m] K. Harada and R. G. Wolfe, *J. Biol. Chem.* **243**, 4131 (1968).
[n] M. Matthew and A. Neuberger, *Biochem. J.* **87**, 601 (1963).
[o] J. A. Watson, M. Fang, and J. M. Lowenstein, *Arch. Biochem. Biophys.* **135**, 209 (1969).
[p] J. P. Glusker, *in* "The Enzymes" (P. D. Boyer, ed.), 3rd ed., Vol. 5, p. 413. Academic Press, New York, 1971.
[q] I. Fridovich, *J. Biol. Chem.* **243**, 1043 (1968).
[r] R. Davies, *Biochem. J.* **37**, 230 (1943).
[s] M. S. Neece and I. Fridovich, *J. Biol. Chem.* **242**, 2939 (1967).
[t] E. F. Kovacs and G. Bot, *Acta Physiol. Acad. Sci. Hung.* **27**, 328 (1965).
[u] W. J. Ray and A. S. Moldvan, *Biochemistry* **9**, 3886 (1970).
[v] Y. L. Huang and K. E. Ebner, *Experientia* **25**, 917 (1969).
[w] O. H. Lowry and C. Passonneau, *J. Biol. Chem.* **244**, 910 (1969).
[x] N. K. Kochetov, E. I. Budowsky, T. N. Druzhinina, N. D. Gabrielyan, I. V. Komlev, Y. Y. Kusov, and V. N. Shibaev, *Carbohydr. Res.* **10**, 152 (1968).
[y] R. W. Gracey and E. A. Noltmann, *J. Biol. Chem.* **243**, 5410 (1968).
[z] P. Meindl and H. Tuppy, *Hoppe-Seyler's Z. Physiol. Chem.* **350**, 1088 (1969).
[aa] D. Carlström, *Biochim. Biophys. Acta* **59**, 361 (1962).
[bb] J. I. Tu, G. R. Jacobson, and D. J. Graves, *Biochemistry* **10**, 1229 (1971).
[cc] M. Calvin, *Fed. Proc., Fed. Am. Soc. Exp. Biol.* **13**, 697 (1954).
[dd] G. E. Lienhard, *Science* **180**, 149 (1973).

Selected Structures from Table II

A.
B.
C.
D.

E. $R_{1,2,3} = H$
F. $R_1 = CH_3$; $R_{2,3} = H$
G. $R_1 = CH_3$; $R_{2,3} = Cl$

H.

I.
J.
K.
L.

In this case, spironolactone is locked into a conformation of similar geometry to the active conformer of aldosterone. Simple isosterism is shown, on the other hand, by the competition between 19-nortestosterone or 17β-estradiol and Δ^5-androstene-3,17-dione in the Δ^5-3-ketosteroid isomerase reaction.[38]

Little has been done with prostaglandin analogs as inhibitors, but a recent report that 9,11-azoprosta-5,13-dienoic acid competes out prostaglandin H_2 (PGH_2) in the thromboxane synthetase reaction[39] appears to involve an isosteric effect and merits attention because of the ability of this inhibitor to prevent platelet aggregation.[39]

Another group of compounds that can be mentioned here comprises the metal ion chelators. These are not inhibitors in the sense of the compounds discussed above since they do not, usually, compete for binding on the surface of the enzyme. They function, instead, by removing required metal ions from solution. As such they are used as general inhibitors of metal ion-dependent enzymes and as probes for metal ion dependency and specificity. A typical application is the use of o-phenanthroline and m-phenanthroline to distinguish Zn^{2+}-dependent reactions.[40] This works on the basis that o-phenanthroline chelates Zn^{2+} whereas m-phenanthroline does not and can be used as a control for nonspecific inhibition. This method has been most recently applied to show that RNA polymerases from *Escherichia coli*[41] and yeast[42] are Zn^{2+} enzymes. Since unsuspected metal ion chelators can create problems in kinetic studies, a separate discussion of this will be given later in the section on complicating factors in inhibition experiments.

The last class of compound to be discussed here is the transition-state analog. This approach is based on the premise that a major factor in accounting for enzymic catalysis is the preferential stabilization of the reaction transition state.[43] That is, if $\Delta G_0^{\ddagger}$, the free energy of formation of the transition state, is reduced on the enzyme surface relative to the value in solution, then the free energy of activation is accordingly reduced and catalysis results. This is illustrated in Fig. 2. In theory, any compound that structurally resembles the transition state of the reaction should function, therefore, as an inhibitor. This depends, however, on the relative

---

[38] R. Jarabak, M. Colvin, S. H. Moolgavkar, and P. Talalag, this series, Vol. 15, p. 642.
[39] R. R. Gorman, G. L. Bundy, D. C. Peterson, F. F. Sun, O. V. Miller, and F. A. Fitzpatrick, *Proc. Natl. Acad. Sci. U.S.A.* **74,** 4007 (1977).
[40] B. L. Vallee, *Adv. Protein Chem.* **10,** 317 (1955).
[41] M. D. Scrutton, C. W. Wu, and D. A. Goldthwait, *Proc. Natl. Acad. Sci. U.S.A.* **68,** 2497 (1971).
[42] T. M. Wandzilak, and R. W. Benson, *Biochemistry* **17,** 426 (1978).
[43] R. Wolfenden, *Nature (London)* **223,** 704 (1969).

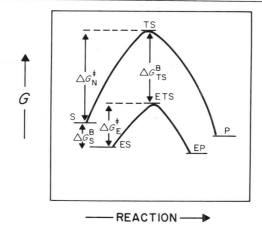

FIG. 2. Energy-reaction coordinate diagram for a hypothetical unireactant enzyme system. $\Delta G_N^\ddagger$ and $\Delta G_E^\ddagger$ refer to the energies of activation for the nonenzymic and enzymic processes, respectively. ΔG_S^B and ΔG_{TS}^B refer, similarly, to the binding energies of substrate and the transition state on the enzyme surface.

binding constants of S and TS (the transition-state analog). If the binding of substrate on the enzyme surface is much greater than that of the transition state analog, then the inhibition will be slight. More commonly, binding of TS is much greater than that of S, and the degree of inhibition may be very large. The action of methionine sulfoximine on glutamine synthetase[44,45] and tripeptide synthetase[6] is a prime example of a tight-binding transition state analog. The use of the transition-state analog approach is not routine, however. For it to be effective, the configuration of the active site in the transition complex must be known. This is not a trivial problem. In fact, one very important use of transition-state analogs is in active site mapping. In this approach, developed by Meister[4-6] and based on work of Wold,[30,46] a series of compounds is prepared and their relative potencies as transition-state analogs is analyzed by computer techniques to give information about the spatial arrangement of the catalytic site during the catalytic step of the reaction. This method will not be covered in detail here, and the reader is referred to Meister's excellent work on this subject.[6] The catalytic site of an enzyme cannot be thought of, however, as a rigid domain of binding but, instead, must be considered as existing in several conformation states. The definition of the geometry

---

[44] R. A. Ronzio, W. B. Rowe, and A. Meister, *Biochemistry* **8**, 1066 (1969).
[45] R. E. Weisbrod and A. Meister, *J. Biol. Chem.* **248**, 3997 (1973).
[46] F. C. Hartman and F. Wold, *Biochim. Biophys. Acta* **141**, 445 (1967).

of the ES and ETS complexes may depend on the presence of S and TS.[6] Similarly, the low activity shown by 5-hydroxy-6-phosphononorleucine as an inhibitor of glutamine synthetase[47] relative to methionine sulfoximine phosphate[44,48] may be related to this phenomenon. As a guide to the reader, some representative transition-state analogs have been cited in the latter portion of Table II.

Complicating Factors in Inhibitor Studies

Several problems, some encountered quite generally in enzymology, can arise during the course of inhibitor studies. These fall into the categories of mutual contamination of substrate and inhibitor and the group of effects giving rise to pseudo-inhibition. These will be discussed and methods of coping with these problems will be examined.

Many substrates and their competitive inhibitors are related structurally and chemically. As such, it should not be too surprising that contamination of substrate by inhibitor and vice versa is a common problem. If a substrate is contaminated by a competitive inhibitor for that substrate, the fractional contamination can be represented by α and $I_c = \alpha S$ where I_c is the amount of inhibitor carried by a given amount of S. The corresponding rate equation (from Table I) becomes

$$\frac{1}{v} = \frac{1}{V_m} + \frac{K_m}{V_m(S)} + \frac{K_m(I + \alpha S)}{V_m K_i(S)} \tag{3}$$

and expanding:

$$\frac{1}{v} = \frac{1}{V_m} + \frac{K_m}{V_m(S)} + \frac{K_m(I)}{V_m K_i(S)} + \frac{K_m \alpha}{V_m K_i} \tag{4}$$

From Eq. (4), it is apparent that the position of the intercept in the double-reciprocal plot will be altered to the value

$$\frac{1}{V_m} + \frac{K_m \alpha}{V_m K_i}$$

If the contamination is unsuspected, then the kineticist will, unaware, arrive at a faulty value for V_m. The simplest way to avoid this trap is to deliberately contaminate S with I at fixed values of α, determine the apparent V_m as a function of α, and extrapolate for the value of V_m when $\alpha = 0$. Clearly if α is large to begin with, such a procedure will yield a nonlinear plot unless very large deliberate contaminations are used. Con-

---

[47] J. A. Todhunter, Ph.D. Thesis, Univ. of California, 1976.
[48] R. A. Ronzio and A. Meister, *Proc. Natl. Acad. Sci. U.S.A.* **59**, 164 (1968).

versely, if the result of such deliberate contamination experiments yields a value for V_m in agreement with that originally determined, the substrate may be inferred to be substantially free of inhibitor.

Alternatively, it is quite possible to have a substrate contamination in the competitive inhibitor. This presents a more difficult problem. Expressing the fractional contamination of S in I by the ratio, the appropriate expression (from Table I) becomes

$$\frac{1}{v} = \frac{1}{V_m} + \frac{K_m}{V_m(S + \alpha I)} + \frac{K_m}{V_m K_i} \cdot \frac{(I)}{(S + \alpha I)} \quad (5)$$

This is, of course, a nonlinear equation, and this type of contamination effect will be manifested as a nonlinearity when $1/v$ is plotted as a function of (I) at constant (S). This is not, therefore, distinguishable from partially competitive inhibition on the basis of $1/v$ vs (I) plots. At very large values of I, this contamination also gives a self-limiting effect, like a partially competitive inhibition, which can be written

$$\lim_{(I) \to \infty} \frac{1}{v} = \frac{1}{V_m}\left(1 + \frac{K_m}{K_i}\right)$$

This value differs from that obtained with the partially competitive inhibitor. For the partial competitive case the limit as (I) became very large was

$$\lim_{(I) \to \infty} \frac{1}{v} = \frac{1}{V_m} + \frac{K_m}{V_m(S)} \quad (6)$$

Thus, in the partially competitive case the limiting value was seen to be the noninhibited rate and was dependent on (S). For the competitive inhibitor contaminated with substrate, the limiting value is independent of (S) and is always larger than the noninhibited rate. These observations should allow the kineticist to establish whether a partially competitive inhibition or a substrate-contaminated inhibitor effect is in operation. For the former, the limiting value of V_m at (I) $\to \infty$ should equal that at (I) = 0. For the latter, at (I) $\to \infty$ $V_m > V_m$ when (I) = 0.

The discussion above concentrated on competitive inhibitors, since these are the inhibitors usually chosen for inhibitor studies. Contamination problems with other classes of inhibitors are, also, less common and usually detectable as deviations from ideal behavior (see Fig. 1). As common practice, the investigator should carefully analyze and purify all substrates and inhibitors used for kinetic studies, and different lots should be tested for agreement of the experimentally determined kinetic constants.

The problem of inhibition not related to binding to the enzyme (pseudo-inhibition) is more general in scope than those problems just discussed. Generally, these are compounds that form complexes with substrates or cofactors, thereby limiting their availability for reaction. Obviously, such inhibitors are of little value for determining the kinetic constants and mechanism of an enzyme. They can, however, serve to shed light on the requirements of an enzymic reaction.[49] As was discussed in the preceding section, metal ion chelators are particularly useful in this vein. Problems arise when the formation of complexes between components of the enzyme assay and substrates or cofactors is not obvious. When a compound is suspected of being an inhibitor, it is necessary to establish by physical methods the nature of binding to the enzyme. A case in point would be the inhibition of yeast alcohol dehydrogenease by o-phenanthroline. Rather than binding to Zn^{2+} on the enzyme, o-phenanthroline may be hydrophobically complexed to the enzyme and act thus, as a true inhibitor in this system[50] rather than a chelator.

Yet another related problem may arise when the investigator is testing crude or only partially purified cell-free fractions for their effect on an enzymic process. It is quite common in such cases to observe inhibitions due to the addition of enzymes which compete for the substrates or cofactors of the reaction in question. Activations can also occur by competition for the reaction products or unsuspected inhibitors in the enzyme assay. Such effects disappear on further purification, but their existence speaks to the dangers of working with crude preparations. Similarly, during the course of enzyme purification, pseudo-inhibitors, competing enzymes, and even real inhibitors are purified away from the enzyme. Thus, during the course of purification the kinetic constants of an enzyme may change considerably. The general practice of relying on recovery of activity as a measure of yield and on increased specific activity as a measure of degree of purification is, therefore, quite liable to error. In the author's own laboratory, this is generally observed in the preparation of RNA polymerases and is reported also by Burgess et al.[51] and Valenzuela et al.[52] for these enzymes. As a result, the reader should be cautious about drawing too many conclusions from work with crude preparations.

Practical

The preceding sections have covered the problems of distinguishing the kinetic types of inhibition, inhibitor constants, selecting useful inhibi-

---

[49] B. L. Vallee and F. L. Hoch, *J. Biol. Chem.* **225**, 185 (1957).
[50] C. Branden, H. Jornvall, H. Eklund, and B. Furugren, in "The Enzymes" (P. D. Boyer, ed.), 3rd ed., Vol. 11, p. 182. Academic Press, New York, 1975.
[51] R. R. Burgess and J. J. Jendrisak, *Biochemistry* **14**, 4634 (1975).
[52] P. Valenzuela, F. Weinberg, G. Bell, and W. J. Rutter, *J. Biol. Chem.* **251**, 1464 (1976).

tors, and dealing with unwanted or unsuspected inhibitors. This section will go briefly into some of the practical aspects of using inhibitors in kinetic studies. Where the discussion overlaps points brought out before or elsewhere in this volume, the reader will be referred to the appropriate sections for greater detail.

Usually, it will be necessary to perform a preliminary determination of K_i unless the value is known reliably from other sources. This will allow a working range of inhibitor concentrations to be established. As discussed above, the inhibitor should be varied over a nonsaturating range of concentrations, and so a value for K_i is needed. For inhibitors that are substrate or product analogs, the K_m value for the substrate or product can often be used as a first approximation to K_i for the preliminary work. Since substrate ought, also, to be nonsaturating, the first K_i value can be determined with (S) = K_m as discussed in the preceding sections. The values of K_i determined in this fashion are usually adequate for use in selecting inhibitor concentration ranges, and much effort need not be expended in this preliminary work. For accurate determinations of K_i, determinations at varying values of (S) are used as discussed in the preceding sections and summarized in Table I.

Once K_i is determined approximately, the kinetic type of inhibition is established. This is done by the protocols given in the first section above and illustrated in Fig. 1. The actual measurement of rates must be such that initial velocities are determined. In practice, this means that the investigator must work in the linear portion of the time course and within the first 10–15% of the reaction's progress. For the purposes of obtaining rates for double-reciprocal plots, it is advisable to actually determine time courses as opposed to single-point assays. This is due to the magnification of experimental error that occurs in a double-reciprocal plot and the resultant need for accuracy in the primary data. Of course, single time points may be used, but then multiple assays must be performed.

The immediate objective of these rate measurements is to obtain plots of $1/v$ vs $1/(S)$ at various values of (I). Note that while reciprocal dilutions of (S) are used (i.e., 1-, 3-, 5-, 7-, and 9-fold) to give even spacing of $1/(S)$ on the plot axis; that (I) plots in a linear space and linear dilutions are suitable (i.e., 0.2, K_i, 0.5 K_i, 1 K_i, 1.5 K_i, 2 K_i). Double-reciprocal plots are constructed, with the data so obtained, to determine the type of inhibition observed (see Fig. 1), and replots, as specified in Table I, are made of the primary plot data to determine accurate K_i values and other parameters of interest. As pointed out above, certain types of inhibition, such as partially competitive inhibition, will have been detected in the preliminary work from $1/v$ vs (I) plots, and appropriate methods to deal with these cases were given.

It should be pointed out that for a suspected uncompetitive inhibition, the author recommends the use of a Hanes plot,[53] which shows $(S)/v$ as a function of (S). This type of plot has a more constant error envelope than the Lineweaver–Burk plot[53] and is, therefore, preferable in many respects. The uncompetitive inhibition will give parallel lines in the double-reciprocal plot (Fig. 1). Moderate deviations from this will be hard to detect and may lie within the error envelope of the plot, which increases considerably as (S) becomes small. In the Hanes plot the uncompetitive inhibition will give a family of lines convergent at the $(S)/v$ axis if $(S)/v$ vs (S) is plotted at varying concentrations of inhibitor. The author has used this device[21] and finds that the diagnosis of uncompetitive inhibition is most reliable with this plotting device.

It is, of course, necessary to establish that the inhibition observed is a true inhibition, not a pseudo-inhibition as described above. Along this line, binding studies to show direct association of inhibitor and enzyme may be helpful. These studies may also point out if there is more than one inhibitor binding site per monomer. Nonspecific inhibitions can sometimes be detected by the use of compounds similar in functionality to the inhibitor but of different structure. Nonspecific inhibitors will generally give inhibitions of a mixed type (see above), and this is also an index of suspicion.[1]

Related to this is the effect of inhibitors with high affinity for metal ions. Such compounds tend to be nucleotide di- and triphosphate analogs and multifunctional compounds, such as dicarboxylic acids or diamines. When such compounds are used, the binding of metal ions from the assay mixture to the added inhibitor may severely distort the results obtained. This is worst for enzymes with sharp metal ion optima or that use metal ion:substrate complexes as the true substrates. Such problems may be avoided simply. From the metal ion binding constant of the inhibitor and the free metal ion concentration of the assay mixture, the fraction of inhibitor that would be complexed with metal ion is calculated. This fractional amount (on a molar basis) of metal ion is then added to the stock inhibitor solution. Upon dilution, the inhibitor will then carry with it the appropriate fraction of bound metal ion and will not perturb the metal ion concentration of the assay. It should be noted that metal ion binding constants are often pH dependent, and this must be taken into consideration.

Since all studies with inhibitors tend to vary as to their objectives and progress, more detailed procedural information than is presented here would be inappropriate. Generally, however, the course of investigation is to determine an approximate K_i, to determine if the inhibition is real,

---

[53] G. N. Wilkinson, *Biochem. J.* **80**, 324 (1961).

and then to determine the kinetic parameters of the inhibition. The theoretical material presented in this chapter is intended as a guide to interpretation of the data obtained in such investigations and as a warning against the more common pitfalls. The practical aspects discussed are points that the author has found to be generally useful and should be incorporated in any inhibitor study. As discussed here, inhibitor studies can be of great utility to the enzymologist and the scope of application is, perhaps, limited only by the resourcefulness of the experimentalist.

[16] Product Inhibition and Abortive Complex Formation

By FREDERICK B. RUDOLPH

The use of products of a reaction to investigate the kinetic mechanism of an enzyme through inhibition studies has become one of the most common tools of the kineticist. Alberty[1] was the first to point out that bisubstrate sequential mechanisms could be differentiated by product inhibition studies. He showed how the order of addition of the substrates could be determined in that type of mechanism, allowing such kinetic analysis to be done easily for the first time. This theory was first applied and expanded by Fromm in studies on ribitol dehydrogenase[2] and lactate dehydrogenase.[3]

Product inhibition experiments are valuable since information about the mechanism and a good deal of insight into regulation of enzyme activity by product levels can be obtained. Such studies are complicated, however, by factors such as abortive complex formation and substrate inhibition, which can make interpretation of results equivocal. This chapter will deal with procedures for product inhibition studies, interpretation of the results, and possible experimental complications.

Nomenclature and Experimental Protocol

The nomenclature suggested by Cleland[4] will be used in this chapter. The experimental protocol for a multisubstrate enzyme involves holding one substrate (or more depending on the number of reactants) constant at

[1] R. A. Alberty, *J. Am. Chem. Soc.* **80**, 1777 (1958).
[2] H. J. Fromm and D. R. Nelson, *J. Biol. Chem.* **237**, 215 (1962).
[3] V. Zewe and H. J. Fromm, *J. Biol. Chem.* **237**, 1668 (1962).
[4] W. W. Cleland, *Biochim. Biophys. Acta* **67**, 173 (1963).

nonsaturating level and varying the substrate being studied in a concentration range around its Michaelis constant. The product is added at several fixed concentrations (including zero). Each substrate and product is evaluated in that manner. Double-reciprocal plots of 1/initial velocity (v) versus 1/substrate (S) are constructed. It must be emphasized that only one product is added at a time to avoid the complication of the reverse reaction. The special case of a single-product reaction will be considered below. In certain cases the fixed substrate concentration may be held at saturating levels.[4] This technique can allow distinctions between mechanisms but is limited by several experimental difficulties, which will be discussed.

There are three general types of inhibitions observed with product inhibitors defined by the effect on the slope and intercept of the double-reciprocal plot. If the slope of the plots in the presence of the product are affected but the intercept is not, the inhibition is competitive relative to the varied substrate. The reciprocal plots are varying product concentrations intersect on the $1/v$ axis, and saturation with the varying substrate will eliminate the inhibition. When the intercepts only are affected, the inhibition pattern is a series of parallel lines at different product concentrations and is termed uncompetitive. If both slopes and intercepts are affected by the presence of product, the reciprocal plots converge to the left of the vertical axis and the inhibition is termed noncompetitive. In this latter case, the convergence point can be above, below, or on the horizontal axis. Several different terms have been used for this type of inhibition (e.g., mixed) depending on the intersection point, but such distinctions are not generally useful and the term noncompetitive will describe the effect of a product where the lines on the double-reciprocal plot converge to the left of the vertical axis, regardless of the position relative to the horizontal axis. The effect of a product inhibitor demonstrating competitive, noncompetitive, and uncompetitive inhibition is illustrated in Fig. 1, panels a, b, and c, respectively. Terms used are defined as follows: v = initial reaction velocity, A = substrate concentration, I = product concentration, K_a = Michaelis constant for A, V_1 = maximum velocity, K_{is} = inhibition constant from the slope term, and K_{ii} = inhibition constant from the intercept term. It should be noted that only K_{is} is defined for competitive inhibition; K_{ii} is defined for uncompetitive inhibition, and noncompetitive inhibition is described by both constants.

The inhibition patterns may be further analyzed by replots of the slopes and intercepts of the reciprocal plots versus the inhibitor concentration. For most products the replots are linear and the inhibition is termed linear competitive, linear noncompetitive, or linear uncompeti-

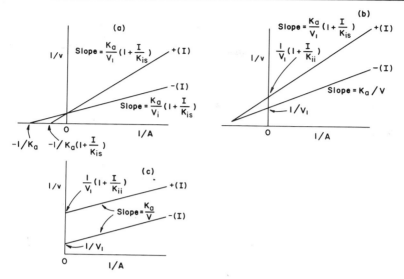

FIG. 1. Double-reciprocal plot of the reciprocal of the initial velocity versus the reciprocal of the substitute concentration in the presence and in the absence of a product. The panels depict competitive inhibition (a), noncompetitive inhibition (b), and uncompetitive inhibition (c). The values of the slopes and intercepts are indicated in each panel.

tive. An inhibition constant can be determined from the slope replot (K_{is}) or the intercept replot (K_{ii}) and is the horizontal intercept on the plot as illustrated in Fig. 2. In addition, either K_{is} or K_{ii} can be calculated directly from either the slopes or intercepts of the plots in Fig. 1.

The majority of inhibitions are linear, but occasionally the slope or intercept replot can be curved as illustrated in Fig. 2. The form of the equation describing this concave upward behavior is[4]

$$\text{Slope or intercept} = a(1 + bI + cI^2) \tag{1}$$

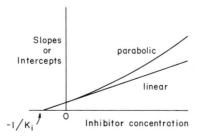

FIG. 2. A replot of the slopes on intercepts from a double-reciprocal plot describing linear or parabolic inhibition by a product.

This describes parabolic inhibition and is due to combination of at least two molecules of inhibitor to the enzyme or, as will be discussed below, to abortive complex formation with products. With noncompetitive inhibition either the slope or intercept or both can be parabolic and is termed S-parabolic I-linear, S-linear I-parabolic, or parabolic noncompetitive, respectively. If the slope and intercepts are described by different functions, the double-reciprocal plots will not intersect in a common point.

Other, more complex, replots have been suggested for various types of inhibitors[4] but are generally not applicable to product inhibition studies. The nonlinearity predicted for parabolic inhibition can be confirmed directly by use of a Dixon plot[5] in which the reciprocal velocity is plotted against the inhibitor concentration at a fixed substrate concentration. A wide range of inhibitor levels can be used, and any curvature should be readily apparent. This plot is not, however, useful for determining the type of inhibition[6,7] even though it has often been used for such purposes.

Theory

One-Substrate Systems. The normal technique in product inhibition studies as formulated by Alberty[1] is to determine the initial velocity of the reaction at varying substrate levels at different fixed levels of one product. For the simplest case, the Michaelis–Menten model

$$\begin{array}{ccccc} & A & & P & \\ & \downarrow & & \uparrow & \\ \overline{\quad k_1 \quad} & \overline{\quad k_2 \quad} & \overline{\quad k_3 \quad} & \overline{\quad k_4 \quad} \\ E & & EA & & E \end{array} \qquad (2)$$

the total equation is

$$v = \frac{V_1 V_2 (A - P/K_{eq})}{V_2 K_a + V_2 A + V_1 P/K_{eq}} \qquad (3)$$

where V_1 and V_2 are the maximum velocities in the forward and reverse directions, K_a and K_p are the Michaelis constants for A and P, and K_{eq} is the equilibrium constant.

In reactions with more than one product the negative term in the rate equation has a common factor that is the mathematical product of all the concentrations of the products. So, for any system where more than one

---

[5] M. Dixon, *Biochem. J.* **55,** 170 (1953).
[6] D. L. Purich and H. J. Fromm, *Biochim. Biophys. Acta* **268,** 1 (1972).
[7] H. J. Fromm, "Initial Rate Enzyme Kinetics." Springer-Verlag, Berlin and New York, 1975.

product is produced, the negative term is zero when only one product is present. In a single-product reaction the negative term in the rate equation does not disappear if the product concentration is initially fixed at some finite value. A double-reciprocal plot for such a reaction will be nonlinear. In itself, such nonlinear behavior is of little diagnostic use and evaluation of binding order for a Uni Uni mechanism is obviously not necessary. Product inhibition does, however, allow a distinction to be made between the normal Uni Uni mechanism [Eq. (2)] and the Iso Uni Uni mechanism

$$\begin{array}{ccccc} & A & & P & \\ & \downarrow k_1 & k_2 & \uparrow k_3 & k_4 \\ E & & EA & & F \underset{k_6}{\overset{k_5}{\rightarrow}} E \end{array} \quad (4)$$

The rate equation for this mechanism is

$$v = \frac{V_1 V_2 (A - P/K_{eq})}{V_2 K_a + V_2 A + V_1 P/K_{eq} + V_2 AP/K_{iip}} \quad (5)$$

where K_{ipp} is an inhibition constant $= (k_3 + k_5)/k_4$. The other terms are as defined for Eq. (3). The only difference is the presence of the $V_2 AP/K_{iip}$ term. Initial rate studies will not distinguish these two mechanisms, since the AP term will not affect the kinetic plots if the initial product concentration is zero. However, the two mechanisms can be differentiated by the use of production inhibition.

If the substrate is held at a saturating level, then Eq. (3) reduces to

$$v = V_1 \quad (6)$$

and Eq. (5) reduces to

$$v = \frac{V_1}{1 + (K_a/K_{iip})(P)} \quad (7)$$

Theoretically, a plot of $1/v$ (at saturating A) against P will be linear for the Iso Uni Uni mechanism and independent of P for the Uni Uni mechanism.[8] Saturation is often difficult to achieve for many substrates, and distinctions will be difficult using such a procedure. To circumvent this problem Darvey has presented a graphical analysis that will afford distinction to be made without requiring saturation by the substrate. The method involves defining a term $\triangle v$ which is $[v_{(P=0)} - v]$ or simply the difference in initial velocity in the presence and in the absence of a given product con-

---

[8] C. Cennamo, *J. Theor. Biol.* **23**, 53 (1969).

centration. For Eq. (3)

$$\Delta v = \frac{K_a P[(V_1 + V_2)A + V_2 K_a]}{K_p(K_a + A)^2 + K_a P(K_a + A)} \quad (8)$$

and for Eq. (5)

$$\Delta v = \frac{K_a P[V_1 A(K_{iip} + K_p A) + V_2 K_{iip}(K_a + A)]}{K_p K_{iip}(K_a + A)^2 + K_a P(K_{iip} + K_p A)} \quad (9)$$

A plot of $1/\Delta v$ versus $1/P$ at a fixed A concentration is linear for both mechanisms. Use of various A concentrations should allow distinction, however. The intercept of the double-reciprocal plot on the $1/P$ axis (I_p) is

$$I_p = \frac{K_a}{K_p(K_a + A)} \quad (10)$$

for Eq. (8) and

$$I_p = \frac{K_a(K_{iip} + K_p A)}{K_p K_{iip}(K_a + A)} \quad (11)$$

for Eq. (9). A plot of $-1/I_p$ versus A is linear for the Uni Uni mechanism and nonlinear for the Iso mechanism. If the experimental data appear linear the result should not be considered absolutely conclusive, as the nonlinearity will depend on the values of individual rate constants and may not be evident in the concentration ranges employed. A useful aspect of the treatment of Darvey[9] is that an estimate can be made of K_{iip}. The reader is referred to the original reference for the procedure.

Fromm[7] has suggested that the two mechanisms can be readily distinguished owing to the fact that in the Iso mechanism the product inhibition cannot be reversed by substrate. This causes an intercept change on the $1/v$ axis of a double-reciprocal plot (versus $1/A$ at different product concentrations including zero). Thus, the lines on the reciprocal plot will intersect at $1/V_1$ for the Uni Uni mechanism and at $(K_{iip} + K_a P)/V_1 K_{iip}$ for the Iso Uni Uni mechanism. This is based on the same principle as the first procedure suggested but allows saturation to be approached by extrapolation. This is the simplest method for distinguishing between these two mechanisms.

One-substrate mechanisms with more than one product are easier to deal with and fortunately more commonly encountered than the previous examples. There are two basic types of such mechanisms.[10] The Ordered

[9] I. G. Darvey, *Biochem. J.* **128**, 383 (1972).
[10] These mechanisms and others to be discussed are illustrated in this volume [3] and the rate equations can be found in this volume [7] or in references cited in footnotes 4 and 7.

Uni Bi mechanism is described by

$$v = \frac{V_1V_2[A - (PQ/K_{eq})]}{V_2K_a + V_2A + V_1K_qP/K_{eq} + V_1K_pQ/K_{eq} + V_1PQ/K_{eq} + V_2AP/K_{eq}} \quad (12)$$

where the terms are analogous to those described for Eq. (3). The rate expression for product inhibition is derived simply by setting all product concentrations except the one of interest to zero. This procedure is applicable to all mechanisms unless additional complexes occur in the presence of the product.

The rate equation in the presence of P (Q = 0) is

$$1/v = \frac{1}{V_1}\left(1 + \frac{P}{K_{ip}}\right) + \frac{1}{V_1}\left(K_a + \frac{V_1K_aP}{V_2K_{eq}}\right)\frac{1}{A} \quad (13)$$

while in the presence of Q (P = 0) it is

$$1/v = \frac{1}{V_1} + \frac{1}{V_1}\left(K_a + \frac{V_1K_pQ}{V_2K_{eq}}\right)\frac{1}{A} \quad (14)$$

These equations predict that P will affect both the slope and intercept of the double-reciprocal plot of $1/v$ versus $1/A$ and be a noncompetitive inhibitor relative to A. Q affects only the slope term and will be a competitive inhibitor relative to A. Both inhibitors will give linear replots. Thus, if the mechanism is Ordered Uni Bi, the product inhibition will define the order of release of the products.

The second type of Uni Bi mechanism involves a random release and binding of products. If the pathway is assumed to be rapid equilibrium where the rate-limiting step is the breakdown of the central complexes, the rate equation will be the same as the Michaelis–Menten equation with K_{ia} substituted for K_a. Either product when present will bind free enzyme

$$E + P = EP, K_{ip} \quad (15)$$

The initial rate equation in the presence of P will be

$$1/v = \frac{1}{V_1} + \frac{K_{ia}}{V_1A}\left(1 + \frac{P}{K_{ip}}\right) \quad (16)$$

The products are symmetrical, so that the term $(1 + Q/K_{ia})$ can replace the $(1 + P/K_{ip})$ term in Eq. (16) to describe inhibition by Q.

The product inhibition patterns for the rapid equilibrium Random Uni Bi mechanism will both be linear-competitive with respect to the substrate, since only the slope term is affected in both cases. This allows distinction between this mechanism and the Ordered Uni Bi mechanism.

When the steady-state assumptions are made for the Random mechanism, the following rate equation is derived[7]:

$$1/v = \frac{1}{V_1}(1 + aP) + \frac{b + cP}{V_1 dA}\left(1 + \frac{P}{K_p}\right) \tag{17}$$

The coefficients a, b, and c are combinations of various rate constants. An analogous equation describes the effect of Q. The inhibition by the products will be noncompetitive with linear intercept replots and parabolic slope replots or S-parabolic I-linear noncompetitive. The product patterns for the Uni Bi mechanisms are summarized in Table I. All three possibilities can be readily differentiated by use of product inhibitors.

Two-Substrate Systems. The original formulation for use of product inhibition was to differentiate between various two-substrate mechanisms.[1] The rate equations describing product inhibition for the four basic bisubstrate mechanisms are as follows:

a. Ordered Bi Bi
with Q present

$$1/v = \frac{1}{V_1} + \frac{K_a}{V_1 A}\left(1 + \frac{Q}{K_{iq}}\right) + \frac{K_b}{V_1 B} + \frac{K_{ia}K_b}{V_1 AB}\left(1 + \frac{Q}{K_{iq}}\right) \tag{18}$$

and with P present

$$1/v = \frac{1}{V_1}\left(1 + \frac{P}{K_{ip}}\right) + \frac{K_a}{V_1 A} + \frac{K_b}{V_1 B}\left(1 + \frac{K_q P}{K_{iq}K_p}\right)$$
$$+ \frac{K_{ia}K_b}{V_1 AB}\left(1 + \frac{K_q P}{K_{iq}K_p}\right) \tag{19}$$

b. Rapid Equilibrium Random Bi Bi
with Q present

TABLE I
PRODUCT INHIBITION PATTERNS FOR UNI BI MECHANISMS[a]

Mechanism	Product	Type of inhibition
Ordered Uni Bi	P	N
	Q	C
Rapid Equilibrium Random Uni Bi	P	C
	Q	C
Steady State Random Uni Bi	P	N[b]
	Q	N[b]

[a] The abbreviations are C (competitive) and N (noncompetitive). The release of products, if ordered, is P first and Q second.
[b] The slope replots from the double-reciprocal plots on parabolic.

$$1/v = \frac{1}{V_1} + \frac{K_a}{V_1 A} + \frac{K_b}{V_1 B} + \frac{K_{ia} K_b}{V_1 AB}\left(1 + \frac{Q}{K_{iq}}\right) \qquad (20)$$

and with P present

$$1/v = \frac{1}{V_1} + \frac{K_a}{V_1 A} + \frac{K_b}{V_1 B} + \frac{K_{ia} K_b}{A_1 AB}\left(1 + \frac{P}{K_{ip}}\right) \qquad (21)$$

c. Theorell–Chance Bi Bi

with Q present

$$1/v = \frac{1}{V_1} + \frac{K_a}{V_1 A}\left(1 + \frac{Q}{K_{iq}}\right) + \frac{K_b}{V_1 B} + \frac{K_{ia} K_b}{V_1 AB}\left(1 + \frac{Q}{K_{iq}}\right) \qquad (22)$$

and with P present

$$1/v = \frac{1}{V_1} + \frac{K_a}{V_1 A} + \frac{K_b}{V_1 B}\left(1 + \frac{P}{K_{ip}}\right) + \frac{K_{ia} K_b}{V_1 AB}\left(1 + \frac{P}{K_{ip}}\right) \qquad (23)$$

d. Ping Pong Bi Bi

with P present

$$1/v = \frac{1}{V_1} + \frac{K_a}{V_1 A} + \frac{K_b}{V_1 B}\left(1 + \frac{P}{K_{ip}}\right) + \frac{K_{ia} K_b P}{V_1 K_{ip} AB} \qquad (24)$$

and with Q present

$$1/v = \frac{1}{V_1} + \frac{K_a}{V_1 A}\left(1 + \frac{Q}{K_{iq}}\right) + \frac{K_b}{V_1 B} + \frac{K_{ib} K_a Q}{V_1 K_{iq} AB} \qquad (25)$$

The product inhibition patterns predicted by these equations are shown in Table II. In all cases these equations predict linear inhibition by products. A choice between the three sequential mechanisms can be readily made if the products act as described above. The more general situation involves the formation of other complexes as discussed in the next section. These effects can limit the utility of product inhibition, but information about the mechanism can always be obtained from such experiments.

Abortive Complexes

In one of the first studies done with product inhibitions to differentiate mechanisms, Fromm and Nelson[2,11] studied the enzyme ribitol dehydrogenase from *Aerobacter aerogenes*. The results of the study were not consistent with the inhibition patterns suggested by Alberty.[1] It was found

---

[11] H. J. Fromm and D. R. Nelson, *Fed. Proc., Fed. Am. Soc. Exp. Biol.* **20**, Abstr. 229 (1961).

TABLE II
PRODUCT INHIBITION PATTERNS FOR BIREACTANT MECHANISMS[a]

Mechanism	Product	Varied substitute			
		A		B	
		Unsaturated, with B	Saturated, with B	Unsaturated, with A	Saturated, with A
Ordered Bi Uni	P	C^b	C	N	—
Ordered Bi Bi	P	$N(N)^{c,d}$	U	$N(N)^e$	N
	Q	C(C)	C	N(N)	—
Theorell–Chance Bi Bi	P	$N(N)^d$	—	$C(C)^e$	C
	Q	C(C)	C	N(N)	—
Ping Pong Tetra Uni	P	$N(N)^e$	—	$C(NL)^f$	C
	Q	C(C)	C	$N(NL)^f$	—
Rapid Equilibrium Ordered Bi Bi	P	—(UC)	—	—(C)	—(C)
	Q	C(C)	—(C)	C(N)	—
Rapid Equilibrium Random Bi Bi	P	C(NC)	—	C(C)	—(C)
	Q	C(C)	—(C)	C(NC)	—

[a] Terms that provide substrate inhibition as indicated in the text are not included [e.g., B/V_1K_{IB} in Eq. (27)].
[b] The abbreviations are C, competitive; N, noncompetitive; U, uncompetitive; and NL, nonlinear. A dash indicates that no inhibition is observed.
[c] Intercept replots versus inhibitor concentration are parabolic.
[d] The patterns in parentheses indicate the effect of abortive complex formation.
[e] Sope replots versus inhibitor concentration are parabolic.
[f] Hyperbolic concave up when binary abortive complexes form.

that inactive complexes called abortive ternary complexes form, causing changes in the actually observed inhibition patterns for a given mechanism.

These ternary abortive complexes are dead-end complexes of the type enzyme–substrate–product. In mechanisms where an ordered sequence of binding occurs, the product in the complex is the one that does not normally bind to the free enzyme. Considerable evidence exists for the formation of such complexes. Fromm[12] observed a NAD–pyruvate–enzyme complex with lactate dehydrogenase by difference spectroscopy, and similar complexes have been observed in fluorescence emission studies on enzyme-bound NADH with alcohol, lactate, and glutamate dehydrogenases.[13,14] Such complexes are now generally recognized with a

[12] H. J. Fromm, *Biochim. Biophys. Acta* **52**, 199 (1961).
[13] G. W. Schwert and A. D. Winer, *Biochim. Biophys. Acta* **29**, 424 (1958).
[14] A. D. Winer, W. B. Novoa, and G. W. Schwert, *J. Am. Chem. Soc.* **79**, 6571 (1957).

large number of enzymes, particularly dehydrogenases, by a variety of techniques including isotope exchange at equilibrium.[15,16] Substrate inhibition is commonly observed with enzymes that form abortive complexes, and it has been suggested that the inhibition is due to these complexes. Discussions about substrate inhibition and abortive complexes can be found in this volume [20] and in the recent review on dehydrogenases by Dalziel.[17] Wong[18] has recently argued that substrate inhibition with lactate dehydrogenase is due to enzyme–pyruvate complex formation, not the enzyme–pyruvate–NADH complex. A more recent study by Burgner *et al.*[19,20] suggested both the ternary abortive complex forms as well as an enzyme (NAD adduct) binary complex. Regardless, the effects to be described here are those occurring when products are present, and the formation of the abortive ternary complexes is well documented.

The effects of abortive complexes on the basic bisubstrate mechanisms will be considered.

a. Ordered Bi Bi

The formation of inactive ternary complexes would be as follows:

$$
\begin{array}{cccccccc}
& A & & B & & P & & Q \\
k_1 \downarrow & k_2 & k_3 \downarrow & k_4 & k_5 \uparrow & k_6 & k_7 \uparrow & k_8 \\
\hline
E & EA & & (EAB) & & EQ & & E \\
& P \rightarrow \big| K_{IP} & & (EPQ) & & B \rightarrow \big| K_{IB} & & \\
& EAP & & & & EQB & &
\end{array} \quad (26)
$$

In the presence of Q, the equation is

$$1/v = \frac{1}{V_1} + \frac{B}{V_1(K_{IB})} + \frac{K_a}{V_1 A}\left[1 + \frac{Q}{K_{iq}}\left(1 + \frac{B}{K_{IB}} + \frac{K_{ia}K_b}{K_a K_{IB}}\right)\right] + \frac{K_b}{V_1 B}$$
$$+ \frac{K_{ia}K_b}{V_1 AB}\left(1 + \frac{Q}{K_{iq}}\right) \quad (27)$$

The effect of Q by Eq. (27) is competitive relative to A as was predicted previously for an ordered mechanism, but the effect relative to B is unusual. The 1/B versus 1/v plot will be hyperbolic concave-up either in

---

[15] E. Silverstein and G. Sulebele, *Biochemistry* **8**, 2543 (1969).
[16] E. Silverstein and S. Sulebele, *Biochemistry* **12**, 2164 (1973).
[17] K. Dalziel, in "The Enzymes" (P. D. Boyer, ed.), 3rd ed., Vol. 11, p. 1. Academic Press, New York, 1975.
[18] C. S. Wong, *Eur. J. Biochem.* **78**, 569 (1977).
[19] J. W. Burgner, G. R. Ainslie, W. W. Cleland, and W. J. Ray, *Biochemistry* **11**, 1646 (1978).
[20] J. W. Burgner and W. J. Ray, *Biochemistry* **17**, 1654 (1978).

the presence or in the absence of Q. This nonlinear effect in the absence of Q is due to the $B/(V_1 K_{IB})$ term, and observation of the substrate inhibition depends on the concentration of B and the magnitude of K_{IB}. Generally, this term is not a major factor in the mechanism. The other term with B in the numerator is $K_a QB/V_1 K_{ia} K_{IB} A$. If the abortive complexes have relatively high dissociation constants, the apparent substrate inhibition will not be observed. This is a case where attempts to saturate an enzyme with a substrate may indicate the existence of the abortive complexes, but such effects are not readily differentiated from inactive EB complexes in an ordered mechanism.

The rate equation for Eq. (26) in the presence of P is

$$1/v = \frac{1}{V_1}\left(1 + \frac{P}{K_{ip}}\right) + \frac{K_a}{V_1 A} + \frac{K_b}{V_1 B}\left(1 + \frac{K_q P}{K_{ia} K_p}\right)\left(1 + \frac{P}{K_{ip}}\right)$$
$$+ \frac{K_{ia} K_b}{V_1 AB}\left(1 + \frac{K_q P}{K_{iq} K_p}\right) \tag{28}$$

Equation (28) predicts that P will be a noncompetitive inhibitor of both substrates A and B as expected for the ordered mechanism, but the expression is second order in P so that the intercept replot for a $1/v$ versus $1/A$ plot is parabolic and the slope replot from the $1/v$ versus $1/B$ is also parabolic. The inhibition by P relative to A is S-linear I-parabolic noncompetitive, and it is S-parabolic I-linear noncompetitive with respect to B.

b. Theorell–Chance Bi Bi

$$
\begin{array}{cccccccc}
 & A & B & & P & & Q & \\
 & \downarrow k_1 & \downarrow k_2 & \searrow k_3 & \nearrow k_4 & & \downarrow k_5 & \uparrow k_6 \\
\hline
E & & EA & & & EQ & & E \\
 & & P \downarrow K_{IP} & B \downarrow K_{IB} & & & & \\
 & & EAP & EBQ & & & &
\end{array} \tag{29}
$$

The rate equation in the presence of P is

$$1/v = \frac{1}{V_1} + \frac{K_a}{V_1 A} + \frac{K_b}{VB}\left(1 + \frac{P}{K_{ip}}\right)\left(1 + \frac{P}{K_{IP}}\right) + \frac{K_{ia} K_b}{V_1 AB}\left(1 + \frac{P}{K_{ip}}\right) \tag{30}$$

where $K_{ip} = k_5/k_4$. The formation of EQB was not included in this equation, but it would have the same form as in Eq. (27). The inhibition by P is different from that observed in the Ordered mechanism in that P is S-parabolic I-linear competitive toward B and S-linear I-parabolic noncompetitive with respect to A.

In the presence of Q the equation is similar to Eq. (27) predicting the same inhibition patterns for both the Ordered and Theorell–Chance mechanisms.

c. Rapid Equilibrium Random Bi Bi

The possible abortive ternary complexes are as shown in Eq. (31).

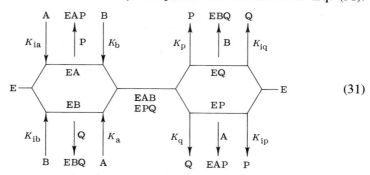

With P present, the rate equation is

$$1/v = \frac{1}{V_1} + \frac{K_a}{V_1 A} + \frac{K_b}{V_1 B}\left(1 + \frac{P}{K_{ip}}\right) + \frac{K_{ia}K_b}{V_1 AB}\left(1 + \frac{P}{K_{ip}}\right) \quad (32)$$

The analogous equation [Eq. (33)] is obtained with Q present.

$$1/v = \frac{1}{V_1} + \frac{K_a}{V_1 A}\left(1 + \frac{Q}{K_{Iq}}\right) + \frac{K_b}{V_1 B} + \frac{K_{ia}K_b}{V_1 AB}\left(1 + \frac{Q}{K_{iq}}\right) \quad (33)$$

These equations predict that P will be competitive and noncompetitive and Q noncompetitive and competitive with B and A, respectively. If abortives do not form, the product inhibition patterns are all competitive as shown above. Examples of both behavior are known and will be discussed below. This behavior, when abortive complexes form, is identical with the Theorell–Chance mechanism when abortives do not form. Such similarity requires other experiments to differentiate these possibilities.

d. Ping Pong Bi Bi

In the Ping Pong mechanism the abortive complexes are not ternary but binary complexes. The interactions including possible abortive complex formations are shown in Eq. (34).

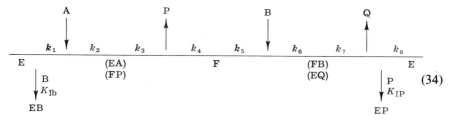

The rate equation describing the presence of P in this scheme [Eq. (34)] is Eq. (35).

$$1/v = \frac{1}{V_1} + \frac{K_a}{V_1 A}\left(1 + \frac{B}{K_{Ib}} + \frac{P}{K_{Ip}}\right) + \frac{K_b}{V_1 B}\left(1 + \frac{P}{K_{ip}}\right)$$
$$+ \frac{K_{ia} K_b P}{V_1 K_{ip} AB}\left(1 + \frac{B}{K_{Ib}} + \frac{P}{K_{Ip}}\right) \tag{35}$$

where $K_{ia} = k_2/k_1$ and $K_{ip} = k_3/k_4$.

The equation is modified in the presence of Q to

$$1/v = \frac{1}{V_1} + \frac{K_a}{V_1 A}\left(1 + \frac{B}{K_{Ib}} + \frac{Q}{K_{iq}}\right) + \frac{K_b}{V_1 B} + \frac{K_{ib} K_a Q}{V_1 K_{iq} AB} \tag{36}$$

where $K_{ib} = k_6/k_5$ and $K_{iq} = k_7/k_8$.

If the $1/v$ versus $1/B$ plot is not hyperbolic concave-up, the B/K_{Ib} term is not significant, and EB is not an important complex kinetically. Under this condition where EP does form, P will be S-parabolic I-linear noncompetitive with respect to both A and B. Q will give the same patterns as described previously, since no Q complexes form besides the normal EQ complex. The predicted effects of abortive complex formation on two-substrate mechanisms (in parentheses) are summarized in Table II compared to inhibition patterns in the absence of abortive complex formation.

The initial observation of abortive complex formation in product inhibition studies was from the study of Fromm and Nelson,[2] where the inhibition by the product, NAD$^+$, on ribitol dehydrogenase was evaluated. It was found that hyperbolic concave-up inhibition resulted. The replots from inhibition studies with ribulose indicated that ribulose was a I-parabolic S-linear noncompetitive inhibitor relative to NAD and I-linear S-parabolic noncompetitive relative to ribitol. These data are consistent only with the Ordered Bi Bi mechanism where the abortive complex EAP forms. Based on these data, it was suggested that the following interactions occurred[5]:

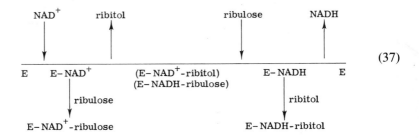

(37)

Saturation Techniques

An additional technique that is used in product inhibition studies is evaluation of the effect of saturation by substrates on the inhibition patterns.[4] In the Ordered Bi Bi mechanism [Eq. (26)], A and Q both combine with E and are competitive relative to each other. All other patterns are noncompetitive. Saturation with A will eliminate all inhibition by Q relative to B. When A is the variable substrate, saturation by B will change the inhibition by P relative to A from noncompetitive to uncompetitive. This is due to conversion of the EA + B = EAB step into an irreversible reaction. The elimination of inhibition by Q at high A will be very difficult to achieve and saturation by B may require a level 100 × the K_m for B before the inhibition is truly uncompetitive. This level may be very difficult to achieve experimentally and may give rise to substrate inhibition effects that preclude analysis.

The effect of saturation by each substrate on the product inhibition patterns for the basic two-substrate mechanisms is summarized in Table II. A noncompetitive pattern may be changed to uncompetitive, but neither competitive nor uncompetitive inhibition can be converted to another type, although the inhibition can disappear with any type under the proper circumstances.

Iso Mechanisms

Isomerization of central (transitory) enzyme forms does not affect the form of the initial rate equation but when a stable enzyme form isomerizes additional terms are present in the rate equation as illustrated in the discussion for the Uni Uni mechanisms [Eq. (4)]. The maximal velocity for an enzyme-catalyzed reaction cannot be greater than any unimolecular rate constant involved in product formation.[21] If the rate constants calculated for a given mechanism have negative values or are smaller than the maximal velocity, it is likely that stable enzymes forms are isomerizing in at least partially rate-limiting steps. In addition, unusual product inhibition patterns[7] and Dalziel ϕ relationships[22] are generated by such isomerizations.

Isomerization of stable enzyme forms does not affect the velocity in the absence of products, so that other types of experiments can define the mechanism and product inhibition can indicate the existence of such forms. The multiple forms of inhibition possible for each mechanism necessitate that care be exercised in interpretation of product inhibition

---

[21] L. Peller and R. A. Alberty, *J. Am. Chem. Soc.* **81**, 5907 (1959).
[22] K. Dalziel, *Acta Chem. Scand.* **11**, 1706 (1957).

data. The possible patterns for a variety of mechanisms with isomerizations steps are listed in Table III. The schemes for these have been described previously and will not be reported here. It should be noted that the Di-Iso mechanism listed in Table III has similar product inhibition patterns as the non-Iso mechanisms and can be distinguished only by evaluation of such parameters as the ϕ relationships.

Three Substrate Systems. A general discussion of the use of product inhibitors for three-substrate enzymes has not appeared, owing likely to the complexity of the systems. The inhibition patterns for a number of Ter Ter mechanisms are presented in Table IV. It is assumed that Products R, Q, and P are derived from substrates A, B, and C, respectively, and that three products are released. This assumption is not generally applicable to

TABLE III
PRODUCT INHIBITION PATTERNS OCCURRING WITH VARIOUS ISO MECHANISMS[a]

Mechanism	Product	Varied substrate		
		A	B	C
Iso Ordered Bi Uni	P	N[b](C)[c]	N	
Mono-Iso Ordered Bi Bi	P	N	N	
	Q	N(C)	N	
Di-Iso Ordered Bi Bi	P	N	N	
	Q	C	N	
Mono-Iso Theorell–Chance	P	N	C	
	Q	N(C)	N	
Di-Iso Theorell–Chance	P	N	C	
	Q	C	N	
Mono-Iso Tetra Uni Ping Pong	P	N	C	
	Q	N(C)	N	
Di-Iso Tetra Uni Ping Pong	P	N	N(C)	
	Q	N(C)	N	
Mono-Iso Uni Uni Bi Bi Ping Pong	P	N	N(C)	N
	Q	U	N	N
	R	C	U	U
Di-Iso Bi Uni Uni Bi Ping Pong	P	N	N	N(C)
	Q	U	U	N
	R	N(C)	N	U
Tri-Iso Hexa Uni Ping Pong	P	N	N(C)	U
	Q	U	N	N(C)
	R	N(C)	U	N

[a] Abortive complex formation was not considered in this analysis. Such effects will not alter the general conclusions.
[b] The abbreviations are C (competitive), N (noncompetitive), and U (uncompetitive).
[c] The patterns in parentheses indicate differences from the analogous non-Iso mechanism.

TABLE IV
Product Inhibition Patterns for Three-Substrate Mechanisms

Class of mechanism[a]	Product inhibitor	Varied substrates		
		A	B	C
I				
Random Ter Ter	P[b]	N[c]	N	C[d]
	Q[b]	N	C	N
	R[b]	C	N	N
II(a)				
Ordered Ter Ter	P[e]	N	N	N[g]
	Q	U[f]	N	NL[i]
	R	C	N[h]	N
II(a′)				
Ordered Theorell–	P[e]	N	N	C
Chance Ter Ter	Q	U	U[j]	U[j]
	R	C	N[k]	N
II(b)				
Random AC (all ternary	P[b]	N	N	C
complexes) Ter Ter	Q	N	C	N
	R[b]	C	N	N
III				
Random C (Ordered B)	P[b]	N	N	C
Ter Ter	Q	N	C	N
	R[b]	C	N	N
IV				
Random A Ter Ter	P	N	U	C
	Q[b]	N	C	N
	R[b]	C	N	N
V				
Random (No EAC)	P[b]	N	N	C
Ter Ter	Q[b]	N	C	N
	R[b]	C	N	N
VI				
Random AB Ter Ter	P	U	U	C
	Q[b]	N	C	N
	R[b]	C	N	N
VII				
Random BC (Equi-	P	U	N	C
librium A) Ter Ter	Q	U	C	N
	R[b]	C	N	N
VIII				
Random BC (Steady	P	U	N	C
State A) Ter Ter	Q	U	C	N
	R[b]	C	N	N
IX				
Equilibrium Ordered	P	U	U	C
Ter Ter	Q	U	C	N[k]
	R[b]	C	N	N

(continued)

TABLE IV (continued)

Class of mechanism[a]	Product inhibitor	Varied substrates		
		A	B	C
X				
Hexa Uni Ping Pong	P	N	C	U
	Q	U	N	C
	R	C	U	N
XI				
Ordered Bi Uni Uni Bi	P	N	N	C
Ping Pong	Q	U	U	N
	R	C	N	U
XII				
Ordered Uni Uni Bi Bi	P	N	C	N
Ping Pong	Q	U	N	N
	R	C	U	U
XIII				
Random Bi Uni Uni Bi	P	N	N	C
Ping Pong	Q	U	U	C
	R	C	C	U
XIV				
Random Uni Uni Bi Bi	P	N	C	C
Ping Pong	Q	U	N	N
	R	C	U	U

[a] It is assumed that the reactions are Ter Ter. Modifications can be made for other cases. All random binding steps are assumed to be rapid equilibrium. A number of these mechanisms were originally described in a report by F. B. Rudolph and F. C. Clayton [Abstr. Southwestern Regional Am. Chem. Soc. Meet. A166 (1974)]. The rate equations for the mechanisms are listed in this volume [7].

[b] If the product binds only to free enzyme, the inhibition would be competitive toward all substrates. This is not likely, since the substrates bind in a random order and binding to an enzyme–substrate complex should be possible unless the product is much larger or otherwise distinguished from its corresponding substrate.

[c] N refers to noncompetitive inhibition toward the substrate indicated.

[d] C refers to competitive inhibition toward the substrate indicated.

[e] Formation of abortive complexes has no effect on the inhibition by this product.

[f] U refers to uncompetitive inhibition toward the substrate indicated.

[g] This inhibition is noncompetitive if EAQC is not formed.

[h] This inhibition is nonlinear if ERB is formed.

[i] This inhibition is nonlinear.

[j] This inhibitor is noncompetitive if EAQ is formed.

[k] This inhibition is nonlinear if ERB is formed.

three-substrate enzymes but will apply to a number of systems including nucleoside triphosphate-linked synthetases.

The information that can be gained from product inhibition studies of three-substrate systems is considerable, as a number of uncompetitive patterns are generated that are very diagnostic of binding order. The effect of saturation can be also considered as was done for two-substrate cases but will not be detailed here. Formation of abortive complexes will be a real problem with that technique in many cases. By use of the saturation technique, all the classes of sequential mechanisms (I–IX) can be potentially distinguished except for I and II. If saturation cannot be achieved, which is quite likely for many enzymes, the information derived is the same as from competitive inhibition studies. This is because in the rapid-equilibrium mechanisms the interactions are potentially the same in both cases. Actually, in the partially random mechanisms II–IX, the effect of saturation by another substrate may be of the same diagnostic value in competitive inhibition studies as in the product inhibition studies. The predicted inhibition patterns for these mechanisms would be the same in either case. Iso mechanisms likely occur in three-substrate systems and would be indicated by unusual product inhibition patterns compared to competitive inhibition results and other effects as discussed in the two-substrate cases.

Elliott and Tipton[22a] have presented a discussion of the expected product inhibition patterns for a limited number of four substrate–three product and four substrate–four product mechanisms and cases where more than one molecule of a substrate is involved in the reaction. They have applied the theory to analysis of the product inhibition of carbamylphosphate synthetase.

Prediction of Inhibition Patterns

Cleland[23] has suggested a set of rules to predict the type of inhibition expected for a steady-state mechanism that has no random sequences unless they are in rapid equilibrium. All the enzyme forms in rapid equilibrium are considered as a single enzyme form for this analysis. The rules will apply to most enzymes, but care must be taken to include possible abortive complex formation in the analysis. The rules can be summarized as follows:

Rule 1. The ordinate intercept of the reciprocal plot is affected by a product which associates reversibly with an enzyme form other than the

---

[22a] K. R. F. Elliott and K. F. Tipton, *Biochem. J.* **141**, 789, 807, 817 (1978).
[23] W. W. Cleland, *Biochim. Biophys. Acta* **67**, 188 (1963).

one with which the variable substrate combines. The effect of the product cannot be eliminated by saturation of the variable substrate.

Rule 2. The slope of the reciprocal plot is affected by a product that associates with an enzyme form that is the same as, or is at least connected by a series of reversible steps to, the enzyme form with which the variable substrate combines. The effect of the product cannot be eliminated by saturation if the two enzyme forms involved are not connected by reversible steps. An irreversible step is one in which a product is released at essentially zero free concentration or where the substrate binding at that step is saturating.

To use these rules, one simply analyzes for slope and intercept effects separately and then combines the results to determine the pattern. If only Rule 1 applies, uncompetitive inhibition is observed. If only Rule 2 is applicable, competitive inhibition is seen. Noncompetitive inhibition results from both rules being involved. If the product adds to only one enzyme form, the effect is always linear; that is, a replot of slopes or intercepts versus concentration for inhibitors, or versus reciprocal concentration for substrates and activators, is a straight line. If the product reacts with two (or more) enzyme forms, as in abortive complex formation, the effects on slopes or intercepts are determined separately for each point of combination, and the combined effects are given by Rule 3.

Rule 3. When combination by a product at two or more points in a reaction sequence gives by Rules 1 or 2, a multiple effect on slopes or intercepts, the resultant effects will be parabolic (or of higher degree for more than two points of combination) if: (a) the effects produced by all combinations are the same; (b) the points of combination are separated in the reaction sequence by reversible steps along which interaction may take place so that an increase in the concentration of the compound specifically causes, as the result of combination at one point, an increase in the steady-state concentration of the enzyme form reacting with the product at the other (or next) point of combination. If no reversible sequences connect the points of combination, or if combination at one point does not affect the steady-state concentrations of the enzyme forms combined with at other points, then the resulting effects are linear. When analyzing intercept effects, it should be remembered that the variable substrate is saturating, and its combination with the enzyme is an irreversible step.

Applications of these rules are illustrated in the original article by Cleland[23] and are useful in analyzing data. Competitive and uncompetitive patterns are especially useful in defining reaction sequences. The formation of multiple complexes causes a noncompetitive inhibitor to be of less diagnostic value.

Practical Aspects

It is necessary to carry out proper assays to measure the true initial velocity of the reaction. Experimental control is considered in this volume [1] and statistical analysis of the initial rate data is detailed in this volume [6]. The assay systems available will influence which products can be used as inhibitors. If the products are measured by use of coupled enzyme assays, an alternative assay will be needed to evaluate the inhibitory effect of the product that is normally determined. This problem can sometimes be circumvented by use of alternative products[4,7,24,25] as discussed below. If the reaction is followed by appearance of a chromophoric product, care must be exercised to avoid introduction of straylight errors at higher concentrations of added measured product. Absorbances above two require care in measurement and interpretation.

Substrate concentrations should be maintained near their K_m levels to avoid abortive complex formation. When saturation by one substrate is attempted, care must be exercised to avoid pH, ionic strength, or other changes in the assay mix. Saturation will be difficult to achieve in many systems owing to such changes and substrate inhibition, which is often induced at high concentrations.

The data are treated as described in the section Nomenclature and Experimental Protocol by construction of double-reciprocal plots, evaluation of the slope and intercept effects and making replots of slope and intercept versus inhibitor comcentration. It is necessary to know both the type of inhibition (i.e., competitive, noncompetitive, or uncompetitive) and whether the inhibition is linear or not. Once this is determined one can determine the likely mechanism and binding order for a given enzyme based on the discussion above or using the predictive approach suggested by Cleland.[23]

The K_{ii} or K_{is} or both should be determined either from the replots or calculation. This value is useful for evaluation of potential regulatory effects by products and allow calculation of individual rate constants. Knowledge of individual steps gives insight into the actual catalytic mechanism and some idea of rate-limiting steps in the reaction sequence. The rate constants can be calculated from experiments in a single direction if the K_{eq} is known.[7]

If the product inhibition patterns do not agree with initial rate and other inhibitor studies, isomerizations may be occurring. This can be determined as illustrated for the two-substrate examples in Table III. This

[24] H. J. Fromm and V. Zewe, *J. Biol. Chem.* **237**, 3027 (1962).
[25] C. C. Wratten and W. W. Cleland, *Biochemistry* **4**, 2442 (1965).

will be especially evident if no competitive product inhibition patterns are observed for the products. This requires, in an Ordered mechanism, that the first substrate and last product bind to different forms of the enzyme.

Examples and Limitations of Product Inhibition Studies

Product inhibition studies have been done with a large number of enzymes, and no attempt will be made to catalog them here. Instead, a few pertinent examples will be given illustrating both the utility and limitations of the procedure. The technique is most useful for determination of the binding order of sequential mechanisms or, in the case of three-substrate systems, the sequential part of Ping Pong mechanisms.

An example of determination of the order of product release for Uni Bi mechanism is illustrated by the study of adenylosuccinate lyase by Bridger and Cohen.[26] AMP is a competitive inhibitor relative to adenylosuccinate, and fumarate is noncompetitive, indicating that product release is ordered with fumarate released first. Similar results were found for adenosine-5-phosphosulfate sulfohydrolyase with AMP released prior to sulfate.[27]

The differentiation between Ordered and Rapid Equilibrium Bi Bi mechanisms should be readily accomplished as indicated in Tables II and IV. The Ordered mechanism should exhibit one competitive and three noncompetitive patterns, and the Random mechanism should have two, three, or four competitive patterns depending on whether the EAP and EBQ abortive complexes form. Generally one would expect a dead-end complex to form between the substrate and the product lacking the piece of the molecule transferred during the reaction. For example, the E–creatine–ADP complex forms for creatine kinase and causes noncompetitive inhibition relative to each molecule by the other.[28] The formation of the complex representing the large substrate and product (e.g., E–ATP–creatine–P) depends on the relative binding affinities and the geometry of the active site. It does occur with creatine kinase but not always with other enzymes.

The behavior of uridine phosphorylase is unusual. The products uridine and phosphate are both competitive with both substrates, uracil and ribose 1-phosphate.[29] The four competitive patterns suggest that the mechanism is Rapid Equilibrium Random with no abortive complexes

[26] W. A. Bridger and L. J. Cohen, *J. Biol. Chem.* **253**, 644 (1968).
[27] A. M. Stokes, W. H. B. Denner, and K. S. Dodgston, *Biochim. Biophys. Acta* **315**, 402 (1973).
[28] J. F. Morrison and E. James, *Biochem. J.* **97**, 37 (1965).
[29] T. A. Krenitsky, *Biochim. Biophys. Acta* **429**, 352 (1976).

formed. This result is what was originally predicted by Cleland[4] for this general mechanism and has been often accepted as being the only patterns observed for a rapid equilibrium mechanism. In reality, it is probably a rare exception.

The more typical behavior for a random mechanism is illustrated by studies on fructokinase.[30] There are two competitive and two noncompetitive patterns consistent with the formation of both abortive complexes. This pattern is identical with that predicted for the Theorell–Chance mechanism, requiring isotope exchange, analysis of initial-rate convergence points, or competitive inhibition studies for differentiation.

An additional complication is illustrated in the product inhibition of muscle phosphofructokinase[31]; ADP is a competitive inhibitor relative to ATP and noncompetitive with respect to fructose 6-phosphate. Fructose diphosphate is noncompetitive relative to both substrates, consistent with an ordered mechanism where ATP binds first to the enzyme. However, other results, such as inhibition by sugar analogs show the mechanism in fact to be random. The noncompetitive inhibition by fructose diphosphate relative to fructose 6-phosphate is apparently due to formation of another abortive complex, E–fructose 6-phosphate–fructose diphosphate, where the 1-phosphate of the diphosphate binds at the γ-phosphate site of the ATP binding site. Using somewhat different experimental conditions, Kee and Griffin[32] and Bar-Tana and Cleland[33] found that fructose diphosphate is competitive with fructose 6-phosphate. The formation of the other complexes depends on the relative concentrations of the substrates and products. An analogous situation exists for yeast hexokinase. If saturating levels of glucose are used, glucose 6-phosphate is a competitive inhibitor with respect to ATP,[34] indicating that it must be binding at the γ-phosphate site.

Engel and Chen[35] have made an extensive study of glutamate dehydrogenase which catalyzes a Bi Ter reaction. All possible combinations of variable substrate and product inhibitor were used for both directions of the reaction. The results could be reconciled with either an ordered or random mechanism by inclusion of abortive complexes. Based on other results, they support a random order mechanism but cannot draw any valid conclusions based on the product inhibition data alone.

Cleland[4] has suggested that in a Steady State Random Bi Bi mecha-

---

[30] F. M. Raushel and W. W. Cleland, *Biochemistry* **16**, 2169 (1977).
[31] R. L. Hanson, F. B. Rudolph, and H. A. Lardy, *J. Biol. Chem.* **248**, 7852 (1973).
[32] A. Kee and C. C. Griffin, *Arch. Biochem. Biophys.* **149**, 361 (1972).
[33] J. Bar-Tana and W. W. Cleland, *J. Biol. Chem.* **249**, 1271 (1974).
[34] K. G. Wettermark, E. Borgland, and S. E. Brolin, *Anal. Biochem.* **22**, 211 (1968).
[35] P. C. Engel and S.-S. Chen, *Biochem. J.* **151**, 305 (1975).

nism inhibition by either product will be hyperbolic noncompetitive toward both substrates. The kinetic mechanism for yeast hexokinase is best described as Steady State Random, and ADP and glucose 6-phosphate are linear noncompetitive inhibitors of both ATP and glucose.[36] In order to determine that such an effect is consistent with the steady-state mechanism, the steady-state equation describing product inhibition of hexolinase was derived. The formation of an abortive complex, EBP, in the presence of product P was assumed. The postulated interactions are

$$EB + P \rightleftharpoons EBP$$
$$EP + B \rightleftharpoons EBP$$

The total rate equation as described for the mechanism was 45 numerator and 672 denominator terms, which include many squared and cubed concentration terms. The rate equation may be reduced by making assumptions concerning magnitude of certain terms to the following form:

$$V = \frac{K_1 AB}{K_2 + K_3 P + K_4 A + K_5 B + K_6 AB + K_7 BP + K_8 ABP} \quad (38)$$

The K's represent various combinations of rate constants. The equation predicts the noncompetitive behavior observed experimentally.

As an additional check the total rate equation was evaluated with a digital computer, and it was found that the product was a linear noncompetitive inhibitor of both substrates whether P was ADP or glucose 6-phosphate. Only when the unimolecular rate constants for the abortive complex formation were much smaller than the outer rate constants, would the inhibition approach competitive. Various combinations of dissociation constants and rate constants were tested and found generally to give similar results. Thus, in a Steady State Random mechanism the product inhibition patterns may all be linear noncompetitive depending on various rate constants. Observation of hyperbolic noncompetitive inhibition will likely by seen only if the initial rate behavior of the steady-state mechamism is also nonlinear.

These results for the various examples given above require that one should exercise care in interpretation of product inhibition data for sequential mechanisms. In a study on carbamate kinase it was found that ADP was a competitive inhibitor relative to ATP and that carbamyl phosphate was noncompetitive relative to carbamate.[37] Based on these data it was suggested that the reaction was ordered with ATP binding first even though binding studies had shown effective binding of carbamate for the

[36] F. B. Rudolph and H. J. Fromm, *J. Biol. Chem.* **246**, 6611 (1971).
[37] M. Marshall and P. P. Cohen, *J. Biol. Chem.* **241**, 4197 (1966).

free enzyme. Observation of only one competitive pattern clearly is not sufficient in itself to confirm an ordered mechanism.

A number of interesting studies have been done on multisubstrate Ping Pong mechanisms where the product inhibition study gave considerable information concerning the mechanism. In a classical Ping Pong Bi Bi mechanism, A and Q combine with free enzyme and B and P bind to the enzyme complex (F) and thus are competitive pairs, and noncompetitive inhibition is observed relative to the other substrate and product. With transcarboxylase, however, pyruvate and propinoyl-CoA are noncompetitive toward each other as are methylmalonyl-CoA and oxaloacetate, and the two keto acids are mutually competitive as are the two CoA thioesters.[38] For the classical Ping Pong mechanism these inhibition patterns would be reversed. The enzyme contains bound biotin that carries the carboxyl group between catalytic centers thus catalyzing the reaction at two separate sites. Cleland[39] has suggested general rules for describing the inhibition observed with multisite mechanisms: (1) When a product combines in the same site as the variable substrate the inhibition is competitive. (2) A product combining at the same site, but in a different region than the variable substrate, will be a noncompetitive inhibitor. (3) A product that combines at a different site from the variable substrate will be an uncompetitive inhibitor unless it can connect the form reacting with the variable substrate into a different form and the site reaction for the substrate is random. Among enzymes that have been suggested to have such multisite Ping Pong mechanisms are pyruvate carboxylase,[40,41] pyruvate dehydrogenase,[42] and fatty acid synthetase.[43]

A special case involving studies with products deals with systems such as maltodextran phosphorylase. The substrate glycogen is also the product, so that addition of phosphate as a product inhibitor causes all substrates and products to be present. This can occur with any substrate that is not significantly altered by the reaction and will serve again as a substrate. Product inhibition studies are not useful in these cases unless more than two products are released. For systems of that type the study would require dealing with the effects of the products on the initial reaction. The mechanism of *Escherichia coli* maltodextran phosphorylase was studied

---

[38] D. B. Northrup, *J. Biol. Chem.* **244**, 5808 (1969).
[39] W. W. Cleland, *J. Biol. Chem.* **243**, 820 (1968).
[40] W. R. McClure, H. A. Lardy, M. Wagner, and W. W. Cleland, *J. Biol. Chem.* **246**, 3579 (1971).
[41] R. E. Barden, C.-H. Fung, M. F. Utter, and M. C. Scrutton, *J. Biol. Chem.* **247**, 1323 (1972).
[42] C. S. Tasi, M. W. Bugett, and L. J. Reed, *J. Biol. Chem.* **248**, 8348 (1973).
[43] S. S. Katiyar, W. W. Cleland, and J. W. Porter, *J. Biol. Chem.* **250**, 2709 (1975).

by Chao et al.,[44] using isotope exchange and competitive inhibitors, but product inhibition was not employed for the above reasons.

Alternative Product Inhibition

If an enzyme can use multiple substrates for the same reaction, it will have alternative products that may be used in inhibition studies. They can afford distinctive kinetic behavior and often allow product inhibition experiments when the normal product is used for a completed enzyme assay or is not usable for some technical reason. Fromm and Zewe[24] used mannose 6-phosphate as an alternative product for yeast hexokinase. It was a noncompetitive inhibitor of both glucose and ATP, indicating that it bound to free enzymes and the enzyme–substrate binary complexes.

Wratten and Cleland[25] have made an extensive study of liver alcohol dehydrogenase using alternative oxidized products. The technique allowed demonstration of abortive ternary complex formation and allowed exclusion of the Theorell–Chance mechanism. The technique seems most useful for this latter purpose, as the presence of central enzyme–substrates complexes are readily detected as described by Cleland.[4]

Concluding Remarks

The use of product inhibition to study kinetic mechanisms has become a widely used technique. Owing to complications with various possible abortive complexes, the data should be interpreted carefully and definite conclusions reached only after use of other approaches as outlined in this volume. It should be emphasized that, if complications are not encountered in such studies, product inhibitors may very effectively define a kinetic mechanism.

Acknowledgment

Supported by Grant 14030 from the National Cancer Institute, DHEW and Grant C-582 from the Robert A. Welch Foundation.

---

[44] J. Chao, G. F. Johnson, and D. J. Graves, *Biochemistry* **8**, 1459 (1969).

[17] The Kinetics of Reversible Tight-Binding Inhibition

By Jeffrey W. Williams[1] and John F. Morrison

A reversible tight-binding inhibitor is one that exerts its reversible inhibitory effect on an enzyme-catalyzed reaction at a concentration compariable to that of the enzyme. Therefore, allowance must be made for the change in the concentration of the free inhibitor that occurs as a result of it undergoing interaction with one or more forms of the enzyme. In this respect tight-binding inhibitors differ from classical inhibitors, which cause inhibition only at concentrations that are considerably greater than the enzyme concentration. Since these two classes of inhibitor differ only in the relative amounts of free and enzyme-bound inhibitor that exist when the total concentration of added inhibitor is in the region of its K_i value, no clear line of demarcation can be drawn between classical and tight-binding inhibitors. Nevertheless, it is apparent when one is dealing with tight-binding inhibitors, as they exhibit characteristic kinetic properties.

Inhibitors that cause irreversible inhibition because they form covalently bonded enzyme–inhibitor complexes exhibit some of the characteristics of tight-binding inhibitors. However, in this chapter consideration will be given only to inhibition that is reversible. The term tight-binding inhibitor will be used to refer to an inhibitor that gives rise to reversible tight-binding inhibition.

Over the past decade there has developed an increased interest in tight-binding inhibition. This interest has been stimulated largely by the finding that some folate analogs are very useful chemotherapeutic agents because they act as very effective and specific inhibitors of dihydrofolate reductase. Much current work is concerned with a search for folate analogs that have selective lethal effects on the dihydrofolate reductase of target cells and tolerable effects on the sensitive enzyme of normal cells.[1a] Concomitant with the applied research, there has been some development of the kinetic theory for tight-binding inhibitors,[2-6] but few studies have

---

[1] Dr. Williams holds a Postdoctoral Fellowship of the American Cancer Society.
[1a] R. L. Blakley, "The Biochemistry of Folic Acid and Related Pteridines." North-Holland Publ., Amsterdam, 1969.
[2] A. Goldstein, *J. Gen. Physiol.* **27**, 529 (1944).
[3] O. H. Strauss and A. Goldstein, *J. Gen. Physiol.* **26**, 559 (1943).
[4] W. W. Ackermann and V. R. Potter, *Proc. Soc. Exp. Biol. Med.* **72**, 1 (1949).
[5] J. M. Reiner, "Behavior of Enzyme Systems." Burgess, Minneapolis, Minnesota, 1959.
[6] L. Easson and E. Stedman, *Proc. R. Soc. London, Ser. B,* **121**, 142 (1936).

been undertaken in sufficient detail to test the theoretical predictions. The future development of tight-binding inhibitors for chemotherapeutic purposes will undoubtedly depend on the application of kinetic techniques that yield quantitative information about the kinetic behavior of the inhibitor. When the structures of tight-binding inhibitors can be correlated with the true dissociation constants for their enzyme–inhibitor complexes, a systematic approach can be made toward the synthesis of more effective inhibitors for a particular enzyme.

In preparing this chapter, the authors have been able to draw only on their own limited experience of tight-binding inhibition and the few reports of other investigators. Hence it has not been possible to evaluate the research in this field and to reach conclusions about the best approaches to use for the study of this type of inhibition. The emphasis will be placed on describing in quantitative terms the different kinetic behavior that may be observed with tight-binding inhibitors. These inhibitors will be referred to as tight-binding or slow-tight-binding inhibitors according to whether the equilibria involving enzyme–inhibitor complexes are established rapidly or slowly. In addition, reference will be made to slow-binding inhibition where the time required to establish equilibrium between enzyme and inhibitor is not short relative to the time-scale of the assay. Inhibitors exhibiting this behavior may or may not be tight-binding inhibitors.

Theory

A quantitative description of tight-binding inhibition cannot be based on the Michaelis–Menten equation, since the assumption that the free inhibitor concentration is equal to the total inhibitor concentration is not valid. Tight-binding inhibitors cause inhibition at concentrations comparable to those at which enzymes are used for steady-state kinetic studies. Consequently the formation of an enzyme–inhibitor complex can result in a considerable reduction in the concentration of added inhibitor, and allowance must be made for this reduction. It is not necessary that the inhibition constant for an inhibitor be very low for differences in the free and total concentrations of inhibitor to become significant. Indeed this situation could arise when an enzyme with a low turnover number acts on a poor substrate. Under these conditions the amount of enzyme required for reaction may well be so high that a significant amount of inhibitor may be enzyme-bound even when the K_i values are of a magnitude usually associated with classical inhibition.

Development of Kinetic Theory

Easson and Stedman[6] were the first to demonstrate the tight-binding inhibition of an enzyme-catalyzed reaction using physostigmine as an inhibitor of cholinesterase. They presented equations that took into account the depletion of the inhibitor, but did not consider the type, or influence, of substrate on the inhibition. Subsequently, the kinetics of the inhibition of enzymes by tight-binding inhibitors were discussed by several authors[2–5] who were primarily concerned with the qualitative aspects of the subject. Only single-substrate reactions were considered, and thus the theory could not generally be applied to multisubstrate reactions. However, the introduction by Ackermann and Potter[4] of the plot of velocity against enzyme concentration at different inhibitor concentrations has proved to be a useful method for detecting tight-binding inhibition.

A general steady-state rate equation that allows for the combination of a tight-binding inhibitor with one or more enzyme forms in any kinetic mechanism was derived by Morrison.[7] The equation is quadratic in v, which represents the true steady-state velocity of the reaction. Cha[8,9] further extended the theory for tight-binding inhibition by considering the situation where the rate of establishment of the equilibrium between free enzyme, tight-binding inhibitor, and the enzyme–inhibitor (EI) complex is slow relative to the rate of conversion of substrate to products. He developed equations to describe this situation and elaborated procedures for determining values for rate constants under conditions where the rate of formation and dissociation of an EI complex is slow. The steady-state and non-steady-state kinetic theories as well as the predictions that follow from the equations describing various models will constitute the main themes of the following section.

Steady-State Kinetic Theory

Under conditions where (a) the various equilibria[10] involving the enzyme, substrates, and inhibitor are attained at a sufficiently rapid rate to permit measurements of steady-state velocities and (b) the amount of enzyme-bound inhibitor forms a significant proportion of the total inhibitor concentration, the rate equation for any kinetic mechanism can be expressed as

$$v^2 + N \left[\frac{1}{\Sigma(N_i/K_i)} + \frac{I_t - E_t}{D} \right] v - \frac{N^2 E_t}{D \Sigma(N_i/K_i)} = 0 \qquad (1)$$

In this equation the term N contains the rate constants that determine the maximum velocity of the reaction together with the concentrations of sub-

---

[7] J. F. Morrison, *Biochim. Biophys. Acta* **185**, 269 (1969).
[8] S. Cha, *Biochem. Pharmacol.* **24**, 2177 (1975).
[9] S. Cha, *Biochem. Pharmacol.* **25**, 2695 (1976).
[10] Unless stated otherwise, no distinction will be made between steady-state and thermodynamic equilibria.

strates; N_i denotes the term(s) in the denominator of the rate equation that represent the form(s) of the enzyme with which the inhibitor reacts, and K_i represents the dissociation constant(s) for the EI complex(es); E_t and I_t denote the total concentrations of enzyme and inhibitor, respectively; D represents the denominator of the rate equation in the absence of inhibitor. Equation (1) is a general steady-state rate equation that describes the inhibition of any enzyme-catalyzed reaction irrespective of the type of inhibition and the magnitude of the inhibition constant(s) in relation to E_t. When I_t is set equal to zero, $v = NE_t/D$ while if $E_t \ll K_i$ [i.e., $E_t \ll D/\Sigma(N_i/K_i)$] then $v = NE_t/(D + I_t\Sigma(N_i/K_i))$.

Plots of Steady-State Velocity (v) as a Function of Total Enzyme Concentration (E_t)

It follows from Eq. (1) that a plot of v against E_t at a fixed concentration of I_t consists of curved and linear regions (Fig. 1). The linear region or asymptote is described by Eq. (2), which predicts that the asymptote will

$$v = N/D\,[E_t - I_t] \qquad (2)$$

cut the abscissa at the point where $I_t = E_t$ and the vertical ordinate at $-v = (N/D)I_t$. Thus a plot of this type allows determination of the concentration of enzyme in an impure preparation, provided that only one molecule of inhibitor reacts at each catalytic site. The vertical intercept together with values for the kinetic parameters associated with the substrates, the concentrations of the substrates, and the concentration of inhibitor can be used to calculate the turnover number of the enzyme. When different fixed concentrations of I_t are used, the asymptotes of plots of v against E_t will consist of a family of parallel lines that cut the horizontal and vertical axes. A secondary plot of horizontal intercepts against I_t will be linear with a slope of plus one and pass through the origin. A similar plot will be obtained for vertical intercepts as a function of I_t, except that this plot will have a slope of minus one.

The initial slope of the curves illustrated in Fig. 1 is described by Eq. (3), but estimates of the values

$$\text{Initial slope} = \frac{N}{D + \Sigma(N_i/K_i)I_t} \qquad (3)$$

for K_i cannot be readily determined by graphical procedures because of the difficulty of measuring the initial slopes. It will be shown later that val-

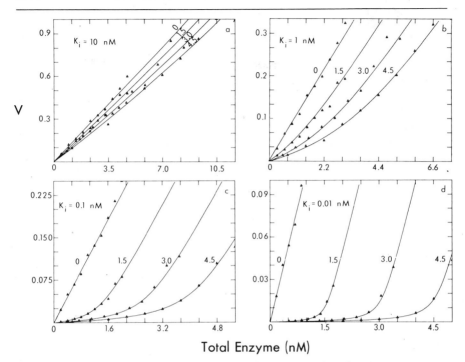

FIG. 1. Theoretical plots of steady-state velocity (v) as a function of total enzyme concentration (E_t) in the absence and in the presence of different concentrations of a reversible tight-binding inhibitor. The figures adjacent to each curve represent the total concentrations (nM) of inhibitor. The points represent simulated data that were generated using E_t concentrations up to that required to give a v/v_o ratio of 0.7 and into which was introduced 5% random scatter.

ues for the kinetic parameters, together with their standard errors, can be obtained by making a least squares fit of the data of the type shown in Fig. 1 to the appropriate equation. The degree of curvature with data conforming to Eq. (1) is a function of the ratios $K_i : E_t$ as well as $E_t : I_t$ (cf. Fig. 1). For the extreme condition where K_i and I_t are very much greater than E_t, curvature disappears so that the plots are linear. That is, the equation for tight-binding inhibition degenerates into one that describes classical inhibition for a reaction conforming to Michaelis–Menten kinetics.

The shapes of the plots of steady-state velocity against E_t at different I_t concentrations are diagnostic of tight-binding inhibition. However, it may be difficult to distinguish between irreversible and tight-binding inhibitors because the two types of inhibitor would give rise to similar plots (cf. Fig. 1d).

Plots of Steady-State Velocity as a Function of Total Inhibitor Concentration

The general shape of the curves that are obtained when the steady-state velocity (v) is plotted against total inhibitor concentration (I_t) is illustrated in Fig. 2. It follows from Eq. (1) that the tangent to the curve at $I_t = 0$ is

$$v = \frac{NE_t}{D[E_t + D/\Sigma(N_i/K_i)]} I_t + \frac{NE_t}{D} \qquad (4)$$

When the tangent is extrapolated to cut the abscissa, it will do so at the point where

$$I_t = E_t + \frac{D}{\Sigma(N_i/K_i)} \qquad (5)$$

It will be noted that the intersection point on the abscissa is not only a function of E_t but also the term $D/\Sigma(N_i/K_i)$, whose magnitude will be determined by the K_i value(s), the substrate concentrations, and the values of the kinetic parameters associated with the substrates. A plot of v against I_t has been used commonly for the determination of E_t on the as-

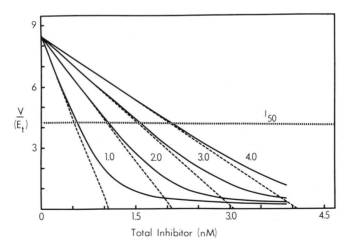

FIG. 2. Theoretical curves for the variation of the steady-state velocity (v) as a function of the total concentration of a reversible tight-binding inhibitor (I_t) at different concentrations of total enzyme (E_t). The figures to the right of each curve represent the total enzyme concentration (nM), and the dashed lines represent the calculated tangents to the curves at the point where $I_t = 0$. The horizontal dashed line indicates how the total inhibitor concentration required to give 50% inhibition (I_{50}) increases as a function of total enzyme concentration.

sumption that the intersection point does give a direct measure of enzyme concentration. But, in any event, there are likely to be difficulties associated with the drawing, by graphical methods, of tangents to curves such as those illustrated in Fig. 2. Difficulties would be minimized when enzyme is used at higher concentrations. It is also apparent from the curves of Fig. 2 that the concentration of tight-binding inhibitor required to give 50% inhibition (I_{50}) depends on the concentration of enzyme. Therefore, I_{50} values for tight-binding inhibitors have no meaning in the absence of information about the concentration of enzyme used.

Plots of the reciprocals of the Steady-State Velocity as a Function of Total Inhibitor Concentration

Plots of v^{-1} against I_t suffer from the same weakness as those described for plots of v against I_t. The equation of the asymptote for such a plot is

$$\frac{1}{v} = \frac{\Sigma(N_i/K_i)}{NE_t} I_t + \frac{D - \Sigma(N_i/K_i)E_t}{NE_t} \tag{6}$$

from which it may be determined that the intersection point on the abscissa, equal to $E_t - D/\Sigma(N_i/K_i)$, underestimates the value for E_t by the same amount as the overestimate from Eq. (5).

Linearization of the General Tight-Binding Inhibition Equation

To facilitate the calculation of values for E_t and K_i, using graphical procedures Henderson[11] has rearranged the general tight-binding inhibition equation [Eq. (1)] into a linear form as

$$\frac{I_t}{1 - (v/v_o)} = \frac{D}{\Sigma(N_i/K_i)} \frac{v_o}{v} + E_t \tag{7}$$

where v_o is the steady-state velocity in the absence of inhibitor and is equal to N/D. Equation (7) predicts that plots of $I_t/(1 - v/v_o)$ against v_o/v would be linear with a slope equal to the value for apparent K_i and intercept equal to E_t. With data obtained at different fixed concentrations of enzyme, the plot would consist of a family of parallel straight lines.

Analysis of Steady-State Velocity Data Obtained with E_t and I_t as Variables

There are several problems associated with the use of the linear form of the equation [Eq. (7)] to determine parameters. The values of

[11] P. J. F. Henderson, *Biochem. J.* **127**, 321 (1972).

$I_t(1 - v/v_o)$ are very susceptible to error unless v_o is well determined. Indeed, if the value of v_o is not sufficiently precise, an artifactual nonlinear plot can be obtained. Further, Eq. (7) is not an explicit function for which the dependent variable can be expressed as a function of the kinetic parameters and independent variables. The errors in the dependent and independent variables are related, and hence a correct weighting factor cannot be assigned and it would be inappropriate to make a least squares fit of the data to Eq. (7), as such a fit would not yield the best values for the kinetic parameters. An attempt has been made to eliminate the problems associated with data analysis,[12] but the computer program devised for this purpose does not, in effect, minimize the residual sum of squares.

The alternative approach to the estimation of values for E_t and apparent K_i is to make a least squares fit of the data to Eq. (1). For this purpose, Eq. (1) can be expressed as

$$v = 1/2R\{[(S + I_t - \alpha E_x)^2 + 4S\,\alpha E_x]^{1/2} - (S + I_t - \alpha E_x)\} \quad (8)$$

where $R = D/N$, $S = D/\Sigma(N_i/K_i)$, and $\alpha E_x = E_t$. Alpha (α) represents the fraction of total protein (E_x) present as enzyme that can react with the inhibitor. If a proportion of the enzyme is catalytically inactive but combines with the inhibitor, that proportion will be included in the estimate for E_t and the turnover number of the enzyme will be less than the true value. R denotes the reciprocal of the steady-state velocity in the absence of inhibitor while S represents the apparent inhibition constant. A computer program has been written for the fitting to Eq. (8) of steady-state velocity data obtained at different fixed concentrations of E_t and I_t.[13] The output gives values for R, S, and α together with values for the standard errors of these parameters. Since

$$\text{Apparent } K_i = D/\Sigma(N_i/K_i) \quad (9)$$

calculation of the true values for K_i requires not only a knowledge of the kinetic parameters in the absence of inhibitor, but also information about the type of inhibition.

Determination of the Type of Inhibition and Calculation of True Values for Inhibition Constants

If a tight-binding inhibitor causes noncompetitive inhibition in relation to the variable substrate, then the value for apparent K_i would be given by

---

[12] P. J. F. Henderson, *Biochem. J.* **135**, 101 (1973).
[13] K. J. Ellis and J. F. Morrison (1977). Unpublished work.

the general expression

$$\text{Apparent } K_i = K_{is} \{[1 + A/K_a]/[1 + A/(K_a K_{ii}/K_{is})]\} \quad (10)$$

where K_a denotes the Michaelis constant for substrate A; K_{is} and K_{ii} refer, respectively, to the inhibition constants associated with the slopes and intercepts of double-reciprocal plots. Equation (10) indicates that for noncompetitive inhibition the apparent K_i value will vary as a hyperbolic function of the substrate concentration. When [A] tends to zero, the slope of the plot will equal K_{is}, whereas when [A] tends to infinity, the slope will be equal to K_{ii}. The plot will be concave-up when $K_{is} > K_{ii}$ and concave-down when $K_{is} < K_{ii}$. If $K_{is} = K_{ii}$, the value of apparent K_i will be independent of the substrate concentration.

With competitive inhibition K_{ii} equals infinity and thus Eq. (10) reduces to

$$\text{Apparent } K_i = K_{is} (1 + A/K_a) \quad (11)$$

so that a plot of apparent K_i against [A] is linear with a vertical intercept equal to K_{is}. With uncompetitive inhibition K_{is} equals infinity and Eq. (10) becomes

$$\text{Apparent } K_i = K_{ii} (1 + K_a/A) \quad (12)$$

Equation (12) shows that uncompetitive inhibition is characterized by the fact that a plot of apparent K_i against $[A]^{-1}$ is linear with an intercept of K_{ii}.

Relationship between Apparent K_i and I_{50} Values

Enzyme inhibitors have been used extensively in many fields for structure–activity studies.[14] In such studies the extent of inhibition is often expressed as I_{50}, which is the concentration of inhibitor required to give 50% inhibition. It has been shown[15] that, when there is negligible depletion of the inhibitor because of its interaction with enzyme, the relationship between K_i and I_{50} is dependent on the type of inhibition. The same is true when the inhibitor is of the tight-binding type, but in addition, the concentration of enzyme enters into the relationship. It follows from Eq. (7) that, when there is 50% inhibition of the reaction ($v = 0.5\ v_0$),

$$I_{50} = E_t/2 + D/\Sigma(N_i/K_i) \quad (13)$$

From this equation it is apparent that the vertical intercept of a plot of I_{50}

---

[14] B. R. Baker, "Design of Active-Site-Directed Irreversible Enzyme Inhibitors." Wiley, New York, 1967.
[15] Y.-C. Cheng and W. H. Prusoff, *Biochem. Pharmacol.* **22**, 3099 (1973).

against E_t would yield a value for apparent K_i. The $E_t/2$ term of Eq. (13) allows for depletion of the inhibitor as a result of binding to the enzyme. If apparent K_i were zero, then 50% inhibition would be achieved when the total inhibitor concentration was equal to half the total enzyme concentration. As the concentration of added enzyme increases, so also does the I_{50} value. Consequently, the concentration of enzyme used to determine I_{50} values for tight-binding inhibitors must always be given. Equation (13) illustrates how the I_{50} value depends on the value of $D/\Sigma(N_i/K_i)$, which in turn is determined by the nature of the inhibition [cf. Eqs. (10)–(12)]. The plots of Fig. 2 also demonstrate how the concentration of inhibitor required to obtain 50% inhibition is a function of the total enzyme concentration.

Steady-State Velocity as a Function of Substrate Concentration at Different Concentrations of Inhibitor

When either E_t or I_t is used as the independent variable, Eq. (1) is applicable irrespective of the type of inhibition and the form(s) of enzyme with which the tight-binding inhibitor combines. However, this is not true when substrate is a variable reactant. The appropriate equation to describe the type of inhibition with respect to a variable substrate can be derived by substituting the appropriate expression for N, N_i, D, and $\Sigma(N_i/K_i)$ into Eq. (1). For purposes of illustration, it will be assumed that a reaction involving two substrates proceeds via their random addition to the enzyme, under rapid equilibrium conditions (Scheme 1) and that a

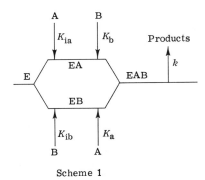

Scheme 1

tight-binding analog of B combines both with the free enzyme and the EA complex with dissociation constants of K_i and K_I, respectively. For this kinetic mechanism

$$N = kAB; \qquad D = K_{ia}K_b + K_aB + K_bA + AB \qquad (14)$$

where K_{ia}, K_a, and K_b are dissociation constants for the reactions illustrated in Scheme 1. Since $K_{ia}K_b$ and K_bA represent the proportions of total enzyme present as free enzyme and EA complex, respectively, it follows that

$$\Sigma\left(\frac{N_i}{K_i}\right) = \frac{K_{ia}K_b}{K_i} + \frac{K_bA}{K_I} \tag{15}$$

Substitution into Eq. (1) of the relationships given in Eqs. (14) and (15) and rearrangement of the resulting equation in double-reciprocal form gives Eq. (16), with A as the variable substrate and Eq. (17) with B as the variable substrate. In both equations v^{-1} is represented by y and $[A]^{-1}$ is denoted by x and $[B]^{-1}$ by z.

$$k^2B^2E_ty^2 - kK_{ia}K_bB\left(1 + \frac{B}{K_{ib}} + \frac{I_t - E_t}{K_I}\right)xy$$

$$- kK_bB\left(1 + \frac{B}{K_b} + \frac{I_t - E_t}{K_I}\right)y - K_{ia}K_b\left(\frac{K_b + B}{K_i} + \frac{K_b + K_aB/K_{ia}}{K_I}\right)x$$

$$- \frac{K_{ia}^2K_b^2}{K_i}\left(1 + \frac{B}{K_{ib}}\right)x^2 - \frac{K_b^2}{K_I}\left(1 + \frac{B}{K_b}\right) = 0 \tag{16}$$

$$k^2A^2E_ty^2 - kK_{ia}K_bA\left[1 + \frac{I_t - E_t}{K_i} + \frac{A}{K_{ia}}\left(1 + \frac{I_t - E_t}{K_I}\right)\right]zy$$

$$- kA(K_a + A)y - \left[(K_a + A)\left(\frac{K_{ia}K_b}{K_i} + \frac{K_bA}{K_I}\right)\right]z$$

$$- \left[K_b^2(K_{ia} + A)\left(\frac{K_{ia}}{K_i} + \frac{A}{K_I}\right)\right]z^2 = 0 \tag{17}$$

For equations having the form of Eqs. (16) and (17), plots of y against x (or z) will yield curves that have an asymptote and a concave-down curved region near the vertical ordinate (Fig. 3). Thus, unless steady-state velocities are determined at relatively high concentrations of the variable substrate, the curvature may be missed. In contrast to the inhibition of reactions conforming to Michaelis–Menten kinetics, the slopes of the asymptotes will not vary as a linear function of I_t. When x is set equal to zero, Eq. (16) reduces to Eq. (18), which shows that the vertical intercepts will

$$k^2B^2E_ty^2 - kK_bB\left(1 + \frac{B}{K_b} + \frac{I_t - E_t}{K_I}\right)y + \frac{K_b^2}{K_I}\left(1 + \frac{B}{K_b}\right) = 0 \tag{18}$$

vary in a nonlinear manner with the total concentration of inhibitor and that their magnitude will be dependent on the fixed concentration of B. The tight-binding inhibition of the reaction (Scheme 1) with respect to A is

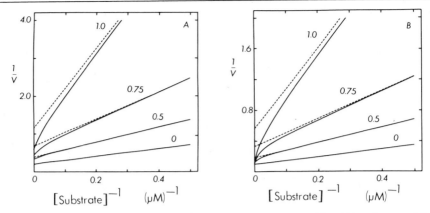

FIG. 3. Theoretical double-reciprocal plots for the noncompetitive (A) and competitive (B) inhibition of a reaction by a tight-binding reversible inhibitor when steady-state conditions apply. The enzyme concentration was 1 nM, K_i was 0.01 nM, and K_I was either 0.2 nM or infinity according to whether the inhibition was noncompetitive or competitive. Velocities were calculated using either Eq. (16) or Eq. (17). The figures adjacent to each curve denote the total inhibitor concentration (nM), and the dashed lines represent calculated asymptotes to the curves.

noncompetitive, as it would be with a classical inhibitor reacting with free enzyme and the EA complex. Similarly, when B is the variable substrate the tight-binding inhibition is competitive. This may be shown by setting z equal to zero in Eq. (16), whereupon Eq. (19) is obtained. This equation shows that the intersection point is independent of the I_t concentration.

$$v = (K_a + A)/VA \qquad (19)$$

It should be noted also that if A were present at a saturating concentration, Eq. (16) would reduce to Eq. (20), which has the same form as that for the tight-binding inhibition of a single substrate reaction. The in-

$$k^2 E_t K_I y^2 - k K_b (K_I + I_t - E_t) z y - k K_I y - K_b z - K_b^2 z^2 = 0 \qquad (20)$$

hibitor would now be able to react only with the EA complex.

While double-reciprocal plots of data conforming to a tight-binding inhibition mechanism may well be useful for determining the nature of the inhibition, the complexity of tight-binding inhibition equations mitigate against their use for determination of values for kinetic parameters.

Inhibition with Slow Establishment of Equilibrium between Enzyme and Enzyme-Inhibitor Complexes

When the equilibrium for the interaction of an inhibitor with any form of enzyme is not established rapidly, there will occur a transient or pre-steady-state phase in a plot of product formation as a function of time. The shape of the progress curve would vary according to whether or not the inhibitor was preincubated with the enzyme (cf. Fig. 4), but generally would be similar to those that have been observed with substrates in classical transient-state kinetic studies.[16] The time scale, however, would be in the minute rather than the millisecond range. A time-dependent interaction between enzyme (E) and inhibitor (I) could involve the rapid formation of an EI complex, which then undergoes a slow isomerization to an EI* complex. Under such circumstances the overall dissociation constant (K_i^*) for the EI* complex is likely to be low, as it would be a function of the K_i for the EI complex as well as of the forward and reverse isomerization constants associated with the interconversion of EI and EI*. Alterna-

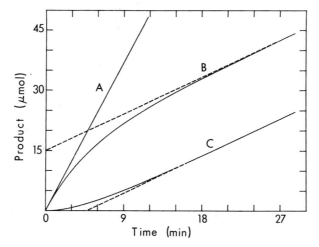

FIG. 4. Theoretical plots of progress curves for an enzyme-catalyzed reaction in the absence (curve A) and in the presence (curves B and C) of a slow-binding inhibitor. Curves B and C were generated using Eq. (24). B is the expected curve when reaction is started by the addition of enzyme; C is the expected curve when reaction is started by the addition of substrate after preincubation of the enzyme with the inhibitor. The dashed lines represent the true steady-state rate (v_s) in the presence of the inhibitor.

[16] K. J. Laidler, "The Chemical Kinetics of Enzyme Action," 2nd ed. Oxford Univ. Press (Clarendon), London and New York, 1973.

tively, the slow establishment of the equilibrium between E, I, and EI could be due simply to the low value of the pseudo-first-order rate constant with respect to enzyme for the formation of EI or of the first-order rate constant for the dissociation of EI. There is, of course, the possibility that the EI complex so formed could undergo a further slow interconversion to EI*. For reactions of this type, the dissociation constants must be low, but they need not be so low that the inhibitors would exhibit the kinetic characteristics of tight-binding inhibitors.

Inhibitors that differ kinetically from classical inhibitors in the relatively slow rate at which they form an EI complex have not been considered previously as a class. Slow-binding inhibition has been observed with studies of the inhibition of hexokinase by chromium–nucleotide complexes[17] of aldehyde dehydrogenase by cyclopropanone hydrate,[18] and of β-galactosidase by D-galactal,[19] but there are few other examples, and this is undoubtedly due to the fact that potential members of this class have not been investigated in detail. Nevertheless, it seems that inhibitors of this type could well be described as slow-binding inhibitors. While some inhibitors that bind strongly to enzymes will exhibit slow establishment of EI equilibrium, this characteristic is insufficient to characterize them as tight-binding inhibitors. The latter term applies only if the formation of an EI complex causes significant reduction in the concentration of added inhibitor.

A discussion of slow-binding inhibition does not rightly belong in a chapter on tight-binding inhibition. But since there has been only limited reference to this topic,[8,9] and as it would be advantageous to arrange experimental conditions to permit some tight-binding inhibitors to be studied as slow-binding inhibitors, brief discussion of the topic would appear to be warranted.

Kinetic Characteristics of Slow-Binding Inhibition

For the reaction

$$\text{E} + \text{I} \underset{k_2}{\overset{k_1}{\rightleftharpoons}} \text{EI} \tag{21}$$

the rates of formation and dissociation of EI would be given by the expressions $k_1[\text{I}_t][\text{E}_t]$ and $k_2[\text{EI}]$, respectively. The formation of EI by a simple bimolecular reaction would be expected to occur at almost diffusion-limited rates, so that k_1 might have a maximum value in the vicinity of $10^9 \, M^{-1} \, \text{sec}^{-1}$. However, EI might arise via the formation of an

---

[17] K. D. Danenberg and W. W. Cleland, *Biochemistry* **14**, 28 (1975).
[18] J. S. Wiseman and R. H. Abeles, *Biochemistry* **18**, 427 (1979).
[19] D. F. Wentworth and R. Wolfenden, *Biochemistry* **13**, 4715 (1974).

initial encounter complex between E and I, which then undergoes a further conformational change. In this event the value of k_1 that is observed might be as low as $10^5\ M^{-1}\ \text{sec}^{-1}$.[20] The relatively slow rate of EI formation when the dissociation constant (K_i) for the EI complex is low can be illustrated well by use of a numerical example. For kinetic studies on enzyme inhibition, an inhibitor is used at concentrations in the region of its K_i value. If $K_i (= k_2/k_1)$ were $10^{-9}\ M$, k_1 were $10^7\ M^{-1}\ \text{sec}^{-1}$, and enzyme concentration were $10^{-10}\ M$, then the pseudo-first-order rate constant with respect to enzyme for EI formation would be $k_1[I_t] = (10^7\ M^{-1}\ \text{sec}^{-1})(10^{-9}\ M) = 10^{-2}\ \text{sec}^{-1}$. Thus the half-time ($t_{0.5}$) for EI formation would be $0.693/(0.01\ \text{sec}^{-1}) = 69.3$ sec, and the slow development of maximum enzyme inhibition would be apparent. It may be calculated that the first-order rate constant for the dissociation of EI (k_2) equals $K_i \times k_1 = (10^{-9}\ M)(10^7\ M^{-1}\ \text{sec}^{-1}) = 10^{-2}\ \text{sec}^{-1}$ so that the $t_{0.5}$ value is also 69.3 sec. Thus the establishment of the equilibrium between E and EI with an inhibitor that possesses a low K_i value may be due slow formation of EI, slow dissociation of EI, or both. A 10-fold decrease in the value of K_i or k_1 would lead to a 10-fold increase in the value for $t_{0.5}$. Under these conditions it is possible that a true steady-state velocity would not be reached on mixing enzyme, inhibitor, and substrate(s) because substrate would be depleted before equilibrium between E and EI was reached. Further, enzyme and inhibitor would now be present in reaction mixtures in comparable concentrations, so that free inhibitor concentration would not be equal to its total concentration.

The conclusion about whether or not an inhibitor acts as a tight-binding inhibitor depends not only on the magnitude of the K_i value, but also on the concentration of enzyme used, this, in turn, depends on its turnover number. If enzyme were present in a reaction mixture at a concentration of 0.1 nM and inhibitor with a K_i value of 1 nM were added at a concentration of 1 nM, then in the absence of substrate the concentration of the EI complex would be about 0.05 nM. As the maximum reduction in inhibitor concentration would be 5%, it could be taken for all practical purposes that the concentrations of free and total inhibitor were identical. This being so, the kinetics of the slow-binding inhibition could be studied and values could be obtained for k_1 and k_2 as outlined below.

Kinetic Analysis of Slow-Binding Inhibition

Cha[8,9] has deomonstrated how the data from progress curves, obtained in the presence of a slow-binding inhibitor and consisting of expo-

---

[20] G. G. Hammes and P. R. Schimmel, *in* "The Enzymes" (P. Boyer, ed.), 3rd ed., Vol. 2, p. 67. Academic Press, New York, 1970.

nential and linear phases (cf Fig. 4), can be utilized to determine rate constants for the formation and dissociation of EI complexes. In deriving the non-steady-state rate equation the simplifying assumption was made that the reactions involving enzyme and substrate occurred under steady-state conditions. For the competitive inhibition by a slow-binding inhibitor of a

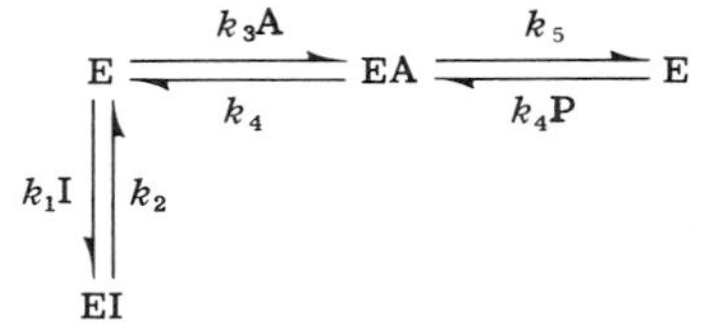

$$
\begin{array}{c}
\mathrm{E} \underset{k_4}{\overset{k_3 \mathrm{A}}{\rightleftharpoons}} \mathrm{EA} \underset{k_4 \mathrm{P}}{\overset{k_5}{\rightleftharpoons}} \mathrm{E} \\
k_1\mathrm{I} \updownarrow k_2 \\
\mathrm{EI}
\end{array}
$$

Scheme 2

single-substrate reaction (Scheme 2) the rate equation was shown to be

$$v = v_s + (v_o - v_s)e^{-kt} \tag{22}$$

where v is the observed velocity at any time, v_o is the initial velocity of the reaction, v_s is the steady-state velocity of the reaction, and $k = k_2(1 + \mathrm{I}/K_i + \mathrm{A}/K_a)/(1 + \mathrm{A}/K_a)$.[21] The equation applies whether the reaction is started by the addition of enzyme to a reaction mixture containing substrate and inhibitor or by the addition of substrate after preincubation of enzyme and inhibitor so as to establish the equilibrium between the free enzyme and the EI complex. In the former case, v_o is the steady-state velocity in the absence of inhibitor, as no EI is present at zero time, and in the latter case, the initial velocity is not a steady-state velocity. (*At this point it must be strongly emphasized that the application of steady-state kinetic theory demands that steady-state reaction velocities be determined. An initial observed velocity may or may not be a steady-state velocity. It has become the practice in this laboratory to refer to steady-state, rather than initial, velocity patterns when such patterns are constructed by using steady-state velocity data.*)

For the special condition that v_s becomes negligibly small compared with v_o, Eq. (22) reduces to

$$v = v_o e^{-kt} \tag{23}$$

so that the slope of a plot of ln (v/v_o) against t yields a value for k. The values for k_1, k_2, and K_i may be calculated as described later.

The integrated form of Eq. (22), which is given as Eq. (24)

$$\mathrm{P} = v_s t - (v_s - v_o)(1 - e^{-kt})/k \tag{24}$$

is a more useful expression, as values for v_s, v_o and k could be determined

[21] S. Cha, *Biochem. Pharmacol.* **25**, 1561 (1976).

by the fitting of data for product (P) formation as a function of time to this equation. Equation (24) predicts that the curve will be concave-up when enzyme is preincubated with inhibitor ($v_o < v_s$) and concave-down when the reaction is initiated by the addition of enzyme ($v_o > v_s$) (Fig. 4). If the reaction rates as t tends to infinity are the same irrespective of the way in which the experiment is conducted, then the rates represent steady-state velocities (v_s), which may be utilized to determine the value for K_i by the usual steady-state kinetic procedures. But additional information, relating to the magnitude of the values for the bimolecular and unimolecular rate constants associated with EI formation, can be obtained from the kinetic data for a slow-binding inhibitor. When $t \to \infty$ Eq. (25) describes the asymptote of Eq. (24).

$$P = v_s t - (v_s - v_o)/k \tag{25}$$

For curves B and C of Fig. 4, the horizontal (t_o) and vertical (P_o) intercepts are described by the relationship given in Eq. (26).

$$t_o = \frac{v_o - v_s}{k} \qquad P_o = \frac{v_s - v_o}{k v_s} \tag{26}$$

These relationships, together with values for v_o, v_s, and the intercepts, allow calculation of k, which is only an apparent value without physical significance. The values for k_1 and k_2 (Scheme 2) may be determined from plots of k as a function of [I] as illustrated in Eq. (27).

$$k = k_2 + \left(\frac{k_1}{1 + A/K_a}\right) I \tag{27}$$

The value for k_2 is given by the vertical intercept of a replot of k as a function of [I], while the values of k_1 and K_i can be calculated from the slope and horizontal intercepts, respectively, using the values for [A] and K_a.

The value of k may also be determined by rearrangement of Eq. (22) to give Eq. (28)

$$\ln\left(\frac{v - v_s}{v_o - v_s}\right) = kt \tag{28}$$

which indicates that the slope of a plot of $\ln[(v - v_s)/(v_o - v_s)]$ against t is equal to k.

Tight-Binding Inhibition with Slow Establishment of Equilibria Involving Inhibitor

The strong interaction of an inhibitor with an enzyme can lead not only to a delay in the establishment of the steady-state velocity, but also to the need to allow for depletion of inhibitor concentration. If the steady-state is reached without any significant change in substrate concentration, then the steady-state velocity data can be analyzed in the same way as when the equilibria involving the tight-binding inhibitor are rapidly established [Eq. (1)]. There is again the potential for utilizing the information contained in the pre-steady-state curve to obtain values for the uni- and bimolecular rate constants associated with the dissociation and formation of EI complexes. For a reaction conforming to Scheme 2, Cha[9] has derived a rate equation that takes into account both the slow establishment of the equilibria involving the inhibitor and the tight-binding inhibition. The equation obtained is

$$v = \frac{v_s + [v_0(1 - \gamma) - v_s]e^{-kt}}{1 - \gamma e^{-kt}} \tag{29}$$

where v, v_0, and v_s are as previously defined and γ and k, for the competitive case, are given by the expressions

$$\gamma = \frac{K_i + E_t + I_t - [(K_i + E_t + I_t)^2 - 4E_t I_t]^{1/2}}{K_i + E_t + I_t + [(K_i + E_t + I_t)^2 - 4E_t I_t]^{1/2}} \quad \text{and}$$

$$k = \frac{k_1[(K_i + E_t + I_t)^2 - 4E_t I_t]^{1/2}}{1 + A/K_a} \tag{30}$$

in which K_i is an apparent inhibition constant whose value varies with the substrate concentration and k is an apparent decay constant value, which is also dependent on substrate concentration. Equation (29) has not been used in connection with the analysis of experimental data, but it has been pointed out that, for reactions described by this equation, a semilog plot of ln v against t would yield a concave curve.[9] Again, the integrated form of Eq. (29), which is given below as Eq. (30a), is a more useful expression for the analysis of tight-binding inhibition data.[22]

$$P = v_s t + \frac{(1 - \gamma)(v_0 - v_s)}{k\gamma} \ln\left(\frac{1 - \gamma e^{-kt}}{1 - \gamma}\right) \tag{30a}$$

Very Slow Tight-Binding Inhibition

If the K_i value for a tight-binding inhibitor were very low then, for reasons similar to those discussed in relation to slow-binding inhibition, the rates of enzyme–inhibitor complex formation and dissociation must be slow compared with the rate of catalysis. This being so, there could be

[22] J. W. Williams, J. F. Morrison, and R. G. Duggleby, *Biochemistry* (in press).

significant depletion of substrate before establishment of the equilibrium of the reaction

$$E + I \underset{k_2}{\overset{k_1}{\rightleftharpoons}} EI$$

Under these circumstances, none of the aforementioned procedures could be applied. In theory, at least, it is possible to write differential equations that describe the rate of formation of EI and the change in substrate concentration as a function of time. These equations could then be used in conjunction with the basic theory for tight-binding inhibition to analyze data by means of the CRICF computer program.[23] However, to date, no investigations along these lines have been made.

An alternative approach to the determination of the K_i value for a very slow tight-binding inhibitor is to measure the rates of EI formation and dissociation. From the ratio of the rate constants, the value of K_i can be determined. As far as EI formation is concerned the inhibition would be considered as being irreversible because of the slow dissociation of EI and under these circumstances, the concentration of free enzyme at any time would be given by the expression

$$E = E_t \cdot e^{-(k_1 I_t)t} \tag{31}$$

Apparent values for the bimolecular rate constant at different fixed inhibitor concentrations would be obtained from the slopes of plots of ln (enzyme activity) against time, and the true value for the bimolecular rate constant would be determined from the slope of the plot of the apparent rate constants against total inhibitor concentration. For these experiments the inhibitor would be varied over a range of concentrations that was high relative to that of the enzyme, so the concentrations of free and total inhibitor could be taken as being equal. The unimolecular rate constant for the dissociation of EI would be determined more directly. The determination would involve complete conversion of enzyme to the EI complex by preincubation with excess inhibitor, reduction of unbound inhibitor to a negligible concentration, and then measurement of the increase in enzymic activity as a function of time. The slope of a plot of ln (enzyme activity) as a function of time would give the value of the unimolecular rate constant.

If different concentrations of a tight-binding inhibitor with the above characteristics were preincubated with an enzyme and the velocity of the reaction then determined as a function of substrate concentration, the inhibition pattern obtained by plotting v^{-1} against [substrate]$^{-1}$ would appear to be simple noncompetitive. That is, all the lines of the plot would

---

[23] J. P. Chandler, D. E. Hill, and H. O. Spivey, *Comp. Biomed. Res.* **5**, 515 (1972).

intersect at the same point on the abscissa. The variation of the vertical intercepts while the K_m value remains constant would be simply a consequence of the very tight-binding inhibitor reducing the amount of free enzyme available for reaction with the substrate. It should be noted that the inhibition pattern would differ from that obtained with a simple noncompetitive inhibitor. With a very tight-binding inhibitor the slopes and vertical intercepts of the double reciprocal plot would vary as a nonlinear function of the total inhibitor concentration. This is apparent from the form of Eq. (32), which describes the type of inhibition under discussion.

$$v = \frac{VA}{(K_a + A)\left\{\dfrac{(K_i + I_t - E_t) + [(K_i + I_t - E_t)^2 - 4I_tE_t]^{1/2}}{2K_i}\right\}} \quad (32)$$

Practical Aspects

From the aforegoing discussion it is apparent that inhibitors of enzyme-catalyzed reactions can be placed in distinct categories although the boundaries between the categories cannot be well defined. The approach to studies with a particular inhibitor will vary according to the category into which it falls. In the first instance, attention should be directed toward determining whether the inhibitor is of the classical or tight-binding type. For a reaction conforming to Michaelis–Menten kinetics, plots of steady-state velocity as a function of enzyme concentration will be linear in both the absence and the presence of a classical inhibitor (cf. this volume [15]). On the other hand, such plots will be nonlinear in the presence of a tight-binding inhibitor (cf. Fig. 1). Tight-binding inhibition should be suspected when a compound, which might be expected to act as a competitive inhibitor, yields an apparently noncompetitive inhibition pattern for which a secondary plot of slopes against inhibitor concentration is nonlinear (cf. Fig. 3B). If the concentration of inhibitor required to cause inhibition is comparable to the enzyme concentration, then tight-binding inhibition will almost certainly be occurring.

When it has been established that a compound acts as a tight-binding inhibitor, it is essential to determine that the inhibition is reversible before any tight-binding inhibition theory is applied. Reversibility may be demonstrated by showing that enzymic activity is recovered on dilution of a mixture of enzyme and inhibitor or by separation of the inhibitor from the enzyme by, for example, passing the mixture through a column of Sephadex.

When it has been shown that a compound acts as a reversible tight-binding inhibitor, it is then necessary to gain information about the rate at which equilibrium is established between an enzyme form and the enzyme–inhibitor complex. Such information is obtained by measuring product formation as a function of time and comparing the results obtained when (a) the reaction is started by the addition of substrate after preincubation of the enzyme with inhibitor and (b) the reaction is started by the addition of enzyme (cf. Fig. 4). When product formation increases as a linear function of time and both experiments yield the same rate of product formation, it may be concluded that all the equilibria involving the inhibitor (and substrate) are established rapidly. The initial rate of the reaction would represent a steady-state rate, and application of the steady-state theory for tight-binding inhibition would be appropriate.

If the equilibria are not established rapidly so that both exponential and linear phases are observed over the time course of the reaction, the results will vary according to the procedure used (cf. Fig. 4). However, if the same steady-state rate is reached with and without preincubation of the inhibitor with the enzyme, then it can be taken that there is no significant depletion of substrate before the equilibria are established. Under these circumstances, the data may be analyzed on the basis of the steady-state kinetic theory for tight-binding inhibition. In practice, it would be advisable to use enzyme at a concentration as low as possible and to initiate the reaction by adding substrate. These precautions would minimize changes in substrate concentration over the period required for attainment of the equilibria. The progress curve data could also be used for determination of the rate constants associated with the formation of the enzyme–inhibitor complex [cf. Eqs. (25)–(30)].

If there is significant depletion of substrate before equilibrium is reached for those reactions involving the tight-binding inhibitor, the steady-state kinetic theory is not applicable. To date, there has been no analysis of data of this type, but investigations are being undertaken currently in this laboratory by Dr. R. G. Duggleby on the use of the CRICF computer program for this purpose. If the time taken to establish the equilibria between inhibitor and the enzyme is very prolonged, it is likely that no reversal of the inhibition will be observed on the addition of substrate to enzyme that has been preincubated with the inhibitor. Under these circumstances, rate constants for the formation and dissociation of the enzyme-inhibitor complex would be determined, as described earlier, and used to calculate the value of the dissociation constant for the enzyme–inhibitor complex.

In the preceding discussion, it has been tacitly assumed that the enzymes subject to tight-binding inhibition have only a single substrate. If the reaction involves two substrates, then studies with a slow-binding inhibitor should involve preincubation of the enzyme with the inhibitor in the presence of each substrate separately as well as in the absence of both substrates. The results of such studies can yield information about the form(s) of enzyme with which the inhibitor reacts and any differential rates of interaction. There is no *a priori* reason why an inhibitor that reacts with free enzyme and an enzyme–substrate complex should do so at the same rates and to the same degree.

Determination of the Type of Inhibition Given by a Tight-Binding Inhibitor when Steady-State Kinetics Apply

The calculation of the dissociation constant for an enzyme–inhibitor complex from the directly determined apparent K_i value requires knowledge of the type of inhibition and the values of the kinetic parameters associated with the substrate(s) of the reaction. The latter values can be determined readily (cf. this volume [6]), but elucidation of the type of inhibition given by a tight-binding inhibitor is not so straightforward.

The data of Fig. 3 indicate that it may be difficult to distinguish between competitive and noncompetitive (or uncompetitive) inhibition by a tight-binding inhibitor. Indeed, if velocities were determined only at lower concentrations of substrate, a competitive inhibitior might well appear to be a noncompetitive inhibitor. Characterization of the inhibition requires use of relatively high concentrations of the variable substrate. For some enzyme-catalyzed reactions there may be no difficulty about using higher substrate concentrations, but with others problems could arise because of substrate inhibition, limited solubility of the substrate, or sensitivity of the enzyme to increases in ionic strength.

An alternative procedure for elucidating the type of inhibition given by a tight-binding inhibitor is (a) to determine the steady-state velocity as a function of total enzyme concerntration (E_t) at changing fixed concentrations of total inhibitor (I_t) while holding substrate at different fixed concentrations; (b) to plot the data at each fixed substrate concentration as described by Henderson[11] [Eq. (7)]; and (c) to replot the slopes of the lines as a function of substrate concentration. In theory at least, this procedure would permit determination of the type of inhibition [Eqs. (10)–(12)]. However, attention has been drawn to the problems associated with the linearized form of the general tight-binding inhibition equation [Eq. (1)], and the amount of information available is not sufficient for conclusions to be reached about the usefulness of the proposed procedure for determining the type of inhibition. It does seem that a tight-binding structural analog of the substrate would combine with the same enzyme form(s) and give rise to the same type of inhibition as would a structural analog that is not a tight-binding inhibitor. Thus, in the absence of evidence to the contrary, it would be reasonable to assume that tight-binding and non-tight-binding structural analogs of the substrate would cause the same type of inhibition with respect to each variable substrate.

Calculation of Enzyme Concentration and Apparent K_i Values

As indicated in an earlier section, the apparent K_i value for a tight-binding inhibitor, as well as the concentration of enzyme, may be determined by the fitting to Eq. (11) steady-state velocity data obtained by varying E_t and I_t at fixed substrate concentration (Fig. 1). Simulation studies show that the most precise estimates of apparent K_i are obtained when most of the data, containing 5% random error, are collected in the curve region of the plot (see the Table). The use of 40 evenly spaced data points to within 50% of the slope of the asymptote yields better values of apparent K_i than when the same number of points are used to within 1% of the asymptote. By contrast, the values for α, respresenting the degree of enzyme purity, and R, denoting the reciprocal of the steady-state rate in the absence of inhibitor, are much less sensitive to the range of E_t concentrations over which the data are collected. It has also become apparent (see the table) that the precision of parameter estimation increases as the apparent K_i value decreases. At high apparent K_i values when the inhibitor is not bound tightly to the enzyme and there is little change in free inhibitor concentration on formation of enzyme–inhibitor complexes, none of the three parameters of Eq. (8) is well determined.

Determination of True K_i Values

For the calculation of true from apparent K_i values, it is necessary to know that kinetic mechanism of the reaction in the absence of the tight-binding inhibitor and the form(s) of enzyme with which the inhibitor reacts. The calculations will vary with the kinetic mechanism and number of enzyme–inhibitor complexes that are formed, but can be illustrted by reference to the Rapid Equilibrium Random mechanism of Scheme 2. In this case

$$S = \frac{D}{\Sigma(N_i/K_i)} = \frac{K_{ia}K_b + K_aB + K_bA + AB}{(K_{ia}K_b/K_i) + (K_bA/K_I)} \qquad (33)$$

Since the relationship for S includes two inhibition constants K_i and K_I, calculation of values for the individual constants requires determination of the value of S at two (or more) nonsaturating concentrations of A. If A were used at a saturating concentration, then no free enzyme would be present for reaction with the inhibitor and the dissociation constant for

ANALYSIS OF TIGHT-BINDING INHIBITION DATA DESCRIBED BY EQ. (8)[a]

Ratio $v:v_0$ at highest E_t concentration used	K_i (nM)	Percent error in		
		R	S	α
0.99	10.0	57.7	62.1	56.0
	1.0	14.0	43.1	12.8
	0.1	4.7	33.3	3.1
	0.01	2.0	38.5	1.0
0.95	10.0	47.0	37.7	45.6
	1.0	9.0	21.5	7.8
	0.1	2.9	17.4	2.0
	0.01	1.4	23.1	0.8
0.90	10.0	49.2	31.1	49.1
	1.0	7.7	15.4	6.5
	0.1	2.6	13.9	1.7
	0.01	1.4	23.0	0.9
0.70	10.0	69.2	25.0	68.8
	1.0	7.0	11.8	6.5
	0.1	2.2	10.2	1.8
	0.01	1.4	16.7	1.1
0.50	10.0	70.7	23.3	90.5
	1.0	6.9	10.1	6.1
	0.1	2.3	9.3	1.9
	0.01	1.5	8.2	1.2

[a] Percentage error in parameter values as a function of the K_i value and the range over which total enzyme concentration is varied. Determinations were made of the total concentration of enzyme (E_t) that was necessary to give the required $v:v_0$ ratio at each of the four inhibitor concentrations given in Fig. 1 with a particular value of K_i. Ten reaction velocities were then calculated for each inhibitor concentration using Eq. (8) and E_t concentrations that started at 10% of that required to give the desired $v:v_0$ ratio and were increased by the same amount. After introduction of random scatter (5%) into the calculated velocities, the 40 data points were fitted to Eq. (8) to obtain estimates for R, S, and α, together with values for their standard errors. The latter are expressed as relative errors.

the reaction of inhibitor with the EA complex (K_I) would be determined directly.

Determination of K_i Values for Slow-Binding, Slow-Tight-Binding, and Very Slow-Tight-Binding Inhibitors

As indicated earlier, slow-binding inhibitors may be of either the classical or tight-binding type, and the distinction can be made by analysis of the steady-state velocity data. These data can be used to determine K_i values for the inhibitor by the well established procedure for classical inhibitors or by the methods outlined above for tight-binding inhibitors. The transient phase data may also be used for calculation of the rate constants associated with the formation of the enzyme–inhibitor complex. With an inhibitor that does not cause tight-binding inhibition, an overall fit to Eq. (24) of data from a single progress curve might be expected to yield a reasonable value for the apparent rate constant, k. However, the precision of the determination would undoubtedly be improved by analysis of multiple progress curves obtained at different concentrations of inhibitor. With single-substrate reactions the value of k, together with the concentration of substrate and its Michaelis constant, would be used to determine rate constants for the formation (k_1) and dissociation (k_2) of the enzyme–inhibitor complex as indicated by Eq. (27).

A similar procedure can be applied to progress curve data for tight-binding inhibition that is described by Scheme 2. Indeed, there has been developed a computer program that gives values for V, K_a, K_i, and k_1.[22] The program is based on members of the STEPT package of routines for function minimization, which was obtained from the Quantum Chemistry Program Exchange at Indiana University at QCPE 307. In connection with the above analysis, it should be noted the γ of Eqs. (29) and (30a) is not an independent parameter, as its value is determined by those for v_o, v_s, E_t, and I_t.[22]

It should be noted that Eq. (24) is based on the premise that the inhibitor undergoes a slow interaction with an enzyme form. If, however, the inhibitor were to undergo a rapid interaction with the enzyme to form an EI complex which subsequently underwent a slow isomerization to form an EI* complex, Eq. (24) would not apply. The values of v_o would not represent steady-state rates for the reaction in the absence of inhibitor, but rather, steady-state rates under conditions where the total enzyme was distributed between all enzyme forms except EI*. It has been demonstrated recently[22] that data conforming to this mechanism of inhibition can be analyzed to yield values for the dissociation constant of the EI complex as well as for the forward and reverse isomerization rate constants associated with the interconversion of EI and EI*. The procedure involved use of the CRICF program[23] to solve by numerical intergration the differential equations that describe the inhibited reaction.

In the above discussion it has been assumed that there is no significant depletion of substrate over the period required to reach steady-state equilibrium. If this assumption does not hold, then a steady-state equilibrium will not be attained and a different procedure must be used for analysis of such progress curve data. While it appears that the CRICF program could be used for this purpose, no tests have so far been made. Because the determination of K_i values for very slow tight-binding inhibitors involves establishment of the equilibrium between enzyme and inhibitor in the absence of substrate, the question of substrate depletion does not arise. But it may be necessary to add one substrate of a multisubstrate reaction in order to observe the formation of an enzyme–inhibitor complex. The true value for the rate constant would be obtained by performing measurements in the presence of a saturating concentration of substrate or by extrapolation of data obtained at different nonsaturating substrate concentrations. The same general approach would have to be used to determine the true rate constant for the dissociation of inhibitor from the enzyme–substrate–inhibitor complex. If a very slow tight-binding inhibitor reacts with the free form of enzyme that catalyzes a multisubstrate reaction, then substrate protection studies may be undertaken in the same way as described for irreversible inhibitors.

Examples of Tight-Binding Inhibition

The inhibition of enzymes by tight-binding inhibitors has received relatively little attention although it has been established that substrate analogs do act as tight-binding inhibitors of several enzymes. Thus there have been demonstrations of the tight-binding inhibition of dihydrofolate reductase by aminopterin and methotrexate,[24,25] of thymidylate synthase by 5-fluoro-UMP,[26] of adenosine deaminase by coformycin,[27,28] of pepsin by pepstatin,[29] and of ATPase by bongkrekic acid.[11] With the increasing interest in the synthesis of transition-state analogs, it is likely that the action of tight-binding inhibitors will receive increasing attention. However, for the most part, the kinetic studies undertaken with enzymes subject to tight-binding inhibition have not been designed specifically to elucidate the kinetic characteristics of the inhibitor. This is probably due to the facts that development of the kinetic theory for tight-binding inhibition has lagged behind the experimental work, that there has been no clear formulation of the approaches which might be used to study this type of inhibition and that there have not been available computer programs which would allow analysis of data obtained with the different classes of tight-

---

[24] F. M. Huennekens, in "Biological Oxidations" (T. P. Singer, ed.), p. 439. Wiley (Interscience), New York, 1968.
[25] M. N. Williams, M. Poe, N. J. Greenfield, J. M. Hirshfield, and K. Hoogsteen, *J. Biol. Chem.* **248**, 6375 (1973).
[26] P. Reyes and C. Heidelberger, *Mol. Pharmacol.* **1**, 14 (1965).
[27] S. Cha, R. P. Agarwal, and R. E. Parks, Jr., *Biochem. Pharmacol.* **24**, 2187 (1975).
[28] R. P. Agarwal, T. Spector, and R. E. Parks, Jr., *Biochem. Pharmacol.* **26**, 359 (1977).
[29] D. H. Rich and E. Sun, in "Peptides" (Proceedings Fifth American Peptide Symposum, (M. Goodman and J. Meinhofer, eds.), p. 209. Wiley, New York, 1977.

binding inhibitor. In view of this situation it is not possible to present an overview of achievements in this important area of enzymology. Instead, some selected examples will be taken to illustrate the results obtained with tight-binding inhibitors belonging to each class under discussion.

Tight-Binding Inhibition under Steady-State Conditions

Studies on creatine kinase have shown that europium–nucleotide complexes function as reversible tight-binding inhibitors of the enzyme under conditions where the reactions involving the inhibitor are fast compared with the time scale of the assays.[13] Two procedures were used to demonstrate that the usual steady-state assumptions are applicable. The first involved measurement of the steady-state rate of the partially inhibited enzyme and then the addition of either EDTA, to reverse the inhibition, or additional Eu^{3+}, to cause further inhibition. The experiments showed that there was no discernible lag in the establishment of the new reaction rate. The second procedure involved preincubation of the enzyme with buffer alone, buffer plus $MgADP^-$, or buffer plus ADP^{3-}, Eu^{3+}, and phosphocreatine. Relatively small samples from each mixture were then added to identical reaction mixtures to initiate the reaction. All reactions proceeded at the same steady-state velocity, and no lags were observed.

When the steady-state velocities for the creatine kinase reaction were determined as a function of the concentration of $MgADP^-$ at different fixed concentrations of Eu^{3+}, the resulting inhibition pattern (Fig. 5) was as expected for tight-binding inhibition that occurs under steady-state conditions. The plots in the presence of Eu^{3+} are nonlinear and consist of curved and linear regions. The downward curvature of these plots toward a common point of intersection on the vertical ordinate strongly suggests that the inhibition is competitive. Such a result is in accord with the idea that a europium–nucleotide would act as a structural analog of $MgADP^-$. The inhibition was noncompetitive with respect to phosphocreatine, and therefore the findings are similar to those observed with non-tight-binding analogs of $MgADP^-$.[30] Inhibition data obtained at two different phosphocreatine concentrations (cf. Fig. 6) were fitted to Eq. (8) to give apparent K_i values that were used to determine that the dissociation constants for the binding of europium–ADP to the free enzyme and the enzyme–phosphocreatine complex were 1.5 nM and 2.2 nM, respectively.

The inhibition of xanthine oxidase by allopurinol and alloxanthine are good examples of the slow development of inhibition.[27] When reactions

---

[30] E. James and J. F. Morrison, *J. Biol. Chem.* **241**, 4758 (1966).

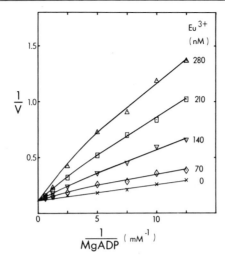

FIG. 5. Tight-binding competitive inhibition of creatine kinase by europium (Eu^{3+}) ions. The data are taken from unpublished work of K. J. Ellis and J. F. Morrison (1977).

were started by the addition of enzyme, it appeared that allopurinol behaves as a classical linear competitive inhibitor with a K_i value of 0.7 μM. However, preincubation of the inhibitor with the enzyme causes a considerable increase in the degree of inhibition and a change in the inhibition pattern. Cha et al.[27] have demonstrated the gradual development of inhibition that occurs when xanthine oxidase is not preincubated with allopurinol and the attainment of essentially the same steady-state rate with and without preincubation of enzyme and inhibitor. Detailed analysis of such data would be of considerable interest. No investigations have yet been made to determine whether allopurinol acts simply as a slow-binding inhibitor or as a slow-tight-binding inhibitor.

The inhibition of adenosine deaminase by EHNA [erythro-9(2-hydroxy-3-nonyl)adenine] also exhibits the characteristics of slow-binding inhibition.[28] Steady-state velocities are reached within a relatively short period, and the data indicate that EHNA behaves as a linear competitive inhibitor with a K_i value of 1.6 nM. This result confirms that the slow development of inhibition can take place in the absence of tight-binding inhibition and provides justification for proposing the class of slow-binding inhibitors. Examination of the ratios of EHNA concentrations to the concentration of adenosine deaminase shows that the values range from 6 to 23. Thus changes in the concentration of free inhibitor on formation of an enzyme–inhibitor complex would be sufficiently small so

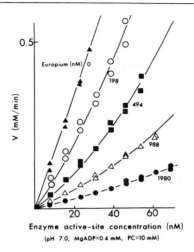

FIG. 6. Plots of steady-state velocity (v) as a function of total enzyme concentration (E_t) in the absence and in the presence of different concentrations of europium ions. The lines are theoretical curves that were drawn using the values of the kinetic parameters obtained by the nonlinear fitting of experimental data to Eq. (8). The results were taken from unpublished work of K. J. Ellis and J. F. Morrison (1977).

that the reaction conforms essentially to Michaelis–Menten kinetics. It is of interest that the slow-binding inhibition of aldehyde dehydrogenase by cyclopropanone hydrate involves dehydration of the inhibitor to cyclopropanone, which then forms a kinetically stable thiohemiketal with the active site of the enzyme.[18] On the other hand, when D-galactal causes slow-binding inhibition of the β-galactosidase reaction, it does so by interacting with the enzyme and undergoing slow conversion to deoxyglucose.[19]

In contrast to the results obtained with EHNA, the inhibition of adenosine deaminase by d-coformycin exhibits a prolonged transient phase which has prevented determination of the K_i value for the inhibitor by analysis of the progress curve data.[28] However, measurements have been made of the rates of formation and dissociation of the enzyme–coformycin complex, and the values obtained were $2.6 \times 10^6 \, M^{-1} \, \text{sec}^{-1}$ and $6.6 \times 10^{-6} \, \text{sec}^{-1}$. Thus the dissociation constant for the enzyme–inhibitor complex is $2.5 \times 10^{-12} \, M$. The magnitude of this value suggests that d-coformycin behaves as a transition-state analog.[31] The inhibitor may, therefore, form an initial enzyme–inhibitor complex that undergoes

[31] R. Wolfenden, *Acc. Chem. Res.* **5**, 10 (1972).

subsequent slow isomerization. Such a mechanism has been demonstrated for the tight-binding inhibition of dihydrofolate reductase by methotrexate.[22] Analysis of the progress curve data has yielded a K_i value of 2.3×10^{-8} M for the dissociation of methotrexate from the E–NADPH–methotrexate complex and values of 5.1 min^{-1} and 0.013 min^{-1} for the forward and reverse isomerization constants, respectively. Thus the overall inhibition constant is 5.8×10^{-11} M. It has also been concluded that the inhibition of pepsin by pepstatin proceeds via a two-step process involving a collision complex that is transformed slowly to a second enzyme–inhibitor complex.[29]

Conclusion

The limited number of detailed studies that have been made on the kinetics of tight-binding inhibition makes difficult the presentation of general conclusions about the mode of action of tight-binding inhibitors and the approaches that should be utilized to elicit this information. In particular, it is not possible to indicate if slow-tight-binding inhibition involves slow interaction of the enzyme and the inhibitor, the slow isomerization of an encounter complex formed rapidly between the enzyme and the inhibitor or a combination of both reactions. Nevertheless, it has become apparent that tight-binding inhibitors fall into three classes that can be described as tight-binding, slow-tight-binding, and very slow-tight-binding according to whether the equilibria involving the inhibitor are established rapidly, slowly, or extremely slow. By contrast, inhibitors giving rise to non-tight-binding inhibition may be of the classical or slow-binding variety, both of which produce linear steady-state inhibition patterns.

Since the experimental approach to the study of the inhibition of an enzyme by a particular inhibitor will vary with the properties of the inhibitor, it is essential at the outset to determine the nature of the inhibition. Some of the procedures that can be utilized for investigations of slow- and tight-binding inhibition have been outlined. It is to be hoped that in the forthcoming years they will be used with a wider range of both enzymes and inhibitors to yield more basic data about each type of inhibition. There is also the need for the more extensive development of computer programs for the analysis of slow- and tight-binding inhibition data. The availability of such programs together with the utilization of automatic data collection equipment to obtain accurate data should permit quantitative interpretation of experimental findings and calculation of binding constants. A knowledge of these values, as well as results from thermodynamic and X-ray crystallographic studies, should lead to a greater insight into the reasons why, for example, methotrexate is such a potent inhibitor of dihydrofolate reductase.

It can be expected that, over the next decade, there will be an increasing interest in the subject of tight-binding inhibition because of the demonstrated importance of tight-binding inhibitors as chemotherapeutic agents. Interest could also be stimulated by the current work on the development of transition-state analogs that must function as tight-binding inhibitors.

The aims of this contribution have been to give some broad outlines of the theory for and approaches to investigations of tight-binding (and slow-binding) inhibition. There can be no doubt that more information could have come from a number of earlier studies with tight-binding inhibitors if a wider range of experimental approaches had been applied. Perhaps the comments made herein will provide some stimulus for the further advancement of the theoretical and practical aspects of tight-binding inhibition.

Acknowledgment

The authors are grateful to Dr. R. G. Duggleby for helpful discussions during the preparation of this manuscript.

[18] Use of Competitive Inhibitors to Study Substrate Binding Order

By HERBERT J. FROMM

Although competitive inhibitors have been used extensively for a variety of reasons in enzyme experiments, their value as tools for making a choice of kinetic mechanism from among possible alternatives was not realized until 1962, when Fromm and Zewe[1] suggested that competitive inhibitors of substrates could be used to differentiate between random and ordered mechanisms. Furthermore, in the latter case, a determination of the substrate binding order could be made from such experiments. This protocol is quite likely the simplest approach for differentiating between Ordered and Random Bi Bi[2] possibilities. In addition, it has the advantage of permitting the kineticist to come to definitive conclusions from studies

[1] H. J. Fromm and V. Zewe, *J. Biol. Chem.* **237**, 3027 (1962).
[2] The nomenclature of Cleland will be used throughout this chapter. See W. W. Cleland, *Biochim. Biophys. Acta* **67**, 104 (1963).

of reactions in a single direction only. Its obvious limitation involves the requirement that a competitive inhibitor be available for each substrate. However, when other initial-rate data are available, a good deal of information concerning the kinetic mechanism may be provided from experiments with only one substrate analog even for Bi and Ter reactant systems.

In this discussion it will be assumed that the competitive inhibitor, which is usually a substrate analog, when bound to the enzyme, will not permit product formation to occur. These inhibitors are then dead-end inhibitors as contrasted with "partial" competitive inhibitors, which when associated with the enzyme, allow formation of product either at a reduced or accelerated rate.[3] In addition it will be assumed that enzyme–inhibitor complex formation occurs rapidly relative to other steps in the reaction pathway.

Theory

One-Substrate Systems

Before describing how competitive inhibitors are used to make a choice of mechanism in the case of multisubstrate systems, some discussion is warranted concerning reversible dead-end inhibition for one-substrate reactions. Scheme 1 illustrates a typical Uni Uni mechanism

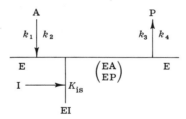

Scheme 1

and how a linear competitive inhibitor enters into the reaction mechanism.

By definition, a competitive inhibitor competes with the substrate for the same site on the enzyme. Identical kinetic results are obtained, however, if the inhibitor and substrate compete for different sites, but where binding is mutually exclusive. Although there is usually a structural similarity between the substrate and the competitive inhibitor, this is not always the case. Finally, competitive inhibition is reversed when the enzyme is saturated with substrate.

[3] H. J. Fromm, "Initial Rate Enzyme Kinetics," p. 86. Springer-Verlag, Berlin and New York, 1975.

The derivation of the rate equation for a competitive inhibitor can be done algebraically or by using any of the more sophisticated procedures described in this volume [4] and [5]. Derivation by the former method is as follows.

The velocity expression is $v = k_3 \left(\dfrac{EA}{EP}\right)$ and the conservation of enzyme equation is $E_0 = E + \left(\dfrac{EA}{EP}\right) + EI$, where E_0 is total enzyme. The dissociation constant for the enzyme–inhibitor complex is taken as $K_{is} = (E)(I)/(EI)$. Thus,

$$E_0 = E + \left(\dfrac{EA}{EP}\right) + \dfrac{(E)(I)}{K_{is}} = (E)\left(1 + \dfrac{I}{K_{is}}\right) + \left(\dfrac{EA}{EP}\right) \tag{1}$$

From the expression $E = (K_a/A)\left(\dfrac{EA}{EP}\right)$, and the equation for initial velocity,

$$v = \dfrac{V_1}{1 + (K_a/A)(1 + I/K_{is})} \tag{2}$$

where K_a is the Michaelis constant, $(k_2 + k_3)/k_1$, and V_1 is the maximal velocity, $k_3 E_0$.

It can be seen from Eq. (1) that the free enzyme component in the conservation of enzyme equation is multiplied by the factor $(1 + I/K_{is})$. A shorthand method for including the effect of a reversible dead-end inhibitor in the rate equation is simply to first identify the enzyme species in the noninhibited rate equation that reacts with the inhibitor, and then to multiply that enzyme form by the proper factor. This will perhaps be somewhat clearer when the rate equation for noncompetitive inhibition is discussed.

Figure 1 illustrates a double-reciprocal plot for an enzyme system in the presence and in the absence of a competitive inhibitor. The rationale for this particular plot comes by taking the reciprocal of both sides of Eq. (2),

$$\dfrac{1}{v} = \dfrac{1}{V_1}\left[1 + \dfrac{K_a}{A}\left(1 + \dfrac{I}{K_{is}}\right)\right] \tag{3}$$

The inhibition described in Eq. (3) is referred to as linear competitive, because a replot of slopes vs I gives a straight line. It will be seen later that competitive inhibition need not be linear.

A number of procedures are currently in vogue for linearizing initial rate plots for inhibitors. The method used most is that illustrated in Fig. 1. Another graphing method, the Dixon plot, is frequently used; however,

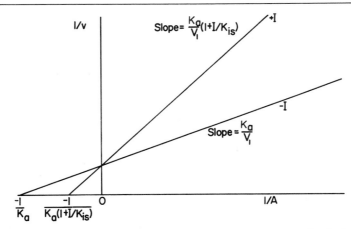

FIG. 1. Plot of 1/initial velocity (v) versus 1/substrate (A) concentration in the presence and in the absence of a linear competitive inhibitor.

it suffers from very serious inherent limitations,[4,5] and its use is not recommended.

Scheme 2 describes linear noncompetitive enzyme inhibition. In this

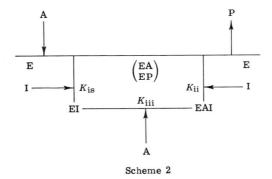

Scheme 2

model the substrate and the inhibitor are not mutually exclusive binding ligands, and complexes of enzyme, inhibitor, and substrate are thus possible.

The rate equation for linear noncompetitive inhibition, in double-reciprocal form, is

[4] D. L. Purich and H. J. Fromm, *Biochim. Biophys. Acta* **268**, 1 (1972).
[5] H. J. Fromm, "Initial Rate Enzyme Kinetics," p. 99. Springer-Verlag, Berlin and New York, 1975.

$$\frac{1}{v} = \frac{1}{V_1}\left[\left(1 + \frac{I}{K_{ii}}\right) + \frac{K_{ia}}{A}\left(1 + \frac{I}{K_{is}}\right)\right] \quad (4)$$

where K_{ia} is the dissociation constant of the EA complex. In the derivation of Eq. (4), it is assumed that the inhibitory complexes are dead-end and that all steps shown in Scheme 2 are in rapid equilibrium relative to the breakdown of the central complex to form product.

In the derivation of Eq. (4) the enzyme forms containing inhibitor must be added to the conservation of enzyme equation. Thus, $E_0 = E + EA + EI + EAI$. From the interactions shown in Scheme 2, it can be seen that EI arises from reaction of the inhibitor and E, whereas EAI *may* arise from interaction of the inhibitor and EA. In deriving Eq. (4), it is then necessary merely to multiply the factor $(1 + I/K_{ii})$ by the EA term of the uninhibited rate equation and the factor $(1 + I/K_{is})$ by the E term of the uninhibited rate expression. The very same result will be obtained if the K_{iii} step, rather than the K_{ii} pathway, is used. This is true because the four dissociation constants are not independent, but are related by the following expression: $K_{is}K_{iii} = K_{ia}K_{ii}$.

A typical plot of $1/v$ versus $1/A$ in the presence and in the absence of the noncompetitive inhibitor is shown in Fig. 2. It can be seen from the graph that the lines converge in the second quadrant. Convergence of the curves may also occur in the third quadrant or on the abscissa, depending upon the relationship between the dissociation constants for the enzyme–inhibitor complexes. The (x, y) coordinates of the intersection

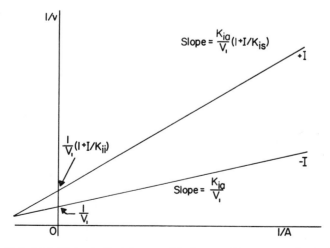

FIG. 2. Plot of 1/initial velocity (v) versus 1/substrate (A) concentration in the presence and in the absence of a linear noncompetitive inhibitor.

point for noncompetitive inhibition are $-K_{is}/K_{ia}K_{ii}$, $(1/V_1)(1 - K_{is}/K_{ii})$. If $K_{is} = K_{ii}$, i.e., the binding of the inhibitor to the enzyme is not affected by the presence of the substrate on the enzyme, the curves will intersect on the 1/A axis. When $K_{is} < K_{ii}$, intersection will be above the abscissa, whereas when $K_{is} > K_{ii}$, intersection of the curves will be in the third quadrant.

It can be seen from Eq. (4) that replots of slopes and intercepts against inhibitor concentration will give linear lines. This type of noncompetitive inhibition is more formally referred to as S-linear (slope) and I-linear (intercept) noncompetitive.

Another type of reversible dead-end enzyme inhibition that will be useful for later discussion is linear-uncompetitive inhibition. Scheme 3 de-

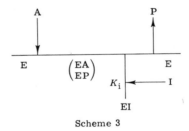

Scheme 3

picts the pathway that illustrates this phenomenon.

The rate expression for linear-uncompetitive inhibition is shown in Eq. (5), where K_i is the dissociation constant

$$\frac{1}{v} = \frac{1}{V_1}\left[\left(1 + \frac{I}{K_i}\right) + \frac{K_a}{A}\right] \tag{5}$$

for the EAI complex.

Figure 3 describes the results of a plot of $1/v$ versus $1/A$ in the presence and in the absence of an uncompetitive inhibitor. It can be seen that linear-uncompetitive inhibition gives rise to a family of parallel lines. It is often difficult to determine whether the lines in inhibition experiments are really parallel. Rather sophisticated computer programs are available that both test and fit *weighted* kinetic data[6] in an attempt to address this problem. A discussion of this point will not be presented here; however, it may be helpful to point out that uncompetitive inhibition may be graphed as a Hanes plot, i.e., A/v versus A. This plot is illustrated in Fig. 4 and has the advantage, relative to the double-reciprocal plot, that the inhibited and uninhibited lines must converge at a common point on the A/v axis.

---

[6] H. J. Fromm, "Initial Rate Enzyme Kinetics," p. 63. Springer-Verlag, Berlin and New York, 1975.

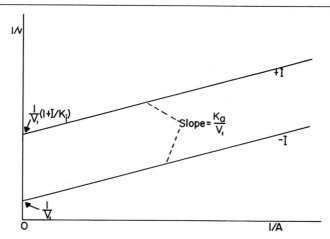

FIG. 3. Plot of 1/initial velocity (v) versus 1/substrate (A) concentration in the presence and in the absence of a linear uncompetitive inhibitor.

This approach does not eliminate the necessity of model fitting and testing, but it does permit the investigator to make a preliminary judgment more easily on the nature of the inhibition.

The last type of reversible dead-end inhibition to be considered in conjunction with unireactant systems is nonlinear inhibition. Nonlinear enzyme inhibition may be obtained from replots of primary double-reciprocal plots as a result of multiple dead-end inhibition, substrate and

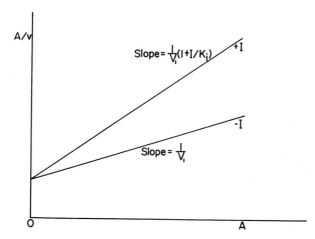

FIG. 4. Plot of substrate (A) concentration divided by initial velocity (v) versus substrate (A) concentration in the presence and in the absence of a linear uncompetitive inhibitor.

product inhibition, partial inhibition, and allostery. This discussion is limited to multiple dead-end inhibition.

When an inhibitor adds reversibly to different enzyme forms (e.g., Scheme 2), inhibition is linear; however, when there is multiple inhibitor binding to a single enzyme form, or to enzyme forms that are sequentially connected, replots of either slopes or intercepts against inhibitor may be nonlinear. If the following equilibrium is added to the simple Uni Uni mechanism of Scheme 1, $EI + I = EI_2$, K_{ii}, the rate equation obtained is

$$\frac{1}{v} = \frac{1}{V_1}\left[1 + \frac{K_{ia}}{A}\left(1 + \frac{I}{K_{is}} + \frac{I^2}{K_{is}K_{ii}}\right)\right] \quad (6)$$

Equation (6) describes parabolic-competitive inhibition. The slope of this equation is

$$\text{slope} = \frac{K_{ia}}{V_1}\left(1 + \frac{I}{K_{is}} + \frac{I^2}{K_{is}K_{ii}}\right) \quad (7)$$

and a plot of slope as a function of I will give rise to a parabola. It is also possible to obtain parabolic-uncompetitive inhibition. In this case only the intercept will be affected by inhibitor. In the event that parabolic noncompetitive inhibition is encountered, either the slope or the intercept, or both, may contain inhibitor terms of greater than first degree.

Nonlinear noncompetitive inhibition may affect slopes and intercepts; in this case it is called S-parabolic I-parabolic noncompetitive inhibition. If the replots of intercept against I are linear, whereas the slope replot is parabolic, the inhibition is called S-parabolic I-linear noncompetitive inhibition.

Parabolic inhibition will also result from interactions of the type:

$$E + I = EI, K_{is}; \quad EI + A = EIA, K_{ii}; \quad EIA + I = EI_2A, K_{iii} \quad (8)$$

Slope and intercept replots versus inhibitor may be of a more complicated nature. Cleland[2] has referred to some of these as 2/1, 3/2, etc., functions. Equation (9) illustrates an example of a S-2/1 function, in which a second-order polynomial is divided by a first-order polynomial.

$$\text{slope} = \frac{K_{ia}(1 + aI + bI^2)}{V_1(1 + cI)} \quad (9)$$

It may be difficult to distinguish a plot of slope versus inhibitor concentration for Eq. (9) from linear replots.

Two-Substrate Systems

Let us now consider the case of a random mechanism to determine how the dead-end competitive inhibitor affects the kinetics of the system.

The rapid-equilibrium random pathway of enzyme and substrate interaction (Random Bi Bi) is illustrated in this volume [3], Scheme A-6. In the case of a competitive inhibitor for substrate A, the inhibitor, I, would participate at every step in the kinetic mechanism in which the substrate A normally reacts. Thus, the following interactions of enzyme with inhibitor might be expected:

$$E + I = EI, K_i; \quad EB + I = EIB, K_{ii}; \quad EI + B = EIB, K_{iii} \quad (10)$$

When the expressions EI and EIB are added to the conservation of enzyme expression, and the rate equation derived for the effect of the competitive inhibitor of substrate A, the following relationship is obtained:

$$\frac{1}{v} = \frac{1}{V_1} + \frac{K_a}{V_1(A)}\left(1 + \frac{I}{K_{ii}}\right) + \frac{K_b}{V_1(B)} + \frac{K_{ia}K_b}{V_1(A)(B)}\left(1 + \frac{I}{K_i}\right) \quad (11)$$

where K_b is the Michaelis constant for substrate B.

When double-reciprocal plots of $1/v$ as a function of $1/A$ are made at different fixed concentrations of inhibitor, only the slope term of the rate expression is altered; i.e.,

$$\text{Slope} = \frac{K_a}{V_1}\left(1 + \frac{I}{K_{ii}}\right) + \frac{K_{ia}K_b}{V_1(B)}\left(1 + \frac{I}{K_i}\right) \quad (12)$$

On the other hand, when B is the variable substrate, double-reciprocal plots at different fixed levels of inhibitor will exhibit increases in both slopes and intercepts.

$$\text{Intercept} = \frac{1}{V_1}\left[1 + \frac{K_a}{A}\left(1 + \frac{I}{K_{ii}}\right)\right] \quad (13)$$

$$\text{Slope} = \frac{1}{V_1}\left[K_b + \frac{K_{ia}K_b}{A}\left(1 + \frac{I}{K_i}\right)\right] \quad (14)$$

Equation (11) predicts then that a dead-end competitive inhibitor for substrate A of the Random Bi Bi mechanism is a noncompetitive inhibitor of substrate B.

If now a dead-end competitive inhibitor for substrate B is used, the following interactions are to be expected:

$$E + I = EI, K_i; \quad EA + I = EAI, K_{ii}; \quad EI + A = EAI, K_{iii} \quad (15)$$

The rate equation for the effect of a dead-end competitive inhibitor of substrate B is described by Eq. (16).

$$\frac{1}{v} = \frac{1}{V_1} + \frac{K_a}{V_1(A)} + \frac{K_b}{V_1(B)}\left(1 + \frac{I}{K_{ii}}\right) + \frac{K_{ia}K_b}{V_1(A)(B)}\left(1 + \frac{I}{K_i}\right) \quad (16)$$

It can be seen from Eq. (16) that a dead-end competitive inhibitor of substrate B will show noncompetitive inhibition relative to substrate A. In summary then, for the rapid-equilibrium Random Bi Bi mechanism, a competitive inhibitor for either substrate will act as a noncompetitive inhibitor for the other substrate. These observations are consistent with the symmetry inherent in the random mechanism. Similar inhibition patterns are to be expected for the rapid-equilibrium Random Bi Uni mechanism.

Very few Random Bi Bi mechanisms are truly rapid-equilibrium random in both directions; however, this condition will be approximated in the "slow direction." When steady-state conditions prevail, i.e., when the interconversion of the ternary complexes is not slow relative to other steps of the kinetic mechanism, it may be supposed that the initial rate plots in double-reciprocal form would not be linear. This is to be expected because of the second-degree substrate terms generated under steady-state conditions; however, Schwert[7] has suggested that the deviation from linearity might be too subtle to discern. A similar point was also made by Cleland and Wratten,[8] and Rudolph and Fromm[9] concluded from computer simulations of the steady-state Random Bi Bi mechanism proposed for yeast hexokinase, that the kinetics approximate the limiting equilibrium assumption. These workers also found that the competitive inhibition patterns proposed for the rapid-equilibrium case would be indistinguishable from the situation in which steady-state conditions prevail.

In the case of the Ordered Bi Bi mechanism, competitive dead-end inhibitors of the first substrate to add to the enzyme give inhibition patterns relative to the other substrate that are distinctively different from the pattern obtained when a competitive dead-end inhibitor of the second substrate is employed. It is this very point that permits the kineticist to make a choice between Random and Ordered Bi Bi mechanisms and allows one to identify the first and second substrates to add in the ordered mechanism.

In the case of the Ordered Bi Bi mechanism, substrate A adds only to free enzyme (this volume [3], Scheme A-3a). By analogy, the competitive dead-end inhibitor should add only to this enzyme form. In addition, it is assumed that the conformation of the enzyme has been distorted enough by the inhibitor so as to preclude addition of substrate B to the enzyme-inhibitor complex.

The competitive dead-end inhibitor for substrate A may react as follows with the enzyme:

[7] G. W. Schwert, *Fed. Proc., Fed. Am. Soc. Exp. Biol.* **13**, Abstr. 971 (1954).
[8] W. W. Cleland and C. C. Wratten, "The Mechanism of Action of Dehydrogenases," p. 103. Univ. Press of Kentucky, Lexington, 1969.
[9] F. B. Rudolph and H. J. Fromm, *J. Biol. Chem.* **246**, 6611 (1971).

$$E + I = EI, \quad K_i = \frac{(E)(I)}{EI} \tag{17}$$

If the conservation of enzyme equation for the Ordered Bi Bi mechanism is modified to account for the additional complex, EI, the initial rate expression is

$$\frac{1}{v} = \frac{1}{V_1} + \frac{K_a}{V_1(A)}\left(1 + \frac{I}{K_i}\right) + \frac{K_b}{V_1(B)} + \frac{K_{ia}K_b}{V_1(A)(B)}\left(1 + \frac{I}{K_i}\right) \tag{18}$$

Equation (18) predicts that the competitive inhibitor for substrate A, the first substrate to add in the ordered mechanism, will be noncompetitive relative to substrate B. On the other hand, for this mechanism, a dead-end competitive inhibitor for substrate B would be expected not to react with free enzyme, but rather with the EA binary complex. This interaction may be described by the following relationship:

$$EA + I = EAI, \quad K_i = \frac{(EA)(I)}{EAI} \tag{19}$$

The kinetic expression obtained when this effect is included in the Ordered Bi Bi mechanism is

$$\frac{1}{v} = \frac{1}{V_1} + \frac{K_a}{V_1(A)} + \frac{K_b}{V_1(B)}\left(1 + \frac{I}{K_i}\right) + \frac{K_{ia}K_b}{V_1(A)(B)} \tag{20}$$

It is quite clear that a dead-end competitive inhibitor for the second substrate will yield uncompetitive inhibition relative to substrate A. This unique inhibition pattern allows a distinction to be made between ordered and random bireactant kinetic mechanisms, and permits determination of the substrate binding order in the former case. These points are summarized in Table I.

Three-Substrate Systems

The kinetic mechanisms for enzymes that utilize three substrates may be divided into Ping Pong and Sequential categories. It is then possible to make a choice of mechanism from among these terreactant systems using dead-end competitive inhibitors for the substrates. The kinetic mechanisms and expected inhibition patterns are illustrated in Table II.

Practical Aspects

Experimentally, it is very important that the fixed, or nonvaried, substrate be held at a subsaturating level, preferably in the region of its Mic-

TABLE I
USE OF DEAD-END COMPETITIVE INHIBITORS FOR DETERMINING
BIREACTANT KINETIC MECHANISMS

Mechanism	Competitive inhibitor for substrate	1/A plot	1/B plot
Random Bi Bi, Ordered-Subsite mechanism[e]	A	C[a]	N[b]
and Random Bi Uni	B	N	C
Ordered Bi Bi	A	C	N[c]
and Theorell–Chance	B	U[d]	C
Ping Pong Bi Bi	A	C	U
	B	U	C

[a] Refers to a double-reciprocal plot that shows competitive inhibition.
[b] Refers to a double-reciprocal plot that shows noncompetitive inhibition.
[c] In the ordered mechanism convergence may be on, above, or below the abscissa; however, the point of intersection with the inhibitor must have the same ordinate as a family of curves in which the other substrate is substituted for the inhibitor.
[d] Refers to a double-reciprocal plot that shows uncompetitive inhibition.
[e] This volume [3], Scheme A-4.

haelis constant. If, for example, when considering Eq. (11), the concentration of substrate A is held very high when B is the variable substrate, it is possible that the intercept increases to be expected in the presence of inhibitor may not be discernible, and the inhibition may appear to be competitive with respect to either substrate. It is important to note that, when replots of slopes and intercepts are made as a function of inhibitor concentration for the type of inhibition illustrated, the plots will be linear.

In studies in which dead-end competitive inhibitors are employed, it is often useful to evaluate the various inhibition constants. This can be done in a number of ways, and a few of the methods that may be used will be illustrated.

It is possible to evaluate either K_i or K_{ii} in Eqs. (11) and (16) from secondary plots of slopes and intercepts versus inhibitor concentration. It can be seen from Eq. (14) that a plot of slope versus I will give a replot in which the slope of the secondary plot is

$$\text{Slope} = \frac{K_{ia}K_b}{V_1 K_i(A)} \tag{21}$$

K_i may also be evaluated by determining the intersection point of the secondary plot on the abscissa; i.e., where slope = 0. In this case

$$I = -K_i \left(1 + \frac{A}{K_{ia}}\right) \tag{22}$$

TABLE II
Competitive Inhibition Patterns for Various Three-Substrate Mechanisms[a]

Mechanism[b]	Competitive inhibitor for substrate	1/A plot	1/B plot	1/C plot
B 2a and 2b:	A	C[c]	N[d,e]	N[g]
Ordered Ter	B	U[f]	C	N[h]
Ter and Ter Bi	C	U	U	C
B 3: Random Ter	A	C	N	N
Ter and Ter Bi*	B	N	C	N
	C	N	N	C
B 4: Random AB*	A	C	N	C[i]
	B	N	C	C[j]
	C	U	U	C
B 5: Random BC*	A	C	N	N
	B	U	C	N
	C	U	N	C
B 6: Random AC*	A	C	N	N
	B	N	C	N
	C	N	N	C
D 1: Hexa Uni Ping Pong	A	C	U	U
	B	U	C	U
	C	U	U	C
D 2: Ordered Bi Uni	A	C	N[k]	U
Uni Bi Ping Pong	B	U	C	U
	C	U	U	C
D 3: Ordered Uni	A	C	U	U
Uni Bi Bi Ping Pong	B	U	C	N[l]
	C	U	U	C
D 4: Random Bi Uni Uni	A	C	N	U
Bi Ping Pong +	B	N	C	U
	C	U	U	C
D 5 and D 6: Random	A	C	U	U
Uni Uni Bi Bi	B	U	C	N
Ping Pong +	C	U	N	C

[a] The various interactions of the competitive inhibitors are presented elsewhere along with the inhibited rate equations (H. J. Fromm, "Initial Rate Enzyme Kinetics," p. 102. Springer-Verlag, Berlin and New York, 1975).
[b] The numbers refer to the mechanisms listed in this volume [3]. *It is assumed that all steps equilibrate rapidly relative to the breakdown of the productive quarternary complex to form products. + It is assumed that all steps equilibrate rapidly relative to the breakdown of the productive ternary complex to form products.
[c] C refers to a double-reciprocal plot that shows competitive inhibition.
[d] N refers to a double-reciprocal plot that shows noncompetitive inhibition.
[e] If EI reacts with B to form EIB, the plots would be nonlinear.
[f] U refers to a double-reciprocal plot that shows uncompetitive inhibition.
[g] If EIB reacts with C to form EIBC, the plots would be nonlinear.
[h] If EAI reacts with C to form EAIC, the plot would be nonlinear.
[i] If EIB reacts with C to form EIBC, the plot would be noncompetitive.
[j] If EIA reacts with C to form EIAC, the plot would be noncompetitive.
[k] If EI reacts with B to form EIB, the plot would be nonlinear.
[l] If FI reacts with C to form FIC, the plot would be nonlinear.

The advantage of using Eq. (22) rather than Eq. (21) is that it is not necessary to evaluate V_1. Presumably, data for A and K_{ia} will be in hand.

The value for K_{ii} can be determined with a knowledge of A and K_a by evaluating the point of intersection on the abscissa of data from Eq. (13). In this case, where the intercept = 0, in the replot

$$I = -K_{ii}\left(1 + \frac{A}{K_a}\right) \tag{23}$$

Methods similar to those described for the Random Bi Bi mechanism can be used to determine the dissociation constants in the case of the Ordered Bi Bi mechanism. For example, K_i can be evaluated from either a slope or intercept versus inhibitor replot using Eq. (18), in which B is the variable substrate. It is of interest to note that the inhibition constant must be the same for this mechanism regardless of whether the determination is made from the slope or intercept. This may or may not be true for the Random pathway, depending upon whether $K_i = K_{ii}$. It will be possible to determine K_i in Eq. (20) readily, using the methods already described.

Examples

The use of dead-end competitive inhibitors for choosing between the Random and Ordered Bi Bi mechanisms has been employed with many enzyme systems. The basic protocol involves segregation of mechanisms into either the Ping Pong or Sequential class from initial rate experiments. After the Sequential nature of the system has been established, the dead-end competitive inhibitors may be used to establish whether the kinetic mechanism is Random or Ordered.

Two examples can be used to illustrate this point. Fromm and Zewe,[1] reported that yeast hexokinase is Sequential when they demonstrated that double-reciprocal plots of $1/v$ versus $1/\text{MgATP}^{2-}$ at different fixed concentrations of glucose converged to the left of the axis of ordinates. From the same data, they observed that, when $1/v$ was plotted as a function of $1/\text{glucose}$ at different fixed concentrations of MgATP^{2-}, the resulting family of curves also converged to the left of the $1/v$ axis. In addition, both sets of primary plots intersected on the abscissa. These investigators also demonstrated that AMP, a competitive dead-end inhibitor for MgATP^{2-}, was a noncompetitive inhibitor with respect to glucose. From these experiments, it was concluded that the kinetic mechanism for yeast hexokinase was either Random Bi Bi or Ordered Bi Bi with MgATP^{2-} as the initial substrate to add to the enzyme. If glucose were to add to hexokinase before MgATP^{2-} in an Ordered Bi Bi mechanism, AMP inhibition would have been uncompetitive with respect to glucose.

The same investigators employed oxalate, a dead-end competitive inhibitor of L-lactate, to help establish the kinetic mechanism of the muscle lactate dehydrogenase reaction.[10] They observed that oxalate was uncompetitive with respect to NAD^+ and concluded from these findings, and other studies, that the kinetic mechanism was Iso-Ordered Bi Bi with the nucleotide substrates adding to the enzyme first.

Dead-end competitive inhibitors have been used recently to study the kinetic mechanism of bireactant systems in which the inhibitor exhibits multiple binding to the enzyme.[11] Tabatabai and Graves[11] established that phosophorylase kinase has a sequential mechanism. They observed that adenyl-5'-yl(β,γ-methylene)diphosphate was a linear competitive inhibitor for ATP and a linear noncompetitive inhibitor relative to a tetradecapeptide, the other substrate. On the other hand, a heptapeptide was found to be nonlinear competitive (S-parabolic) with respect to the tetradecapeptide and nonlinear noncompetitive (I-parabolic, S-parabolic) relative to ATP. From these findings, it was concluded that the kinetic mechanism for phosphorylase kinase was Random Bi Bi. Their rationalization of the nonlinear inhibition data was as follows: From the linear inhibition data provided by the ATP analog, the mechanism was either Random Bi Bi or Ordered Bi Bi with ATP as the first substrate to add to the enzyme. In the case of the ordered mechanism with multiple binding of the heptapeptide, if A and B are ATP and tetradecapeptide, respectively, then

$$EA + I = EAI, K_i \quad \text{and} \quad EAI + I = EAI_2, K_{ii} \qquad (24)$$

and thus,

$$\frac{1}{v} = \frac{1}{V_1}\left[1 + \frac{K_a}{A} + \frac{K_b}{B}\left(1 + \frac{I}{K_i} + \frac{I^2}{K_i K_{ii}}\right) + \frac{K_{ia}K_b}{(A)(B)}\right] \qquad (25)$$

It was possible to exclude the ordered mechanism for phosphorylase kinase from consideration, because the inhibition pattern relative to ATP was not nonlinear uncompetitive (I-parabolic).

In the case of the Random Bi Bi mechanism, multiple binding by the inhibitor would be expected to occur as follows:

$$E + I = EI, K_i; \quad EI + I = EI_2, K_{ii};$$
$$EA + I = EAI, K_{iii}; \quad EAI + I = EAI_2, K_{iv};$$
$$EI + A = EIA, K_v; \quad EI_2 + A = EI_2A, K_{vi} \qquad (26)$$

and thus,

---

[10] V. Zewe and H. J. Fromm, *Biochemistry* **4**, 782 (1965).
[11] L. Tabatabai and D. J. Graves, *J. Biol. Chem.* **253**, 2196 (1978).

$$\frac{1}{v} = \frac{1}{V_1}\left[1 + \frac{K_a}{A} + \frac{K_b}{B}\left(1 + \frac{I}{K_{iii}} + \frac{I^2}{K_{iii}K_{iv}}\right)\right.$$
$$\left. + \frac{K_{ia}K_b}{(A)(B)}\left(1 + \frac{I}{K_i} + \frac{I^2}{K_{iii}K_{iv}}\right)\right] \quad (27)$$

The initial rate results were consistent with Eq. (27), and Tabatabai and Graves[11] concluded that the kinetic mechanism for phosphorylase kinase is rapid-equilibrium Random Bi Bi.

Wolfenden[12] and Lienhard[13] have recently outlined how transition state or multisubstrate (sometimes referred to as geometric) analogs may be used to provide information on the chemical events that occur during enzymic catalysis. If it were possible to design an inactive compound that resembles the transition state, this analog would be expected to bind very tightly to the enzyme. In theory, a good deal of binding energy when enzyme and substrate interact is utilized to alter the conformation of both the enzyme and substrate so that proper geometric orientation for catalysis is provided between enzyme and substrate. Therefore, some of this binding energy is conserved because the multisubstrate analog more closely resembles the transition state than does the substrate. It certainly does not follow that, if the inhibition constant is lower than the dissociation constant for enzyme and substrate, the inhibitor is a transition-state analog. There are many examples in the literature where competitive inhibitors bind more strongly to enzymes than do substrates and yet are clearly not transition-state analogs.

These suggestions may be formalized by considering the following two reactions

$$E + A = EA', \quad K_1 = 10^{-7} M$$
$$EA' = EA, \quad K_2 = 10^4$$

for the overall reaction

$$E + A = EA, \quad K_{ia} = 10^{-3} M$$

The first reaction represents the thermodynamically favorable process of enzyme-substrate binding. The second reaction may be taken to be the enzyme-induced distortion of both the substrate and enzyme leading to the transition state.

From the perspective of kinetics, the use of geometric analogs may not be clear-cut in the case of multisubstrate systems. The analog should bind free enzyme, and, in theory, for a two-substrate system, the analog and

---

[12] R. Wolfenden, *Acc. Chem. Res.* **5**, 10 (1972).
[13] G. E. Lienhard, *Science* **180**, 149 (1973).

substrate should not be able to bind to the enzyme simultaneously. This situation is difficult to check experimentally because it is not easy to determine whether, for example, 50% of the enzyme has substrate bound and the other 50% of the enzyme is associated with both substrate and analog and analog alone. For example, the kinetic pattern for a random mechanism where the analog can bind free enzyme and the EB complex will be the same as that for an ordered mechanism in which the analog binds exclusively to free enzyme.

It becomes fairly clear when considering the effect of multisubstrate or geometric analogs on the kinetics of bireactant enzyme systems that only in the case of the rapid-equilibrium Random Bi Bi mechanism may one obtain unequivocal results, and then only under certain circumstances. Consider, for example, the interaction of the analog and enzyme in an Ordered Bi Bi mechanism. The inhibitor will bind enzyme and will not permit the addition of the second substrate. Thus, the analog will act like any other competitive inhibitor of substrate A for this mechanism (see Table I); i.e., it will be a noncompetitive inhibitor of substrate B. Therefore, the inhibition patterns provided by geometric analogs are identical to those to be expected for dead-end competitive inhibitors of the first substrate of the Ordered Bi Bi mechanism.

In the case of the Random Bi Bi mechanism, multisubstrate analogs may indeed give unique inhibition patterns, and this observation has been used to provide support for the Random Bi Bi mechanism for muscle adenylate kinase.[14,15] The multisubstrate analog used to test this theory with adenylate kinase was p^1,p^4-di(adenosine-5') tetraphosphate (AP_4A).[14] Recently, it was shown that AP_5A binds even more strongly to the enzyme than does AP_4A.[15]

When considering the Random Bi Bi mechanism, the geometric analog should bind exclusively to free enzyme. This binding should effectively preclude binding of substrates A and B, and thus only the E term of the rate equation will be affected by the analog I. The rate expression is therefore,

$$\frac{1}{v} = \frac{1}{V_1} + \frac{K_a}{V_1(A)} + \frac{K_b}{V_1(B)} + \frac{K_{ia}K_b}{V_1(A)(B)}\left(1 + \frac{I}{K_i}\right) \qquad (28)$$

where K_i is the dissociation constant of the enzyme–multisubstrate inhibitor complex.

Equation (28) indicates that the multisubstrate analog will function as a competitive inhibitor for both substrates. This effect is unique to the

---

[14] D. L. Purich and H. J. Fromm, *Biochim. Biophys. Acta* **276**, 563 (1972).
[15] G. E. Lienhard and I. I. Secemski, *J. Biol. Chem.* **248**, 1121 (1973).

Random mechanism and suggests that the inhibitor bridges both substrate binding pockets.

If the inhibitor binds only at one substrate site in either the Random or the Ordered Bi Bi cases, or if, for the latter mechanism, the compound does resemble the transition state and substrate B does not add, inhibition patterns will be competitive and noncompetitive relative to the two substrates. Thus, it will not be possible to differentiate between these two mechanisms based upon these inhibition patterns, nor will it be possible to determine whether the inhibitor is really a transition-state analog in the Ordered mechanism. In the case of the Random pathway, the enzyme–inhibitor complex will permit binding of one substrate and the enzyme–substrate complex will allow analog to bind. In summary then, only the unique inhibition pattern illustrated by Eq. (28) allows one to use multisubstrate analogs to unambiguously differentiate between kinetic mechanisms.

Frieden[16] has presented a rapid-equilibrium Ordered Bi Bi-subsite mechanism (this volume [3], Scheme A-4) that cannot be differentiated from the rapid-equilibrium Random Bi Bi mechanism from studies of product inhibition, substrate analog inhibition, isotope exchange at equilibrium, and the Haldane relationship.[16,17] Multisubstrate analogs can be used to make this differentiation, along with other kinetic procedures.[17]

In the case of Frieden's mechanism,[16] the transition-state analog should bind at both the active site and the subsite for B, i.e., the EB site. This will result in modification of a number of terms in the rate equation. The dead-end complexes to be expected are EI_1, EI_2, EI_1I_2, and EAI_2, where I_1 and I_2 represent binding of the transition-state analog at the active and subsites, respectively. The resulting rate expression is described by Eq. (29), in which K_i, K_{ii}, K_{iii}, and K_{iv} represent dissociation constants for EI_1, EI_2, EI_1I_2, and EAI_2 complexes, respectively.

$$\frac{1}{v} = \frac{1}{V_1}\left[1 + \frac{K_{ia}K_b}{K_{ib}(A)} + \frac{K_b}{B}\left(1 + \frac{I}{K_{iv}}\right) \right.$$
$$\left. + \frac{K_{ia}K_b}{(A)(B)}\left(1 + \frac{I}{K_i} + \frac{I}{K_{ii}} + \frac{I^2}{K_iK_{iii}}\right)\right] \quad (29)$$

Equation (29) indicates that the multisubstrate analog acts like a competitive inhibitor relative to substrate B and as a noncompetitive inhibitor with respect to substrate A. In addition, replots of the slopes from these primary plots versus inhibitor concentration will yield a concave-up

---

[16] C. Frieden, *Biochem. Biophys. Res. Commun.* **68**, 914 (1976).
[17] H. J. Fromm, *Biochem. Biophys. Res. Commun.* **72**, 55 (1976).

parabola. This analysis indicates that a differentiation can be made between the Ordered Bi Bi-subsite and the Random Bi Bi mechanisms.

Limitations

When considering competitive substrate inhibitors, the possibility is automatically excluded that the inhibitor may bind to an enzyme–product complex. In the case of the rapid-equilibrium Random Bi Bi mechanism, a competitive inhibitor for substrate B could in theory bind the EQ complex; however, this complex occurs after the rate-limiting step and is not part of the kinetic equation. Similarly, although this binary complex is kinetically important in the Ordered Bi Bi case, if an EQI complex did form, inhibition would be noncompetitive rather than competitive relative to substrate B. Under these conditions, the approach would not be a viable technique, and another inhibitor should be sought.

Huang[18] has suggested that it is not possible to differentiate between Ordered and Random Bi Bi mechanisms if in the Ordered case an inhibitor for B binds free enzyme and the EA complex to form complexes EI and EAI. This will be true only when all steps in the kinetic mechanism equilibrate rapidly relative to the breakdown of the productive ternary complex to form products. This mechanism is easily separated from those under discussion here by its unique double-reciprocal plot patterns in the absence of inhibitors.[19]

It should be pointed out that the dead-end competitive inhibitors cannot be used to differentiate between normal and Iso mechanisms. Nor can they be used to make a choice as to whether ternary complexes are kinetically important in Ordered mechanisms.

The competitive substrate inhibitors cited above have been referred to as "dead-end" inhibitors.[2] The question arises as to what happens if the inhibitors are not of the dead-end type, i.e., if the enzyme-inhibitor complexes of the ordered mechanism act in a manner similar to those analogous complexes in the random mechanism. This possibility was considered by Hanson and Fromm.[20] If in the Ordered mechanism, the EI complex permits substrate B to add, the additional reaction would be

$$EI + B = EIB, \quad K_{ii} \qquad (30)$$

and Eq. (18) would be modified as shown in Eq. (31)

---

[18] C. Y. Huang, *Arch. Biochem. Biophys.* **184**, 488 (1977).
[19] H. J. Fromm, "Initial Rate Enzyme Kinetics," p. 38. Springer-Verlag, Berlin and New York, 1975.
[20] T. L. Hanson and H. J. Fromm, *J. Biol. Chem.* **240**, 4133 (1965).

$$\frac{1}{v} = \frac{1}{V_1} + \frac{K_a}{V_1(A)}\left[1 + \frac{I}{K_i} + \frac{(I)(B)}{K_iK_{ii}} + \frac{K_{ia}K_b(I)}{K_aK_iK_{ii}}\right] + \frac{K_b}{V_1(B)}$$
$$+ \frac{K_{ia}K_b}{V_1(A)(B)}\left(1 + \frac{I}{K_i}\right) \quad (31)$$

Inhibition relative to substrate A would of course be competitive; however, a $1/v$ versus $1/B$ plot would show concave-up hyperbolic inhibition. This effect is obviously readily distinguishable from the case in which a dead-end binary complex is formed.

Concluding Remarks

The use of competitive substrate inhibitors for studying kinetic mechanisms has received wide attention in recent years. If one is interested in determining the kinetic mechanism for an enzyme, this protocol should be used immediately after a determination is made as to whether the initial rate kinetics are Sequential or Ping Pong. Theoretically, using competitive substrate inhibitors to study the sequence of enzyme and substrate binding is far less ambiguous than either substrate or product inhibition or isotope exchange experiments. The protocol is no more complicated than initial velocity experiments in which inhibitors are not used.

The procedure does suffer from certain limitations, as do most kinetic methods. For example, it will not permit one to differentiate between Iso and conventional mechanisms, nor can it be used to choose between Theorell–Chance mechanisms and analogous mechanisms involving long-lived central complexes. However, on balance, using dead-end competitive inhibitors is a powerful tool for studying enzyme kinetics.

Acknowledgments

This work was supported by research grants from the National Institutes of Health (NS 10546) and the National Science Foundation (PCM 77-09018).

[19] Use of Alternative Substrates to Probe Multisubstrate Enzyme Mechanisms

By CHARLES Y. HUANG

The alternative substrate is a useful tool for differentiating kinetic models, whose potential as a mechanistic probe has not been adequately exploited. It is ideal for studying enzyme mechanisms since it presumably undergoes the same catalytic events as the substrate, yet its presence gives rise to alternative reaction pathways. Wong and Hanes[1] first pointed out that the effect of introducing alternative substrates into an enzyme

[1] J. T. Wong and C. S. Hanes, *Can. J. Biochem. Physiol.* **40**, 763 (1962).

system could be utilized to deduce the order of substrate addition. In this chapter, several current methods of using the alternative substrate will be described. Generally speaking, they can be classified into two types. One approach is to determine the kinetic parameters for the substrate and the alternative substrate separately and compare the numerical values; the other is to evaluate the graphical patterns obtained in the presence of both substrate analogs. The latter approach can be carried out by measuring the rate of formation of either one single product or the common product (or the summation of two correspondent products). In addition, the alternative substrate can be held at a fixed level or maintained at a constant ratio to the substrate.

The Numerical Method

Let us use the four usual two-substrate mechanisms as examples: the Theorell–Chance, the Ordered, the Rapid Equilibrium Random, and the Ping Pong mechanisms. The substrates will be denoted by A and B, and the products by P and Q. The alternative substrate and products will be represented by A', B' and P', Q', respectively. Consider the Theorell–Chance mechanism shown in Scheme 1a:

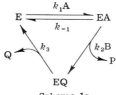

Scheme 1a

The reciprocal initial velocity equation is

$$\frac{E_o}{v} = \frac{1}{k_3} + \frac{1}{k_1 A} + \frac{1}{k_2 B} + \frac{k_{-1}}{k_1 k_2 AB} \tag{1}$$

or, using the ϕ expressions of Dalziel, it has the general form of

$$\frac{E_o}{v} = \phi_o + \frac{\phi_A}{A} + \frac{\phi_B}{B} + \frac{\phi_{AB}}{AB} \tag{2}$$

When the alternative substrate B' is used in place of B (Scheme 1b),

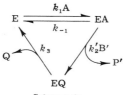

Scheme 1b

we have

$$\frac{E_o}{v} = \frac{1}{k_3} + \frac{1}{k_1 A} + \frac{1}{k_2' B'} + \frac{k_{-1}}{k_1 k_2' AB'} \qquad (3)$$

Comparison of Eqs. (1) and (3) reveals that ϕ_o, ϕ_A, and ϕ_{AB}/ϕ_B remain unchanged.

If A' is used instead of A (Scheme 1c),

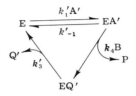

Scheme 1c

all the coefficients are different:

$$\frac{E_o}{v} = \frac{1}{k_3'} + \frac{1}{k_1' A'} + \frac{1}{k_4 B} + \frac{k_{-1}'}{k_1' k_4 A' B} \qquad (4)$$

For the ordered mechanism (Scheme 2),

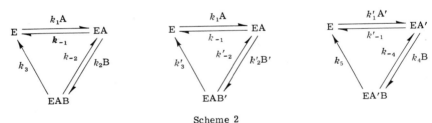

Scheme 2

the reciprocal rate equations are

$$\frac{E_o}{v} = \frac{1}{k_3} + \frac{1}{k_1 A} + \frac{k_{-2} + k_3}{k_2 k_3 B} + \frac{k_{-1}(k_{-2} + k_3)}{k_1 k_2 k_3 AB} \qquad (5)$$

$$\frac{E_o}{v} + \frac{1}{k_3'} + \frac{1}{k_1 A} + \frac{k_{-2}' + k_3'}{k_2' k_3' B'} + \frac{k_{-1}(k_{-2}' + k_3')}{k_1 k_2' k_3' AB'} \qquad (6)$$

and

$$\frac{E_o}{v} = \frac{1}{k_5} + \frac{1}{k_1' A'} + \frac{k_{-4} + k_5}{K_4 k_5 B} + \frac{k_{-1}'(k_{-4} + k_5)}{k_1' k_4 k_5 A' B} \qquad (7)$$

It can be seen that ϕ_A and ϕ_{AB}/ϕ_B remain constant when B' is the alternative substrate.

For the Rapid Equilibrium Random mechanism (Scheme 3),

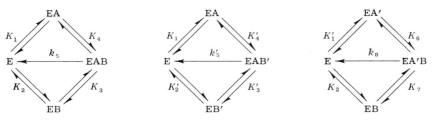

Scheme 3

the rate equations are

$$\frac{E_o}{v} = \frac{1}{k_5}\left(1 + \frac{K_1 K_4}{K_2 A} + \frac{K_4}{B} + \frac{K_1 K_4}{AB}\right) \quad (8)$$

$$\frac{E_o}{v} = \frac{1}{k_5'}\left(1 + \frac{K_1 K_4'}{K_2' A} + \frac{K_4'}{B'} + \frac{K_1 K_4'}{AB'}\right) \quad (9)$$

and

$$\frac{E_o}{v} = \frac{1}{k_8}\left(1 + \frac{K_1' K_6}{K_2 A'} + \frac{K_6}{B} + \frac{K_1' K_6}{A'B}\right) \quad (10)$$

In this case, when B′ is used, ϕ_{AB}/ϕ_B is unaltered; when A′ is used, ϕ_{AB}/ϕ_A is unaltered.

For the Ping Pong mechanism shown in Scheme 4

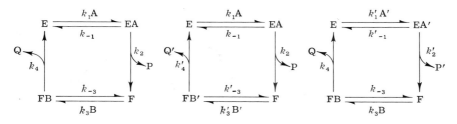

Scheme 4

the rate equations are given by

$$\frac{E_o}{v} = \frac{k_2 + k_4}{k_2 k_4} + \frac{k_{-1} + k_2}{k_1 k_2 A} + \frac{k_{-3} + k_4}{k_3 k_4 B} \quad (11)$$

$$\frac{E_o}{v} = \frac{k_2 + k_4'}{k_2 k_4'} + \frac{k_{-1} + k_2}{k_1 k_2 A} + \frac{k_{-3}' + k_4'}{k_3' k_4' B'} \quad (12)$$

and

$$\frac{E_o}{v} = \frac{k_2' + k_4}{k_2' k_4} + \frac{k_{-1}' + k_2'}{k_1' k_2' A'} + \frac{k_{-3} + k_4}{k_3 k_4 B} \quad (13)$$

TABLE I
INVARIANT ϕ COEFFICIENTS PREDICTED FOR A' AND B'

Mechanism	Alternative substrate	
	A'	B'
Theorell–Chance	None	ϕ_0, ϕ_A, ϕ_{AB}/ϕ_B
Ordered	None	ϕ_A, ϕ_{AB}/ϕ_B
Rapid Equilibrium Random	ϕ_{AB}/ϕ_A	ϕ_{AB}/ϕ_B
Ping Pong	ϕ_B	ϕ_A

ϕ_A is unchanged when B' is used, and ϕ_B is invariant when A' is used. Although the Ping Pong mechanism usually can be easily identified by the characteristic parallel double-reciprocal plots, it is necessary to make certain that the parallel patterns are not due to certain Rapid Equilibrium Random mechanisms in which the two substrates are highly antagonistic or synergistic such that the 1/AB term becomes unimportant. Alternative substrates are most useful in differentiating these two mechanisms.

The predicted invariant ϕ coefficients for the four models presented above are summarized in Table I. These predicted results have been shown by Wong,[2] who also considered variations of these four mechanisms. In theory, calculation of ϕ_0, ϕ_A, ϕ_B, ϕ_{AB}/ϕ_A, and ϕ_{AB}/ϕ_B allows one to identify the mechanism and substrates A and B (in Theorell–Chance and ordered mechanisms) in most instances. Using these parameters, one can also generate theoretical curves to verify the experimental results obtained in the presence of both A and A' or B and B'. In practice, this approach requires rather precise data. Furthermore, an invariant ϕ_0 may not indicate a Theorell–Chance mechanism if $k_3 = k_3'$ (see Scheme 2) in the ordered mechanism. Like most kinetic tools, it should be used in conjunction with other techniques.

The Graphical Methods

When an alternative substrate is introduced into a two-substrate enzyme system, in general a common product and two alternative products are formed. For instance,

$$A + B \text{ and } B' \longrightarrow P \text{ and } P' + Q \qquad (14)$$

$$A + B \text{ and } B' \longrightarrow P + Q \text{ and } Q' \qquad (15)$$

If the formation of one of the alternative products in mesured, say, dP/dt

---

[2] J. T.-F. Wong, "Kinetics of Enzyme Mechanisms," p. 100. Academic Press, New York, 1975.

in Eq. (14), the effect of the alternative substrate, B', is inhibition because of the competitive nature of B and B'. If the formation of the common product, dQ/dt (or $dP/dt + dP'/dt$) in Eq. (14), is followed, the result may be inhibition or activation or both, depending on the relative K_m and specific V of the two cosubstrates. We shall refer to these two methods as the one-product approach and the common-product approach.

The One-Product Approach

With this method, the alternative substrate, A' or B', is held at several fixed levels while the concentration of A or B is varied. The proper levels of the alternative substrate can be determined by its V/K_m ratio compared with the V/K_m ratio for the substrate. The four common two-substrate mechanisms, again, will serve as our examples. For the Theorell–Chance mechanism, when the B-B' pair is present, the reaction sequence is given in Scheme 5.

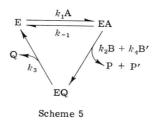

Scheme 5

For $v = dP/dt$, the rate equation is

$$\frac{E_0}{v} = \frac{1}{k_3} + \frac{1}{k_1 A} + \frac{(k_3 + k_4 B')}{k_2 k_3 B} + \frac{(k_{-1} + k_4 B')}{k_1 k_2 AB} \tag{16}$$

Thus the inhibition patterns are: B' competitive with B, noncompetitive with A.

When A' is the added alternative substrate (Scheme 6),

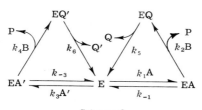

Scheme 6

for $v = dQ/dt$

$$\frac{E_o}{v} = \frac{1}{k_5} + \frac{1}{k_1 A} + \frac{1}{k_2 B}$$
$$+ \frac{1}{k_1 k_2 AB}\left[k_{-1} + \frac{k_3(k_6 + k_4 B)(k_{-1}k_4 + k_2 k_5 B)A'}{k_5 k_6(k_{-3} + k_4 B)}\right] \quad (17)$$

In this case, A' is competitive with A, buit *nonlinear* noncompetitive with B. A number of conditions, however, may result in negligible curvilinearity, the simplest one being $k_{-3} \simeq k_6$ or $k_4 B \gg k_{-3}, k_6$.

The alternative substrate inhibition patterns in the ordered mechanism are similar to those of the Theorell–Chance. From Scheme 7

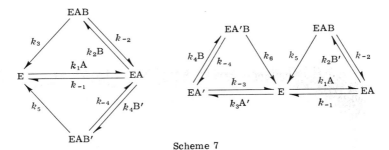

Scheme 7

for the B-B' pair, we have

$$\frac{E_o}{v} = \frac{1}{k_3} + \frac{1}{k_1 A} + \frac{(k_{-2} + k_3)}{k_2 k_3 B}\left(1 + \frac{k_4 B'}{(k_{-4} + k_5)}\right)$$
$$+ \frac{(k_{-2} + k_3)}{k_1 k_2 k_3 AB}\left(k_{-1} + \frac{k_4 k_5 B'}{(k_{-4} + k_5)}\right) \quad (18)$$

and for the A-A' pair,

$$\frac{E_o}{v} = \frac{1}{k_5} + \frac{1}{k_1 A} + \frac{k_{-2} + k_5}{k_2 k_5 B} + \frac{1}{k_1 k_2 k_5 AB}\left[k_{-1}(k_{-2} + k_5)\right.$$
$$\left. + \frac{k_3(k_{-1}k_{-2} + k_{-1}k_5 + k_2 k_5 B)(k_{-4} + k_6 + k_4 B)A'}{k_{-3}(k_{-4} + k_6) + k_4 k_6 B}\right] \quad (19)$$

These equations predict that B' is competitive with B, noncompetitive with A; A' is competitive with A, nonlinear noncompetitive with B. The fact that the simultaneous presence of A and A' in ordered mechanisms gives rise to nonlinear or higher-degree profile for B has been shown by Wong and Hanes,[1] Rudolph and Fromm,[3] and Ricard et al.[4] Ricard et al.

[3] F. B. Rudolph and H. J. Fromm, *Biochemistry* **9**, 4660 (1970).
[4] J. Ricard, G. Noat, C. Got, and M. Borel, *Eur. J. Biochem.* **31**, 14 (1972).

also pointed out situations where the nonlinearity became undetectable. Under those conditions, a choice of mechanism is possible provided the Dalziel coefficients fit certain diagnostic rules.[4] These rules are similar to those presented in Table I.

For the Rapid Equilibrium Random case, when both B and B' are present (Scheme 8), we have

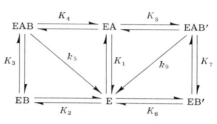

Scheme 8

$$\frac{E_o}{v} = \frac{1}{k_5}\left[1 + \frac{K_1 K_4}{K_2 A} + \frac{K_4}{B}\left(1 + \frac{B'}{K_8}\right) + \frac{K_4}{AB}\left(K_1 + \frac{B'}{K_6}\right)\right] \quad (20)$$

The predicted inhibition patterns are: B' competitive with B, noncompetitive with A. Similar patterns are obtained with the A-A' pair: A' is competitive with A, noncompetitive with B.

The alternative substrate inhibition patterns in the Ping Pong mechanism are identical to those of the Rapid Equilibrium Random case. Scheme 9 yields a rate equation that is similar in form to Eq. (20).[3]

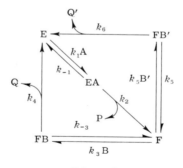

Scheme 9

$$\frac{E_o}{v} = \frac{1}{k_2} + \frac{1}{k_4} + \frac{(k_{-1} + k_2)}{k_1 k_2 A} + \frac{(k_{-3} + k_4)}{k_3 k_4 B}\left[1 + \frac{k_5(k_2 + k_6)B'}{k_2}\right]$$
$$+ \frac{k_5 k_6(k_{-1} + k_2)(k_{-3} + k_4)B'}{k_1 k_2 k_3 k_4(k_{-5} + k_6)AB} \quad (21)$$

It is clear from the equations presented above that the predicted alter-

TABLE II
Alternative Substrate Inhibition Patterns in the One-Product Approach

Mechanism	Alternative substrate pair	Varied substrate[a]	
		A	B
Theorell–Chance	AA'	C	Nonlinear (N)
	BB'	N	C
Ordered	AA'	C	Nonlinear (N)
	BB'	N	C
Rapid Equilibrium Random	AA'	C	N
	BB'	N	C
Ping Pong	AA'	C	N
	BB'	N	C

[a] Symbols used: C, competitive; N, noncompetitive.

native substrate inhibition patterns in the one-product approach are invariably competitive with the correspondent substrate and noncompetitive with the other substrate (Table II). This approach is useful only in identifying ordered substrate addition when nonhyperbolic profiles are seen. Linear double-reciprocal plots, however, cannot be used to exclude ordered mechanisms. Olive et al.[5] and Rudolph and Fromm[3] have applied this method to the study of glucose-6-phosphate dehydrogenese from *Leuconostoc mesenteroides* and equine liver alcohol dehydrogenase. Rudolph and Fromm[6] also extended the theory to three-substrate enzyme systems.

The Common-Product Approach

Zewe et al.[7] first studied the mechanism of yeast hexokinase by measuring the formation of both glucose 6-phosphate and fructose 6-phosphate using glucose and fructose as alternative substrates (the sum of glucose 6-phosphate + fructose 6-phosphate should equal the amount of ADP formed). Theoretical treatment of this technique has been provided by Fromm[8] and by Darvey.[9] A major problem with this approach, in which the alternative substrate is held at fixed levels in the usual manner, is that most of the predicted double-reciprocal plots are nonlinear (Table III). The nonlinear patterns make the data treatment difficult and may not provide useful diagnostic clues.

[5] C. Olive, M. E. Geroch, and R. Levy, *J. Biol. Chem.* **246**, 2047 (1971).
[6] F. B. Rudolph and H. J. Fromm, *Arch. Biochem. Biophys.* **147**, 515 (1971).
[7] V. Zewe, H. J. Fromm, and R. Fabino, *J. Biol. Chem.* **239**, 1625 (1964).
[8] H. J. Fromm, *Biochim. Biophys. Acta* **81**, 413 (1964).
[9] I. G. Darvey, *Mol. Cell. Biochem.* **11**, 3 (1976).

TABLE III
Graphical Patterns in the Common-Product Approach with the Alternative Substrate Present at Fixed Levels

Mechanism	Alternative substrate pair	Varied substrate[a,b]	
		A	B
Theorell–Chance	AA'	Nonlinear (C)	Nonlinear (N)
	BB'	N	Nonlinear (C)
Ordered	AA'	Nonlinear (C)	Nonlinear (N)
	BB'	N	Nonlinear (C)
Rapid Equilibrium Random	AA'	Nonlinear (C)	N
	BB'	N	Nonlinear (C)
Ping Pong	AA'	Nonlinear (C)	U
	BB'	U	Nonlinear (C)

[a] Symbols used: C, competitive; N, noncompetitive; U, uncompetitive.
[b] The predicted inhibition patterns for nonlinear plots are given in parentheses. However, nonlinear plots may not allow a distinction between competitive and noncompetitive inhibition to be made graphically.

A different approach has been developed by Huang,[10] which leads to linearization of most of the double-reciprocal plots, yet retains the nonhyperbolic features of the $1/v$ vs $1/B$ plot in ordered substrate addition when the A-A' pair is present. In this procedure, the substrate:alternative substrate ratio is kept constant such that when the substrate is varied the alternative substrate is also varied but maintains a constant ratio with substrate. The graphical patterns obtained at several different fixed alternative substrate:substrate ratios form the basis for mechanism differentiation. Secondary intercept plots are also utilized as a diagnostic aid.

As an example, consider the Theorell–Chance mechanism for the BB' pair shown in Scheme 5. When $v = dQ/dT$ (or $dP/dt + dP'/dt$)

$$\frac{E_o}{v} = \frac{1}{k_3} + \frac{1}{k_1 A} + \frac{1}{k_2 B + k_4 B'} + \frac{k_{-1}}{k_1(k_2 B + k_4 B')A} \quad (22)$$

If B' is held constant, the E_o/v vs $1/B$ plot will be nonlinear. By maintaining B and B' at a constant ratio, say $B' = \beta B$, Eq. (22) becomes

$$\frac{E_o}{v} = \frac{1}{k_3} + \frac{1}{k_1 A} + \frac{1}{(k_2 + k\beta)B} + \frac{k_{-1}}{k_1(k_2 + k_4\beta)AB} \quad (23)$$

Double-reciprocal plots with respect to $1/B$ or $1/A$ now become linear. The plots obtained at various ratios (β's) will generate a family of lines intersecting to the left of the E_o/v axis in the $1/A$ plot and on the E_o/v axis in the $1/B$ plot. We shall call these patterns noncompetitive and competi-

[10] C. Y. Huang, *Arch. Biochem. Biophys.* **184,** 488 (1977).

tive, respectively, for the sake of convenience. It should be noted, however, that an increase in the ratio may result in either activation or inhibition. The ratio, β, replaces the usual alternative substrate concentration as the parameter to be varied in this approach.

With the AA' pair, the rate equation for the Theorell–Chance mechanism (of Scheme 6) has the form

$$\frac{E_o}{v} = \frac{k_1k_6(k_5 + k_2B)(k_{-3} + k_4B) + k_3k_5(k_{-1} + k_2B)(k_6 + k_4B)\alpha}{k_5k_6[k_1k_2(k_{-3} + k_4B) + k_3k_4(k_{-1} + k_2B)\alpha]B}$$
$$+ \frac{(k_{-1} + k_2B)(k_{-3} + k_4B)}{[k_1k_2(k_{-3} + k_4B) + k_3k_4(k_{-1} + k_2B)\alpha]AB} \quad (24)$$

in which $\alpha = A'/A$. The predicted patterns are: α noncompetitive with A and nonlinear noncompetitive with B. The noncompetitive plot may intersect to the left or to the right of the ordinate, and the nonlinear plot may concave upward or downward or appear linear.[10]

For the ordered mechanism (Scheme 7), with B and B' present, we have

$$\frac{E_o}{v} = \frac{k_2(k_{-4} + k_5) + k_4(k_{-2} + k_3)\beta}{k_2k_3(k_{-4} + k_5) + k_4k_5(k_{-2} + k_3)\beta} + \frac{1}{k_1A}$$
$$+ \frac{(k_{-2} + k_3)(k_{-4} + k_5)}{[k_2k_3(k_{-4} + k_5) + k_4k_5(k_{-2} + k_3)\beta]B}$$
$$+ \frac{k_4(k_{-2} + k_3)(k_{-4} + k_5)}{k_1[k_2k_3(k_{-4} + k_5) + k_4k_5(k_{-2} + k_3)\beta]AB} \quad (25)$$

Equation (25) predicts that β is noncompetitive with both A and B. One salient feature of this case is the expression for the intercepts on the ordinate of an E_o/v vs $1/B$ plot

$$I_{1/B} = \frac{k_2(k_{-4} + k_5) + k_4(k_{-2} + k_3)\beta}{k_2k_3(k_{-4} + k_5) + k_4k_5(k_{-2} + k_3)\beta} + \frac{1}{k_1A} \quad (26)$$

Consequently, if experiments at each β are done at several levels of A, a secondary $I_{1/B}$ vs $1/A$ plot should yield parallel lines ("uncompetitive"). This secondary plot makes a distinction between ordered and random mechanism possible when nonlinear E_o/v vs $1/B$ plots with the AA' substrate pair are not observed.

Rate equations for the other situations are given below: AA' pair, ordered mechanism:

$$\frac{E_o}{v} = \frac{\phi_1\phi_2 + \phi_3\phi_4\alpha}{(\phi_5 + \phi_6\alpha)B} + \frac{\phi_1\phi_2}{(\phi_5 + \phi_6\alpha)AB} \quad (27)$$

where

$$\phi_1 = k_{-3}(k_{-4} + k_6) + k_4 k_6 B$$
$$\phi_2 = k_1[(k_{-2} + k_5) + k_2 B]$$
$$\phi_3 = k_{-1}(k_{-2} + k_3) + k_2 k_5 B$$
$$\phi_4 = k_3[(k_{-4} + k_6) + k_4 B]$$
$$\phi_5 = k_1 k_2 k_5 \phi_1$$
$$\phi_6 = k_3 k_4 k_6 \phi_3$$

BB' pair, rapid equilibrium random (cf. Scheme 8; rate equation for the AA' pair is identical in form):

$$\frac{E_o}{v} = \frac{K_1 K_4 K_8}{k_5 K_8 + k_9 K_4 \beta}\left[\frac{1}{K_1}\left(\frac{1}{K_4} + \frac{\beta}{K_8}\right) + \left(\frac{1}{K_2} + \frac{\beta}{K_6}\right)\frac{1}{A} + \frac{1}{K_1 B} + \frac{1}{AB}\right] \quad (28)$$

BB' pair, Ping Pong mechanism (cf. Scheme 9; rate equation for the AA' pair has the same form):

$$\frac{E_o}{v} = \frac{k_3(k_2 + k_4)(k_{-5} + k_6) + k_5(k_2 + k_6)(k_{-3} + k_4)\beta}{k_2 k_3 k_4(k_{-5} + k_6) + k_2 k_5 k_6(k_{-3} + k_4)\beta} + \frac{k_{-1} + k_2}{k_1 k_2 A}$$
$$+ \frac{(k_{-3} + k_4)(k_{-5} + k_6)}{[k_3 k_4(k_{-5} + k_6) + k_5 k_6(k_{-3} + k_4)\beta]B} \quad (29)$$

All the primary and secondary graphical patterns are summarized in Table IV.

TABLE IV
GRAPHICAL PATTERNS PREDICTED BY THE COMMON-PRODUCT,
CONSTANT-RATIO ALTERNATIVE SUBSTRATE METHOD

		Varied substrate[a]			
		A		B	
Mechanism	Alternative substrate pair	1/A plot	Intercept plot[b]	1/B plot	Intercept plot[b]
Theorell–Chance	AA'	N	Nonlinear (N)	Nonlinear (N)	(N)[c]
	BB'	N	C	C	—
Ordered	AA'	N	Nonlinear (N)	Nonlinear (N)	(N)[c]
	BB'	N	N	N	U
Rapid Equilibrium Random	AA'	N	N	N	N
	AA'	N	N	N	N
Ping Pong	AA'	N	U	U	N
	BB'	U	N	N	U

[a] Symbols used: C, competitive; N, noncompetitive; U, uncompetitive.
[b] An intercept plot refers to a plot of intercepts on the ordinate (obtained at different constant levels of the nonvaried substrate) vs the reciprocal of its nonvaried substrate.
[c] In theory, the intercept plots are linear, but such plots would be impracticable if the primary plots were nonlinear.

If alternative substrates for both A and B are available, the mechanisms listed in Table IV can all be differentiated since the four sets of graphical patterns are different. When all the double-reciprocal plots appear to be linear noncompetitive, the secondary intercept plots can be utilized to make a choice between the ordered and random mechanisms, since the BB' pair in the ordered case should produce an uncompetitive intercept plot. It should be noted, however, that, although a noncompetitive intercept plot is indicative of a random mechanism, an uncompetitive secondary plot may not be used to rule it out. It can be shown that if $k_5 K_2 K_8$ happens to equal $k_9 K_4 K_6$ [see Eq. (28)], the intercept plot will yield parallel lines because the coeefficient for the $1/A$ term becomes $K_1 K_8 / k_9 K_6$ and is independent of the $B'/B (=\beta)$ ratio. In this case, one should examine the ϕ coefficients according to Table I to determine the correct mechanism.

The constant-ratio method can be extended to three-substrate enzyme systems.[10] For example, for the three-substrate Ordered and Rapid Equilibrium Random mechanisms, when one of the substrates is held constant, the predicted graphical patterns for the two remaining substrates in the presence of their alternative substrates are identical with those of the two-substrate cases presented in Table IV. Also, the method is not limited to studying the order of addition of substrates. Huang and Kaufman[11] have applied the constant-ratio alternative substrate technique to differentiate models proposed for the enzyme concentration-dependent activity of phenylalanine hydroxylase.

Experimental Procedures for the Constant-Ratio Alternative-Substrate Approach

Suppose we want to study the effect of B' on the graphical patterns of A and B (arbitrarily designated). The first step is to determine the K_m and V for B' so that the appropriate range of B' and B ratios can be estimated. For example, if the K_m's for B' and B are 0.5 mM and 0.1 mM, respectively, then a $B'/B(=\beta)$ of 5 should make B' roughly equal in strength relative to B. However, if B' has a V twice as high as the V for B, a β of 2.5 should make B' about as potent as B. The proper ratios can be calculated, therefore, according to the relative V/K_m values of the alternative substrates. The steps described below can then be followed:

1. Prepare four stock solutions, all of which have the same concentration of B, but each has a different concentration of B' such that the ratio $B'/B = \beta_1, \beta_2, \ldots \beta_4$ is varied, e.g., $\beta_1 = 0$, $\beta_2 = 1$, $\beta_3 = 2.5$, $\beta_4 = 4$.

[11] C. Y. Huang and S. Kaufman, *J. Biol. Chem.* **248**, 4242 (1973).

2. To obtain data for the E_o/v vs $1/B$ plot, carry out separate experiments in which each stock solution is varied (by addition of different amounts to a set total volume) in the presence of a fixed, nonsaturating concentration of A. Since a common product is to be measured in all experiments, the amounts of product formed may increase when β is increased. Thus appropriate adjustments must be made in fixed-time assays. In making double-reciprocal plots, one should consider only the concentration of B (not B + B'!) in the calculation of $1/B$. The data will yield four lines corresponding to β_1, β_2, β_3, and β_4.

3. To obtain data for E_o/v vs $1/A$, carry out separate experiments in which the concentration of A is varied with each of the β_1, β_2, β_3, and β_4 stock solutions held at a nonsaturating level (at the same B concentration). The E_o/v vs $1/A$ plot will generate four lines corresponding to the four β values.

4. If intercept plots are needed to differentiate a mechanism, e.g., when A is the varied substrate, repeat the experiments as in step 3 at several levels of B, using the same β_1, β_2, β_3, and β_4 stock solutions. The intercepts obtained from the E_o/v vs $1/A$ plots are replotted against $1/B$, which should yield four lines, each corresponding to one of the four β values.

Concluding Remarks

The alternative substrate method has two limiting factors. One of the factors is its availability, which is shared by other tools utilizing structural analogs, such as the competitive inhibitor, the transition-state analog, and, to some extent, the product. The other limitation is a suitable assay in the measurement of the common product. This limitation also exists in product inhibition studies if product formation, rather than substrate disappearance, is measured. Furthermore, the alternative substrate approach only provides information regarding the substrate addition in the direction of the reaction. However, although product inhibition patterns allow one to detect enzyme isomerization and abortive complex formation, a choice of mechanism is not always possible because of these complications. Consequently, various approaches often need to be employed simultaneously in order to deduce the correct kinetic model. In this regard, the alternative-substrate approach is as important as other more frequently used techniques.

Concerning the several procedures available in the alternative-substrate method, the numerical approach, because of the demand on its precision, is probably more useful as a support for the graphical approach. The one-product approach is useful only when nonlinear patterns indicative of the Theorell–Chance or ordered mechanisms are seen. The common-product, constant-ratio alternative-substrate approach has wider application because of the distinctive patterns generated by different mechanisms. Even when only one alternative substrate is available, several mechanisms still can be readily differentiated. As can be seen from Table IV, an N,C or an N,U pattern would indicate Theorell–Chance or Ping Pong, respectively. An N,N pattern with an uncompetitive intercept plot would indicate an ordered mechanism. A nonlinear plot would mean that the substrate analogs are the AA' pair in one of the ordered cases. Knowledge of substrate A makes it possible to differentiate among the Theorell–Chance, Iso-Theorell-Chance, Ordered, and Iso-Ordered models by product inhibition patterns. In the case of one alternative substrate yielding linear noncompetitive primary and secondary plots, with additional clues from product inhibition patterns and/or ϕ relationships, all the usual two-substrate mechanisms can be distinguished. It is clear, therefore, that the alternative substrate method is most useful when used in combination with other techniques.

[20] Substrate Inhibition

By W. Wallace Cleland

When a substrate concentration is varied at fixed levels of other reactants, one expects the kinetics to follow the usual equation:

$$v = VA/(K + A) \tag{1}$$

where A is substrate concentration, and K the Michaelis constant. As A increases, v should increase asymptotically to the value V. In some cases, however, v increases only to a certain point, and then decreases at higher levels of A; this is called substrate inhibition. Such inhibition can arise for several reasons. First, if the substrate combines as a dead-end inhibitor with an enzyme form with which it is not supposed to react, it will cause linear (or total) substrate inhibition in which the rate goes to zero at infinite substrate concentration. Second, if high levels of substrate cause an altered order of addition of reactants, or in any other way generate an altered reaction pathway, one observes a partial (or hyperbolic) substrate inhibition, in which infinite substrate gives a finite, but reduced rate. Third, the substrate may combine at an allosteric site and cause either total or partial substrate inhibition. Fourth, levels of substrate above the middle millimolar range (above 10 mM, say) may cause nonspecific inhibition as a result of increased ionic strength, or a higher level of toxic counterions.

Substrate inhibition is in most cases a nonphysiological phenomenon (an exception, the allosteric substrate inhibition by MgATP of phosphofructokinase, which is relieved by AMP or MgADP as a balancing allosteric activator), and is thus seen in the normal physiological direction of a reaction only at levels of substrate considerably above those found in the cell. In the nonphysiological direction, however, substrate inhibition is much more common. In this direction we are using what are normally the products of the physiological reaction as the substrates, and such products are almost always present in the cell at levels low enough not to cause product inhibition, and thus well below their Michaelis constants. As a result there has been no evolutionary pressure to prevent combinations in dead-end fashion with dissociation constants equal to or higher than the Michaelis constants, and such combinations, which lead to prominent substrate inhibitions, are common.

While substrate inhibition can be a nuisance to a kineticist trying to determine kinetic patterns, it also is an excellent tool for deducing mechanism, at least where the inhibition constants are low enough for the inhibitory effects to be specific ones, and in this chapter we will present first the theory, including some examples of how deductions concerning mechanism can be made, and then discuss methods for analysis of the results.

Theory

Total Substrate Inhibition

Let us consider first the result of the combination of a substrate as a dead-end inhibitor. This normally produces linear substrate inhibition (that is, $1/v$ will become linear with A at high enough A levels, and infinite A produces zero rate), unless two molecules of substrate combine with the same enzyme form in dead-end fashion (in which case one gets parabolic inhibition), or there are cooperative effects between subunits (in which case $1/v$ vs A may be a more complex function with a linear asymptote). To analyze the substrate inhibition one varies one of the other substrates (see below for the case where there is only one substrate), and notes whether the inhibition shows up on slopes, intercepts, or both of the standard reciprocal plots ($1/v$ vs $1/A$). When the effect is on slopes only, it is called competitive; on intercepts only, uncompetitive; and on both slopes and intercepts, noncompetitive. These patterns allow deductions to be made concerning mechanism.

Substrate Inhibition in Ping Pong Mechanisms

In Ping Pong mechanisms of the type

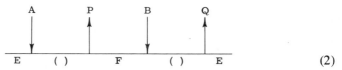
(2)

the enzyme oscillates between two stable enzyme forms, E and F. For a transaminase, these are pyridoxal-enzyme and pyridoxamine-enzyme, and for nucleoside diphosphate kinase, they are free enzyme and phosphoenzyme. In the diagram above, A is shown reacting only with E, and B with F. Since E and F are so similar, however, it is reasonable to expect B to have some affinity for E as well as F, and if the active site in F is not too full for adsorption of A, for A to have some affinity for F. Such combinations will probably not be very prominent in the physiological direction of the reaction, but may be very much so in the reverse direction. Combination of B with E multiplies the K_aB term in the denominator of the rate equation by $(1 + B/K_{Ib})$, where K_{Ib} is the dissociation constant of B from EB, and causes competitive substrate inhibition when A is varied:

$$v = \frac{VAB}{K_aB(1 + B/K_{Ib}) + K_bA + AB} \quad (3)$$

As can be seen from Fig. 1, the intercepts drop with increasing B and reach saturation in normal manner, but the lines, instead of being parallel as expected for the Ping Pong mechanism, show slopes increasing linearly with B. If A combines with F as well, the K_bA term is multiplied by $(1 + A/K_{Ia})$, where K_{Ia} is the dissociation constant of A from FA. If both

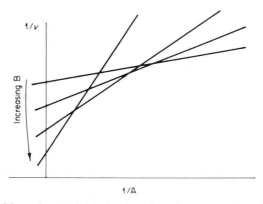

FIG. 1. Competitive substrate inhibition in a Ping Pong mechanism. A replot of slopes will be linear with B. Data points not shown for clarity.

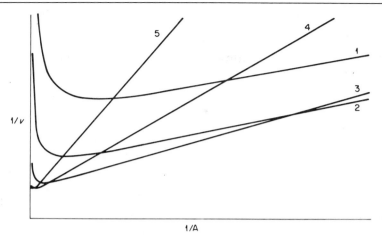

FIG. 2. Double competitive substrate inhibition in a Ping Pong mechanism. The concentration of B increases from 1 to 5. Data points not shown for clarity.

combinations produce strong substrate inhibition, one gets double competitive substrate inhibition, which is plotted in Fig. 2. The pattern looks messy, but can be easily fitted to give the kinetic constants (see this volume [6]).

Competitive substrate inhibition does not show up as nonlinear reciprocal plots if both substrates are varied together in constant ratio, and one must be careful when performing such experiments to determine maximum velocities to correct for any substrate inhibition known to occur. This can be seen from Eq. (3) (assuming the $K_b A$ term to be multiplied by $(1 + A/K_{Ia})$) by letting $B = xA$, where x is a constant. We now get:

$$v = \frac{VA/y}{(K_a + K_b/x)/y + A} \qquad (4)$$

where

$$y = 1 + \frac{K_a x}{K_{Ib}} + \frac{K_b}{x K_{Ia}} \qquad (5)$$

It is clear that the apparent maximum velocity for this plot, with A and B varied together, must be multiplied by y to get the true V. The correction factor can minimized by picking

$$x = (K_b K_{Ib}/K_a K_{Ia})^{1/2} \qquad (6)$$

but unless $K_{Ia} \gg K_a$, and $K_{Ib} \gg K_b$, the correction cannot be ignored. For the application of this theory, see Garces and Cleland.[1]

[1] E. Garces and W. W. Cleland, *Biochemistry* **8**, 633 (1969).

Substrate inhibitions other than competitive will be rare in Ping Pong mechanisms. Even in multisite Ping Pong Mechanisms, one expects (and sees[2,3]) competitive substrate inhibitions.

Substrate Inhibition in Sequential Mechanisms

Substrate inhibition patterns for sequential mechanisms can be competitive, uncompetitive, or even noncompetitive, depending on the mechanism. In a random mechanism, one does not expect substrate inhibition except where one substrate shows some affinity for the site of another, and this should result in competitive substrate inhibition. An example is myokinase, where ADP will combine at the MgADP site, but not vice versa.[4] Another example is competitive substrate inhibition (possibly parabolic) of CO_2 (or more likely, the accompanying bicarbonate) vs TPNH in TPN isocitrate dehydrogenase. This inhibition is noncompetitive vs ketoglutarate, showing that the CO_2 (or bicarbonate) combines as a dead-end inhibitor at the nucleotide site.[5]

In ordered mechanisms such as

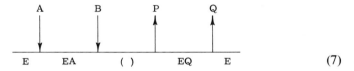 (7)

EA and EQ are often very similar (E–DPN and E–DPNH, for instance), and it is not unreasonable for B to have some affinity for EQ as well as for EA. If the combination of B with EQ prevents release of Q, the result is linear uncompetitive substrate inhibition by B when A is varied. In this pattern the slopes decrease with increasing B as expected, but the intercepts decrease at first, and then increase again at higher levels of B (Fig. 3). The intercept replot for such a pattern is shown in Fig. 4. This type of substrate inhibition is not as common as one might expect, largely because combination of B with EQ may not prevent release of Q (see e.g., Schimerlik and Cleland[6]). If release of Q is only slowed down by B, one sees partial substrate inhibition (see below).

One must be cautious about ascribing uncompetitive substrate inhibi-

---

[2] D. B. Northrop, *J. Biol. Chem.* **244,** 5808 (1969).
[3] S. S. Katiyar, W. W. Cleland, and J. W. Porter, *J. Biol. Chem.* **250,** 2709 (1975).
[4] J. C. Khoo and P. J. Russell, Jr., *J. Biol. Chem.* **245,** 4163 (1970).
[5] M. L. Uhr, V. W. Thompson, and W. W. Cleland, *J. Biol. Chem.* **249,** 2920 (1974); D. B. Northrop and W. W. Cleland, *J. Biol. Chem.* **249,** 2928 (1974).
[6] M. I. Schimerlik and W. W. Cleland, *Biochemistry* **16,** 565 (1977).

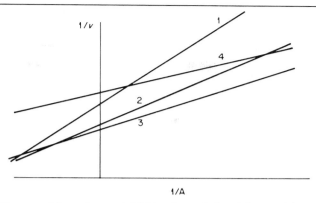

FIG. 3. Uncompetitive substrate inhibition in an Ordered Sequential mechanism. The slope replot will appear normal, but the intercept replot will be a hyperbola (see Fig. 4). Substrate concentration increases from 1 to 4. Data points not shown for clarity.

tion to combination of B with EQ, however, since combination of B with the central complexes will also produce uncompetitive substrate inhibition. While such combination seems at first glance unlikely, it seems to be the explanation for the strong uncompetitive substrate inhibition by ketoglutarate seen with TPN isocitrate dehydrogenase. This substrate inhibition, which is linear up to 5000 times the Michaelis constant for ketoglutarate, and is uncompetitive vs either CO_2 or TPNH, was shown not to result from combination with E-TPN and prevention of TPN release by a double inhibition experiment.[5] With CO_2 and TPNH held constant, $1/v$ was plotted vs ketoglutarate (using levels well up in the substrate inhibi-

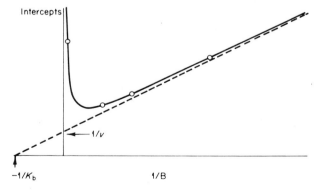

FIG. 4. Intercept replot for data such as those in Fig. 3. If the intercepts are plotted against B, instead of $1/B$, the shape is the same, but the horizontal intercept of the asymptote gives the substrate inhibition constant.

tion region so linear $1/v$ vs ketoglutarate plots would be obtained) at several fixed levels of TPN. The resulting parallel pattern showed the absence of the term in the denominator of the rate equation in (ketoglutarate)^2(TPN) which should be present if ketoglutarate combined in dead-end fashion with E–TPN. Possibly a second molecule of ketoglutarate forms Schiff's bases with lysines exposed during the conformation change that converts the initial central complex into one that is catalytically active. This type of double-inhibition experiment should always be done for uncompetitive substrate inhibitions to determine whether the combination is really with EQ, or with central complexes.

Mechanism of Substrate Inhibition When There Is Only One Substrate

In the cases described above, one varies the noninhibitory substrate to deduce the type of substrate inhibition, but if there is only one substrate this is not possible. One can envision two mechanisms for substrate inhibition when there is only one substrate. First, if product release is ordered, as is common for some hydrolytic enzymes where the reaction is really Ping Pong with water as the second substrate, the substrate could react with EQ in dead-end fashion [Eq. (8)]

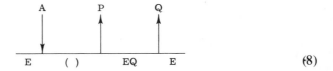
(8)

The more classical explanation is that two molecules of substrate each get part of themselves stuck in part of the active site (that is, we get a dead-end EA$_2$ complex). These mechanisms can be told apart by running a product-inhibition experiment in which A is varied over a wide concentration range (including the substrate-inhibition region) at several fixed levels of P. If the kinetics of A at each P level are fitted to the usual equation for substrate inhibition

$$v = \frac{VA}{K + A + A^2/K_i} \qquad (9)$$

both $1/V$ and K/V will show the expected linear dependence on P (that is, P is expected to be a noncompetitive product inhibitor). For the mechanism involving dead-end combination of A with EQ, $1/VK_i$ will not be a function of P, whereas for the mechanism where a dead-end EA$_2$ complex forms, $1/VK_i$ will also be a linear function of P. This technique has not yet been used, despite the large number of hydrolytic enzymes that show substrate inhibition.

Partial Substrate Inhibition

When infinite substrate concentrations still result in a finite rate, one has partial substrate inhibition, and high substrate concentrations are causing an alteration in the reaction pathway. This can result from allosteric combination of the substrate, but more commonly represents some randomness in the mechanism. For example, with glutamate dehydrogenase, ketoglutarate shows uncompetitive substrate inhibition vs ammonia that is partial and represents a 4-fold reduction in the rate of release of TPN when ketoglutarate is present.[7] With liver alcohol dehydrogenase, an even more complicated pattern is observed. Primary alcohols, or good substrates like cyclohexanol, show partial substrate inhibition vs DPN caused by combination with E–DPNH (acyclic secondary alcohols react so slowly that there is no E–DPNH present in the steady state, and thus they do not show substrate inhibition).[8] The substrate inhibition is noncompetitive, the intercept effect resulting from slower release of DPNH from E–DPNH–alcohol than from E–DPNH, while the slope effect results from a lower V/K value for combination of DPN with E–alcohol than with free enzyme. The presence of E–alcohol at very low DPN concentrations shows randomness in the mechanism, but the slower V/K value for combination of DPN with E–alcohol shows that the normal order of combination with DPN adding first is preferred.

Note the contrast to the partial uncompetitive pattern seen for ketoglutarate in the glutamate dehydrogenase case; the lack of a slope effect with glutamate dehydrogenase results from the ordered combination of TPNH, ketoglutarate, and ammonia, so that at low ammonia levels ammonia is combining with the same E–TPNH–ketoglutarate complex regardless of whether ketoglutarate is at high or low levels.

It should be noted that the same mechanisms that cause partial substrate inhibition can result in partial activation if the values of rate constants are different. For example, with liver alcohol dehydrogenase high levels of cyclohexanol show partial slope inhibition, but partial intercept activation when DPN is varied.[8] The activation of the intercepts arises because DPNH is released more rapidly from E–DPNH–cyclohexanol than from E–DPNH, while the partial inhibition of slopes again shows that DPN reacts with free enzyme faster than with E–cyclohexanol.

Induced Substrate Inhibition

This substrate inhibition is not seen normally, but is induced by the presence of a dead-end inhibitor that can combine in place of a reactant in

---

[7] J. E. Rife, Ph.D. Dissertation, Univ. of Wisconsin, Madison, 1978.
[8] K. Dalziel and F. M. Dickinson, *Biochem. J.* **100**, 34 and 491 (1966).

an ordered mechanism. It is thus highly diagnostic for ordered mechanisms. In the following example [Eq. (10)]

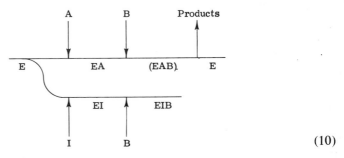

(10)

the presence of I induces substrate inhibition by B that is competitive vs A (infinite A keeps I from combining with the enzyme, and thus overcomes the effect). The effect is seen only in Steady State Ordered mechanisms and is absent in a Rapid Equilibrium mechanism, and absent (or only a partial inhibition; see below) in a Random mechanism. For the inhibition to be total, I must not be able to dissociate from EIB until B does. The reason for the substrate inhibition is that EIB is a dead-end complex, while (EAB) continuously breaks down to products. Thus at high B, all EI that forms is converted to EIB and stays there, while (EAB) regenerates E, and in turn EI. After a few reaction cycles, all of the enzyme is EIB.

The rate equation for this situation is obtained by multiplying the terms corresponding to free enzyme ($K_{ia}K_b$ and K_aB) in the denominator of the rate equation by $[1 + (I/K_i)(1 + B/K_{Ib})]$, where K_i is the dissociation constant of I from EI, and K_{Ib} is the dissociation constant of B from EIB. The resulting equation can be written in reciprocal form

$$\frac{V}{v} = 1 + \left(\frac{K_a}{A} + \frac{K_{ia}K_b}{AB}\right)\left(1 + \frac{I}{K_i}\right) + \frac{K_{ia}K_bI}{AK_{Ib}K_i} + \frac{K_aBI}{AK_{Ib}K_i} \quad (11)$$

The last term is the one causing the substrate inhibition, and it is clear that infinite A overcomes the inhibition, and a finite level of I must be present to induce it.

An interesting example of this situation is thymidylate synthetase where 5-bromo-dUMP (a competitive inhibitor of dUMP) induces substrate inhibition by methylene tetrahydrofolate that is competitive vs dUMP.[9] This pattern, plus other kinetic evidence, shows clearly that the kinetic mechanism is ordered (dUMP adds first), and methylene tetrahydrofolate keeps dUMP, Br-dUMP, or F-dUMP from dissociating (this was directly demonstrated with F-dUMP[9]). In this case it appears that

[9] P. V. Danenberg and K. D. Danenberg, *Biochemistry* **17**, 4018 (1978).

Br-dUMP and F-dUMP actually begin to act as substrates, but the reaction halts at the point where the proton replaced by Br or F would have to be removed.

Induced substrate inhibition is not always competitive. When there are three substrates, as in Eq. (12),

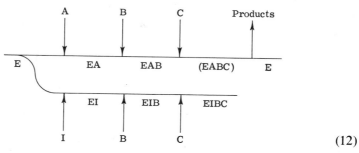

(12)

the presence of I induces substrate inhibition by B that is competitive vs A, but uncompetitive vs C. The absence of any slope effect in the latter pattern results from the fact that at low C levels the paths between E and EAB and EIB are at equilibrium, and the ratio of EIB to EAB is constant and dependent only on the ratio of I to A. If C also combines as shown, I will induce substrate inhibition by C that is competitive vs A, but uncompetitive vs B (at low B there is no EIB for C to react with).

If I is competitive with B

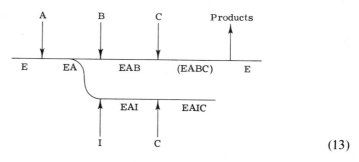

(13)

the presence of I will induce substrate inhibition by C that is competitive vs B, but uncompetitive vs A (at low A, there is no EA for I to react with).

Partial Induced Substrate Inhibition

If binding of B in Eq. (10) causes tighter binding of I, but does not prevent its release from EIB, one will observe partial induced substrate inhibition by B, which is competitive vs A. Such behavior is characteristic of random mechanisms with a preferred order of combination of reactants, but which are not rapid equilibrium ones. With yeast hexokinase, for instance, lyxose, which is competitive vs glucose, induces substrate inhibition by MgATP, which is partial and competitive vs glucose.[10] The inhibition arises from the complex E–MgATP–lyxose, in which there is very synergistic binding of nucleotide and sugar (lyxose binds weakly to free enzyme). The dissociation constant of MgATP from a binary complex with enzyme is 4–5 mM, although the Michaelis constant (which is for reaction of MgATP with E–glucose) is about 0.1 mM. Very little E–MgATP–lyxose thus forms at K_m levels of MgATP, but by the time MgATP reaches 5 mM, this complex is present in large amounts. Since lyxose can dissociate and be replaced by glucose, however, the substrate inhibition is partial.

Methods of Data Analysis

In the Theory section above we have discussed the various patterns one may observe in substrate-inhibition experiments, and now we must come to grips with how to analyze the experimental data. When one has varied a noninhibitory substrate at fixed levels of the one giving the substrate inhibition, one first makes reciprocal plots ($1/v$ vs $1/A$) for the varied substrate and ideally fits each set of data to Eq. (1) (see this volume [6]). The slopes and intercepts of each line are then replotted vs the reciprocal of the inhibitory substrate to determine which one (or both) shows the inhibitory effect (see Fig. 4). Once the probable overall pattern is determined, all data points should be fitted to the appropriate overall equation (see this volume [6]).

There are times, however, when one has varied only one substrate concentration (or there is only one substrate), and one must have a way to analyze substrate inhibition data directly. For data showing total substrate inhibition and appearing to fit Eq. (9), the best procedure is to fit the data to this equation (see this volume [6]), but graphical solutions for the constants are useful for preliminary analysis.

When the data are plotted in double-reciprocal form, the curve is a hyperbola with one asymptote lying on the vertical axis and the other lying symmetrically on the other side of the curve (see Fig. 5). Although the necessary analytical geometry is shown in Fig. 5 for analyzing this curve, this is not the best plot for graphical analysis. The normal error is to draw the asymptote too close to the curve, rather than leave the required space; most people, given the curve in Fig. 5, would get a V only 2/3 the correct

---

[10] K. D. Danenberg and W. W. Cleland, *Biochemistry* **14,** 28 (1975).

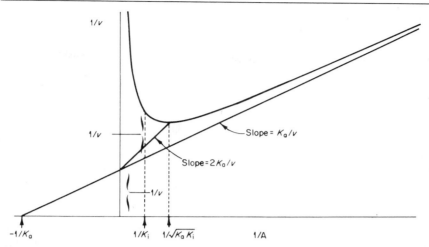

FIG. 5. Linear substrate inhibition by the variable substrate [data fit Eq. (9)]. Data points not shown for clarity.

value, and a K_m about 60% of the correct one. If one is to use this plot, the asymptote must be adjusted until the spacings between the curve and the two asymptotes are similar (it helps to fold the paper so that the two asymptotes coincide; the two halves of the curve should then also coincide). When things are correct, the slope of the line between the vertical intercept of the asymptote and the minimum point of the curve is twice that of the asymptote. The $1/K_i$ is then where the curve lies $1/V$ above the asymptote; alternatively, one can determine K_i from the coordinate of the minimum point $(1/(K_a K_i)^{1/2})$ and the value of K_a.

Another method of plotting is v vs log A, which gives a symmetric bell-shaped curve. When A_m, A_1, and A_2 are the values of A at the maximum point, and at the two points where velocity is half that at the maximum, and v_m is the velocity at the maximum point, we have

$$K_i = A_1 + A_2 - 4A_m \tag{14}$$
$$K_a = A_m^2/K_i \tag{15}$$
$$V = v_m[1 + (K_a/K_i)^{1/2}] \tag{16}$$

If the data fit Eq. (9) well, one should be able to fold the paper along a vertical line at A_m, and both limbs of the curve should coincide. This is a useful test for partial substrate inhibition, or for functions more complex than Eq. (9).

Probably the best graphical procedure is that of Marmasse.[11] In this

[11] C. Marmasse, *Biochim. Biophys. Acta* **77**, 530 (1963).

plot, we define

$$\alpha = A/A_m \tag{17}$$

and plot $1/v$ vs $(\alpha + 1/\alpha)$

$$1/v = [(K_a/K_i)^{1/2}/V](\alpha + 1/\alpha) + 1/V \tag{18}$$

Since

$$A_m = (K_a K_i)^{1/2} \tag{19}$$

one combines the apparent Michaelis constant from this plot with the value of A_m by multiplication to give K_a and by division to give K_i.

This plot has the great advantage that if A_m is not correctly picked the curve is not a straight line, but a hyperbola in which all points where $A < A_m$ lie on one limb, and all points where $A > A_m$ lie on the other (see Fig. 6). If $x = A_m/(\text{apparent } A_m)$, where (apparent A_m) was the initial estimate of A_m used to construct the plot, the asymptotes to the hyperbola will cross on the vertical axis at $1/V$, and the ratio of their slopes is x^2. The value of x is then used to get a newer estimate of A_m, and the plot is made

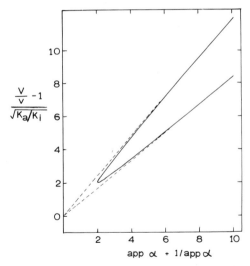

FIG. 6. Plot of substrate inhibition data fitting Eq. (9) by the method of Marmasse. The chosen value of A_m is too small by 1.2. All data for levels of A below A_m are on the upper limb, and data for levels of A above A_m are on the lower limb; the ratio of the slopes of the low A and high A asymptotes (1.44 here) is used to determine the correction factor for A_m. When the chosen value of A_m is too large, the low A limb is the lower one, and the ratio of asymptote slopes is thus lower than 1. When the plot is made with $1/v$ on the vertical axis, as would normally be the case, the asymptotes will converge at $1/V$.

again. When the two limbs of the curve lie on top of one another, one has the correct value of A_m, and analysis is straightforward. Note that the expression $(\alpha + 1/\alpha)$ has a minimum value of 2, and thus some extrapolation is always involved in using this plot. Failure to fit Eq. (9) is obvious here, since the points on one limb will not fit the curve properly. It should be emphasized once again that all graphical procedures are for preliminary analysis only, and that the proper statistical analysis and curve fitting as described in this volume [6] should be used for final analysis.

Author Index

Numbers in parentheses are reference numbers and indicate that an author's work is referred to although the name is not cited in the text. Numbers in italics show the page on which the complete reference is listed.

A

Abbasnezhad, M., 32(*l*), 35
Abeles, R. H., 399(*e*), 401, 450, 461(18), 465(18)
Ackerman, E., 338(*ee*), 339, 358
Ackermann, W. W., 437, 439(4)
Achmatowicz, B., 399(*k*), 402
Adams, E. Q., 184
Adams, M., 394, 396(15), 398(15)
Agarwal, R. P., 462, 464(27,28), 465(28)
Ahmed, A. I., 338, 340, 352(8)
Ainslie, G. R., 421
Ainsworth, S., 198
Akerlof, G., 354
Alber, T., 340, 352(9)
Alberding, N., 368
Albert, A., 296, 297(13)
Alberty, R. A., 43, 50, 139, 155, 172, 178, 180, 190, 194, 200, 201(14,16), 206(14, 16), 229, 230(70), 253, 254(23), 298, 300(13), 304(13), 305, 312(18), 321(18), 411, 414, 418(1), 419, 425
Albrecht, H. P., 395
Aldersley, M. F., 206
Ali, M., 33(*v*), 35
Allen, C. R., 369
Allende, C. C., 52(49), 53
Allende, J. E., 52(49), 53
Allison, R. D., 12, 13
Altman, C., 58, 65, 84
Anderson, A. W., 32(*o*), 35
Anderson, B. M., 396, 398(26)
Anderson, C. M., 396, 398(26)
Anfinsen, C. B., 396, 398(23)
Angelides, K. J., 338, 340, 342, 344(17), 345(17), 346(17), 347, 348(17), 350, 351(10,28), 352(10)
Appleman, J. R., 297, 304(16), 318(16), 320(16), 321(16), 322(16)
Arion, W. J., 51(31), 52(31), 53
Asensio, C., 32(*b,d*), 33(*d*), 34
Ashour, A. L. E., 373

Asp, N., 33(*p*), 35
Atkins, G. L., 176, 177, 178, 179, 180(14)
Atkinson, M. R., 394, 395(8), 398(8)
Austin, R. H., 368
Avigad, G., 37

B

Babior, B. M., 399(*d*), 401
Bada, J. L., 249, 251(21)
Bagnara, A. S., 325
Bailey, J. M., 371, 372
Baker, B. R., 396, 399(28), 445
Baker, R. H., Jr., 50
Baldessarini, R. J., 36
Balko, B., 362
Ballou, C. E., 396, 399(30), 405(30)
Balney, C., 338
Balny, C., 338, 340, 357, 366(13), 368
Banerjee, R., 338(*ff*), 339, 368
Banyasz, J. L., 260, 292(13)
Barbour, B. H., 397, 400(33)
Barden, R. E., 52(52), 53, 435
Barker, R., 371
Barnard, M. L., 245
Barnett, L. B., 229, 230(70)
Barrett, M. J., 35, 178
Bar-Tana, J., 373, 433
Barwell, C. J., 22
Baxendale, J. H., 241
Beardall, A. J., 239
Beeson, K., 368
Bell, G., 408
Bell, J. E., 51(45), 52(45), 53
Bell, R. M., 9, 42
Benaglia, A. E., 255
Bender, M. L., 236, 238
Benkovic, P. A., 374, 376, 377
Benkovic, S. J., 370, 371, 374, 375, 376, 377, 378
Benson, R. W., 404
Berger, L., 299
Berger, R. L., 362

Bergmeyer, H. U., 22, 31
Bernard, M., 368
Berthou, J., 352
Beyer, T. A., 51(45), 52(45), 53
Bhacca, M. S., 371
Biedermann, W., 300
Bielski, B., 338
Blair, J., 285
Blake, C. C. F., 231
Blakley, R. L., 437
Blaustein, M. P., 310
Bloomfield, V., 50, 200, 201(16), 206(16)
Bodourian, S. H., 33(s), 35
Bonnichsen, R. K., 240, 241
Boots, M. R., 399(i), 402
Boots, S. G., 399(i), 402
Borel, M., 260, 264(16), 292(16), 492, 493(4)
Borgland, E., 433
Borgmann, U., 246, 247(18), 248(20), 250, 251(18,19), 252
Bot, G., 400(t), 402
Botts, J., 239
Boulter, D., 394, 396(13), 398(13)
Boyer, P. D., 51(40), 52(40), 53, 76, 138, 157, 206, 230(30), 287
Brady, D. R., 33(cc), 35
Branden, C., 408
Brant, D. A., 229, 230(70)
Bredig, G., 241
Bridger, W. A., 51(11,22), 52(22), 53, 158, 432
Briggs, G. E., 55
Briggs, T. N., 304
Britton, H. G., 79
Brocklehurst, K., 195, 201(8), 202(8), 218(21)
Brolin, S. E., 433
Brook, A. J. W., 369
Brown, A. J., 42
Brownstone, Y. S., 33(v), 35
Bruice, T. C., 231
Buchanan, J. M., 157
Budowsky, E. I., 401(x), 402
Bugett, M. W., 435
Bullard, W. P., 378
Bulos, B., 51(26), 52(26), 53
Bundy, G. L., 401(39), 404
Bunting, P. S., 166, 175(5), 246, 254, 256
Burgess, R. R., 408
Burgner, J. W., 421
Burk, D., 141
Burnell, J. N., 32(g), 35
Burton, K., 297, 300(17), 303, 310(17), 312(17), 329
Butler, J. A. V., 237, 244
Byrnes, K., 32(l), 35

C

Cabiby, E., 236
Caldin, E. F., 369
Calvin, M., 401(cc), 402
Canva, J., 338(bb), 339
Carlström, D., 401(aa), 402
Cary, L. W., 371
Casey, E. J., 244
Caughey, W. S., 399(j), 402
Cavalieri, R. L., 31
Cavallini, D., 34(dd), 35
Cedar, H., 52(51), 53
Cennamo, C., 415
Cha, S., 71, 224, 439, 450(8,9), 451, 452, 454, 462, 464(27)
Chaimovich, H., 52(49), 53
Chamberlin, M. J., 395
Chan, S. S., 368
Chance, B., 240, 241, 338, 368, 372
Chandler, J. P., 297, 304(16), 318(16), 320(16), 321(16), 322(16), 455
Chandrasekhar, K., 394, 396(15), 398(15)
Chang, M., 31
Chao, J., 436
Chapman, B. E., 311
Chapman, H. F., 362
Chase, J. F. A., 227, 229
Chen, D. T. Y., 240
Chen, S.-S., 433
Cheng, Y.-C., 445
Chesne, R. B., 397, 400(33)
Chirikjian, J. G., 52(50), 53
Christensen, J. J., 318
Christian, G. D., 34(ee), 35
Chung, T., 31
Clayton, F. C., 151, 427(a), 428
Cleland, W. W., 9, 43, 51(13,17,30,34,35, 44), 52(4,17,30,34,35,44), 53, 80, 106, 111, 116, 138(5), 139, 143, 144, 147, 155, 158, 179, 206, 211(29), 229, 230, 258, 272, 291(6,8), 328, 373, 386, 395, 398(19,20), 411, 412(4), 413(4), 414(4), 416(4), 421, 425(4), 429, 431(4), 433,

AUTHOR INDEX

435, 432, 450, 467, 474, 476, 485(2), 503, 504, 505(5), 510
Clochard, A., 352
Cohen, L. H., 51(11), 52
Cohen, L. J., 432
Cohen, P. P., 51(14), 52(14), 53, 434
Cohn, M., 259, 294, 308, 311, 315(46)
Cohn, W. E., 299
Colvin, M., 401(38), 404
Connolly, T. N., 225, 226(59), 285, 298
Cooney, P. A., 33(*aa*), 35
Cornish-Bowden, A., 22(8), 23, 28, 29, 30, 37, 38, 39, 40(8), 41, 141, 159, 206, 208, 223, 325
Coward, J. K., 32(*n*), 35
Cozzona, P., 352
Cruikshank, W. H., 219
Cuatrecasas, P., 395, 396, 398(23)
Cunningham, B. A., 376

D

Dahlquist, F. W., 233
Dahlqvist, A., 33(*p*), 35
Dalziel, K., 37, 50, 51(39), 52(39), 53, 139, 143, 144, 152, 153, 155(4,13), 157, 206, 387, 388(3), 421, 425, 507
Damodaran, N. P., 395
Danenberg, K. D., 450, 508, 510
Danenberg, P. V., 508
Daron, K. S., 399(*h*), 400(*h*), 402
Darvey, I. G., 173, 416, 494
Davidsohn, H., 183
Davies, C. W., 306
Davies, R., 400(*r*), 402
Dawson, R. M. C., 225
Debey, P., 338(*gg,hh*), 339, 341, 366, 368
deMaine, M. M., 374, 375
De Moura, J., 305
Dempsey, B., 20, 287
Denner, W. H. B., 432
Dennis, D. T., 52(47), 53, 293
De Pamphilis, M. L., 395, 398(20)
Dickinson, F. M., 508
Diggs, D., 36
Dinovo, E. C., 206, 230(30)
Dixon, H. B. F., 186, 195, 201(8), 202(8), 214(3), 218(21), 232(5a)
Dixon, M., 141, 180, 196, 197, 198(10), 214(10), 414, 424(5)
Dodgston, K. S., 432

Douzou, P., 337, 338(*bb, dd, ff, gg, hh*), 339, 340, 341, 343, 350, 351(29), 353(3, 5,7), 354, 355, 356(33), 357, 358(37), 360, 363, 366(13), 368, 369(7)
Druzhinina, T. N., 401(*x*), 402
Dubravac, S. A., 9
Duggleby, R. G., 52(47), 53, 173, 174, 177(13), 179(13), 180(13), 181, 293, 454
Duley, J., 34(*hh*), 35
Dunlop, R. B., 32(*h*), 35
Dunn, B., 231
Duprè, S., 34(*dd*), 35
Durst, R. A., 305
Dygert, S., 370

E

Easson, L. H., 437, 438
Easterby, J., 22(6), 23, 26, 28, 29, 30, 39
Ebner, K. E., 272, 400(*v*), 402
Eisenhardt, R. H., 362, 372
Eisenstein, L., 368
Eisenthal, R., 141
Eklund, H., 408
Elliott, D. C., 225
Elliott, K. R. F., 429
Elliott, W. H., 225
Ellis, K. J., 303, 310(30), 444, 463(13), 464, 465
Ellison, W. R., 147, 148(11)
Engel, P. C., 51(39), 52(39), 53, 157, 433
Engers, H. D., 51(22), 52(22), 53
Engle, J. L., 371
Ethier, M., 239
Evans, M. G., 241
Evenson, L., 310, 312(17)

F

Fabino, R., 494
Fang, M., 400(*o*), 402
Farror, W. W., 158
Fastrez, J., 205
Federici, G., 34(*dd*), 35
Ferguson, D. A., 34(*jj*), 35
Fernley, N. H., 178
Fersht, A. R., 196, 202, 205, 210, 213, 220(35)
Findlay, D., 211

Fink, A. L., 337, 338, 340(6), 342, 343, 344(17), 345(16,17), 346(16,17), 347, 348(17), 349(6,11,16,17,21,22,23), 350, 351(6,10,22,27,28), 352(4,8,10,27), 364, 366(14)
Fishbein, R., 374, 375, 376, 378
Fisher, R. R., 32(h), 35
Fishman, P. H., 371, 372
Fitzpatrick, F. A., 401(39), 404
Flossdorf, J., 76
Folk, J. E., 51(32), 52(32), 53
Foster, R. J., 161, 162, 169(4), 170(4), 171(4), 173(4), 177(4), 180(4)
Franks, F., 341
Frauenfelder, H., 368
Freed, S., 338
Frey, C. M., 260, 292(13)
Frey, W. A., 375
Fridovich, I., 400(q,s), 402
Frieden, C., 12, 37, 153, 158, 484
Fromm, H. J., 5, 7, 12, 13(12), 14, 25(3), 43, 50, 51(16,37), 52(16,21,37), 53, 65, 66, 84(3), 85, 141, 142(5), 146, 147, 148(11), 151, 152, 155(5), 156(5), 157, 223, 258, 264, 266(22), 292, 411, 414, 416, 418(7), 419, 420, 424, 425(7), 431(7), 434, 436, 467, 468, 470, 472, 476, 479, 480, 481, 483, 484, 485, 492, 493(3), 494
Fujioka, M., 157
Fung, C.-H., 52(52), 53, 435
Furugren, B., 408
Fyfe, J. A., 32(i), 35

G

Gaboriaud, R., 358
Gabrielyan, N. D., 401(x), 402
Garbers, D. L., 51(43), 52(43), 53, 158
Garces, E., 51(30), 52(30), 53, 258, 259(8), 292(8)
Garfinkel, D., 106
Garfinkel, L., 106
Gass, J. D., 394, 405(4,5)
Gatica, M., 52(49), 53
Geeves, M. A., 338
George, P., 295, 298(12), 308, 309(6), 312(6), 315(6), 316(6)
Gerhards, E., 397, 400(36)
Geroch, M. E., 494
Giacomello, A., 32(k), 35
Gibian, H., 397, 400(36)

Gibson, O. H., 372
Gibson, Q. H., 362
Gilbert, H. D., 32(c), 34
Ginsburg, A., 398(c), 401
Glaid, A. J., 51(15), 52(15), 53
Glantz, R. R., 244
Glusker, J. P., 400(o), 402
Gnosspelius, G., 33(bb), 35
Goldberg, D. M., 34(ii), 35
Goldstein, A., 437, 439(2,3)
Goldstein, B. N., 55, 62, 65, 84(2), 85
Goldthwait, D. A., 404
Gomori, G., 225
Good, D., 368
Good, N., 338, 340, 349(11)
Good, N. E., 225, 226, 285, 298
Goray, G. R., 371
Gorbunoff, M. J., 213
Gorman, R. R., 401(39), 404
Got, C., 260, 292(16), 492, 493(4)
Graces, E., 503
Gracey, R. W., 401(y) 402
Graham, N., 338, 368
Graves, D. J., 401(bb), 402, 436, 481, 482
Gray, G. R., 371
Green, T. R., 32(o), 35
Greenfield, N. J., 462
Grey, B., 338, 343, 364
Griffin, C. C., 433
Griffith, G. W., 397, 399(32)
Gross, M., 51(32), 52(32), 53
Grossman, G., 327
Guerrieri, P., 34(dd), 35
Guilbault, G. G., 35
Guinard, S., 338(ii), 339
Gulbinsky, J. S., 258, 292(6)
Gunsalus, I. C., 358, 368, 399(h), 400(h), 402
Gunter, C., 238
Gupta, R. K., 371, 372, 374(10), 378
Gutfreund, H., 22
Gutowsky, M. S., 372

H

Hakala, M. T., 51(15), 52(15), 53
Haldane, J. B. S., 55, 155, 207
Halford, S. E., 215
Hammes, G. G., 253, 254(23), 287, 451
Han, Y. W., 32(o), 35
Handler, P., 51(26), 52(26), 53
Hanes, C. S., 65, 141, 142(7), 145, 486, 492

Hansen, J. N., 206, 230(30)
Hansen, R. J., 33(2), 35
Hanson, R. L., 51(23), 52(23), 53, 156, 376, 433
Hanson, T. L., 485
Harada, K., 400(*m*), 402
Hardman, J. G., 51(43), 52(43), 53, 158
Hare, J., 397, 400(34)
Harris, R. C., 33(*s*), 35
Hartley, B. S., 218, 219(50)
Hartman, F. C., 405
Hartree, E. F., 371
Hastings, J. W., 338, 340, 366(13)
Hatch, M. D., 36
Hegazi, M., 376
Heidelberger, C., 396, 398(25), 462
Heinke, H., 157
Heinle, H., 51(41), 52(41), 53
Hellerman, L., 399(*j*), 402
Henderson, J. F., 51(24), 52(24), 53
Henderson, P. J. F., 439, 444
Henri, V., 161
Hersh, L. B., 51(12,25), 52(25), 53
Hershey, J. W. B., 394, 395(9), 398(9)
Hess, B., 22(5), 23, 370, 376, 378
Hess, G. P., 220
Heyde, E., 258
Heyde, M. E., 311
Hill, D. E., 455
Hill, R. L., 51(45), 52(45), 53
Himoe, A., 220
Hinberg, I., 236
Hines, R. H., 51(42), 52(42), 53, 157
Hinkle, P. M., 238
Hinz, H., 33(2), 35
Hirshfield, J. M., 462
Hoberman, H. D., 76
Hoch, F. L., 408
Hochachka, P. W., 250
Hoffman, G., 33(*x*), 35
Hofstee, B. H., 141
Hogenkamp, H. P. C., 396, 399(29)
Holme, D., 34(*ii*), 35
Holmes, R. S., 34(*hh*), 35
Holzer, H., 33(2), 35
Homer, R., 338, 348, 352
Hoogsteen, K., 462
Horgan, D. J., 32(*f*), 34
Hori, M., 51(24), 52(24), 53
Houston, L. L., 33(*cc*), 35
Howard, A., 338
Hsu, R. Y., 51(34), 52(34), 53

Hu, A., 311
Huang, C. Y., 65, 66(6), 80(6), 485, 495, 496(10), 498(10)
Huang, Y. L., 400(*v*), 402
Huennekens, F. M., 462
Hui Bon Hoa, G., 338(*hh,ii*), 339, 353, 354, 356(33), 357, 358(37), 369
Hunkapiller, M. W., 212
Hurst, J. K., 287
Hurst, R. O., 85, 87
Hutchinson, H. D., 33(*w*), 35

I

Illiano, G., 395
Inagami, T., 212
Infante, J. P., 266, 270(24), 292(24)
Irwin, R., 52(50), 53
Isbell, H. S., 371
Isemura, T., 370
Izawa, S., 225, 226(58,59), 285, 298
Izatt, R. M., 318

J

Jacobson, G. R., 401(*bb*), 402
Jallon, J. M., 308, 315(46)
James, E., 51(20), 52(20), 53, 266, 269, 292(23,27), 432, 463
Janson, C. A., 147, 395, 398(19)
Jarabak, R., 401(38), 404
Jayaram, H. N., 33(*aa*), 35
Jayaram, S., 33(*aa*), 35
Jeffery, P. D., 22(7), 23, 29(7), 30(7), 39(7)
Jencks, W. P., 51(25), 52(25), 53, 202, 225, 232(19)
Jendrisak, J. J., 408
Jennings, R. R., 169
Johansen, G., 104
Johnson, G. F., 436
Johnson, L. N., 231
Jollès, P., 352
Jones, G. H., 395
Jones, K. F., 225
Jonsson, B., 254
Jornvall, H., 408
Joyce, B. K., 51(42), 52(42), 53, 157

K

Kachmar, J. F., 138
Kaiser, E. T., 338, 340
Kaplan, H., 200, 206(18), 218, 219

Karahasonoglu, A. N., 36
Karplus, S., 213
Katiyar, S. S., 435, 504
Katz, A. J., 297, 304(16), 318(16), 320(16), 321(16), 322(16)
Kaufman, S., 237, 238, 244, 498
Kayne, F. J., 258, 279(3)
Kee, A., 433
Keilin, D., 371
Kellerman, G. M., 32(*e*), 34
Kendrick, N. C., 310
Khoo, J. C., 504
King, E. L., 58, 65, 84
Kinsella, J. E., 266, 270(24), 292(24)
Kirby, A. J., 206
Kirsch, J. F., 238
Kitz, R., 216
Kleeman, J. E., 5
Kleinschuster, J. J., 374
Kneifel, H. P., 270
Knowles, J. R., 206, 210
Kobashi, K., 397, 400(34)
Kochetov, N. K., 401(*x*), 402
Köchli, H., 34(*gg*), 35
Koerber, B. M., 178, 180
Koeppe, R. E., II, 212, 213(39)
Koerner, T. A. W., Jr., 371, 373
Kohn, M. C., 106
Komai, T., 397, 400(34)
Komlev, I. V., 401(*x*), 402
Koshland, D. E., 9, 42
Kovacs, E. F., 400(*t*), 402
Krenitsky, T. A., 32(*i*), 35, 432
Krug, F., 395
Krupka, R. M., 200, 206(15)
Kuby, S. A., 258, 278(10)
Kuchel, P. W., 22(7), 23, 29, 30(7), 39
Kula, M., 76
Kumar, A., 34(*ee*), 35
Kun, E., 399(*k*), 402
Kusov, Y. Y., 401(*x*), 402
Kyd, J., 325

L

Ladunski, M., 238
Laidler, K. J., 166, 175(5), 200, 206(16,15,18), 235, 236, 239, 240, 244, 245, 246, 247(18), 248(20), 250(19), 251(18,19), 252, 254, 256, 449
Lancaster, P. W., 206
Landskroener, P. A., 240
Lardy, H. A., 51(23,34), 52(23,34), 53, 156, 270, 376, 433, 435
Larroque, C., 357
Layani, M., 338
Lazdunski, C., 238
Lee, H. A., 399(*e*), 401
Lee, H.S., 200, 206(17)
Legallais, V., 338, 368
Leigh, J. S., 368
Le Peuch, C., 338, 340, 366(13)
Le Tourneau, D., 305
Levy, M., 255
Levy, R., 494
Li, E. H., 298, 305(19)
Li, E. L.-F., 297, 304(16), 318(16), 320(16), 321(16), 322(16)
Li, H. C., 157
Li, T. K., 394, 396(11), 398(11)
Liddle, G. W., 397, 401(37)
Lienhard, G. E., 7, 401(*dd*), 402, 482, 483
Lineweaver, H., 141
Linn, C.-S., 52(50), 53
Lister, A. J., 397, 400(35)
Lombard, A., 338(*ff*), 339, 368
Lonberg-Holm, K. K., 362, 372
Low, P. S., 249, 251
Lowe, G., 231
Lowenstein, J. M., 51(18), 52(18), 53, 400(*o*), 402
Lowry, O. H., 222, 400(*w*), 402
Lueck, J. D., 147, 148
Lum, C. T., 32(*l*), 35
Lumry, R., 104, 244
Lynn, M., 310, 312(17)

M

McClure, W. R., 22, 26, 27, 30, 31(3), 37, 38, 39, 40, 41, 270, 435
MacGibbon, A., 338
McGillivray, I. H., 256
Madsen, N. B., 51(22), 52(22), 53
McKinley-McKee, J. S., 394, 396(12), 398(12)
McPherson, A., 394, 396(14,15), 398(14,15)
Magnusdottir, K., 338, 349
Mahler, H. R., 50
Mair, G. A., 231
Maier, V. P., 338(*cc*), 339
Makarewicz, W., 396, 398(24)

AUTHOR INDEX

Makinen, M. W., 337, 338, 340, 352(4)
Malakhov, A. A., 34(*ff*), 35
Marco, E. J., 31
Marco, R., 31
Markham, G. D., 51(38), 52(38), 53
Markley, J. L., 213
Marmasse, C., 511
Marshall, M., 51(14), 52(14), 53, 434
Martell, A. E., 226, 285, 299(9), 310, 312(49), 315, 317, 318(9)
Martell, A. R., 295, 312(9)
Massey, V., 51(27), 52(27,48), 53, 190, 194, 201(4), 229(4)
Mathias, A. P., 211
Mathysse, S., 36
Matsuoka, Y., 51(19), 52(19), 53
Matthew, M., 400(*n*), 402
Maurel, P., 338, 342, 345(17a), 353, 354, 356(33), 357
Mayhew, S. G., 36
Meindl, P., 401(*z*), 402
Meister, A., 394, 397, 399(32), 401(44,48), 405(4,5,6), 406(6,44)
Melchior, N. C., 298, 312
Meloche, H. P., 378
Meyers, T. C., 394, 395(7), 398(7)
Michaelis, L., 183
Michaels, G., 52(53), 53
Midelfort, C. F., 371, 372, 374(10), 378
Mildvan, A. S., 266, 270(26), 294, 371
Miller, O. V., 401(39), 404
Miller, R. L., 32(*i,j*), 35
Milner, Y., 52(53), 53
Miwa, I., 33(*r*), 35
Moffatt, J. G., 395
Moffet, F. J., 158
Mohan, M. S., 260, 292(14), 295, 298(8), 305(7,8), 308(7,8,38), 312(7), 317
Moldvan, A. S., 400(*u*), 402
Monk, B. C., 32(*e*), 34
Monroe, R. E., 394, 395(9), 398(9)
Moolgavkar, S. H., 401(38), 404
Moon, T. W., 246, 247(18), 248(20), 250(19), 251(18,19), 252
Morales, M. F., 239, 244, 256
Morley, C. G. D., 400(*l*), 402
Morris, D. L., 394, 396(12), 398(12)
Morrison, J. F., 17, 51(20), 52(20), 53, 173, 174, 177(13), 179(13), 180(13), 181, 258, 264, 266(20, 21), 269, 272, 281(21), 287(20), 292(21,23,27), 294,
300, 303, 310(30), 324(1), 432, 439, 444, 454, 463(13), 464, 465, 466(22)
Morton, D. P., 320
Mourad, N., 51(29), 52(29), 53
Murray, A. W., 394, 395(8), 398(8)

N

Nakatani, Y., 157
Nanninga, L. B., 308
Natelson, S., 33(*y*), 35
Neece, M. S., 400(*s*), 402
Nelson, D. R., 411, 419, 424
Neuberger, A., 400(*n*), 402
Neurath, H., 237, 238, 244
Newbold, R. P., 32(*f*), 34
Newman, P. F. J., 176, 180(14)
Nichol, L. W., 22(7), 23, 29(7), 30(7), 39(7)
Nicholson, J. F., 33(*s*), 35
Niemann, C., 161, 162, 169(4), 170(4), 171(4), 173(4), 177(4), 180(4)
Nimmo, I. A., 176, 177, 178, 179, 180(14)
Ning, J., 5, 52(21), 53
Noat, G., 260, 264(16), 292(16), 492, 493(4)
Noda, L., 259
Noltmann, E. A., 258, 278(10), 401(*y*), 402
Nordlie, R. C., 51(16,31), 52(16,31), 53
Norlund, T. M., 368
North, A. C. T., 231
Northrop, D. B., 51(33), 52(33), 53, 111, 138(5), 435, 504, 504(5)
Novoa, W. B., 399(*f*), 402, 420

O

O'Brien, W., 34(*kk*), 35
O'Connell, E. L., 379
Ogston, A. G., 264, 266(20), 287(20)
Ohta, T., 394, 395(10), 398(10)
Oikawa, T. G., 396, 399(29)
O'Leary, M. H., 111, 138(5)
Olive, G., 494
Oliver, L. T., 32(*a*), 34
Orsi, B. A., 51(35), 52(35), 53
O'Sullivan, W. J., 260, 264, 266(20,21), 281(21), 287(20), 292(21), 294, 295, 299, 304(19a), 305(19a), 310(19a), 311, 312(19a), 318(10), 327
Ottolenghi, P., 200, 206(13)
Ouellet, L., 239, 244, 256
Ozand, P. T., 36

P

Pace, J., 256
Pana, C., 241
Pantaloni, C., 338(ii), 339
Parikh, I., 395
Park, G. S., 241
Park, W. D., 32(h), 35
Parks, P. C., 220
Parks, R. E., Jr., 51(29), 52(29), 53, 462, 464(27,28), 465(28)
Parsons, S. M., 5, 9, 210, 213(36), 232, 233, 320
Passonneau, C., 400(w), 402
Passonneau, J. S., 32(m), 35, 222
Peacock, D., 394, 396(13), 398(13)
Pearson, D. J., 227
Pederson, T., 358
Peller, L., 50, 200, 201(14), 206(14), 425
Pentchev, P. G., 371, 372
Pentz, L., 226
Perrin, D. D., 20, 260, 263, 266, 287, 295, 304(19a), 305(19a), 310(19a), 312(19a)
Perutz, M. F., 368
Pesce, M. A., 33(s), 35
Peterson, D. C., 401(39), 404
Petsko, G. A., 338, 340, 352(9)
Phillips, D. C., 231
Phillips, R. C., 295, 298(12), 308, 309(6), 312(6), 315, 316, 318
Philo, R. D., 178, 180
Pilgrim, W. W., 371
Plowman, K. M., 51(44), 52(44), 53, 158
Popjak, G., 396
Porter, J. W., 435, 504
Potter, V. R., 437, 439(4)
Poznanskaya, A. A., 34(ff), 35
Preller, A., 36
Prusoff, W. H., 445
Purich, D. L., 5, 7, 8, 9(5), 12, 13(12), 14, 24(3), 43, 52(21), 53, 258, 264, 266(22), 292, 395, 398(21,22), 410(21), 414, 470, 483

Q

Qibson, Q. H., 51(27), 52(27), 53

R

Rabin, B. R., 211
Raftery, M. A., 210, 213(36), 232, 233
Rao, B. D. N., 259
Rathbun, W. B., 32(c), 34
Ratzlaff, R. W., 310
Raushel, F. M., 373, 432
Ray, W. J., 303, 400(u), 402, 421
Rechnitz, G. A., 260, 292(14), 295, 298(8), 305(7,8), 308(7,8,38), 312(7), 317
Redfield, A. G., 9
Reed, G. H., 51(38), 52(38), 53, 311
Reed, J. K., 51(28), 52(28), 53
Reed, L. J., 435
Regenstein, J., 225
Reichel, K. B., 395, 398(21), 410(21)
Reiner, J. M., 437, 439(5)
Renard, M., 202
Reyes, P., 396, 398(25), 462
Reynolds, A. H., 368
Reynolds, C. H., 394, 396(12), 398(12)
Rhoads, D. G., 51(18), 52(18), 53
Ricard, J., 492, 493(4)
Rich, D. H., 462, 466(29)
Richard, J., 260, 264(16), 292(16)
Richards, J. H., 212
Rietz, E. B., 35
Rife, J. E., 507
Rimai, L., 311
Roberts, D. V., 30
Roberts, R. M., 17
Robillard, G., 213
Robinson, R. A., 306
Rock, M. K., 222
Rogers, K. S., 399(i), 402
Ronzio, R. A., 401(44,48), 405, 406(44)
Roscelli, G. A., 303
Rose, I. A., 370, 371, 372, 374(10), 377, 379
Rosenblum, L., 33(aa), 35
Rossmann, M., 394, 396(15), 398(15)
Rossotti, F. J. C., 296, 300(14), 304(14)
Rossotti, H., 296, 300(14), 304(14)
Rottenburg, D. A., 32(m), 35
Rowe, W. B., 401(44), 405, 406(44)
Rudolph, F. B., 12, 13, 14, 51(23,37,43,47), 52(23,37,43,46), 53, 151, 156, 157, 158, 258, 376, 427(a), 428, 433, 434, 476, 492, 493(3), 494
Rupley, J. A., 213
Russell, M. E., 338
Russell, P. J., Jr., 504
Rutman, R. J., 295, 298(12), 308, 309(6), 312(6), 315(6), 316(6)
Rutter, W. J., 408
Rytting, J. H., 318

S

Sabater, B., 32(b,d), 33(d), 34
Sable, H. Z., 31
Saika, A., 372
Saint Blanchard, J., 352
Salas, M., 378
Salerno, C., 32(k), 35
Sarkar, S., 394, 395(10), 398(10)
Sarma, V. R., 231
Saronio, C., 368
Sayce, I. G., 263, 304, 305
Schack, P., 181
Schimerlik, M. I., 51(17), 52(17), 53, 504
Schlesinger, M. J., 215
Schmidt, C. L. A., 161, 180
Schmidt, D. E., 210, 215(34), 216
Schønheyder, F., 180
Schonbaum, G. R., 236
Schray, K. J., 370, 374, 376, 377, 378
Schubert, J., 308
Schulman, R. G., 213
Schwartz, J. H., 52(51), 53
Schwartzenbach, G., 300
Schwert, G. W., 51(15), 52(15,53), 53, 174, 237, 238, 244, 399(f,g), 402, 420, 476
Scopes, R. K., 258, 311
Scrutton, M. C., 52(52), 53, 435
Seargent, E. P., 296, 297(13), 300(13), 304(13)
Sebastian, J., 32(b,d), 33(d), 34
Secemski, I. I., 7, 483
Segal, H. L., 138
Selwyn, M. J., 178, 180, 181, 182
Sharma, V. S., 260, 305
Sharrook, M., 368
Sherwin, J. F., 33(y), 35
Shibaev, V. N., 401(x), 402
Shiga, T., 338
Shindler, J. S., 207
Shiner, V. J., 50
Siano, D. B., 142
Siewers, I. J., 374, 376
Sillen, L. G., 226, 285, 295, 299(9), 312(9), 318(9)
Silverman, D. N., 206
Silverstein, E., 51(36), 52(36), 53, 421
Simms, H. S., 232
Simon, L. N., 394, 395(7), 398(7)
Singh, R. M. M., 225, 226(59), 285, 298
Sireix, R., 338(bb,dd), 339, 350, 351(29)
Smallcombe, S. H., 212
Smiley, J. D., 399(j), 402
Smith, E., 292
Smith, E. I., 244
Smith, R. M., 298, 305, 312(18), 321(18)
Smithers, G. W., 299
Snoke, J. E., 238, 244
Snyder, G. H., 213
Solinas, S. P., 34(dd), 35
Sols, A., 378
Somaro, G. N., 249, 250, 251(21)
Sorenson, T. S., 181
Sorenson, L. B., 368
Spector, T., 32(i,j), 35, 462, 464(28), 465(28)
Sperling, J., 213
Spivey, H. O., 297, 304(16), 318(16), 320(16), 321(16), 322(16), 455
Spradlin, J. E., 370
Spring, T. G., 397, 399(31)
Srere, P. A., 51(19), 52(19), 53
Stadtman, E. R., 398(c), 401
Stadtman, T. C., 400(l), 402
Stedman, E., 437, 438
Steffens, J. J., 374, 376
Stein, R., 36
Stevenson, K. J., 218, 219(50)
Stewart, J. A., 200, 206(17)
Stokes, A. M., 432
Stokes, R. H., 306
Storer, A., 22(8), 23, 28, 29, 30, 37, 38, 39, 40(8), 41, 223, 325
Strauss, O. H., 437, 439(3)
Strickland, S., 52(48), 53
Strother, G. K., 338(ee), 339, 358
Stroud, R. M., 212, 213(39)
Stuehr, J. E., 260, 292(13), 304
Sturtevant, J. M., 212
Sulebele, G., 421
Sun, E., 462, 466(29)
Sun, F. F., 401(39), 404
Susi, H., 213
Swenson, C. A., 371
Swoboda, B. E. P., 51(27), 52(27), 53
Sykes, B. D., 213
Symons, R. H., 295

T

Tabatabal, L., 481, 482
Tai, H., 36
Takagi, T., 370
Talalag, P., 401(38), 404

Tammann, G., 234
Tappel, A. L., 338(cc), 339
Taqui Khan, M. M., 310, 312(49), 315, 317
Tasi, C. S., 435
Tate, S. S., 394, 405(6), 406(6)
Tenhunen, R., 36
Thach, R. E., 394, 395(10), 398(10)
Theorell, H., 240, 241
Thomas, J. A., 370
Thompson, R. E., 297, 304(16), 318(16), 320, 321(16), 322
Thompson, V. W., 504, 505(5)
Thorneley, R. N. F., 260, 292(15)
Thornton, E. R., 226
Tildon, J. T., 36
Tipton, K. F., 178, 181, 195, 207, 429
Timasheff, S. N., 213
Tischler, M. E., 32(h), 35
Toda, H., 370
Todhunter, J. A., 8,9(5), 13, 395, 398(21), 406, 410(21)
Tovey, K. C., 17
Townsend, J., 311
Travers, F., 338(dd), 339, 350, 351(29), 353, 354, 355, 356(33), 358, 369
Trenthan, D., 338
Trosper, T., 310, 312(17)
Tsai, A., 338
Tsernoglou, D., 340, 352(9)
Tu, C. K., 206
Tu, J. I., 401(bb), 402
Tubbs, P. K., 227, 229
Tume, R. K., 32(f), 34
Tuppy, H., 401(z), 402

U

Uhr, M. L., 300, 504, 505(5)
Ureta, T., 36
Utter, M. F., 52(52), 53, 435

V

Valenzuela, P., 408
Vallee, B. L., 394, 396(11), 398(11), 404, 408
Vernon, G. A., 231
Villet, R. H., 157
Viñuela, E., 378
Vogel, A. I., 299, 300(26)

Vogt, M., 36
Volkenstein, M. V., 55, 62, 65, 84(2), 85
Voll, R. J., 373
Volman, D. H., 338
von Berneck, R. M., 241
von Wartburg, J. P., 34(gg), 35

W

Wagner, M., 435
Walaas, E., 308, 309, 318(44)
Waley, S. G., 190
Walker, A. C., 161, 180
Wall, M. C., 244
Walter, C., 173, 178
Walton, J. H., 241
Wandzilak, T. M., 404
Wassnik, J. H., 36
Watanabe, S., 310, 312(17)
Watson, J. A., 400(o), 402
Watts, D. C., 258, 285(5)
Webb, B. C., 8,9(5)
Webb, E. C., 141, 180, 196, 197(10), 198(10)
Webb, J. L., 384, 397(1), 410(1)
Weber, J., 338
Wedler, F. C., 51(40), 52(40), 53, 157
Weibel, M. K., 33(q), 35
Weil, M. H., 397, 400(33)
Weiland, O. H., 33(x), 35
Weinberg, F., 408
Weisbrod, R. E., 405
Weiss, L., 33(x), 35
Wendel, A., 51(41), 52(41), 53, 157
Wennerstrom, H., 254
Wentworth, D. F., 450, 465(19)
Westheimer, F. H., 210, 215(34), 216
Wettermark, K. G., 433
Whatley, F. R., 32(g), 35
Whitaker, D. K., 212
Wiese, A. C., 305
Wilczek, J., 36
Wilchek, M., 396, 398(23)
Wildi, E., 349
Wilkinson, G. N., 105, 106, 116(2), 410
Williams, J., 241
Williams, J. F., 173
Williams, J. W., 454, 461(22), 466(22)
Williams, M. N., 462
Willison, K. R., 260, 292(15)

Wilson, C. S., 35
Wilson, I. B., 216, 236
Winer, A. D., 399(g), 402, 420
Winget, G. D., 225, 226(59), 285, 298
Wingler, P. W., 229
Winter, W., 225, 226(59), 285, 298
Wiseman, J. S., 450, 461(18), 465(18)
Wold, F., 396, 399(30), 405
Wolfe, R. G., 400(m), 402
Wolfenden, R., 401(43), 404, 450, 465(19), 482
Wong, C. S., 421
Wong, J. T., 56, 65, 486, 490, 492
Wood, H. G., 52(53), 53
Wratten, C. C., 51(13), 52, 431, 436, 476
Wu, C. W., 404
Wu, F. Y., 32(n), 35
Wurster, B., 22(5), 23, 370, 376, 378

Y

Yagil, G., 76
Yakovlev, V. A., 34(ff), 35
Yakusheva, M. I., 34(ff), 35
Yamamura, K., 338, 340
Yasmineh, W. G., 32(l), 35
Yip, B. P., 51(46), 52(46), 53, 158
Yonetani, T., 368
Younathan, E. S., 371, 373
Yuan, B., 36

Z

Zell, T., 213
Zewe, V., 411, 431, 436, 467, 480, 481, 494
Ziegler, D. N., 33(w), 35
Zyskind, J. W., 142

Subject Index

A

Abortive complex formation, 43, 419–424, 432–436, *see also* Dead-end inhibition
  chymotrypsin, 205
  limitation, 432–436
  pH effects, 205
Acetaldehyde dehydrogenase (acylating), *see* Coenzyme A-linked aldehyde dehydrogenase
Acetate:CoA ligase, *see also* Acetyl-CoA synthetase
  mechanism, 158
Acetate kinase
  cold denaturation, 9
  equilibrium constant, 5
  inhibitor, 398
  metal–ion binding, 275–278
  purine nucleoside diphosphate kinase activity, 8
Acetate thiokinase, *see* Acetyl-CoA synthetase
Acetoacetate, analog, 400
Acetoacetate decarboxylase
  inhibitor of, 400
  perturbed lysine pK, 210
Acetokinase, *see* Acetate kinase
Acetopyruvate, 400
Acetylacetone, 400
Acetyl-CoA, assay, 33
Acetyl-CoA synthetase, assay, 33, *see also* Acetate:CoA ligase
ϵ-N-Acetyllysine, 400
3-Acetylnicotinamide adenine dinucleotide, 31
Acetyl-L-tyrosine, 180, 238
Acetyl-L-tyrosylhydroxamate, 180
Acid phosphatase, progress curve, 180
Aconitase, inhibitor, 400
Activated complex theory, 235
Activation
  buffer, 226
  fructose 1,6-bisphosphate, 7
  isocitrate, 293
  metal-ion, 258, 270–279, 287–289
  nucleotide, 293

Activation entropy, *see* Entropy of activation
Acyl activating enzyme, *see* Acetyl-CoA synthetase
Adenine phosphoribosyl transferase, mechanism, 51
Adenosine deaminase
  auxilliary enzyme, as, 32, 36
  coupled assay, 32
  slow-binding inhibition, 464–466
  tight-binding inhibition, 462
Adenosine-5′-diphosphate
  assay, 32
  metal–ion complex, 261, *see also* metal–nucleotide complex
    calcium, 319
    europium, 303
    magnesium, 314, 315, 319
    manganese, 319
    potassium, 312, 319
    sodium, 312, 319
  product inhibition, 279–280
Adenosine diphosphate ribose, inhibitor, 394, 396, 398
Adenosine 3′-monophosphate, 398
Adenosine 5′-monophosphate
  analog, 398
  assay, 32
  metal–ion complex, 319
Adenosine 5′-monophosphate deaminase
  coupled enzyme assay, in, 6
  inhibitor, 396
Adenosine 5′-phosphosulfate kinase, assay, 32
Adenosine 5′-phosphosulfate sulfohydrolase, product inhibition, 432
Adenosine phosphotransferase, assay, 32
Adenosine triphosphatase, *see also* Myosin adenosine triphosphatase
  assay, 32
  inhibitor, 394, 398, 462
  metal–nucleotide complex, and, 264–265
  tight-binding inhibition, 462
Adenosine 5′-triphosphate
  acetate kinase stabilization, 9

activation, metal ions, 270–271
assay, 32, 36
dead-end inhibition by, 279–280
inhibition by, 37, 269, 276–277, 328–329
ionization effect on metal binding, 259–261
metal–ion complex, 267–269, 305–318
substrate inhibition, 501
tetrapropyl ammonium bromide, interaction with, 298
vanadate ion contamination, 10
Adenosylhomocysteinase, coupled enzyme assay, in, 36
S-Adenosyl-L-homocysteine, assay, 32
S-Adenosylmethionine
assay, 32
tritium labeled, 36
Adenylate cyclase, inhibition, 395
Adenylate kinase
contaminant in phosphotransferases, 7
coupled assay, in, 6, 32
inhibition, 7, 398, 401, 483
mass action ratio, 18
mechanism, 18
quenching problem, 16
substrate, metal–ion complex, 259
Adenylosuccinate lyase
mechanism, 51
product inhibition, 432
Adenylosuccinate synthetase
coupled enzyme assay, 32
mechanism, 51, 157
(4-Adenylyl)-1-butylcobalamine, 399
5′-Adenylylimidodiphosphonate
inhibition by, 398
structure, 403
5′-Adenylyl-β,γ-methylenediphosphonic acid, 481
ADP, see Adenosine 5′-diphosphate
Albumin
enzyme stabilizing agent, 10
inhibition, 37
Alcohol dehydrogenase
abortive complex, 420
alternative product inhibition, 436
alternative substrate effect, 494
coupled assay, use in, 34
cryoenzymology, 338
inhibitor, 398, 408
isotope effect, 110

mechanism, 51
partial substrate inhibition, 507
Aldehyde, assay, 34
Aldehyde dehydrogenase, see also Coenzyme A-linked aldehyde dehydrogenase
slow-binding inhibition, 450, 465
Aldolase
anomeric specificity, 374
direct kinetic determination of, 378
coupled assay, use in, 34
Aldomet, see α-Methyldihydroxyphenylalanine
Aldose-1-epimerase, assay, 33
Aldosterone, 401, 404
receptor, 401
Alkaline phosphatase
coupled assay, 34, 36
cryoenzymology, 338
kinetic parameters, 238
pH effects on, 215
N-Alkylnicotinamide, 396, 398
Allopurinol, slow-binding inhibition by, 463–464
Allosterism,
phosphotransferases, 293
PRPP:ATP phosphoribosyltransferase, 9
substrate inhibition and, 43, 500–501
Alloxanthine, slow-binding inhibition by, 463–464
Alternative product inhibition, 436
Alternative substrate, 10, 145–146, 236, 372–374, 486–500
graphical method, 490–499
common-product approach, 494–498
constant-ratio approach, 498–499
one-product approach, 491–494
numerical method, 487–490
Amino acid, assay, 33
Amino acid residues
acetylation, 218
analogs, 397, 399–400
dielectric constant effects, 211–212
direct titration, 212–214
heats of ionization, 209–211
identification, 209–220
pK values, 209–210
D-Amino-acid oxidase
assay, 34
cryoenzymology, 338

SUBJECT INDEX

L-Amino-acid oxidase, assay, 34
3-Aminoenolpyruvate-2-phosphate, 397, 399
β-Aminoglutaryl-L-aminobutyrate, 397, 399
δ-Aminolevulinate synthetase, inhibitor, 400
Aminomalonate, 400
Aminopterin, 396, 399
  tight-binding inhibition by, 462
Ammonia, assay, 34
Ammonium sulfate, inhibition by, 37
AMP, see Adenosine 5'-monophosphate
Δ^5-Androstene-3,17-dione, 401, 404
2,5-Anhydro-D-glucitol, 373
2,6-Anhydro-D-glucitol, 373
2,5-Anhydro-D-glucitol 6-phosphate
  fructokinase, with, 373
  phosphofructokinase, with, 373–374
2,5-Anhydro-D-mannitol, 373
2,6-Anhydro-D-mannitol, 373
2,5-Anhydro-D-mannitol 6-phosphate, 373
2,5-Anhydro-D-mannose, 373
Anomeric specificity, 14, 370–379
  direct kinetic determination of, 372, 374–378
  generation, by, 378–379
  substrate analog, by, 372–374
Apoflavodoxin, 36
Arginine kinase, 258, 292
Argininosuccinate lyase, assay, 33, 34
Aromatic-L-amino-acid decarboxylase, see DOPA decarboxylase
Arrhenius theory, 235, 240–247
Arsenate, 19
Arsenazo III, 310
Arylamine acetyltransferase, assay, 33
Asparagine synthetase, 52
Asparaginase, assay, 33, 34
Aspartate
  assay, 33
  chemical modification, 213, 232
  heat of ionization 209
  perturbed pK, 233
  pK value, 209
  titration, 213
Aspartate aminotransferase, 33, 181
ATP, see Adenosine 5'-triphosphate
ATPase, see Adenosine 5'-triphosphatase
ATP-citrate lyase
  assay, 33

  inhibition, 400
ATP lyase, inhibition, 398
Auxiliary enzyme assay, see Coupled enzyme assay
9,11-Azoprosta-5,13-dienoic acid, 401–404

B

Beer–Lambert relation
  spectrophotometric assay, 31
  stray light effects on, 15
Beer's law, see Beer–Lambert relation
Benzoyl-L-arginamide, 237
Benzoly-L-phenylalanine ethyl ester, 238
Benzoyl-L-tyrosinamide, 237, 244
Benzoyl-tyrosine ethyl ester, 238, 244
Benzoyl-L-tyrosylglycinamide, 237
BES, see Buffer compound
Billirubin, 36
Billiverdin, 36
Billiverdin reductase, 36
Bi (random) uni uni ping pong, 52
Bireactant systems,
  alternative substrate effects, 486–500
  competitive inhibition, 474–478
  constant ratio method, 146
  general rate equation, 143
  graphical analysis, 143–148
  ping ping 47–48, 51
  product inhibition, 418–419
  progress curves, 173–174
  replotting analysis, 143–146
  sequential, 51
  stoichiometry, 11
  substrate saturation, 11–12, 504–506
N,N-Bis(2-hydroxyethyl)-2-aminoethane sulfonic acid, see Buffer compound
Bisubstrate systems, see Bireactant systems
Bi uni uni bi ping pong, mechanism, 52
Bi uni uni uni ping pong, mechanism, 52
Bongkrekic acid, tight-binding inhibition by, 462
Borate buffers, 19–20, 359
Briggs–Haldane equation, 55–57
Bromoacetyl carnitine, 229
5-Bromo-2'-deoxyuridine 5'-monophosphate, induced substrate inhibition by, 508–509

Buffer compound
  activation, 226
  capacity of, 4
  chelation effects, 225–226
  choice of, 19–20, 285
  dielectric constant, effect on pK, 226
  dilution, 20
  equilibrium constant, effects on, 18
  heavy water and, 226, 227
  ionic strength, 226, 227
  metal–ion effects, 287, 298–299
  stability constant measurements, 298–299

C

Cacodylate buffer, 19, 355, 357, 359–360
Calcium ion, 319
  ATP complex, 310, 318
  spectral determination, 301
Carbamate kinase, 51
  product inhibition, 434–435
Carbamoyl-phosphate synthetase, product inhibition, 429
Carbobenzoxy-L-glutamyl-L-tyrosine, 244
Carbobenzoxyglycyl-L-leucine, 244
Carbobenzoxyglycyl-L-tryptophan, 244
Carbobenzoxy-L-phenylalanine, 244
Carbohydrate, see specific carbohydrate
  analogs, 400–401
  coupled assay of, 32
Carbonic anhydrase, 206
Carbon monoxide, complexes, with hemoglobin, 368
Carboxypeptidase
  cryoenzymology, 338
  energy of activation, 243–245
  kinetic constants, 244
2-Carboxy-D-ribitol 1,5-diphosphate, 401
Carnitine acetyltransferase
  chemical modification, 229
  pH kinetics, 227–229
Catalase
  activation energy, 241
  coupled enzyme assay, use as, 32, 33
  cryoenzymology, 338
  diffusional control, 253
  kinetic parameters, 240
Catechol-O-methyltransferase, assay, 32
Cellobiose, assay, 32
Cellulase, assay, 32

Central complexes, 44
Cha derivation method, 71–75, 224
Chelex, 284
Choline kinase
  magnesium activation, 270, 292
  mechanism, 292
Cholinesterase, inhibitor of, 439
Chromium–nucleotide complexes, 280, 395, 398, 450
Chymotrypsin
  abortive complex, 205
  chemical modification, 219
  competitive labeling, 219
  cryoenzymology, 338
  dielectric constant effects on, 212
  energy of activation, 243–245
  infrared spectrum, 212
  inhibition, 180
  kinetic parameters, 237, 238, 244
  nuclear magnetic resonance properties, 212
  perturbed pK values, 210, 220
  progress curve analysis, 180
Chymotrypsinogen, 244
N-trans-cinnamoylimidazole, kinetic parameters, chymotrypsin, 238
Citrate, analog, 400
Citrate cleavage enzyme, see also ATP:citrate lyase
  mechanism, 51, 158
Citrate synthase, mechanism, 51
Cleland nomenclature, see Nomenclature
Cobalamine, analog, 396, 399
Coenzyme A-linked aldehyde dehydrogenase, 13–14, see also aldehyde dehydrogenase
  lag elimination, 9
Coenzyme A synthetase, mechanism, 158
Coenzyme A transferase, mechanism, 51
d-Coformycin, tight-binding inhibition by, 462, 465
Combined equilibrium-steady state derivation method, 71–75
Competitive binding, nonproductive, 372–374
Competitive inhibition, 10, 385–392, see also specific compound; specific enzyme
  adenosine 5′-diphosphate, 281
  adenosine 5′-triphosphate, 271–277
  complications, 406–408

SUBJECT INDEX

computer program, 114–115, 126–127, 137
constant ratio to substrate, 10
coupled enzyme assay and, 40
cryoenzymology, in, 345
equation, 10, 468–470
fumarate hydratase, 229
metal-nucleotide complex, 268–269, 292, 328–329
pH effects on, 214
product effects, 412–413
statistical analysis, 110
substrate binding order, 467–486
  limitation, 485–486
  practical aspects, 477–485
  theory, 468–477
    one-substrate, 468–474
    three-substrate, 477, 479
    two-substrate, 474–478
tight-binding, 445
Competitive labeling, 218–219
Competitive reactions
  kinases, 6, 10
  transglycosylases, 6
Computer-assisted derivations, 84–103, *see also* Equation
  metal–ion studies, 289–291, 330–336
  program description, 85–90
  statistical analysis, 111–138
Conformation
  cryosolvent effects, 346–348
  enzyme, 9
  substrate, 14
  temperature effects, 239, 348
Congo red, 35
Constant ratio methods
  alternative substrate, 498–499
  substrates, 146
Continuous assays, 4, 15, 16, 22–42
Convergence point method, 147–148
Cooperativity
  false, 6–7
  metal ions, 266
  proton binding, 195–198, 232
  PRPP:ATP phosphoribosyltransferase, 9
Coupled enzyme assay, 22–42, *see also* specific enzyme
  analysis, 22–30
  assumptions, 23
  dehydrogenase, 31–32

fluorescence assay, 37
isozymes, 30
kinase, 5–6
kinetic theory, 22–30
lag time, 23–29
levels, 23, 26, 28, 37–39
model, 22–30
pH effects on, 225
practical aspects, 30–39
precautions, 39–41
progress curve, 177
stop-time, 36
Creatine kinase
  abortive complex formation, 432
  competitive inhibition, 292
  coupled enzyme assay, 32
  metal ion activation, 258, 263–264, 275–278, 292, 327–329
  pH effect on, 280–283
  tight-binding inhibition of, 463
Creatine phosphate, *see* Phosphoryl creatine
Cryoenzymology, 336–370
  cryosolvent
    catalytic effect, 344–346
    choice of, 341–343
    dielectric constant, 354
    electrolyte solubility, 355–356
    enzyme stability, 344
    pH measurements, 357–358
    preparation, 358–361
    viscosity effects, 358
  intermediate detection, 349–350
  mixing techniques, 361–362
  rapid reaction techniques, 367–369
  temperature
    control, 363–367
    effect on catalysis, 348–349
    effect on enzyme structure, 348
    measurement, 365
  theory, 340–352
Cryosolvent, *see* Cryoenzymology; specific compound
Cyclase inhibitor, 398
$3'$-$5'$-Cyclic adenosine monophosphate, analogue, 395
Cyclohexanol
  isotope effect with, 110
  partial substrate inhibition, 507
Cyclopropanone hydrate, slow-binding inhibition by, 450, 465
Cystamine aminoaldehyde, 34

Cysteine
  heat of ionization, 209
  pK value, 209
  ultraviolet spectral titration, 213
Cytochrome oxidase, cryoenzymology, 338, 368
Cytochrome P-450, cryoenzymology, 338, 366, 368

D

Dalziel relationship, see phi relationships
Davies equation, 306–307
Dead-end inhibition, see also Abortive complex formation; Competitive inhibition
  multiple, one-substrate system, 473–474
  practical aspects, 477–480
  substrate inhibition, in, 500–501
  two-substrate system, 474–478
Decay
  enzyme activity, 8–9
  substrate purity, 11
Dehydrogenase, see specific enzyme
  abortive complex, 420–421
  alternative substrate, 31
  coupled enzyme assays with, 31–32
  inhibition of, 5, 394, 396
  isomerization mechanism, 50
  NAD$^+$-dependent, 15
Denaturation, thermal
  prevention, 9
  cryoenzymology, 343–344
5'-Deoxyadenylylcobalamine, see Cobalamine
2-Deoxy-2,3-dehydro-N-acetylneuramic acid, 401
5-Deoxyfructose 1,6-bisphosphate, 374
Deoxythymidine 3',5'-disulfate, 25, 396, 398
Deoxythymidine 3'-fluorophosphate, 396, 398
Deoxythymidine 3'-p-nitrophenylphosphate, 398
Deoxythymidine 5'-phosphate, 396, 398
5'-Deoxythymidylcobalamine, 399
Deoxythymidyl 3',5'-diphosphate, 398, 403
2'-Deoxyuridine 5'-monophosphate, assay, 32
5'-Deoxyuridylylcobalamine, 399
Depolymerase, 14

Determinate method, equation derivation, 56–58
P^1,P^5-Di(adenosine-5')-pentaphosphate, adenylate kinase inhibition by, 7, 401, 483
P^1,P^4-Di(adenosine-5')-tetraphosphate, adenylate kinase inhibition by, 7, 401, 483
Diamine oxidase, assay, 34
O-Dianisidine, 34
Dibasic acid, ionization theory, 184–189
2',7'-Dichlorofluorescein, in enzyme assays, 34
Dichloromethotrexate, inhibition by, 396, 399, 403
Dielectric constant
  effect on pK value, 211–212, 226, 340
  low temperature, 354–355
Diffusion effects, 205–207, 252–254
Dihydrofolate
  analog, 396, 398
  coupled assay, in, 32
  structure, 403
Dihydrofolate reductase
  cryoenzymology, 338
  inhibitor of, 396, 399
  tight-binding inhibitor, 437, 462, 466
Dihydrofolic acid, see Dihydrofolate
2,2'-(1,8-Dihydroxy-3,6-disulfonaphthylene-2,7-bisazo)bis-benzenearsonic acid, see Arsenazo III
Dihydroxyphenylalanine, 397, 400
Di-iso bi uni uni, product inhibition, 426
Di-iso ordered bi bi, product inhibition, 426
Di-iso tetra uni ping pong, product inhibition, 426
Di-iso Theorell–Chance
  mechanism, 52
  product inhibition, 426
Dilution of enzyme samples, 10
Dimethylarsinic acid, see Cacodylate
Dimethylformamide, 341
  altered ionization properties, 352–357
  cryosolvent properties, 338
Dimethylsulfoxide, 341–343, 346
  altered ionization properties, 353–357
  cryosolvent properties, 338
  inhibition by, 397, 400
Diol dehydrase, inhibition, 399
Dioxane, dielectric effects on pK value, 211–212

Dissociation constant, *see* specific compound; *see also* Ionization constant
  amino acid residue, 200–210
  bireactant mechanism, 144
  group, 186–189
  hexokinase, 5
  molecular, 186–189, 195
  point of convergence method, 146–147
  PRPP:ATP phosphoribosyltransferase, 5
Dithioerythritol, 9
Dithiothreitol, 9
Dithizone, metal extractions with, 283–284, 300
Dixon plot, 469–470
DOPA decarboxylase, inhibitor, 397, 400
Dopamine-β-hydrolase, assay, 36

E

Edee-Hofstee plot, *see* Hofstee plot
Eisenthal–Cornish–Bowden plot, 141–142
Elastase
  chemical modification, 218–219
  competitive labeling, 218–219
  coupled enzyme assay, 33
  cryoenzymology, 338
  pH study, 218–219
Elastin, assay, 33
Electrode, ion-specific, *see* ion-specific electrode
Electrolyte, supporting
  cryoenzymology, 355–356
  stability constants and, 317, 321
Electron spin resonance, stability constant determination, use in, 311
Enolase
  coupled assays, with, 33
  inhibitor, 399
  pyruvate kinase contamination, 7
Enthalpy, lactate dehydrogenase reaction, 248–252
Enthalpy of dissociation, *see* Heat of ionization
Entropy, 234–246
  enzyme inactivation and, 254–257
  lactate dehydrogenase, 248–252
Entropy of activation, 234–246
  enzyme inactivation, 254–257
  hydrogen peroxide decomposition, 240–241
Enzyme, *see also* specific enzyme
  concentration, 14, 21
  coupled assays, in, 23, 30
  determination from progress curves, 175
  dilution, 10
  metal-activated, 257–294
  pH effects, 183–234
  purity, 6–10
  side reactions, 8
  solubility, cryosolvents, in, 345
  stability, cryosolvents, in, 344, 346–348, 360–361
  temperature effects, 8, 234–257
Equation, initial velocity
  assumption, 54
  derivation, 54–75
    Cha method, 71–75
    combined equilibrium–steady state treatment, 71–75
    comparison of methods, 69–70
    computer-assisted, 84–106
    determinant method, 56–58, 85
    Fromm systematic method, 65
    King and Altman method, 58–62
    rapid equilibrium treatment, 70–71
    steady state treatment, 55–70
    systematic method, 65–69
    Volkenstein and Goldstein method, 62–64
  progress curve, 159–180
Equation, isotope exchange, derivation, 76–83
  Cleland derivation method, 80–83
  equilibrium assumption, 76–79
  steady state assumption, 79–83
Equilibrium constant, 4, 5, *see also* Mass action ratio
  acetate kinase, 5
  buffer effects on, 18
  determination, 18–19
  hexokinase, 5
Equilibrium exchange equations, *see* Equation, isotope exchange
Equilibrium ordered, *see* Ordered bi bi, rapid equilibrium; Ordered ter ter, rapid equilibrium
Eriochrome Black T, magnesium indicator, 299
Error, *see* Standard error; Statistical analysis
17-β-Estradiol, 401, 404
Ethanol, cryoenzymology, 338, 341–345, 353–355

Ethanolamine ammonia-lyase, inhibition, 399
Ethylene glycol, 341
 altered ionization constants, and, 356–357
 cryosolvent properties, 338, 353–355
N-Ethylmorpholine
 buffer properties, 285, 298
 purification, 285
Europium-nucleotide complex, tight-binding inhibitor, as, 463–465
Exchange equations, see Equations, isotope exchange

F

Fatty acid synthetase, product inhibition, 435
Flash photolysis, subzero temperature, 368
Flavin mononucleotide, assay, 36
Flavodoxin, apo form, 36
Flavokinase, assay, 36
Fluorescence determination, stability constants, of, 309–310
Fluorocitrate, 400
5-Fluoro-deoxyuridine 5'-monophosphate, induced substrate inhibition, 508–509
1-Fluoro-2,4-dinitrobenzene, 219
α-Fluoroglutarate, 25, 397, 399
5-Fluorouridine 5'-triphosphate, 398, 462
FMN, see Flavin mononucleotide
Formate dehydrogenase, inhibitor, 398
Formylglycinamide ribonucleotide aminotransferase, mechanism, 157–158
Formyltetrahydrofolate synthetase, mechanism, 51
Free energy of activation, see Gibbs free energy of activation
Freezing points, cryosolvents, 353–354
Frieden method, kinetics of terreactant systems, 12–13, 153
Fromm derivation method, 69
Fromm method, kinetics of terreactant systems, 12–14, 150–152
β-D-Fructofuranosidase, assay, 33, see also Invertase
Fructokinase
 assay, 32
 product inhibition, 433
 specificity, 373
D-Fructose, 373
Fructose bisphosphatase, anomeric specificity, 374–376
Fructose 1,6-bisphosphate
 activation of pyruvate kinase, 7
 configuration, 371
 inhibitor, 374, 400
 mutarotation, 372
Fructose 6-phosphate
 assay, 32
 configuration, 371
 inhibitor, 374
Fumarase, see also Fumarate hydratase
 coupled enzyme assays, use in, 33
 progress curve analysis, 180
Fumarate, assay, 33
Fumarate hydratase, see also Fumarase
 exchangeable proton, 206, 229–230
 pH study, 229–231
Furylacryloylimidazole, papain substrate, 238

G

D-Galactal, slow-binding inhibitor, 450, 465
Galactokinase, activation, 258, 292
Galactose, assay, 33
Glactose dehydrogenase, assay, 33
Glactose 1,6-diphosphate, inhibitor, 400
Galactose 6-phosphate, inhibitor, 401
Galactose 1-phosphate uridylyltransferase, assay, 33
β-Glactosidase
 cryoenzymology, 338
 inhibitor, 401
 slow-binding inhibition, 450, 465
β-D-Galactoside, analog, 401
Galactosyl transferase, mechanism, 51
Gauss–Newton method, 106–108
Gel filtration method, stability constant determination, 310
Geometric analog, see Multisubstrate analogue
GDP, see Guanosine 5'-diphosphate
Gibbs free energy, 236–246
 enzyme inactivation, 254–257
 lactate dehydrogenase, 248–252
Gibbs energy of activation, 235–246
 enzyme inactivation, 251–257
 hydrogen peroxide decomposition, 240–241
Glucokinase
 assay, 37, 38
 metal ion effects, 325
 Michaelis constant, 38

SUBJECT INDEX

1,5-Gluconolactone, 401
Glucosamine 6-phosphate, 401
D-Glucose
 anomeric forms, 371, 372
 assay, 33
 hexokinase, 5
 mutarotation, 372
Glucose 1,6-bisphosphate, analog, 400
Glucose oxidase
 assay, 32, 33
 cryoenzymology, 338
 mechanism, 51
 specificity, 371
Glucose 6-phosphatase, mechanism, 51
Glucose 1-phosphate
 assay, 33
 metal–ion complex, 311, 319
Glucose 6-phosphate
 analogue, 397, 400, 401
 configuration, 371
 glucokinase, inhibition, 38
 hexokinase, inhibition, 5
 inhibitor, 401
 metal–ion complex, 319
 radioactive, 36
Glucose-6-phosphate dehydrogenase
 alternative substrate inhibition, 494
 coupled assay, 32, 33, 38
 glucose effect, 40
 lag time in assays, 37
 specificity, 378–379
Glucose phosphate isomerase, assay, 32
β-1,4-Glucosidase, assay, 32
Glutamate, analog, 13, 397, 399
Glutamate dehydrogenase
 abortive complex, 420
 coupled assay, 34
 inhibition, 399, 433, 507
 mechanism, 51, 157
Glutamate mutase, inhibitor, 399
Glutamic-alanine transaminase, mechanism, 51
Glutamine synthetase
 assay, 6
 false cooperativity, 6
 inhibition, 401, 405, 406
 mechanism, 51, 157
 substrate saturation, 12–13
γ-Glutamylcyclotransferase, inhibitor, 399
γ-Glutamylcysteine synthetase
 coupled assay, 32
 inhibitor, 399
 mechanism, 51, 158
γ-Glutamyl-phosphate, analog, 401
γ-Glutamyltransferase, assay, 6–7
Glutarate, inhibitor, as, 397, 399
Glutathione, analog, 397, 399
Gluthathione synthetase, mechanism, 51, 157
Glyceraldehyde 2-phosphate, analog, 396–399
Glyceraldehyde 3-phosphate, analog, 396
Glyceraldehyde-3-phosphate dehydrogenase
 cryoenzymology, 338
 inhibition, 399
 mechanism, 51
 NAD^+-specific mechanism, 52
Glycerol, cryosolvent properties, 338, 353
Glycerol dehydratase, assay, 34
Glycerol 1-phosphate, 319
Glycerol 3-phosphate dehydrogenase, assay, 34
Glycine
 analog, 400
 cryosolvent buffer, 355, 359
Glycogen synthetase, assay, 32
Glycolytic intermediates, see specific compound
 assay, 32, 33
 isosteric compounds, 399
Glycylglycine, buffer, 285, 298
Glyoxylate, analog, 396, 399, 400
GMP, see Guanosine 5'-monophosphate
GMP kinase, see Guanosine 5'-monophosphate kinase
GMP synthetase, see Guanosine 5'-monophosphate synthetase
Graphical analysis, 142–155, see also plotting methods
 bireactant mechanism, 143–148
 pH kinetics, 194–198
 point of convergence method, 147–148
 progress curve, 177–180
 terreactant mechanism, 148–153
Group dissociation constant, see Dissociation constant
Guanosine 5'-diphosphate, assay, 32
Guanosine 5'-monophosphate, assay, 32
Guanosine 5'-monophosphate kinase, assay, 32
Guanosine 5'-monophosphate synthetase, assay, 32
Guanylate cyclase, mechanism, 51, 158

H

Haldane relation, 18, 155
Hanes plot, 141, 145
Heavy metal
  inhibition, 283
  removal, 283–284
Heavy water effect, ionization constants, 226–227
Heme oxygenase, assay, 36
Hemoglobin
  carbon monoxide complex with, 368
  cryoenzymology, 338, 368
HEPES, buffer, 285
Henri equation, 160
Hexa uni ping pong, 48, 52
  graphical analysis, 148–155
  ϕ relationship, 149
  product inhibition, 428
Hexokinase
  activation, 258, 264, 292
  alternative product inhibition, 436
  alternative substrate inhibition, 145, 494
  assay, 32, 36, 37, 40
  competitive inhibition, 292, 476, 480
  inhibition, 5, 348
  mechanism, 51
  partial induced substrate inhibition, 510
  product inhibition, 433, 434
  slow-binding inhibitor, 450
  specificity, 378, 379
  steady state assumption, 476
Hexose configuration, 370
Hill equation, progress curve analysis, 175
Histidine
  chemical modification, 229
  competitive labeling, 219
  direct titration, 213
  heat of ionization, 209
  inhibitor, PRPP:ATP phosphoribosyltransferase, 9
  pK value, 209, 230
  proton magnetic resonance, identification by, 213
Histidine ammonia lyase, 161–162
  progress curve analysis, 180
Histidinol, assay, 33
Histidinol dehydrogenase, assay, 33
Histidinolphosphate phosphatase, assay, 33
Hofstee plot, 141–142

Hydrogen ion concentration, see pH
Hydrogen peroxide
  assay, 34
  catalase, substrate, 240
  kinetic parameters, decomposition, 240–241
Hydroxycitrate, inhibition, 400
N-2-Hydroxyethylpiperazine-N'-2-ethane sulfonic acid, see HEPES
γ-Hydroxy-α-ketoglutarate, 400
Hydroxymalonate, inhibitor, 400
$erythro$-9(2-Hydroxy-3-nonyl) adenine, slow binding properties, 464
5-Hydroxy-6-phosphononorleucine, inhibition, glutamine synthetase, 406
β-Hydroxypropionate 3-phosphate, analog, 397, 399
9-Hydroxyprostaglandin dehydrogenase, assay, 36
8-Hydroxyquinoline
  fluoresence of magnesium complex, 309
  spectrum of magnesium complex, 301, 302
  spectrum, calcium complex, 301
  stability constant measurement, 297–303
Hypoxanthine-guanine phosphotransferase, assay, 32
Hysteresis, see lag time

I

Imidazole, effect on equilibrium constants, 18
Inactivation, temperature, 234, 254–257, see also Temperature
Infrared spectroscopy
  chymotrypsin, 212
  configuration determination by, 371
  direct titration by, 213
Initiation of reactions, 3
Integrated rate equation, 159–163
Intercept replot, 13
  bireactant mechanism, 143–146
  terreactant mechanism, 150–152
Intermediate, enzyme
  alkaline phosphatase, 215
  analog, 406
  cryoenzymic detection, 349–352
  glycolytic, isosteric analog, 399
  transition state theory, 404–405

SUBJECT INDEX

Inosine, assay, 32
Invertase, K_m dependence on pH, 207–208, see also β-D-Fructofuranosidase
Ion-exchange method, stability constant determination by, 308–309
Ionic strength
  buffer effect, 226
  cryoenzymology, 355–356
  dissociation constant, effect on, 260
  lysozyme, effect on, 233
  metal–ion complexes, effect on, 307–308, 316–317, 321
  substrate inhibition, 500–501
Ionization
  amino acid residue, 207, 209–219, 227–234
  direct titration method, 212–214
  heat of, 209–211
Ion-specific electrode, 305–308
Irreversible reaction, 23
Isocitrate dehydrogenase
  activation, 293
  isotope effect, 111
  substrate inhibition, 504–506
Isocitrate lyase, inhibitor of, 399, 400
Isomerization mechanism, 50–52, 425, see specific enzyme mechanism
Iso ordered bi bi mechanism, 481
  point of convergence method, 147–148
  product inhibition, 426
Iso ordered bi bi Theorell–Chance mechanism, 147, 148, 426
Iso ordered bi uni mechanism, 426
Iso ping pong mechanism, product inhibition, 426
Isotope effect
  alcohol dehydrogenase, 110
  computer program, 138
  isocitrate dehydrogenase, 111
  primary, 17
  secondary, 17
Isotope exchange equations, see Equations, isotope exchange
Iso uni uni bi bi ping pong mechanism, 426
Iso uni uni mechanism, 415–416
Isozyme
  detection, 8
  coupled assay, 24, 31
Itaconate, inhibitor action, 400
Iterative fitting procedure, 106

J

Joule heating, temperature-jump method, 368

K

2-Keto-3-deoxygluconate aldolase, anomeric specificity, 378
α-Ketoglutarate, substrate inhibition by, 505, 507
Δ^5-3-Ketosteroid isomerase, inhibitor, 401, 404
Kinase, see specific phosphotransferase; see specific kinase
Kinetic reaction mechanism, see Mechanism, kinetic reaction
King–Altman method, 58–62

L

Lactase, assay, 33
Lactate dehydrogenase,
  abortive complex formation, 420–421
  assay, 5, 6, 32, 33, 376
  dead-end competitive inhibition, 481
  inhibitor, 398, 399, 411
  mechanism, 51
  product inhibition, 411
  progress curve analysis, 181
  substrate inhibition, 421
  temperature dependence, 246–252
Lactate 2-phosphate, inhibitor action, 396, 399
Lactoglobin, subzero temperature effect, 338
Lag time, see also Transient time
  coupled assays, in, 23, 38–39
  glucose-6-phosphate dehydrogenase, 37, 38
  inhibition, 9, 40
Least squares fitting method, 103–104
  computer program, 111–138
  evaluation, 109–111
  iterative fitting, 106
  nonlinear, 104–105
  surface, 106–109
  weighting, 105
Leucyl-tRNA synthetase, mechanism, 52
Linearity of initial rate, 4, 140
  coupled assays, 30, 72

stray-light effects on, 15
enzyme activation effect on, 10
Lipase, pancreatic, inactivation thermodynamics, 256
Lipoamide dehydrogenase, mechanism, 51
Luciferase
  cryoenzymology, 338
  intermediate, 366
Lysine
  analog inhibitor, 397, 400
  heat of ionization, 209
  mutase inhibitor, 397, 400
  perturbed pK value, 210
  pK value, 209
  titration, magnetic resonance, 213
Lysozyme
  chemical modification, 213
  cryoenzymology, 338
  difference spectral titration, 332, 333
  inhibitor, 401
  perturbed pK value, 210, 233
  pH study, 231–234
Lyxose, hexokinase inhibition, 510

M

Magnesium ion
  ADP complex, 314, 315, 379
  ATP complex, 308, 311
  enzyme interaction, 258–294
  indicator, 299
  ion-specific electrode, by, 306
  nucleotide interaction, 258–299
  purification, 283
  spectra of 8-hydroxyquinoline complex, 301, 302
  stability constant value, 308, 311
  standardization, 284
Malate dehydrogenase, assay, 33
Malonate, inhibitor action, 400
Maltodextrin phosphorylase, product inhibition difficulties, 435
Manganese
  enzyme interaction, 258–294
  inhibition, 266
  nucleotide complex, 258–294, 311, 318, 319
  purification, 283
  stability complex, 311, 318, 319
  standardization, 283
Mannitol 1-phosphate, inhibitor, 401

Mannokinase, assay, 32, 33
Mannose 6-phosphate
  analog, 401
  assay, 33
Mannose 6-phosphate isomerase
  assay, 33
  inhibitor, 401
Mapping active site, 405–406
Marmasse plot, 511–513
Mass action ratio, 5, 18, *see also* Equilibrium constant
Maximum velocity
  bireactant mechanism, 144
  competitive inhibitor, with, 10
  coupled assay, 24–30, 37, 41
  cryosolvent effect on, 344–346
  metal–ion interaction, 257–294
  point of convergence method, 146–147
  progress curve measurement, 159–183
  substrate analog, with, 372–374
  temperature dependence, 348–349
  terreactant mechanisms, 148–153
2-Merceptoethanol, 9
Metal ion
  analysis, 289–292, 325–336
  buffer interactions, 19–20, 226, 285, 287
  chelation, 226, 404
  enzyme activation, 257–294
  equilibrium constant dependence, 18–19
  experimental conditions, 287–289, 325–329
  inhibition, 279–280
  purification, 283–284
  reaction mixture preparation, 289, 325–339
  removal from solutions, 284, 299–300
  stability constant measurements, 295–319
  standardization, 283–284
Methanol
  cryosolvent properties, 341, 353–355
  interaction with papain, 347
  pH changes, 356–357, 360
Methionine adenosyltransferase, assay, 36
Methionine sulfoximine phosphate, inhibitor, glutamine synthetase, 401, 405, 406
Methotrexate
  inhibitor action, 396, 399, 403
  tight-binding interaction, 462, 466
5-(1-Methoxy)ribosylcobalamine, inhibitor action, 399

SUBJECT INDEX

Methyl-2-acetamido-2-dexyglucose, inhibitor, lysozyme, 233
Methyl-DL-α-chloro-β-phenylpropionate, inhibitor, chymotrypsin, 244
Methylcobalamine, inhibitory analog, 399
α-Methyldihydroxyphenylalanine, inhibitory analog, 397, 400
Methyl-DOPA, see α-Methyldihydroxyphenylalanine
Methylene diphosphonate nucleotide analogue, 398, 403
β,γ-Methylene ATP, see 5'-Adenylylmethylene diphosphonate
Methyl β-D-fructofuranoside 6-phosphate, anomeric analog, 374
β-Methylglutamate, inhibitor action, 397, 399
γ-Methylglutamate, inhibitor action, 397, 399
Methyl hydrocinnamate, substrate, chymotrypsin, 244
3-Methyl-5-oxoproline, inhibitor action, 397, 399
2-Methyl-2,4-pentanediol, cryosolvent, 353
Methyl-D-β-phenyllactate, kinetic constants, 244
Methyl-L-β-phenyllactate, kinetic parameters, chymotrypsin, 238, 244
4-Methyl umbelliferyl phosphate, substrate, alkaline phosphatase, 215
Mixed inhibition, see also Noncompetitive inhibition
  pH, effect of, 214
  progress curve, with, 164, 170–171
Modification, chemical, see specific compound; specific enzyme
  competitive labeling, 218–219
  pH, effect of, 215–219
  titration, with, 212–214
Molecular dissociation constant, see Dissociation constant, molecular
Monoamine oxidase, assay, 34
Mono-iso ping pong bi bi mechanism, scheme, 52
Multisubstrate analog, 482–484
Murein glycan, analog, 401
Mutarotase
  coupled enzyme assay of, 33
  coupling enzyme, use as, 32
Mutarotation, 370–372
  D-fructose 1,6-bisphosphate, rate constants, 372

D-glucose, phosphate-catalyzed, 372
Myokinase, see Adenylate kinase
Myosin adenosine triphosphatase
  activation energy, 243–245
  cryoenzymology, 338
  inactivation
    energy of activation, 256
    entropy of activation, 256
  inhibitor, 398
  kinetic constants, 244
  temperature, solvent and pressure study, 239

N

NADH, see Nicotinamide adenine dinucleotide
NAD$^+$-linked dehydrogenase, see specific dehydrogenase
NADPH-dependent biliverdin reductase, see Biliverdin reductase, NADPH-dependent
Neuraminidase
  assay, 33
  inhibitor of, 401
Newton method, see Gauss–Newton method
Nicotinamide adenine dinucleotide
  analog, 398
  coupled enzyme assay, in, 32, 34
  instability, 222
  molar extinction coefficient, 15
  stability constant determination, 310
p-Nitroaniline, coupled enzyme assay, in, 33
p-Nitrophenyl phosphate, substrate, alkaline phosphatase, 238
p-Nitrophenyl trimethylacetate, substrate, chymotrypsin, 236
Nomenclature
  Alberty, 43
  Cleland, 43–44, 139
    bireactant, 143
    product inhibition, 411–414
  Dalziel, 50, 139
    bireactant, 143
  mechanism, 43–44
Noncompetitive inhibition, 385–392, see specific compound; specific enzyme
  computer program, 115, 128–129, 137
  coupled enzyme assay, and, 40
  experimental design, 394

graph, 389
metal-ion, 292
one-substrate, 470–472
partial, 385
pH effect of, 214
product, 412–414
progress curve, in, 164, 170–171
rate expression, 393
tight-binding, 470–472
Nonlinear least squares, 104–105
19-Nortestosterone, inhibitor, 401, 404
Nuclear magnetic resonance
  ^{13}Carbon
    aspartic acid, titration, 213
    configuration, carbohydrate, determination, 371–372
    cryoenzymology, in, 352
    lysine, titration, 213
  Proton
    chymotrypsin, 212
    cryoenzymology, in, 346–347, 352
    histidine, titration, 213
    tyrosine, titration, 213
Nuclease, inhibitor, 396, 398
Nucleoside diphosphate kinase
  acetate kinase activity, 8
  activation, 258, 259, 292
  coupled enzyme assay of, 6
  mechanism, 51
  product inhibition, 10–11
  substrate inhibition, 502–503
Nucleoside diphosphorylase, assay, 32
3′-Nucleotidase, assay, 32
Nucleotide, see specific nucleotide
  activation, 293
  analog, 398
  chromium, complex with, 280
  coupled assay of, 32
  inhibition, 258, 264, 271—275
  metal–ion complex, 257–294, 323–327
  phosphate, inhibitor, 398
  phosphonate, inhibitor, 398
  phosphoramidate, inhibitor, 398
  purification, 299
  resolution, 17
  stability constants, 311–319

O

Octapamine, assay, 36
One-substrate system, 11, see also Uni uni mechanism; see specific mechanism

Ordered bi bi ternary complex mechanism,
  abortive complex interaction, 421–422
  alternative substrate effect, 146, 487–490, 492
  competitive inhibitor method, 467, 476–478
  computer program, 114, 126
  example, 51, 242–243
  graphical analysis, 144–145
  inhibition patterns, 390–391
  multisubstrate analog, 482–484
  phi relationship, 143
  point of convergence analysis, 147–148
  product inhibition, 418, 420, 432–433
  rate equation, 45
  scheme, 45
  subsite mechanism, 45, 478
Ordered bi bi Theorell–Chance mechanism, 144–145
  abortive complex interactions, 422–423
  alternative substrate effects, 487–490, 492, 494–497
  phi relationships, 143
  point of convergence analysis, 147–148
  product inhibition, 419–420
  scheme, 45
Ordered bi ter mechanism, 45, 51, 142
Ordered bi uni mechanism, 45, 51
Ordered bi uni uni ping pong mechanism
  graphical analysis, 148–155
  phi relationships, 149
  product inhibition, 428
  replotting methods, 150–152
  scheme, 48
  substrate saturation, 154–155
Ordered mechanism, see specific mechanism
Ordered ter ter mechanism,
  example, 51, 157
  phi relationships, 149
  product inhibition, 427
  rapid equilibrium, 148–155
  scheme, 46
Ordered ter ter Theorell–Chance mechanism
  example, 157, 158
  graphical analysis, 148–155
  phi relationships, 149
  product inhibition, 427
  scheme, 45
  substrate saturation, 154–155

Ordered uni bi mechanism, *see also* Ordered bi uni mechanism
  example, 51, 142, 432
  product inhibition, 416–418
  rate expression, 417
  substrate inhibition, 506
Ordered uni uni bi bi ping pong mechanism
  graphical analysis, 148–155
  phi relationships, 149
  product inhibition, 428
  replot, 150–152
  scheme, 49
  substrate saturation, 154–155
Organic solvent, *see* specific compound; cryoenzymology
L-Ornithine, inhibitor action, 400
Oxalate, inhibitor action, 399, 481
Oxaloacetate, assay, 33, 36
Oxalomalate, inhibitor action, 400
Oxamic acid, inhibitor action, 396, 399
5-Oxoprolinase, inhibitor, 399
2-Oxopropane sulfonate, inhibitor action, 400
Oxygen
  coupled enzyme assay, in, 32
  cryoenzymology, 366, 368

P

Papain
  cryoenzymology, 338, 347
  kinetic parameters, 238
PEP, *see* Phosphoenolpyruvate
PEP carboxykinase, *see* Phosphoenolpyruvate carboxykinase
Pepsin
  activation energy, 243–245
  kinetic constants, 244
  tight-binding inhibition, 462, 466
Pepstatin, tight-binding inhibition, 462
Periodic sampling method, 15–16
Peroxidase
  coupled assay, 34
  cryoenzymology, 338
  temperature study, 240
Peroxisomal oxidase, assay, 34
pH
  buffer, 19–20, 225–227
  computer program, 105
  cryosolvent changes in, 345–346, 356–360

effect, studies,
  complications
    abortive complex formation, 205
    change in rate limiting step, 201–203
    multiple intermediates, 199–200
    multiple isomerizations, 200–201, 203–205
  dielectric constant dependence, 211–212, 226
  example, 214, 215–219, 227–234
  identification of amino acid residue, 209–219
  independence of K_m, 207–209
  limitation, 219–220
  metal–ion interactions, 323–327
  practical aspects, 220–224
  theory, 183–219
m-Phenanthroline, inhibitor action, 404
o-Phenanthroline, inhibitor action, 404, 408
Phenol, assay, 34
Phenylalanine hydroxylase, alternative substrate study, 498
Phenylethanolamine N-methyltransferase, assay, 36
Phenyl pyruvate, inhibitor action, 180
Phenyl urea, inhibitor action, 400
Phi relationships, 153–155, *see also* Nomenclature, Dalziel
  bireactant mechanism, 143–146
  isomerization mechanism, 50
  terreactant mechanism, 148–153
Phosphate
  buffer, 285, 298–299, 355–357
  catalyst for mutarotation, 372
  inhibitor action, 180
  metal–ion interaction, 319
  ribonucleic acid, assay, 36
Phosphoarginine, *see* Phosphorylarginine
Phosphocreatine, *see* Phosphorylcreatine
Phosphoenolpyruvate, assay, 36
Phosphoenolpyruvate carboxykinase, assay, 33, 36
Phosphofructokinase
  anomeric specificity, 373–374
  assay, 376
  contamination, 7
  initial rate properties, 156
  mechanism, 51
  product inhibition, 433
  substrate analogue, 373–374

substrate inhibition, 501
Phosphoglucomutase
  coupled enzyme assay, 33
  inhibitor, 400
6-Phosphogluconate, inhibitor action, 397, 400
6-Phosphogluconate dehydrogenase
  coupled enzyme assay, 33
  initial rate properties, 157
Phosphoglucose isomerase
  anomeric specificity, 377
  equilibrium constant, 378
1-Phosphoglucuronate, inhibitor action, 397, 400
3-Phosphoglycerate
  assay, 33
  inhibitor action, 396, 399
  metal-ion interactions, 319
3-Phosphoglycerate kinase
  activation, 258
  assay, 33
2-Phosphoglycolate, inhibitor analog, 401
Phosphomevalonic acid, 318
Phosphomolybdate, phosphate assay, 36
Phosphoramidate, nucleotide derivative, 398
N^1-Phosphoribosyl-ATP, 5
5-Phosphoribosyl-α-1-pyrophosphate
  analysis, 304
  metal-ion interaction, 294, 297, 318, 320-322
  purification, 299
Phosphorylarginine, metal-ion interaction, 319
Phosphorylase b, mechanism, 51
Phosphorylase kinase, inhibition, 481-482
Phosphorylcreatine, metal-ion interaction, 319, 329
Phosphotransferase, see specific enzyme
Physostigmine, inhibitor action, 438
Ping pong mechanism, see also specific enzyme mechanism
  abortive complex interaction, 423-424
  bireactant systems, 47-48, 51
  computer program, 114, 125-126
  constant ratio method, 146, 493-494
  dead-end competitive inhibition, 478
  graphical analysis, 145-146
  modified mechanism, 45, 51
  phi relationships, 143
  product inhibition, 419, 420, 435
  scheme, 48
  substrate inhibition, 145-146, 487-490, 493-497, 502-504
  terreactant system, 13, 52, 148-155
Piperazine-N,N'-bis(2-ethanesulfonic acid), buffer, 285
Plotting methods, see also Graphical analysis
  bireactant systems, 148-153
  computer approach, 109-138
  constant ratio method, 146
  definition, 139-140
  graphical analysis, 142-155
  pH profile, 194-198
  progress curve, 177-180
  statistical analysis, 103-138
  terreactant system, 148-153
  tight-binding inhibitor, 440-443
Point of convergence method, 147-148
Polymerase, inhibitor, 14, 398, 404, 408
Polynucleotide kinase, 14
Potassium ion, stability constant, 305-306, 308, 312, 318, 319
Preincubation method, 3, 9
  coupled enzyme assay, in, 40
  enzyme stability, 220-221
  slow-binding inhibitor, 449, 453
Prephenate dehydratase
  inhibition, 180
  progress curve analysis, 180
Pressure effect, 239
Product inhibition, 4, 5, 10-11, 411-436, see also specific enzyme mechanism
Progress curve
  activation effect, 174-175
  analysis, 140, 159-183
  enzyme concentration, 175-177
  limitations, 181-183
  plotting, 177-180
  practical methods, 175-177
  reversible inhibitor action, 163-175
  reversible reaction, 171-175
  simulation study, 180
  theory, 159-162
Prostaglandin, analog, 401, 404
Protein kinase, 14
Proteolysis, prevention of, 9
Proton, 205-207, see also pH
  hindered equilibration, 206
  metal-ion interactions, 318, 319

SUBJECT INDEX

PRPP, *see* 5-Phosphoribosyl-α-1-pyrophosphate
Purity
  enzyme, 6–10
  metal, 283–285
  radioisotope, 16–18
  substrate, 10–11, 17
3,5-Pyridine dicarboxylate, inhibitor action, 397, 399
Pyrophosphate
  assay, 34
  metal–ion complex, 299, 319
Pyrophosphate:fructose 6-phosphate phosphotransferase, assay, 34
Pyrophosphomevalonic acid, metal–ion complex, 318
Pyruvate
  analog, 399
  assay, 33
  kinetic constants, 247
Pyruvate carboxylase
  activation, 270
  mechanism, 52
  product inhibition, 435
Pyruvate dehydrogenase, product inhibition, 435
Pyruvate kinase
  activation, 7, 258, 270, 279
  contaminating activity, 7
  coupled assay, 32, 36, 56, 376
  dead-end inhibition, 279–280
  product inhibition, 279
Pyruvate phosphate kinase, mechanism, 52

Q

Quenching method
  efficiency, 16–18
  radioactivity, 17
  rapid mixing, 374–378

R

Radiolysis, substrate deterioration, 17
Radiometric assay, 16–18
Raman spectroscopy, stability constant determination by, 311
Rapid reaction technique, cryoenzymology
  flash photolysis, 368
  mixing, 369
  subzero temperature, 367–369
  temperature jump, 368–369
Random A B, random Q R, bi uni uni bi ping pong mechanism, *see also* Bi uni uni bi ping pong
  graphical analysis, 148–155
  phi relationships, 149
  product inhibition, 428
  replot, 150–152
  saturation, effect of, 154–155
  scheme, 49
Random AB, random QR ter ter mechanism
  example, 51, 158
  graphical analysis, 148–155
  phi relationships, 149
  product inhibition, 427
  replot, 150–152
  saturation, effect of, 154–155
  scheme, 46
Random AC, random PR ter ter mechanism
  example, 157
  graphical analysis, 148–155
  phi relationships
    ordered B, 149
    ternary complexes present, all, 149
  product inhibition, 427
  replot, 150–152
  saturation, effect of, 154–155
  scheme, 47
Random A ter ter mechanism
  phi relationships, 149
  product inhibition, 427
Random BC, random PQ ter ter mechanism
  example, 51, 157, 158
  graphical analysis, 148–155
  phi relationships
    equilibrium A, 149
      product inhibition, 427
    steady state A, 149
      product inhibition, 427
  product inhibition, 427
  replot, 150–152
  saturation, effect of, 154–155
  scheme, 46
Random BC, uni uni bi bi ping pong mechanism, *see also* Two-site bi (random) bi (random) uni uni ping pong
  graphical analysis, 148–155

phi relationships, 149
product inhibition, 428
replot, 150–152
saturation, effect of, 154–155
scheme, 49
Random bi bi mechanism
  alternative substrate analysis, 487–490, 493–497
  competitive inhibitor analysis, 467, 474–478
  example, 51, 481
  multisubstrate analog, with, 483–484
  point of convergence method, 147–148
  product inhibition, 423, 432–434
  rapid equilibrium
    abortive complexes, with, 423
    alternative substrate, 146
    example, 227, 280–283, 432–433
    phi relationships, 143
    problems, 158
    product inhibition, 418–420
  scheme, 45
  substrate inhibition, 504–506, 508
Random bi uni mechanism, *see also* Random uni bi mechanism
  dead-end competitive inhibition, 478
  scheme, 45
Random bi uni uni bi ping pong mechanism, *see* Random AB, random QR bi uni uni bi ping pong mechanism
Random C (ordered B) ter ter mechanism, *see also* Random AC, random PR ter ter mechanism
  phi relationships, 149
  product inhibition, 427
Random mechanisms, *see* specific mechanism; *see also* Sequential mechanisms
  metal–ion binding, 275, 277, 286
  point of convergence method, 147–148
  substrate inhibition, 43
Random ter ter mechanism
  example, 51, 157
  graphical analysis, 148–155
  phi relationships,
    no EAC complex, 149
      product inhibition, 427
    random A, 149
  product inhibition, 427
  replot, 150–152
  saturation, effect of, 154–155
  scheme, 46

Random uni bi mechanism, *see also* Random bi uni mechanism
  rapid equilibrium
    equation, 417
    product inhibition, 417–418
  scheme, 45
  steady state
    equation, 418
    product inhibition, 418
Random uni uni bi bi ping pong mechanism, *see* Random BC uni uni bi bi ping pong mechanism
Rapid equilibrium mechanism, *see* specific mechanism
Rapid equilibrium random-ordered bi bi mechanism
  derivation, 71–75
  scheme, 71
Rapid equilibrium ordered bi bi mechanism, 145, 158, *see also* Ordered bi bi mechanism
  example, 51
  phi relationships, 143
Rate constants
  alkaline phosphatase, 238
  chymotrypsin, 237–238
  mutarotation, 370–379
  papain, 238
  separation, 236
  trypsin, 237
Residual least squares, 110–112
Reversible inhibition, 383–411, see also specific inhibition type
  practical aspects, 408–411
  theory, 384–408
    characterization, 385–391
    classes, 385–391
    complications, 406–408
    constant determination, 391–392
Ribitol dehydrogenase
  abortive complex, 419–420, 424
  mechanism, 51
  product inhibition, 411
Riboflavin, coupled enzyme assay, in, 36
Ribonuclease
  cryoenzymology, 338
  dielectric constant, effect of, 211
  quenching problems, 16
  stop-time coupled enzyme assay of, 36
Ribulose 1,5-diphosphate carboxylase, inhibitor of, 401

Ricin
  kinetic parameters, 255
  temperature inactivation, 255–256

S

Saccharopine dehydrogenase, mechanism, 157
Sequential mechanisms, *see also* specific enzyme mechanism
  alternative substrate approach, 486–500
  bireactant systems, 51, 143–148
  computer program, 114, 123–124
  constant ratio method, 146
  terreactant systems, 51, 148–155
Sialic acid, analogue, 401
Side reaction, 8, 14–15, 40
Simulated progress curve, 180
Single-point assay, 4
Slope replot, *see also* Graphical analysis
  bireactant mechanisms, 143–146
  example, 13
  terreactant mechanisms, 150–152
Slow-binding inhibition, 449–453
  analysis, 451–453
  characteristics, 449–453
  kinetic constants, determination, 460–462
Sodium ion, nucleotide complexes, 311–312, 318–319
Solvent effect, 236, 239
α-L-Sorbofuranose, substrate analog, 373
Specific activity, 21
Specificity
  anomeric configuration, 370–379
  substrate, 14, 370–379
Spectrophotometric assay
  coupled enzyme systems, 31–32
  reaction linearity, 15
  stray light limitation, 15
  temperature control, 20–21
Spironolactone, inhibitor, 397, 401, 404
Stability
  enzyme, 6–10
  metal-ion complex, *see* specific complex; *see* Metal ion
  pH, 220–221
  substrate, 10–11, 17, 221–222
Statistical analysis, 14–18, 21, 103–138
  iterative fitting procedure
    computer program, 111–138

  evaluation, 109–111
  surfaces, 106–109
  least squares method
    assumptions, 103
    nonlinear fit, 104–105
    weighting procedure, 105
  progress curve, 159–183
Steady state assumption, *see also* Equations, initial rate
  coupled assay system, 22–30
  experimental verification, 4–19
Steroid, analog, 401
Stoichiometry, 11
Stopping procedure, *see* Quenching method
Structural effect, 236
Substrate, *see also* Alternative substrate
  activation, progress curve analysis, 174–175
  binding order, competitive inhibitor method, 467–487
  concentration range, 11–14, 223–224
  conformation, 14, 370–379
  dead-end complex, 229–280, 288, 372–374
  dehydrogenase assay, 31–33
  inhibition
    analysis, 510–513
    computer program, 113–114, 137
    partial, 507–509
    progress curve, 165–166
    random mechanism, 43
    theory, 501–510
  ionization, 222–223
  metal–ion complex, 223–293
  pH properties, 221–224, 257–294
  purity, 10–11, 17, 221–222, 407
  regeneration, 5–6, 22–42
  resolution, 16–17
Subtilisin, cryoenzymology, 338
Succinate, analogue, 400
Sucrose phosphorylase, side reaction, 8
Sulfhydryl reagent, *see* specific compound
Systematic derivation method, 65–69

T

D-Tartronate semialdehyde phosphate, inhibitor action, 396, 399
Temperature, *see also* cryoenzymology
  activation energy, 236–246

activation entropy, 236–246
buffer dependence, 226
control, 20–21, 363–367
diffision effect, 252
dissociation constant dependence, 260
enzyme activation, 254–257, 343
equilibration, 3, 21
lysozyme, 233
metal–ion stability constant, and, 299, 315–316, 321
multisubstrate system, 246–252
rate constant separation, 236
subzero
  measurement, 365
  column chromatography, and, 366–367
Terreactant enzyme systems, *see* specific mechanism
alternative substrate approach, 494, 498
competitive inhibitor method, 477–479
graphical analysis, 148–155
phi relationships, 149–155
ping pong case, 13, 48–52
product inhibition procedure, 426–429
replot method, 150–152
sequential, 12–14, 45–47, 51
substrate level, 12–14
TES, buffer, 285
Tetraalkylammonium ion, metal–ion interaction, 284, 298
Tetrahydrofolate synthetase, mechanism, 157
Tetrapropylammonium bromide, metal-ATP complex, 298
Theorell–Chance mechanism, *see* specific enzyme; specific mechanism
Thermal equilibration, 3, 20–21
Thiamine 5′-triphosphate, metal-ion complex, 318
β-D-Thiogalactoside, inhibitor action, 401
Thiourea, inhibitor action, 397, 400
Threonyl-tRNA synthetase, mechanism, 52
Thromboxane synthetase, inhibitor, 401, 404
Thymidylate synthase
coupled assay, 32
induced substrate inhibition, 508–509
inhibitor, 398
tight-binding ligand, 462
Tight-binding inhibitor, 462–466

practical considerations, 456–462
theory, 438–456
Titration
difference, 233
direct, 212–214
metal–ion complex, 297–305
substrate, 222
Transcarboxylase
mechanism, 51
product inhibition, 435
Transglutaminase, mechanism, 51
Transglycosylase
competitive reaction, 6
sucrose phosphorylase, side reaction, 8
Transient time, *see also* Lag time
coupled enzyme assay, 28, 38–39, 41
kinase, assay, 38
Transition state analogs, 392, 401, 404–406, 462, 465, 482
Tri-N-acetyl chitotriose, inhibitor action, 401
Tricarboxylic acid pathway, *see* specific enzyme; specific compound
Triethanolamine, buffer, 285–298
Triethyloxonium ion, lysozyme modification, 213, 232
Triosephosphate isomerase
assay, 34
inhibitor, 401
Tripeptide synthetase, inhibitor, 405
Tris-(hydroxymethyl)aminomethane
cryosolvent, buffer, 355, 359
equilibrium constant, effect on, 18
metal-ion complex, 285, 298
temperature effect on, 226
N-Tris-(hydroxymethyl)methyl-3-aminopropanesulfonic acid, *see* TES
Trypsin
coupled assay, 33
cryoenzymology, 338
energy of activation, 256
entropy of activation, 256
kinetic parameters, 237–238, 244
Two-site ping pong mechanism, 48, 49, 51, 52
Tyrosinase, coupled enzyme assay, 34
Tyrosine
heat of ionization, 209
pK value, 209
titration, 213

SUBJECT INDEX

U

UDP, see Uridine 5'-diphosphate
UDP-glucose 4-epimerase, inhibitor, 401
Ultraviolet spectral titration, 213
dUMP, see 2'-Deoxyuridine 5'-monophosphate
Uncompetitive inhibition, 385–392, see specific compound; specific enzyme
    computer program, 115, 127–128, 137
    design, 394
    graphical analysis, 389, 393, 410
    one-substrate, 472
    pH effect, 214
    product inhibition, 412–413
    progress curve analysis, 164, 170
    tight-binding interaction, 445
Uni uni mechanism
    competitive inhibition, 468–470
    equation, 56, 57, 59, 70, 414
    noncompetitive inhibition, 470–472
    product inhibition, 414–418
    uncompetitive inhibition, 472–473
Urea, 244, 397, 400
Urease
    activation energy, 243–245
    inhibitor, 397, 400
    kinetic constant, 244
Uric acid, coupled enzyme assay, 32
Uridine 5'-diphosphate, assay, 32
Uridine 5'-diphosphate-3-deoxyglucose, inhibitor analogue, 401
Uridine diphosphate-glucose, analog, 401
Uridine phosphorylase, product inhibition, 432–433
Uridine 5'-triphosphate, analog, 398
5'-Uridylylcobalamine, inhibitor action, 399

V

Vanadate ion, ATPase inhibitor, 10
Viscosity, cryosolvent, 353, 358, 361–362

X

Xanthine oxidase
    coupled assay, 32
    cryoenzymology, 338
    slow-binding inhibitor, 463–464